The **Intext** Series in BASIC MATHEMATICS
under the consulting editorship of

>RICHARD D. ANDERSON
Louisiana State University

>ALEX ROSENBERG
Cornell University

Modern Analytic Geometry

Modern Analytic Geometry
THIRD EDITION

W. K. Morrill (deceased)
Formerly Professor of Mathematics
The Johns Hopkins University

Samuel M. Selby
Department of Mathematics
Hiram College

Wendell G. Johnson
Department of Mathematics
Hiram College

Intext Educational Publishers
COLLEGE DIVISION of Intext
Scranton   San Francisco   Toronto   London

ISBN 0-7002-2413-0

Copyright © 1972, 1964, 1951, International Textbook Company

All rights reserved. No part of the material protected by this copyright notice may be reproduced or utilized in any form or by any means, electronic or mechanical, including photocopying, recording, or by any informational storage and retrieval system, without written permission from the copyright owner. Printed in the United States of America by The Haddon Craftsmen, Inc., Scranton, Pennsylvania. Library of Congress Catalog Card Number: 76-185821.

Preface

Twenty years ago, when the first edition of this text was published, it was not common to use vector methods in teaching analytic geometry at the introductory level. In the preface of the first two editions, Professor W. K. Morrill acknowledged Professor F. D. Murnaghan's work in first introducing vectors in the teaching of analytic geometry at the Johns Hopkins University. This work at Johns Hopkins and the writing of the first edition by Professor Morrill had an important influence upon the use of vector procedures in analytic geometry.

In this third edition the authors' aim has been to retain the flavor and spirit of Professor Morrill's work while introducing some contemporary notation and terminology. Some portions have been rewritten, new sections and many new exercises have been added. The motivation for some of the new material has come from suggestions by teachers who used the first two editions. Chapter 1 has been significantly changed by adding sections on sets, relations and functions. This should provide the background for those instructors who find it helpful to use the set notation and terminology. Also, some additional topics in algebra and trigonometry have been included.

The significant features of the earlier editions have been retained, some of which are the following.

(1) The use of vector methods is perhaps the most significant feature of the book. The use of these methods makes it easy to proceed from 2-space to 3-space. They will be extremely useful for the student in his study of the calculus and the physical sciences.

(2) The study of locus problems and parametric equations is found throughout several chapters, rather than being allocated to one particular chapter.

(3) Examples and illustrations are constructed to emphasize the procedures involved as well as to increase the students' understanding and ability to reason.

(4) Matrix techniques play an increasingly significant role in mathematics curricula and can conveniently be used in analytic geometry. A brief introduction to matrix theory is found in Appendix A. This affords the instructor an opportunity to use these techniques if he so desires. However, the material in the text can be covered without the use of matrices. Determinants are used throughout the text where convenient, and sufficient reference material for them is included in Chapter 1.

(5) There is flexibility in the way the text can be used. The material is adaptable for a one term course or a more extensive course in analytic geometry. It could well serve as a text for a 3 to 5 semester hour college level course or a year's course at the senior high school level. As a guide for a one term course, Chapters 2, 3, 4, 5, 7, and 8 might be covered. Another combination which has been used successfully is Chapters 2, 3, 9, 10, 11, 4, 5, and 6, in that order. Certain sections which may be omitted are starred in the Contents.

As a gesture of respect and admiration for Professor W. K. Morrill, the authors of this third edition wish to dedicate it to his revered memory.

The authors wish to express appreciation to Mr. Charles J. Updegraph, Mathematics Editor of Intext, for his cooperation and kindness throughout the preparation and production of this third edition. They also wish to thank Mrs. Charles Adams for the typing and handling of the details required in the preparation of the manuscript.

It is hoped that this third edition will experience the same acceptance as the earlier editions.

Samuel M. Selby
Wendell G. Johnson

Hiram, Ohio
January, **1972**

Contents

I. ANALYTIC GEOMETRY IN THE PLANE

Chapter 1. Introduction . . . 3

1–1. Set notation
1–2. Special sets
1–3. Set operations
1–4. Linear equation in one unknown
1–5. The quadratic equation $ax^2 + bx + c = 0$
1–6. Exponents
1–7. Logarithms
1–8. Determinants
1–9. Solution of simultaneous systems of linear equations
1–10. Systems with quadratic equations
1–11. Inequalities and absolute value
1–12. Number intervals
1–13. Relations and functions
1–14. Conditional and identity equations
1–15. Degrees, radians and arc length
1–16. The trigonometric functions
1–17. The functions of 0°, 30°, 45°, 60°, 90° and related measures
1–18. Important identities
1–19. Reduction formulas
1–20. Relations for triangles
1–21. Inverse trigonometric functions

Chapter 2. The Point and Plane Vectors . . . 39

2–1. Introduction
2–2. The rectangular coordinate system
2–3. Symmetry
2–4. Projections
2–5. Scalar components of a segment
2–6. Distance between two points
2–7. Direction cosines of a segment
2–8. Plane vectors
2–9. Angle between two vectors
2–10. Parallel and perpendicular vectors
2–11. The coordinates of a point that divides a segment in a given ratio
2–12. The mid-point of a segmen
2–13. Complementary vectors
2–14. Area of a triangle and the bar product of two vectors

x CONTENTS

Chapter 3. The Straight Line . . . 79

3–1. Direction numbers and direction cosines of a line
3–2. The slope of a line
*3–3. Parametric equations of a line
3–4. Equation of a line in direction number form
3–5. The general linear equation $ax + by + c = 0$
3–6. Angle between two lines
3–7. Point-slope form of equation of a line
3–8. Intercept form of equation of a line
3–9. Slope intercept form of equation of a line
3–10. Normal form of equation of a line
3–11. Lines parallel and perpendicular to a given line
3–12. Distance from a line to a point
3–13. Intersecting lines
*3–14. Three concurrent lines
*3–15. Bisectors of angles
*3–16. Sets of lines
*3–17. Pencils of lines

Chapter 4. The Circle . . . 127

4–1. Standard equation
4–2. The general equation of a circle
4–3. Circle determined by three conditions
4–4. Symmetry
*4–5. Tangents to a circle from an external point
*4–6. Length of the tangent from a point outside a circle to the circle
*4–7. Radical axis
*4–8. Equation of the tangent to the circle $x^2 + y^2 = r^2$ at the point $P_1(x_1, y_1)$ on the circle
*4–9. Equation of the tangent to the circle $x^2 + y^2 = r^2$ when the slope of the tangent is known
*4–10. Pencil of circles
*4–11. Parametric equations of the circle

Chapter 5. The Conics . . . 162

*5–1. The conic as a section of a cone
5–2. General definition of a conic
5–3. Explanation of terms
*5–4. A geometric construction of the parabola
5–5. Simple equations of the parabola
5–6. The latus rectum
*5–7. Parabolic arch
*5–8. Parametric equations of the parabola
5–9. Applications of parabolic curves
5–10. Simple equations of the ellipse
5–11. Explanation of terms for ellipse
5–12. The focal radii
5–13. Discussion of the equation $Ax^2 + Cy^2 + F = 0$, where A and C are of like sign and are not equal to zero
5–14. The latus rectum of an ellipse
*5–15. Construction of an ellipse
*5–16. Parametric equations of the ellipse
5–17. Applications of the ellipse
5–18. Simple equations of the hyperbola
5–19. Explanation of terms for hyperbola
5–20. Asymptotes of a hyperbola
5–21. The focal radii
5–22. Discussion of the equation $Ax^2 + Cy^2 + F = 0$, where A and C are unlike in sign and different from zero
5–23. The latus rectum of the hyperbola
5–24. Conjugate hyperbolas
5–25. Equilateral hyperbola
*5–26. Construction of the hyperbola
5–27. Applications of the hyperbola
*5–28. Parametric equations of a hyperbola
5–29. Summary

* May be omitted in a brief course.

Chapter 6. Transformation of the Axes . . . 217

    6–1. Introduction
    6–2. Translation of the axes
    6–3. Simplification of the equation $Ax^2 + Cy^2 + Dx + Ey + F = 0$, where both A and C are not zero
    6–4. The standard equations of the conics
    6–5. Rotation of the axes
    6–6. Simplification of the equation $Ax^2 + Bxy + Cy^2 + Dx + Ey + F = 0$, where $B \neq 0$
    6–7. Equilateral hyperbola
    *6–8. Invariants of the second-degree equation
    *6–9. The characteristic $4AC - B^2$ and the discriminant Δ
    *6–10. General transformation of coordinates
    *6–11. The conic through five points
    *6–12. Use of set symbolism for conics (Supplementary)

Chapter 7. Polar Coordinates . . . 261

    7–1. Introduction
    7–2. Polar coordinates of a point
    7–3. The relation between polar coordinates and cartesian coordinates
    7–4. The line
    7–5. The circle
    7–6. Intercept points
    7–7. Symmetry
    7–8. Extent
    7–9. Sketching curves representing $\rho = f(\theta)$
    7–10. The conic
    *7–11. Locus problems
    *7–12. Intersection of curves in polar coordinates
    *7–13. Common curves in polar coordinates

Chapter 8. Transcendental and Other Curves . . . 295

    8–1. The graph of $y = x^n$ and other algebraic curves
    8–2. Discussion and sketching or curves in rectangular coordinates
    8–3. The exponential curves
    8–4. The logarithmic curve
    8–5. The trigonometric curves
    *8–6. Curves represented by $y = a \sin(bx + \phi)$, where a, b, and ϕ are constants
    *8–7. Curves represented by other equations involving trigonometric functions
    *8–8. The graphs of the inverse trigonometric functions
    *8–9. Parametric equations
    *8–10. The ovals of Cassini
    *8–11. The hyperbolic functions

II. ANALYTIC GEOMETRY OF SPACE

*Chapter 9. The Point and Space Vectors . . . 335

    9–1. The point in space
    9–2. Projections
    9–3. Scalar components of a segment
    9–4. Length, or magnitude, of a segment
    9–5. Direction cosines of a segment
    9–6. Space vectors
    9–7. Cosine of the angle between two vectors
    9–8. The coordinates of a point that divides a segment in a given ratio
    9–9. The mid-point of a segment

* May be omitted in a brief course.

Chapter 10. The Plane . . . 356

- 10–1. An equation of a plane
- 10–2. Parallel and perpendicular planes
- 10–3. Intercept equation of a plane
- 10–4. Vector product of two vectors
- 10–5. The vector product: A summary
- 10–6. Alternating product, or scalar triple product, of three vectors
- 10–7. Vector triple products
- 10–8. Determining a plane satisfying three conditions
- 10–9. Distance from a plane to a point
- *10–10. Angle between two planes
- *10–11. Pencils of planes

Chapter 11. The Straight Line in Space . . . 381

- 11–1. Direction numbers and direction cosines
- 11–2. Parametric equations of a line
- 11–3. Symmetric equations of a line
- 11–4. The general equations of a line
- 11–5. Intersecting planes
- 11–6. Three homogeneous linear equations
- *11–7. Parametric equations of a plane

Chapter 12. Surfaces and Curves . . . 396

- 12–1. Introduction
- 12–2. Surfaces of revolution
- 12–3. Cylindrical surfaces
- 12–4. Conical surfaces
- 12–5. Quadric surfaces
- 12–6. The spheres
- 12–7. The ellipsoids
- 12–8. The elliptic paraboloids
- 12–9. The hyperboloids
- 12–10. The hyperbolic paraboloid
- 12–11. Ruled surfaces
- 12–12. Curves in space
- 12–13. Sketching the curve of intersection of two surfaces
- 12–14. Transformation of the axes
- 12–15. Non-rectangular coordinate systems

Appendix A. Brief Introduction to Matrix Theory . . . 435

Appendix B. Answers to Selected Exercises . . . 449

Index . . . 479

* May be omitted in a brief course.

Modern Analytic Geometry

Analytic Geometry in the Plane

Introduction

1

The purpose of this introduction is to help the student recall some of the important results of algebra and trigonometry which are applied in analytic geometry. It is suggested that the course in analytic geometry be begun by a hurried review of this introductory chapter.

It may be necessary to elaborate in certain sections, depending upon the background of the students. Since this is a text in analytic geometry, the amount of material included in this review chapter is limited and is presented here for reference rather than teaching purposes. This should be borne in mind when determining how to use the chapter.

SETS

1–1. Set notation

A set, aggregate, or collection of objects is an important concept in mathematics. The objects of the set are called its elements. A particular set is determined precisely if there is a specified way to decide which elements do or do not belong to it as members.

A common way to specify a set is by listing the elements. For example, if the elements of a set are a, b, and c, then we enclose them in braces { }, and list them as $\{a, b, c\}$

A second way of specifying a set is by use of the set-builder notation. For example, a set A may be described as

$$A = \{x \mid x \text{ is an even, positive integer less than } 10\}$$

This set could also be designated by listing the elements, namely,

$$A = \{2, 4, 6, 8\}$$

In words, the set-builder notation asserts that A is the set of all x such that x is an even, positive integer less than 10. The bar, $\mid$, is read "such that." Following the bar, precise information is given as to what properties the elements x must satisfy.

As another example, the symbol

$$\{(x,y) \mid x \text{ and } y \text{ are real numbers}\}$$

reads "the set of all ordered pairs* (x,y) such that x and y are real numbers."

The notation, $a \in A$, simply means that a is an element of set A. For example, $1 \in \{1,2,3\}$. The notation $a \notin A$ means that a is not an element of set A.

Two sets are equal if they contain the same elements. For example, $\{a,b,c\} = \{c,b,a\}$. The order in which the elements are listed is of no importance in describing a set, and only distinct elements are listed. The repetition of distinct elements does not change the set. For example, $\{a,a,b,c\} = \{a,b,c\}$.

1–2. Special sets

The *universal set* U is the working set from which all the elements under discussion are chosen. In most of the work in this text, the elements of the sets will be real numbers or ordered pairs of real numbers; consequently, the universal set will be the real number system. It is usually clear from the context of a problem what the universal set is; therefore, it is customary not to state it explicitly.

The *empty*, or *null* set, denoted by ϕ is the set which contains no elements. For example, $\{x \mid x \text{ is a negative integer greater than } 3\} = \phi$.

1–3. Set operations

The *union* of two sets, $A \cup B$ (read "A union B") is the set of all elements that are in A or in B, i.e., $x \in A \cup B$ provided $x \in A$ or $x \in B$.

In this definition the word "or" is used in the inclusive sense, that is, $x \in A$ or $x \in B$ means x belongs to A, or x belongs to B, or to both. The symbol $\vee$ is commonly used in place of the inclusive "or." Thus $x \in A \vee x \in B$ means $x \in A$ or $x \in B$. (Fig. 1–1a)

The *intersection* of two sets, $A \cap B$ (read "A intersection B") is the set of all elements that are in both A and B, i.e., $x \in A \cap B$ provided $x \in A$ and $x \in B$.

The symbol $\wedge$ is commonly used in place of the word "and" in statements such as $x \in A$ and $x \in B$. Thus the notation $x \in A \wedge x \in B$ means x belongs to A and x belongs to B. (Fig. 1–1b)

The *complement* of a set A, denoted by A', is the set of elements not in A. That is, the set A' consists of all the elements in the universal set that are not in A, namely $A' = \{x \mid x \notin A\}$. (Fig. 1–1c).

The set A is said to be a *subset* of the set B provided every element of A is also an element of B. The notation $A \subseteq B$ means A is a subset

---

* An ordered pair (a,b) simply means that the order in which the elements a and b appear is important. (a,b) and (b,a) are two different ordered pairs provided $a \neq b$.

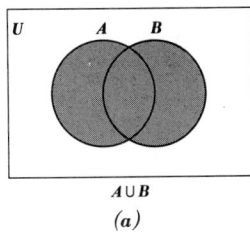
$A \cup B$
(a)

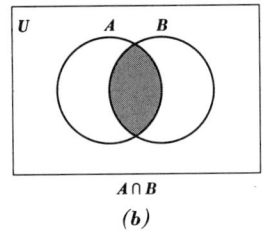
$A \cap B$
(b)

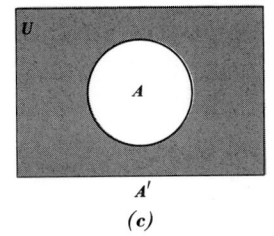
A'
(c)

FIG. 1-1

of B. The set A is said to be a *proper subset* of B provided A is a subset of B and B contains at least one element which is not in A. The notation $A \subset B$ means A is a proper subset of B.

Since the empty set ϕ contains no elements, we agree to write $\phi \subseteq A$ for any set A. The empty set ϕ is a subset of itself and a proper subset of any nonempty set. Thus $\phi \subseteq \phi$ and $\phi \subset A$, provided A is not the empty set.

The *cartesian product* of two sets A and B is the set of all ordered pairs (a,b) such that $a \in A \wedge b \in B$. The notation for the cartesian product is $A \times B$. It should be observed that $A \times B$ does not equal $B \times A$. (See Example 1-1 for $C \times D$ and $D \times C$.)

Let R be the set of all real numbers. We now consider $R \times R$, the cartesian product of R with itself.

$$R \times R = \{(x,y) \mid x \in R \text{ and } y \in R\}$$

This is the set of all ordered pairs (x,y) where x and y are real numbers. Since each point in the plane is represented by an ordered pair (x,y)*, the cartesian product $R \times R$ is commonly used to represent the totality of all points in the plane.

EXAMPLE 1-1. Consider the universal set $U = \{1,2,3,4,5,6,7,8,9,10\}$ and the sets:

$A = \{1,2,3,4,5,6\}$ $\qquad$ $B = \{5,6,7,8\}$
$C = \{3,4,5\}$ $\qquad$ $D = \{8,9,10\}$

The following statements are true for these sets.

$A \cap B = \{5,6\}$, $A \cup B = \{1,2,3,4,5,6,7,8\}$, $A \cap D = \phi$, $A' = \{7,8,9,10\}$, $C \subset A$
$A' \cap B = \{7,8\}$, $A \subset D'$
$C \times D = \{(3,8),(3,9),(3,10),(4,8),(4,9),(4,10),(5,8),(5,9),(5,10)\}$
$D \times C = \{(8,3),(8,4),(8,5),(9,3),(9,4),(9,5),(10,3),(10,4),(10,5)\}$
$C \times C = \{(3,3),(3,4),(3,5),(4,3),(4,4),(4,5),(5,3),(5,4),(5,5)\}$

---

* This is discussed in some detail in Chapter 2.

ANALYTIC GEOMETRY IN THE PLANE

EXERCISES

1. Indicate by appropriate diagrams, such as in Fig. 1–1, the following sets:
   (a) $A' \cap B$
   (b) $A \cap (B \cup C)$
   (c) $(A \cap B) \cup C$
   (d) $(A \cap B)' \cup C$
   (e) $(A \cap B) \cap C$
   (f) $(A \cap B) \cap C'$

2. Use diagrams, such as in Fig. 1–1, to verify:
   (a) $(A \cup B)' = A' \cap B'$
   (b) $(A \cap B)' = A' \cup B'$
   (c) $A \cup (B \cap C) = (A \cup B) \cap (A \cup C)$
   (d) $A \cap (B \cup C) = (A \cap B) \cup (A \cap C)$

3. Let $U = \{a,b,c,d,e,f,g\}$; $A = \{a,b,c\}$, $B = \{c,d\}$, and $C = \{d,e,f,g\}$. Find:
   (a) A'
   (b) $A \cap C$
   (c) $A \times B$
   (d) $B \times A$
   (e) $B' \cup C$
   (f) $(B \cap C) \cup A'$
   (g) $U \cap A'$
   (h) $(A \cup B)'$

4. Write an expression representing the shaded portion in each of the following diagrams.

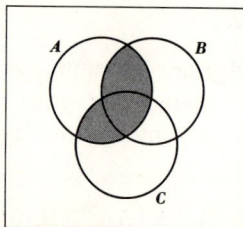 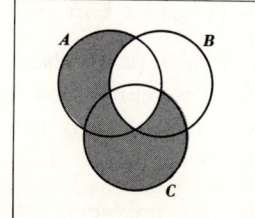 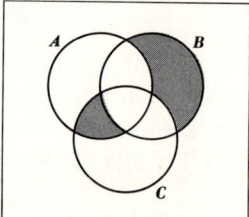

ALGEBRA

1–4. Linear equation* in one unknown

To solve a linear equation in one unknown, collect all terms involving the unknown, say x, on one side of the equality sign and the constant terms (terms not involving the unknown) on the other side. Then multiply both sides of the equation by the multiplicative inverse, or the reciprocal, of the coefficient of the unknown x, provided this coefficient is not zero. In the equation, $ax - b = 0$, add b to both sides to obtain

---
* A linear equation is an equation of the first degree in the unknowns. For example, $ax + by + c = 0$ is a linear equation in x and y and, as will be shown later, represents a line graph (hence the term linear).

$ax = b$, where a and b are constants and $a \neq 0$. Then multiplying both sides by $\frac{1}{a}$, we have

$$x = \frac{b}{a} \tag{1-1}$$

1-5. The quadratic equation $ax^2 + bx + c = 0$

The solutions (or roots) of the quadratic equation $ax^2 + bx + c = 0$, where $a \neq 0$, are given by the formula

$$x = \frac{-b \pm \sqrt{b^2 - 4ac}}{2a} \tag{1-2}$$

This formula gives the two roots

$$x_1 = \frac{-b + \sqrt{b^2 - 4ac}}{2a} \quad \text{and} \quad x_2 = \frac{-b - \sqrt{b^2 - 4ac}}{2a} \tag{1-3}$$

The student would do well to learn this result in terms of the meanings of the coefficients a, b, and c. That is, he should know a as the coefficient of the second-degree term, b as the coefficient of the linear term, and c as the constant term. These coefficients are determined after the equation is written in the standard form $ax^2 + bx + c = 0$.

The quantity $(b^2 - 4ac)$ is called the *discriminant* of the quadratic equation and enables one to determine directly whether the roots are real and distinct, real and equal, or complex (and distinct). Hence, when a, b, and c are real numbers, the roots x_1 and x_2 are real and distinct if $(b^2 - 4ac) > 0$; x_1 and x_2 are real and equal if $(b^2 - 4ac) = 0$; and x_1 and x_2 are complex and distinct if $(b^2 - 4ac) < 0$.

The following relations exist among the coefficients a, b, and c and the roots x_1 and x_2:

$$\begin{aligned} x_1 + x_2 &= \frac{-b}{a} \\ x_1 x_2 &= \frac{c}{a} \end{aligned} \tag{1-4}$$

EXAMPLE 1-2. Find the roots of the equation

$$2x^2 - 5x - 4 = 0$$

Solution. Using (1-3) we obtain

$$x_1 = \frac{-(-5) + \sqrt{(-5)^2 - 4 \cdot 2(-4)}}{2 \cdot 2}$$

and

$$x_2 = \frac{-(-5) - \sqrt{(-5)^2 - 4 \cdot 2(-4)}}{2 \cdot 2}$$

Hence,

$$x_1 = \frac{5 + \sqrt{57}}{4} \quad \text{and} \quad x_2 = \frac{5 - \sqrt{57}}{4}$$

These roots are real and distinct.

8   ANALYTIC GEOMETRY IN THE PLANE

The quadratic formula is useful in solving for one unknown explicitly in terms of the other when they are related by an equation such as $xy^2 + 3yx^2 - 5 = 0$. This equation is of degree 2 in y, hence we can solve for y by using (1–3). Here a, b, and c are x, $3x^2$, and -5 respectively.

$$y_1 = \frac{-3x^2 + \sqrt{9x^4 + 20x}}{2x} \quad \text{and} \quad y_2 = \frac{-3x^2 - \sqrt{9x^4 + 20x}}{2x}$$

EXERCISES

1. Verify equations (1–4) by computing $x_1 + x_2$ and $x_1 x_2$ where x_1 and x_2 are given in (1–3).

2. Solve each of the following linear equations for x:

   (a) $2x - 3 + x - \dfrac{1}{4} = x - \dfrac{x}{2} + 2 - \dfrac{3}{4}$ (d) $2ax + b = -ax + c,\ a \neq 0$

   (b) $\dfrac{x}{2} - \dfrac{1}{3} + \dfrac{2}{5} = 3 - \dfrac{x}{4}$ (e) $rx - 2 = sx + t,\ r - s \neq 0$

   (c) $2x + \dfrac{3}{5} - \dfrac{x}{3} - \dfrac{x}{4} = x - \dfrac{1}{5} + 4$ (f) $\dfrac{x}{a} + ab = \dfrac{x}{b},\ a \neq b \neq 0$

3. Solve each of the following quadratic equations:

   (a) $x^2 - 3x + 4 = 0$ (d) $x^2 + 5x = 0$
   (b) $2x^2 - 3x - 1 = 0$ (e) $3x^2 + 2x = 6$
   (c) $3x^2 + 2x - 5 = 0$ (f) $x^2 - x = 3$

4. Using the discriminant, determine whether the roots of each of the following equations are real or complex:

   (a) $x^2 - 2x + 1 = 0$ (d) $2x^2 - 3x + 5 = 0$
   (b) $x^2 - 2x - 1 = 0$ (e) $9x^2 - 6x + 1 = 0$
   (c) $x^2 - 2x + 2 = 0$ (f) $x^2 + 3x - 4 = 0$

5. Determine k in the equation $x^2 + kx - 4 = 0$ if the sum of the roots is 2.

6. Determine k in the equation $x^2 - 3x + 2k = 0$ if the product of the roots is 6.

7. Determine k in the equation $kx^2 + 2x - 3 = 0$: (a) if the sum of the roots is $-\dfrac{1}{4}$; (b) if the product of the roots is $\dfrac{3}{7}$.

8. Find the roots of a quadratic equation if the sum of the roots is 5 and the product of the roots is 6.

9. Solve for y in terms of x:

   (a) $x^3 y^2 - 3xy + 6x^2 - 2 = 0$ (c) $9(x+2)^2 + (y-1)^2 = 4$
   (b) $2x^2 + 3xy - 2y^2 = 0$ (d) $4x^2 + 4x = (2y-1)^2$

10. Solve for x in terms of y for 9(b), 9(c), and 9(d).

1–6. Exponents

The meanings of various types of exponents are shown by the following typical expressions in which $a \neq 0$ and n, p, and q are positive or negative integers:

$$a^0 = 1 \tag{1-5}$$

$$a^{-n} = \frac{1}{a^n} \tag{1-6}$$

$$a^{\frac{p}{q}} = \sqrt[q]{a^p} = (\sqrt[q]{a})^p \tag{1-7}$$

provided $a > 0$ when q is even.

Examples are: $3^0 = 1$; $2^{-3} = \frac{1}{2^3}$ $7^{2/3} = \sqrt[3]{7^2} = (\sqrt[3]{7})^2$.

The basic laws applied in performing operations involving exponents follow:

1. *When multiplying numbers with the same base, add the exponents.* Thus,

$$a^m a^n = a^{m+n} \tag{1-8}$$

So $2^3 \cdot 2^2 = 2^5$ and $3^{1/2} \cdot 3^{1/3} = 3^{5/6}$.

2. *When dividing numbers with the same base, subtract the exponents.* Thus,

$$\frac{a^m}{a^n} = a^{m-n} \tag{1-9}$$

So $\frac{3^5}{3^2} = 3^3$ and $\frac{2^{1/2}}{2^{1/3}} = 2^{1/6}$.

Other relations follow:

$$(a^m)^n = a^{mn} \tag{1-10}$$

So $(2^3)^2 = 2^6$ and $(3^{1/2})^3 = 3^{3/2}$.

$$a^m b^m = (ab)^m \tag{1-11}$$

So $3^2 2^2 = 6^2$.

$$\frac{a^m}{b^m} = \left(\frac{a}{b}\right)^m \tag{1-12}$$

So $\frac{5^3}{2^3} = \left(\frac{5}{2}\right)^3$.

$$\sqrt[n]{ab} = \sqrt[n]{a} \cdot \sqrt[n]{b}, \quad \sqrt[n]{\frac{a}{b}} = \frac{\sqrt[n]{a}}{\sqrt[n]{b}}, \quad a > 0 \text{ and } b > 0 \text{ when } n \text{ is even.} \tag{1-13}$$

So $\sqrt{7 \cdot 11} = \sqrt{7} \cdot \sqrt{11}$ and $\sqrt[3]{\frac{3}{5}} = \frac{\sqrt[3]{3}}{\sqrt[3]{5}}$.

The equalities in (1–13) are used in the process called *rationalizing the denominator* of a fraction. For example,

$$\frac{1}{\sqrt{3}} = \frac{1}{\sqrt{3}} \cdot \frac{\sqrt{3}}{\sqrt{3}} = \frac{\sqrt{3}}{3} \quad \text{and} \quad \frac{1}{\sqrt[3]{2}} = \frac{1}{\sqrt[3]{2}} \cdot \frac{\sqrt[3]{4}}{\sqrt[3]{4}} = \frac{\sqrt[3]{4}}{\sqrt[3]{8}} = \frac{\sqrt[3]{4}}{2}$$

The symbol $\sqrt{a}$, $a > 0$, represents the *positive* square root, which is called the *principal* square root of a. For example, $\sqrt{4} = 2$, not ± 2, and $\sqrt{(-5)^2} = \sqrt{25} = 5$. Hence, $\sqrt{x^2} = x$ when x is positive and $\sqrt{x^2} = -x$ when x is negative.

EXERCISES

1. Use the laws of exponents to evaluate each of the following: $2^2 2^3$; $3^4 3^2$; $\frac{5^2}{5^3}$; $\frac{3^5}{3}$; 3^0; $(2^2)^3$; $(2)(3^0)$; $\frac{1}{6x^0}$; $\left(\frac{2}{3}\right)^3$; $\sqrt[3]{8^2}$; $\sqrt{\frac{4}{9}}$; $\sqrt[4]{\frac{1}{16}}$; $(32)^{2/5}$; $(3^4)^2$.

2. Write the following with positive exponents and simplify: 3^{-1}; 4^{-2}; $\frac{1}{5^{-1}}$; $\left(\frac{2}{3}\right)^{-2}$; $(2^{-2})^{-1}$; $(3^{-1}2)^0$; $16^{-3/4}$; $2^{-1}+3^{-1}$; $(2^{-2}-3^{-2})^{-1}$; $\frac{1}{\sqrt{3^{-2}}}$; $\frac{3^0}{3^{-1}}$; $27^{-2/3}$; $(a+b)^{-1}$; $a^{-1}+b^{-1}$; $\frac{1}{a^{-1}+b^{-1}}$.

3. Write the following with positive exponents and simplify (assume all the letters represent positive real numbers):
$\sqrt{a^4b^3}$; $\sqrt{20a^5}$; $\sqrt[3]{24x}$; $\sqrt{(-4)^2a^2b^3}$.

4. Rationalize the denominator:
$\frac{3}{\sqrt{2}}$; $\frac{4}{\sqrt{a}}$, $a > 0$; $\frac{3}{\sqrt[3]{3}}$; $\frac{b}{\sqrt[3]{2a^2}}$, $a \neq 0$.

1–7. Logarithms

If $N = a^b$, where a is a positive number other than 1 and N is positive, b is called the logarithm of N to the base a. We write this as follows:

$$\log_a N = b \tag{1-14}$$

The two equations

$$N = a^b \text{ and } \log_a N = b \tag{1-15}$$

are equivalent, the first is in exponential form and the second is in logarithmic form.

EXAMPLE: To solve $\log_5 x = 3$ for x, we first put the equation in exponential form, which is $x = 5^3$. Hence, $x = 125$.

EXAMPLE: To solve $\log_x 8 = 3$, again we use the exponential form, $x^3 = 8$, from which we get $x = 2$.

EXAMPLE: To solve $8^x = b$ for x, use the logarithmic form to obtain $x = \log_8 b$. If $b = 64$, then $x = \log_8 64 = 2$.

A logarithm, therefore, is an exponent associated with a particular base. Logarithms are important in applications since they simplify arithmetical computations. They reduce a problem of multiplication to one of addition, a problem of division to one of subtraction, and a problem of determining a power or extracting a root to one of multiplication or division.

The basic theorems on logarithms are expressed by the following relations:

$$\log_a a = 1 \tag{1-14}$$

$$\log_a 1 = 0 \tag{1-15}$$

$$\log_a (MN) = \log_a M + \log_a N \tag{1-16}$$

$$\log_a \frac{M}{N} = \log_a M - \log_a N \tag{1-17}$$

$$\log_a M^k = k \log_a M \tag{1-18}$$

where k is any real number.

$$\log_b N = \frac{\log_a N}{\log_a b} \tag{1-19}$$

EXERCISES

1. Write each of the following in logarithmic form: $2^3 = 8$; $\sqrt{9} = 3$; $2^{-2} = \frac{1}{4}$; $16^{3/4} = 8$.

2. Write each of the following in exponential form: $\log_2 16 = 4$; $\log_9 3 = \frac{1}{2}$; $\log_2 \frac{1}{2} = -1$; $\log_3 1 = 0$

3. Solve each of the following equations:
    (a) $\log_x 27 = 3$
    (b) $\log_x 3 = \frac{1}{2}$
    (c) $\log_x 4 = \frac{2}{3}$
    (d) $x = \log_{10} 0.1$
    (e) $x = \log_2 \frac{1}{8}$
    (f) $x = \log_{64} 16$
    (g) $\log_{10} x = 3$
    (h) $\log_{16} x = -\frac{3}{4}$
    (i) $\log_a x = \frac{2}{3}$

4. Show that $a^{\log_a x} = x$.

5. Show that $\log_b b^x = x$.

6. Prove the theorems on logarithms given by equations (1–14) to (1–18).

7. Use logarithms to evaluate each of the following:
    (a) $(8.56)(3.47)(198)$
    (b) $(2.86) \div (33.2)$
    (c) $(3.74)^5$
    (d) $(0.00323)^{3/7}$
    (e) $\sqrt[3]{976 \times 10^5}$
    (f) $[(17.3)(0.0453)^2] \div (0.253)$

Note: Problem 7 is inserted for practice in the use of a common (base 10) logarithm table. Even though computers are now usually available, practice in the art of doing arithmetic is still essential.

1–8. Determinants*

A determinant, as used in this text, is a number associated with a square array of elements, each of which is a number. The symbol commonly used to denote the determinant of a square array is

$$\begin{vmatrix} a_{11} & a_{12} & \ldots & a_{1n} \\ a_{21} & a_{22} & \ldots & a_{2n} \\ \vdots & \vdots & & \vdots \\ a_{n1} & a_{n2} & \ldots & a_{nn} \end{vmatrix}$$

The vertical bars enclosing the elements are a symbol for the determinant, which is a number associated with the array. Each element in the array has a position; the two subscripts give that position. The first subscript gives the row and the second gives the column. For example, a_{34} is the element in the third row and the fourth column. To find the value of a determinant, we proceed as follows:

1. Form all possible products of elements with each product consisting of exactly one element from each row and each column.

2. Arrange the elements of each product in such a way that the row subscripts appear in their natural order. Interchange the column subscripts two at a time until they are in their natural order. If the number of interchanges is even, prefix a plus sign to the term; if the number of interchanges is odd, prefix a minus sign to the term.

3. The algebraic sum of all these terms is the *value* of the determinant.

EXAMPLE 1–3. To illustrate the procedure, we evaluate the following determinant.

$$\begin{vmatrix} a_{11} & a_{12} & a_{13} \\ a_{21} & a_{22} & a_{23} \\ a_{31} & a_{32} & a_{33} \end{vmatrix}$$

Solution. Step 1. The six possible terms consisting of products of exactly one element from each row and each column are: $a_{11}a_{22}a_{33}$, $a_{11}a_{23}a_{32}$, $a_{12}a_{21}a_{33}$, $a_{13}a_{22}a_{31}$, $a_{12}a_{23}a_{31}$, $a_{13}a_{21}a_{32}$.

Step 2. We now find the sign associated with each term. A plus sign is associated with $a_{11}a_{22}a_{33}$ since 0 interchanges are required to put the column subscripts in natural order. A minus sign is associated with $a_{11}a_{23}a_{32}$ since 1 interchange is required. One interchange is required for $a_{12}a_{21}a_{33}$ and $a_{13}a_{22}a_{31}$, and two are required for $a_{12}a_{23}a_{31}$ and $a_{13}a_{21}a_{32}$.

* For a more elaborate discussion of determinants and proofs of the theorems, see any reputable text on college algebra.

We then have $+ a_{11}a_{22}a_{33}, - a_{11}a_{23}a_{32}, - a_{12}a_{21}a_{31}, - a_{13}a_{22}a_{31}, + a_{12}a_{23}a_{31},$ and $+ a_{13}a_{21}a_{32}.$

Step 3. The value of the determinant is $a_{11}a_{22}a_{33} - a_{11}a_{23}a_{32} - a_{12}a_{21}a_{33} - a_{13}a_{22}a_{31} + a_{12}a_{23}a_{31} + a_{13}a_{21}a_{32}.$

The *order* of a determinant is the number of rows (or columns) in the determinant. The determinant in Example 1-3 is a third order determinant.

There are n^2 elements in a determinant of order n.

There are $n!$ (read n factorial*) terms in the expansion of a determinant of order n. So for a third-order determinant there are six terms; for a fourth-order determinant there are twenty-four terms; and so on.

A *co-factor* of an element is $(-1)^{p+q}$ times the determinant remaining when the row and column that contain the element have been deleted. Here p is the number of the row and q is the number of the column containing the element. For example, let it be required to find the co-factor of a_{23} in the third-order determinant

$$\begin{vmatrix} a_{11} & a_{12} & a_{13} \\ a_{21} & a_{22} & a_{23} \\ a_{31} & a_{32} & a_{33} \end{vmatrix}$$

The element a_{23} is in row 2 and in column 3. Therefore, the co-factor of a_{23} is

$$(-1)^{2+3} \begin{vmatrix} a_{11} & a_{21} \\ a_{31} & a_{32} \end{vmatrix}$$

We shall denote the co-factors by capital letters. So the co-factor of a_{23} is A_{23}, that of a_{32} is A_{32}, etc.

We state without proof some very important and useful theorems on determinants. These theorems are illustrated with determinants of orders two and three. The student is advised to check each of the results with the expansion of the determinants.

Theorem 1: The value of a determinant is equal to the algebraic sum of the products of the elements of any row (or column) and their corresponding co-factors. For example,

$$\begin{vmatrix} a_{11} & a_{12} \\ a_{21} & a_{22} \end{vmatrix} = a_{11}A_{11} + a_{12}A_{12} = a_{21}A_{21} + a_{22}A_{22} \qquad (1\text{-}22)$$

Either form reduces to $a_{11}a_{22} - a_{12}a_{21}$.

---

*$n! = (n)(n-1)(n-2) \ldots (4)(3)(2)(1); 5! = (5)(4)(3)(2)(1) = 120; 6! = (6)(5)(4)(3)(2)(1) = 720; 0! = 1$ by definition.

A third order determinant can be evaluated in any one of six ways. Thus,

$$\begin{vmatrix} a_{11} & a_{12} & a_{13} \\ a_{21} & a_{22} & a_{23} \\ a_{31} & a_{32} & a_{33} \end{vmatrix} = \begin{matrix} a_{11}A_{11} + a_{12}A_{12} + a_{13}A_{13} \\ a_{21}A_{21} + a_{22}A_{22} + a_{23}A_{23} \\ a_{31}A_{31} + a_{32}A_{32} + a_{33}A_{33} \\ a_{11}A_{11} + a_{21}A_{21} + a_{31}A_{31} \\ a_{12}A_{12} + a_{22}A_{22} + a_{23}A_{23} \\ a_{13}A_{13} + a_{23}A_{23} + a_{33}A_{33} \end{matrix} \qquad (1\text{--}23)$$

The student should check to see that each gives the correct result. For example,

$$a_{11}A_{11} + a_{12}A_{12} + a_{13}A_{13} = (-1)^{1+1}a_{11}\begin{vmatrix} a_{22} & a_{23} \\ a_{32} & a_{33} \end{vmatrix} + (-1)^{1+2}a_{12}\begin{vmatrix} a_{21} & a_{23} \\ a_{31} & a_{33} \end{vmatrix}$$

$$+ (-1)^{1+3}a_{13}\begin{vmatrix} a_{21} & a_{22} \\ a_{31} & a_{32} \end{vmatrix}$$

$$+ a_{11}(a_{22}a_{33} - a_{23}a_{32}) - a_{12}(a_{21}a_{33} - a_{23}a_{31})$$

$$+ a_{13}(a_{21}a_{32} - a_{22}a_{31})$$

$$= a_{11}a_{22}a_{33} - a_{11}a_{23}a_{32} - a_{12}a_{21}a_{33}$$

$$+ a_{12}a_{23}a_{31} + a_{13}a_{21}a_{32} - a_{13}a_{22}a_{31}$$

Theorem 2: If rows and columns are interchanged, the value of the determinant is unchanged.

$$\begin{vmatrix} a_{11} & a_{12} & a_{13} \\ a_{21} & a_{22} & a_{23} \\ a_{31} & a_{32} & a_{33} \end{vmatrix} = \begin{vmatrix} a_{11} & a_{21} & a_{31} \\ a_{12} & a_{22} & a_{32} \\ a_{13} & a_{23} & a_{33} \end{vmatrix} \qquad (1\text{--}24)$$

Theorem 3: If any two rows (or columns) are interchanged, the value of the determinant changes sign.

$$\begin{vmatrix} a_{11} & a_{12} & a_{13} \\ a_{21} & a_{22} & a_{23} \\ a_{31} & a_{32} & a_{33} \end{vmatrix} = - \begin{vmatrix} a_{31} & a_{32} & a_{33} \\ a_{21} & a_{22} & a_{23} \\ a_{11} & a_{12} & a_{13} \end{vmatrix} \qquad (1\text{--}25)$$

Theorem 4: If two rows (or columns) are alike, the determinant is zero.

Theorem 5: If all the elements of any row (or column) have a common factor, this factor may be placed outside the determinant symbol.

$$\begin{vmatrix} ka_{11} & ka_{12} & ka_{13} \\ a_{21} & a_{22} & a_{23} \\ a_{31} & a_{32} & a_{33} \end{vmatrix} = k \begin{vmatrix} a_{11} & a_{12} & a_{13} \\ a_{21} & a_{22} & a_{23} \\ a_{31} & a_{32} & a_{33} \end{vmatrix} \qquad (1\text{--}26)$$

Theorem 6: If the elements of any two rows (or columns) are proportional,* the determinant is zero.

$$\begin{vmatrix} a_{11} & a_{12} & a_{13} \\ ka_{11} & ka_{12} & ka_{13} \\ a_{31} & a_{32} & a_{33} \end{vmatrix} = 0$$

Theorem 7: If each of the elements of any column (or row) is expressed as the sum of two elements, then the determinant is equal to the sum of two determinants, as follows:

$$\begin{vmatrix} a_{11}+b_{11} & a_{12} & a_{13} \\ a_{21}+b_{21} & a_{22} & a_{23} \\ a_{31}+b_{31} & a_{32} & a_{33} \end{vmatrix} = \begin{vmatrix} a_{11} & a_{12} & a_{13} \\ a_{21} & a_{22} & a_{23} \\ a_{31} & a_{32} & a_{33} \end{vmatrix} + \begin{vmatrix} b_{11} & a_{12} & a_{13} \\ b_{21} & a_{22} & a_{23} \\ b_{31} & a_{32} & a_{33} \end{vmatrix} \quad (1\text{-}28)$$

Theorem 8: If the products of a constant and the elements of one row (or column) are added to the corresponding elements of another row (or column), the value of the determinant is not altered.

$$\begin{vmatrix} a_{11} & a_{12} & a_{13} \\ a_{21} & a_{22} & a_{23} \\ a_{31} & a_{32} & a_{33} \end{vmatrix} = \begin{vmatrix} a_{11}+ka_{21} & a_{12}+ka_{22} & a_{13}+ka_{23} \\ a_{21} & a_{22} & a_{23} \\ a_{31} & a_{32} & a_{33} \end{vmatrix} \quad (1\text{-}29)$$

Theorem 9: The sum of the products of the elements of any row (or column) and the co-factors of the elements of any other row (or column) is zero.

This last theorem says, for example, that in the case of a third order determinant,

$$a_{11}A_{12} + a_{21}A_{22} + a_{31}A_{32} = 0$$

EXAMPLE 1-4. Find the value of

$$\begin{vmatrix} 1 & -1 & 3 \\ 2 & 0 & 1 \\ -3 & 1 & 5 \end{vmatrix}$$

Solution. We will use Theorem 1 and will choose the second row since this row contains one zero.

$$\begin{vmatrix} 1 & -1 & 3 \\ 2 & 0 & 1 \\ -3 & 1 & 5 \end{vmatrix} = (-1)^{2+1} \cdot 2 \begin{vmatrix} -1 & 3 \\ 1 & 5 \end{vmatrix} + (-1)^{2+2} \cdot 0 \begin{vmatrix} 1 & 3 \\ -3 & 5 \end{vmatrix}$$

$$+ (-1)^{2+3} \cdot 1 \begin{vmatrix} 1 & -1 \\ -3 & 1 \end{vmatrix} = -2(-8) + 0 - 1(-2) = 18$$

---

* The elements a_{11}, a_{12}, a_{13} are proportional to the elements a_{21}, a_{22}, a_{23} when there is a number k such that $a_{21} = ka_{11}$, $a_{22} = ka_{12}$, and $a_{23} = ka_{13}$ or there is a number l such that $a_{11} = la_{21}$, $a_{12} = la_{22}$, and $a_{13} = la_{23}$.

The evaluation of determinants of order higher than the second can be simplified by using Theorems 5 and 8. The object is to transform the given determinant into an equivalent* one having all but one of the elements in any one row (or column) equal to zero. The determinant can then be expressed as a determinant of the next lower order by Theorem 1.

We now outline a procedure by which this can be accomplished.

PROCEDURE

1. Choose a row (or column) as an operating row (or column). When possible choose a row (or column) which contains a $+1$ or -1, (rows 1, 3, or 4 or columns 1, 3, or 4 in Example 1–5).

2. The operating row (or column) (row 3 in Example 1–5) is then used to transform the given determinant into one with all but one zero in a particular column (or row). Note that only one operating row (or column) is used in any one transformation.

3. The operating row (or column) is not changed, but reproduced exactly in the transformed determinant. (Exception: a common factor can be removed from an operating row (or column) if the determinant is multiplied by the factor.)

4. The row (or column) being operated upon is replaced by the sum of that row (or column) and a constant times the operating row (or column). The row (or column) being operated upon is not multiplied by any factor except 1, (rows 1, 2, and 4 in Example 1–5).

When row or column operations are used in the evaluation of determinants, the operations used should be clearly stated. This is illustrated in the following two examples. In Example 1–5, $4R_3 + R_1$ opposite row 1 means "add each element in row 1 to four times the corresponding element in row 3, then place the result in row 1." In Example 1–6, the $\frac{1}{2} C_1$ below column 1 means "multiply each element in column 1 by $\frac{1}{2}$." Since a 2 is now factored out of column 1, the determinant must be multiplied by 2. (Theorem 5.)

EXAMPLE 1–5

$$\begin{vmatrix} 4 & -3 & 1 & 2 \\ 2 & 3 & -2 & -3 \\ -1 & -2 & 2 & 1 \\ 1 & 2 & -3 & 1 \end{vmatrix} \begin{matrix} 4R_3 + R_1 \\ 2R_3 + R_2 \\ R_3 \\ R_4 + R_1 \end{matrix} = \begin{vmatrix} 0 & -11 & 9 & 6 \\ 0 & -1 & 2 & -1 \\ -1 & -2 & 2 & 1 \\ 0 & 0 & -1 & 2 \end{vmatrix}$$

---

* Here equivalent determinant means one which has the same value as the given one.

$$= (-1)(-1)^4 \begin{vmatrix} -11 & 9 & 6 \\ -1 & 2 & -1 \\ 0 & -1 & 2 \end{vmatrix} \begin{matrix} -11R_2 + R_1 \\ R_2 \\ 0R_2 + R_1 \end{matrix} = - \begin{vmatrix} 0 & -13 & 17 \\ -1 & 2 & -1 \\ 0 & -1 & 2 \end{vmatrix}$$

$$= -(-1)(-1)^3 \begin{vmatrix} -13 & 17 \\ -1 & 2 \end{vmatrix} = -(-26 + 17) = 9$$

EXAMPLE 1-6

$$\begin{vmatrix} -6 & 3 & 4 & -7 \\ 2 & -5 & 2 & 0 \\ 0 & 4 & -7 & 3 \\ 4 & 2 & 5 & 2 \end{vmatrix} = 2 \begin{vmatrix} -3 & 3 & 4 & -7 \\ 1 & -5 & 2 & 0 \\ 0 & 4 & -7 & 3 \\ 2 & 2 & 5 & 2 \end{vmatrix} \begin{matrix} 3R_2 + R_1 \\ R_2 \\ 0R_2 + R_3 \\ -2R_2 + R_4 \end{matrix}$$

$\frac{1}{2} C_1$

$$= 2 \begin{vmatrix} 0 & -12 & 10 & -7 \\ 1 & -5 & 2 & 0 \\ 0 & 4 & -7 & 3 \\ 0 & 12 & 1 & 2 \end{vmatrix}$$

$$= 2 \begin{vmatrix} 0 & -12 & 10 & -7 \\ 1 & -5 & 2 & 0 \\ 0 & 4 & -7 & 3 \\ 0 & 12 & 1 & 2 \end{vmatrix} = 2(1)(-1)^3 \begin{vmatrix} -12 & 10 & -7 \\ 4 & -7 & 3 \\ 12 & 1 & 2 \end{vmatrix} = -2 \begin{vmatrix} -132 & 10 & -27 \\ 88 & -7 & 17 \\ 0 & 1 & 0 \end{vmatrix}$$

$$\begin{matrix} -12C_2 & C_2 & -2C_2 \\ + & & + \\ C_1 & & C_3 \end{matrix}$$

$$= (-2)(1)(-1)^5 \begin{vmatrix} -132 & -27 \\ 88 & 17 \end{vmatrix} = 2(-2244 + 2376) = 264$$

EXERCISES

1. Evaluate each of the following determinants:

   (a) $\begin{vmatrix} 2 & -1 \\ 3 & 4 \end{vmatrix}$ (b) $\begin{vmatrix} 1 & -3 \\ -1 & -2 \end{vmatrix}$ (c) $\begin{vmatrix} 2 & -1 \\ -4 & 2 \end{vmatrix}$ (d) $\begin{vmatrix} a & b \\ a^2 & b^2 \end{vmatrix}$

2. Evaluate each of the following determinants:

   (a) $\begin{vmatrix} 1 & -1 & 1 \\ 2 & 3 & -1 \\ -1 & 3 & 2 \end{vmatrix}$ (b) $\begin{vmatrix} 2 & -3 & 1 \\ -1 & 4 & 2 \\ -2 & -1 & 3 \end{vmatrix}$

   (c) $\begin{vmatrix} 1 & 2 & -1 \\ 3 & 0 & 5 \\ 1 & -4 & 7 \end{vmatrix}$ (d) $\begin{vmatrix} 1 & 2 & -5 \\ 0 & 6 & 1 \\ 0 & 0 & -2 \end{vmatrix}$

18 ANALYTIC GEOMETRY IN THE PLANE

3. Evaluate each of the following determinants:

(a) $\begin{vmatrix} -1 & 1 & 2 & -1 \\ 3 & -2 & 1 & 4 \\ 1 & 2 & -1 & 1 \\ -2 & 3 & 1 & -2 \end{vmatrix}$
(b) $\begin{vmatrix} 2 & -1 & 3 & 5 \\ 1 & -2 & 1 & 3 \\ -1 & 3 & 2 & -1 \\ -2 & -2 & -3 & 1 \end{vmatrix}$

(c) $\begin{vmatrix} 1 & -2 & 3 & 1 \\ 4 & 1 & -6 & 5 \\ 0 & 4 & -1 & 0 \\ -2 & 1 & -3 & 2 \end{vmatrix}$
(d) $\begin{vmatrix} 2 & 0 & -3 & 1 \\ 5 & -2 & 1 & 2 \\ 1 & -1 & 5 & 0 \\ 3 & 2 & -22 & 3 \end{vmatrix}$

4. Construct a determinant satisfying the requirements of each of the nine theorems and expand each one.

5. Show that

$$\begin{vmatrix} a & b & c \\ a^2 & b^2 & c^2 \\ a^3 & b^3 & c^3 \end{vmatrix} = abc(a-b)(b-c)(c-a)$$

1–9. Solution of simultaneous systems of linear equations

We start by solving two linear equations in two unknowns, x and y.

$$\begin{cases} a_1 x + b_1 y = c_1 \\ a_2 x + b_2 y = c_2 \end{cases} \quad (1\text{–}30)$$

Multiplying the first equation by a_2, the second by a_1 and subtracting, we obtain

$$(a_2 b_1 - a_1 b_2) y = a_2 c_1 - a_1 c_2$$

Now solve for y, assuming $a_2 b_1 - a_1 b_2 \neq 0$.

$$y = \frac{a_2 c_1 - a_1 c_2}{a_2 b_1 - a_1 b_2} = \frac{a_1 c_2 - a_2 c_1}{a_1 b_2 - a_2 b_1} \quad (1\text{–}31)$$

In like manner, eliminating y, we obtain

$$x = \frac{b_2 c_1 - b_1 c_2}{a_1 b_2 - a_2 b_1} \quad (1\text{–}32)$$

We now observe that the denominator of (1–31) and of (1–32) is the value of the $\begin{vmatrix} a_1 & b_1 \\ a_2 & b_2 \end{vmatrix}$, the determinant formed by the coefficients in the two given equations. The numerators can also be expressed as determinants. Hence,

$$x = \frac{\begin{vmatrix} c_1 & b_1 \\ c_2 & b_2 \end{vmatrix}}{\begin{vmatrix} a_1 & b_1 \\ a_2 & b_2 \end{vmatrix}}, \quad y = \frac{\begin{vmatrix} a_1 & c_1 \\ a_2 & c_2 \end{vmatrix}}{\begin{vmatrix} a_1 & b_1 \\ a_2 & b_2 \end{vmatrix}}, \text{ providing } \begin{vmatrix} a_1 & b_1 \\ a_2 & b_2 \end{vmatrix} \neq 0 \quad (1\text{–}33)$$

Equation (1–33) illustrates Cramer's Rule for the case of two linear equations in two unknowns.

Cramer's Rule: In a system of n linear equations in n unknowns, each unknown can be expressed as a quotient of two determinants. The denominator is the determinant of coefficients of the unknowns, and the numerator is the determinant obtained from the denominator by replacing the column of coefficients of that unknown by the column of constant terms on the right side of the equality sign in each equation.

This rule assumes that the determinant of coefficients is not zero. The method does not apply otherwise.

EXAMPLE 1–7. Solve the equations

$$2x - 3y + 6 = 0 \text{ and } 3x + 4y - 2 = 0$$

Solution. We first write the equations with the constant terms on the right.

$$\begin{cases} 2x - 3y = -6 \\ 3x + 4y = 2 \end{cases}$$

From equation (1–33),

$$x = \frac{\begin{vmatrix} -6 & -3 \\ 2 & 4 \end{vmatrix}}{\begin{vmatrix} 2 & -3 \\ 3 & 4 \end{vmatrix}} = \frac{-24 + 6}{8 + 9} = \frac{-18}{17}; \quad y = \frac{\begin{vmatrix} 2 & -6 \\ 3 & 2 \end{vmatrix}}{\begin{vmatrix} 2 & -3 \\ 3 & 4 \end{vmatrix}} = \frac{4 + 18}{17} = \frac{22}{17}$$

Therefore, $x = \dfrac{-18}{17}$ and $y = \dfrac{22}{17}$.

Check: Substitute these values for x and y in the original equations and we have:

$$2\left(\frac{-18}{17}\right) - 3\left(\frac{22}{17}\right) = \frac{-36 - 66}{17} = -\frac{102}{17} = -6, \text{ and}$$

$$3\left(\frac{-18}{17}\right) + 4\left(\frac{22}{17}\right) = \frac{-54 + 88}{17} = \frac{34}{17} = 2$$

The work thus far in this section has been with two equations in two unknowns. Cramer's Rule applies to systems of n linear equations in n unknowns. We now give an example of three equations in three unknowns.

Let it be required to solve the following three equations for x, y, and z:

$$a_{11}x + a_{12}y + a_{13}z = d_1$$
$$a_{21}x + a_{22}y + a_{23}z = d_2 \qquad (1\text{--}34)$$
$$a_{31}x + a_{32}y + a_{33}z = d_3$$

20 ANALYTIC GEOMETRY IN THE PLANE

Let A denote the determinant of the coefficients of the unknowns arranged in a definite order as in (1–34). Hence,

$$A = \begin{vmatrix} a_{11} & a_{12} & a_{13} \\ a_{21} & a_{22} & a_{23} \\ a_{31} & a_{32} & a_{33} \end{vmatrix}$$

By Cramer's Rule, provided $A \neq 0$, the solution is

$$x = \frac{\begin{vmatrix} d_1 & a_{12} & a_{13} \\ d_2 & a_{22} & a_{23} \\ d_3 & a_{32} & a_{33} \end{vmatrix}}{A} \tag{1–35}$$

$$y = \frac{\begin{vmatrix} a_{11} & d_1 & a_{13} \\ a_{21} & d_2 & a_{23} \\ a_{31} & d_3 & a_{33} \end{vmatrix}}{A} \tag{1–36}$$

$$z = \frac{\begin{vmatrix} a_{11} & a_{12} & d_1 \\ a_{21} & a_{22} & d_2 \\ a_{31} & a_{32} & d_3 \end{vmatrix}}{A} \tag{1–37}$$

EXAMPLE 1-3. Solve the following equations for x, y, and z, and check the results:

$$x + y - z - 1 = 0$$
$$x - 2y + z + 3 = 0$$
$$2x - y + 2z - 2 = 0$$

Solution. By equation (1–35),

$$x = \frac{\begin{vmatrix} 1 & 1 & -1 \\ -3 & -2 & 1 \\ 2 & -1 & 2 \end{vmatrix}}{\begin{vmatrix} 1 & 1 & -1 \\ 1 & -2 & 1 \\ 2 & -1 & 2 \end{vmatrix}}$$

Using Theorem 8 of Section 1–5, we obtain:

$$x = \frac{\begin{vmatrix} 0 & 0 & -1 \\ -2 & -1 & 1 \\ 4 & 1 & 2 \end{vmatrix}}{\begin{vmatrix} 0 & 0 & -1 \\ 2 & -1 & 1 \\ 4 & 1 & 2 \end{vmatrix}} = \frac{-\begin{vmatrix} -2 & -1 \\ 4 & 1 \end{vmatrix}}{-\begin{vmatrix} 2 & -1 \\ 4 & 1 \end{vmatrix}} = \frac{-2}{-6} = \frac{1}{3}$$

Similarly,

$$y = \frac{\begin{vmatrix} 1 & 1 & -1 \\ 1 & -3 & 1 \\ 2 & 2 & 2 \end{vmatrix}}{\begin{vmatrix} 1 & 1 & -1 \\ 1 & -2 & 1 \\ 2 & -1 & 2 \end{vmatrix}} = \frac{\begin{vmatrix} 0 & 0 & -1 \\ 2 & -2 & 1 \\ 4 & 4 & 2 \end{vmatrix}}{-6} = \frac{-\begin{vmatrix} 2 & -2 \\ 4 & 4 \end{vmatrix}}{-6} = \frac{8}{3}$$

$$z = \frac{\begin{vmatrix} 1 & 1 & 1 \\ 1 & -2 & -3 \\ 2 & -1 & 2 \end{vmatrix}}{\begin{vmatrix} 1 & 1 & -1 \\ 1 & -2 & 1 \\ 2 & -1 & 2 \end{vmatrix}} = \frac{\begin{vmatrix} 1 & 0 & 0 \\ 1 & -3 & -4 \\ 2 & -3 & 0 \end{vmatrix}}{-6} = \frac{-12}{-6} = 2$$

The solution is $x = \frac{1}{3}$, $y = \frac{8}{3}$, and $z = 2$.

Check: Substituting the results in the respective equations, we have:

$$\frac{1}{3} + \frac{8}{3} - 2 - 1 = 0$$

$$\frac{1}{3} - 2\left(\frac{8}{3}\right) + 2 + 3 = 0$$

$$2\left(\frac{1}{3}\right) - \frac{8}{3} + 4 - 2 = 0$$

EXERCISES

1. Solve each of the following equations for the unknown, and check the results:

   (a) $\begin{vmatrix} 1 & x \\ 2 & -3 \end{vmatrix} = 0$

   (b) $\begin{vmatrix} 1 & -1 & x \\ 2 & 1 & -1 \\ 1 & 4 & x \end{vmatrix} = 0$

   (c) $\begin{vmatrix} x & 2 & 1 \\ -1 & x & 0 \\ 1 & 1 & -1 \end{vmatrix} = 0$

2. Solve each of the following pairs of equations for x and y, and check the results:

   (a) $2x - 3y + 5 = 0$
   $x + 2y - 3 = 0$

   (b) $3x + 4y - 6 = 0$
   $2x - 3y + 1 = 0$

3. Solve each of the following systems of equations for x, y, and z, and check the results:

   (a) $2x - y + z = 4$
   $x + 2y - z = 1$
   $x - 6y + 3z = -2$

   (b) $x - 3y - 2z - 1 = 0$
   $2x - y + 3z = 0$
   $3x + 4y - z - 11 = 0$

   (c) $3x + y - z = 2$
   $x - y + 2z = 1$
   $2x + 3y - 2z = 4$

4. Solve each of the following systems of equations for x, y, z, and w, and check results:

(a) $\quad x + y + z = 1$
$\quad\quad 2y - 2z - w = -7$
$\quad\quad x + y - z = -3$
$\quad\quad x + z + w = 2$

(b) $\quad x + 3y - z + 2w = -4$
$\quad\quad 2x - y - 3z + 2w = -1$
$\quad\quad x - 2y - z + 3w = 8$
$\quad\quad x + y - z + 3w = 2$

1–10. Systems with quadratic equations

Some methods for solving systems of equations containing one or more quadratic equations are illustrated in the following examples.

EXAMPLE 1–9. Solve the following pair of equations for x and y.

$$x^2 + y^2 = 25$$
$$x + y = 1$$

Solution. Solve the linear equation for y and then substitute for y in the quadratic equation.

$$y = 1 - x$$
$$x^2 + (1-x)^2 = 25$$
$$2x^2 - 2x - 24 = 0$$
$$x^2 - x - 12 = 0$$
$$(x-4)(x+3) = 0$$
$$x = 4 \text{ or } x = -3$$

When $x = 4$, $y = 1 - 4 = -3$ and when $x = -3$, $y = 1 - (-3) = 4$. Hence the solutions are (4,3) and (−3,4).

EXAMPLE 1–10. Solve the following pair of equations for x and y.

$$x^2 + 2y^2 = 9$$
$$3x^2 + 5y^2 = 25$$

Solution. Multiply the first equation by 3 and subtract the second from the result.

$$3x^2 + 6y^2 = 27$$
$$3x^2 + 5y^2 = 25$$

Subtracting, we get

$$y^2 = 2$$

which yields $y = \sqrt{2}$ or $y = -\sqrt{2}$. Substitute the values of y in one of the original equations to get the corresponding values of x. The set of solutions is

$$\{(\sqrt{5}, \sqrt{2}), (-\sqrt{5}, \sqrt{2}), (\sqrt{5}, -\sqrt{2}), (-\sqrt{5}, -\sqrt{2})\}$$

EXERCISES

1. Solve the following systems of equations and check your answers.

   (a) $y^2 = 2x$
   $3x - y = 4$

   (b) $x^2 + y^2 = 16$
   $x + y = 4$

   (c) $4x^2 + 9y^2 = 36$
   $3x + 2y = 0$

   (d) $x^2 - 4y = 0$
   $x - 2y = 1$

   (e) $3y^2 - 5xy = 27$
   $x + y = 3$

2. Solve the following systems of equations and check your answers.

   (a) $7x^2 - 4y^2 = 47$
   $3x^2 + 2y^2 = 35$

   (b) $7x^2 + 2y^2 = 52$
   $5x^2 - 3y^2 = -16$

   (c) $5x^2 + 2y^2 = 13$
   $4x^2 - 7y^2 = 19$

   (d) $6xy + y^2 = 28$
   $8x^2 + y = 4$

   (e) $x^2 + y^2 = 26$
   $xy = 5$

1–11. Inequalities and absolute value

The real number a is less than a real number b provided there is a positive number x such that $a + x = b$. If a and b are associated with points on a number line, the point associated with a is to the left of the point associated with b. See Fig. 1–2. In symbols, we write $a < b$,

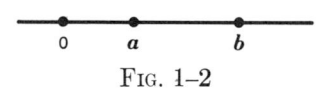

FIG. 1–2

where the symbol $<$ means *is less than*.

If $a < b$, then b is greater than a, or is to the right of a on the number line. The symbol $>$ means *is greater than*. Hence, $a < b$ and $b > a$ are different ways of expressing the relationship between a and b.

When we want to state that a is less than or equal to b, or a is greater than or equal to b, we write

$$a \leq b \text{ or } a \geq b$$

The following are important properties of inequalities:
1. If $a < b$, then $a + c < b + c$
2. If $a < b$ and $c > 0$, then $ac < bc$
3. If $a < b$ and $c < 0$, then $ac > bc$.

EXAMPLE 1–11. Solve the inequalities

   (a) $2x + 3 < 5$

   (b) $x + 5 \geq 3x - 7$

Solution. To solve an inequality in x means to find the set of real numbers x which satisfy the inequality.

(a) $2x + 3 < 5$
  $2x < 2$, adding -3 to each side
  $x < 1$, multiplying each side by $\frac{1}{2}$.

(b) $x + 5 \geq 3x - 7$
  $-2x \geq -12$, adding $-3x$ and -5 to each side
  $x \leq 6$, multiplying both sides by $-\frac{1}{2}$

These say that $\{x \mid 2x + 3 < 5\} = \{x \mid x < 1\}$ and that $\{x \mid x + 5 \geq 3x - 7\} = \{x \mid x \leq 6\}$.

The *absolute value* of the real number x, denoted by $|x|$, is defined as follows:

$$|x| = \begin{cases} x \text{ if } x \geq 0 \\ -x \text{ if } x < 0 \end{cases} \tag{1-38}$$

For example, $|3| = 3$ and $|-3| = 3$ since $-(-3) = 3$. The absolute value of a number is always nonnegative.

If a and b are real numbers then

1. $|a + b| \leq |a| + |b|$
2. $|a| - |b| \leq |a - b|$
3. $|ab| = |a| \cdot |b|$
4. $\left|\dfrac{a}{b}\right| = \dfrac{(a)}{(b)}$
5. $|a| = \sqrt{a^2}$

EXERCISES

1. *Explain* carefully whether the following are true or false; correct the ones which are false: $-3 < -4$; $4 > -5$; $-7 < -2$; $-1 < 0$; $|-3-7| < |-3| + |-7|$; $|-4+5| < |-4| + |5|$; $|-5| - |2| < |-5 - (2)|$; $|-3 \cdot 5| = |-3| \, |5|$.

2. Evaluate the following: $|-6| = $; $|2| = $; $|-3| + |-4| = $; $|-2| \cdot |-1| = $; $|4| - |7| = $; $|5| - |-4| = $; $|-2| + |-6| = $; $|-5| - |-8| = $.

3. Go over equation (1-38) carefully and test it by assigning x the following values: $3, -1, 4, 0, -5, -7, -6, 2, -2/3, -\pi$.

4. Determine when the equality sign holds in
$$|a + b| \leq |a| + |b|$$

5. Determine when the equality sign holds in
$$|a| - |b| \leq |a - b|$$

6. Solve the following inequalities:
   (a) $3x - 2 < 5$
   (b) $2 - 3x \geq 5x + 2$
   (c) $6 - x < x + 17$
   (d) $x(x-1) > 0$
   (e) $(x-1)^2 > 0$
   (f) $x^2 - x - 6 > 0$

7. Solve the following equations:
   (a) $|x| = 2$
   (b) $|x - 2| = 3$
   (c) $|-2 - x| = 4$
   (d) $|2x| = 6$
   (e) $|x - (-2)| = 7$
   (f) $|-3x| = 15$

1–12. Number intervals

Let a and b be two real numbers such that $a < b$. The set of all real numbers x between a and b, is usually written in the form $a < x < b$. This latter form, called a continued inequality, is two inequalities; it says that a is less than x and x is less than b. This set of real numbers is called an *open interval*. If the set includes a and b, it is called a *closed interval*. The notation (a,b) is used to designate an open interval and $[a,b]$ is used to designate a closed interval. The notation (a,b) is sometimes confused with the concept of an ordered pair. The context in which this symbol is used determines the manner in which it is to be interpreted.

$$\text{Open interval } (a,b) = \{x \mid a < x < b\}$$

$$\text{Closed interval } [a,b] = \{x \mid a \leq x \leq b\}$$

Following this notation, $[a,b$ and $a,b]$ are called half-open intervals.

$$\text{Half-open on the right } [a,b) = \{x \mid a \leq x < b\}$$

$$\text{Half-open on the left } (a,b] = \{x \mid a < x \leq b\}$$

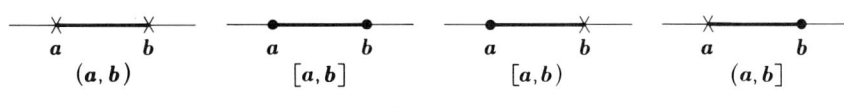

Fig. 1–3

It is convenient to have a notation so that the set of real numbers defined by $\{x \mid x > a\}$ can be expressed as a number interval. By definition,

$$(a, \infty) = \{x \mid x > a\}, \quad (-\infty, a) = \{x \mid x < a\}, \tag{1-39}$$
$$(-\infty, \infty) = \{x \mid x \text{ is a real number}\}$$

The number interval $(-\infty, \infty)$ is the entire real number line. The symbol ∞ is not a number, it is merely a symbol which is to be interpreted as defined in (1–39) when used to designate number intervals.

EXAMPLE 1–12. Graph the set of points contained in each of the following intervals:

(a) $[-2,1)$ (b) $[-1,0] \cup (1,3]$ (c) $[-1,1] \cap [0,2)$ (d) $[2, \infty)$

Solution.

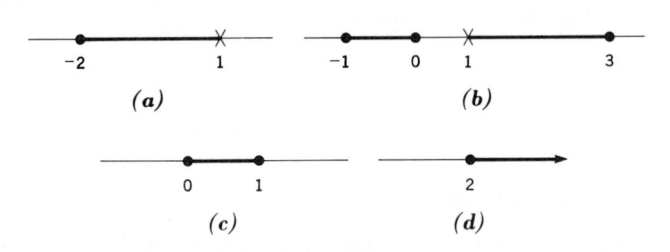

Two important number intervals defined by inequalities when $a > 0$ are:

$$\{x \mid |x| < a\} = \{x \mid -a < x < a\} = (-a,a)$$
$$\{x \mid |x| > a\} = \{x \mid x < -a \text{ or } x > a\} = (-\infty, -a) \cup (a, \infty)$$

EXERCISES

1. Graph each of the following intervals:
    (a) $[-2,2)$,   (b) $(-\infty, -2]$,   (c) $[-1,2]$

2. Graph each of the following sets of real numbers:
    (a) $[0,1] \cup (2,3)$
    (b) $[0,1]'$*
    (c) $(2,\infty)'$
    (d) $[-2,0) \cap (0,2]$
    (e) $(-2,2)'$
    (f) $(-\infty,-2] \cup (2,\infty)$
    (g) $(-2,0) \cap (0,\infty)$

3. Graph the solution of each of the following inequalities and express the solution in terms of number intervals.
    (a) $|x| < 2$
    (b) $|x| \geq 1$
    (c) $|x-2| < 1$
    (d) $|-x| < 3$
    (e) $x-1 > 0$
    (f) $2x \leq 4$
    (g) $|3x+5| < 2$
    (h) $|5-3x| \leq 4$

1–13. Relations and functions

In analytic geometry, the elements x and y in an ordered pair (x, y) are real numbers. A set of ordered pairs is called a *relation*. Hence, a relation is a subset of $R \times R$, the set of all ordered pairs of real num-

---

* Recall that A' is the complement of set A. Thus $[0,1]'$ is the set of all real numbers not in the interval $[0,1]$, which can be expressed $(-\infty,0) \cup (1,\infty)$.

bers. The *graph* of a relation is the set of all points in the plane whose coordinates are the members of the ordered pairs in the relation. The symbols x and y are called *variables*; x the *independent* variable and y the *dependent* variable.

The *domain* of a relation is the set of all first members, and the *range* is the set of all second members in the set of ordered pairs of the relation.

Consider the relation $\{(x,y) \mid x = 1 \text{ and } y \in R\}$. This is the set of all ordered pairs having 1 as the first member with no restriction on the second member. They are all of the form $(1,y)$ where y can be any real number. The graph is shown in Fig. 1–4. The domain of this relation is

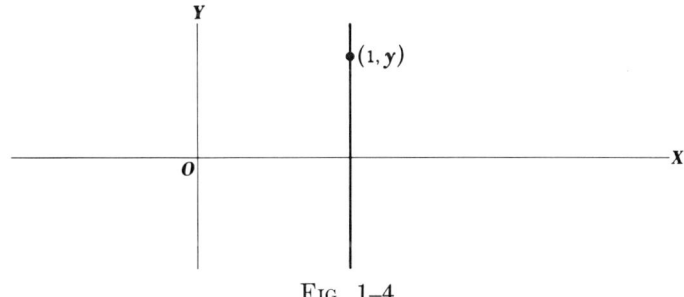

Fig. 1–4

the set $\{1\}$ and the range is R, the set of all real numbers.

A *function* is a set of ordered pairs where each first member is associated with exactly one second member. Hence, a function is a special type of relation, one in which each number in the domain is associated with exactly one number in the range. The relation shown in Fig. 1–4 is not a function. However, the set $\{(x,y) \mid y = x\}$ is a function. This is the set of all ordered pairs with the same number as first and second numbers. The graph is shown in Fig. 1–5.

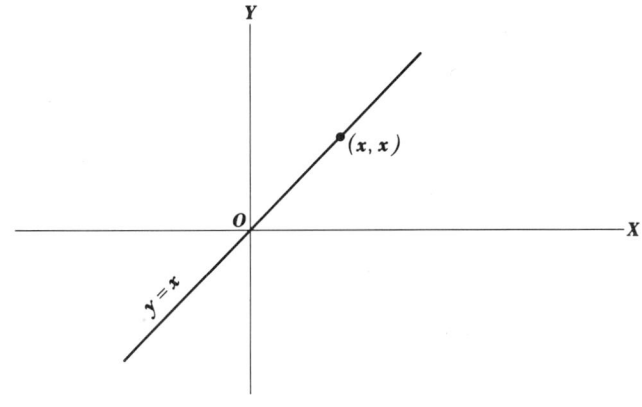

Fig. 1–5

The domain is the set R and the range is the set R.

A function is commonly denoted by f; that is, the symbol f represents the set of ordered pairs of the function. In the above example,

$$f = \{(x, y) \mid y = x\} \tag{1-40}$$

The symbol $f(x)$, (read "f of x") represents the number y associated with x in the set of ordered pairs of f. This is expressed by the equality $y = f(x)$. Thus each member of the ordered pair of the function may be written as (x,y) or $(x,f(x))$. The relationship between x and y defined in (1-40) can be written as $f(x) = x$. Examples are $f(2) = 2$ and $f(-4) = -4$, which yield the ordered pairs $(2,2)$ and $(-4,-4)$. In fact, all ordered pairs of f are of the form (x,x).

In general, we write

$$f = \{(x,y) \mid y = f(x)\} \tag{1-41}$$

This distinguishes between the symbols f and $f(x)$; f is the set of ordered pairs of the function and $f(x)$ is the value of the function for a specified x.

We wish to emphasize the difference between a function and a relation. A function is a relation but a relation is not necessarily a function. A function f is a relation such that for each x in the domain of f there is associated exactly one y in the range of f.

EXAMPLE 1-14. Graph the relation $S = \{(x,y) \mid x = 1 \wedge -1 \leq y \leq 1\}$ and state the domain and range of S.

Solution.

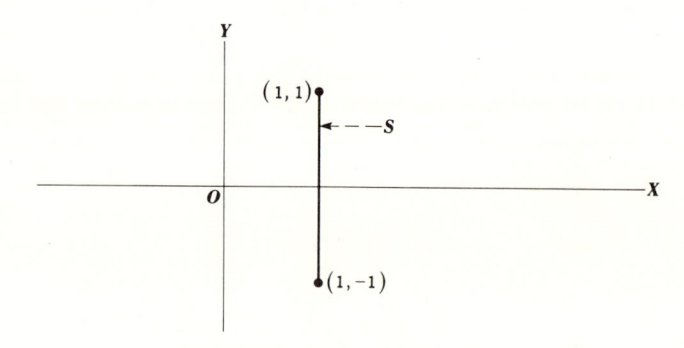

The graph of S is the set of points on the indicated line. The domain is the set $\{1\}$ and the range is the set $\{y \mid -1 \leq y \leq 1\}$ which can be stated as the interval $[-1,1]$.

EXAMPLE 1-15. Graph the relation $S = \{(x,y) \mid 0 < x < 2 \wedge -1 < y < 1\}$ and state the domain and range of S.

Solution.

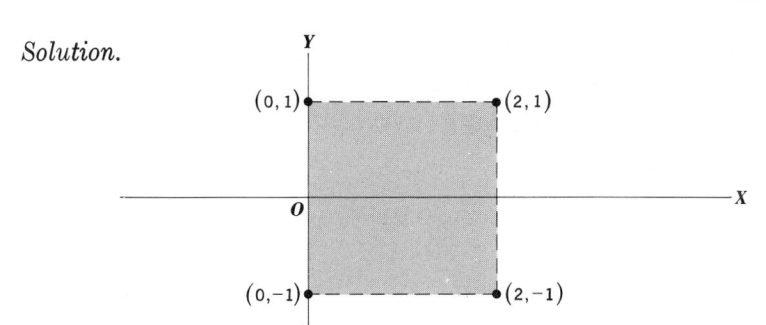

The graph is the shaded portion, not including the boundary lines. The domain is the interval (0,2) and the range is the interval (−1,1).

EXAMPLE 1–16. Graph the function $f = \{(x,y) \mid y = 2x\}$ and state the domain and range of f.

Solution.

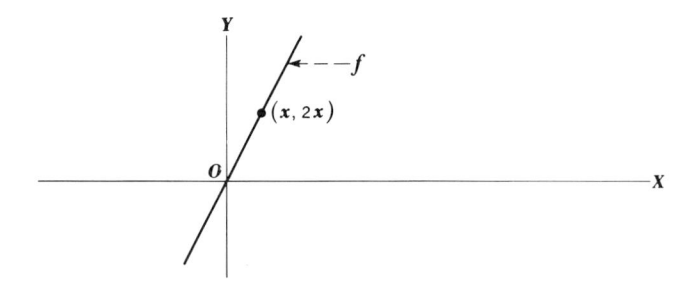

This is the set of points of the form $(x,2x)$. We could also write this as $(x,f(x))$, where $f(x) = 2x$. The domain and range are R.

EXAMPLE 1–17. For the function

$$f = \{(x,y) \mid y = x^2 - 2\}$$

Compute $f(0)$, $f(3)$ and $f(a)$.

Solution. $f(x) = x^2 - 2$, therefore $f(0) = 0 - 2 = -2$, $f(3) = 3^2 - 2 = 7$ and $f(a) = a^2 - 2$.

When a function is defined by an equation $y = f(x)$ (in Example 1–17, $y = x^2 - 2$), y is said to be *explicitly* defined in terms of x. The equation $x^2 + y^2 = 25$, for example, defines a relationship between the variables x and y, but does not give y explicitly in terms of x. In such an equation y is said to be defined *implicitly* in terms of x when we consider y

the dependent and x the independent variables. The equation $x^2 + y^2 = 25$ defines the relation S, where

$$S = \{(x,y) \mid x^2 + y^2 = 25\}$$

Solving the equation for y, we obtain

$$y = \sqrt{25 - x^2} \text{ and } y = -\sqrt{25 - x^2} \qquad (1\text{--}42)$$

There are two y values for each x value; hence, the relation S is not a function. However, each of the equations in (1–42) defines y explicitly in terms of x and each represents a function.

EXERCISES

1. Determine if the following relations are also functions. State the domain and range.
   (a) $\{(1,2), (2,5), (4,-2), (5,0)\}$
   (b) $\{(1,2), (5,2), (4,-2), (6,6)\}$
   (c) $\{(1,2), (1,1), (4,5), (5,0)\}$
   (d) $\{(1,1), (2,1), (3,1), (4,1)\}$
   (e) $\{(x,y) \mid y = x^2\}$
   (f) $\{(x,y) \mid y^2 = x\}$
   (g) $\{(x,y) \mid y = 1\}$
   (h) $\{(x,y) \mid |x| = 1\}$
   (i) $\{(x,y) \mid x^2 = y^2\}$

2. If $f = \{(x,y) \mid y = f(x) = 3x\}$, find
   (a) $f(2)$ (d) $f(a)$
   (b) $f(-2)$ (e) $f(a + h)$
   (c) $f(0)$ (f) $f(a + h) - f(a)$

3. For the function in Exercise 2, is $f(a + 2)$ the same as $f(a) + f(2)$?

4. If $f = \{(x,y) \mid y = f(x) = x^2 + 2\}$, find
   (a) $f(2)$ (d) $f(a)$
   (b) $f(-2)$ (e) $f(a + h)$
   (c) $f(0)$ (f) $f(a + h) - f(a)$

5. For the function in Exercise 4, is $f(a + 2)$ the same as $f(a) + f(2)$?

6. The set of all points on a circle of radius one with center at $(0,0)$ is a relation. Is it also a function? What are the domain and range of this relation?

7. Graph each of the following and state the domain and range of each.
   (a) $\{(x,y) \mid x \in [0,1] \text{ and } y \in [-2,2]\}$
   (b) $\{(x,y) \mid x \geq 0 \text{ and } 0 < y < 1\}$
   (c) $\{(x,y) \mid |x| < 2 \text{ and } |y| \geq 1\}$
   (d) $\{(x,y) \mid x^2 < 4 \text{ and } y = 1\}$

8. Graph each of the following functions and state the domain and range of each.
   (a) $f = \{(x,y) \mid y = 0\}$
   (b) $f = \{(x,y) \mid y = |x|\}$
   (c) $f = \{(x,y) \mid x = 2y\}$
   (d) $f = \{(x,y) \mid y = -x\}$

9. Find functions which are implicitly defined by the following equations, treating x as the independent variable.
   (a) $x^2 + y = 0$
   (b) $4x + y^2 = 0$
   (c) $x^2 + y^2 = 1$
   (d) $xy = 1$
   (e) $y^2 - 4xy + 4x^2 = 0$
   (f) $y^2 + 2xy + 4 = 0$

   Hint: Solve for y in terms of x and see what functions are defined, as in (1–42).

10. Find the functions which are explicitly defined by the equations in Exercise 9, treating y as the independent variable.

    Hint: In this case solve for x in terms of y.

1–14. Conditional and identity equations

An equation is merely a statement of equality between two expressions. In most of the material in this book, the expressions involve symbols representing numbers. It is important to distinguish between two types of equations.

1. The identity equation

An identity, or identity equation, is one in which the two members of the equation are equal for all permissible values of the variable. For example,

$$(x+y)(x-y) = x^2 - y^2, \quad \frac{1}{x} + \frac{1}{x-1} = \frac{2x-1}{x(x-1)}$$

are both identities. The first is a true statement for all values of the variables x and y. The second is true for all permissible values of x, which excludes the values 0 and 1. The right and left members of an identity are equal for all values of the variables for which the members are defined.

Identities are of particular importance in trigonometry. For example, $\sin^2 \theta + \cos^2 \theta = 1$ for all values of θ. This is an identity. The sign "≡" is often used in place of "=" to emphasize the fact that this equation is an identity. Most often it is clear from the context and "=" is used without making this distinction.

2. The conditional equation

A conditional equation is an equation where the right and left members are not equal for all values of the variables for which they are defined. For example,

$$2x + 5 = 9, \quad x(x-1) = 0$$

are conditional equations. The first becomes a true statement if we replace x by 2, and the second becomes true if we replace x by 0 or 1.

A conditional equation is sometimes called an "open statement." To solve the equation means to find values of the variable (or variables) for which the open statement becomes a true statement.

EXERCISES

Determine which of the following are identities and which are conditional equations.

1. $2(x - 3) = x + 1$
2. $\sqrt{x^2 + y^2} = x + y$
3. $2(x + 1) + 5 = 9 - 2(1 - x)$
4. $\dfrac{1}{2x} + \dfrac{1}{3x} = \dfrac{1}{6x}$
5. $\dfrac{x^2 - 9}{x + 3} = x - 3$
6. $2x(x - 2) + 1 = x(x - 2)$
7. $x^3 - 1 = (x - 1)(x^2 + x + 1)$

TRIGONOMETRY

1–15. Degrees, radians and arc length

A *degree* is the measure of a central angle of a circle that is subtended by an arc equal to $\dfrac{1}{360}$ of the circumference of the circle.

A *radian* is the measure of a central angle of a circle that is subtended by an arc equal to the radius of the circle.

The angle θ in Fig. 1–6a has a radian measure of one radian, since the

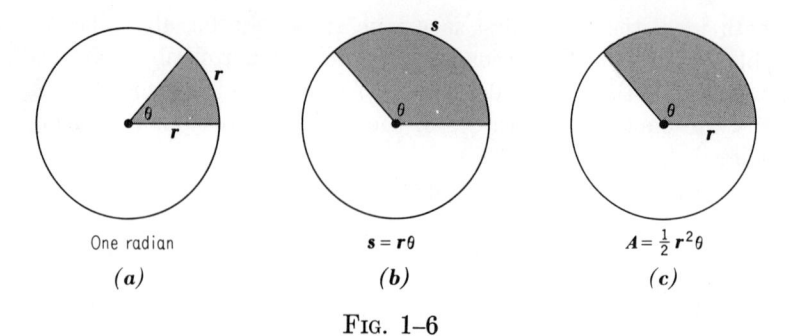

One radian
(a)

$s = r\theta$
(b)

$A = \tfrac{1}{2} r^2 \theta$
(c)

Fig. 1–6

arc subtended has a length r, the radius of the circle. The angle θ in Fig. 1-6b has a radian measure of $\frac{s}{r}$, that is, $\theta = \frac{s}{r}$ or $s = r\theta$, where s is the length of the arc subtended by θ. Fig. 1-6c shows a sector of the circle with central angle θ. If θ is the radian measure of the central angle, the area is given by $A = \frac{1}{2}r^2\theta$.

Since the length of the arc corresponding to an angle of 360° is $2\pi r$, the radian measure of the same angle is $\frac{2\pi r}{r} = 2\pi$.

$$\pi \text{ radians} = 180°, \quad 1° = \frac{\pi}{180} = 0.0174529 \text{ (approximately)}$$

1-16. The trigonometric functions

The trigonometric functions are defined in terms of an angle in *standard position*. We introduce a rectangular coordinate system in the plane by two perpendicular lines which divide the plane into four sections, called quadrants. The point of intersection of the two lines is the origin and the lines are called axes. An angle is in standard position when one side, called the *initial* side, coincides with the positive half of the horizontal axis and the vertex coincides with the origin. The other side of the angle is called the *terminal* side. The angle is measured from the initial side in a counterclockwise direction.

Let P be a point on the terminal side of the angle and let x and y be the abscissa (directed distance from the y axis) and the ordinate (directed distance from the x axis) of the point P. We shall call the distance from the origin to the point P simply the *distance* and shall denote it by r. Fig. 1-7 shows a typical point P with the corresponding angle θ, and the distances x, y, and r.

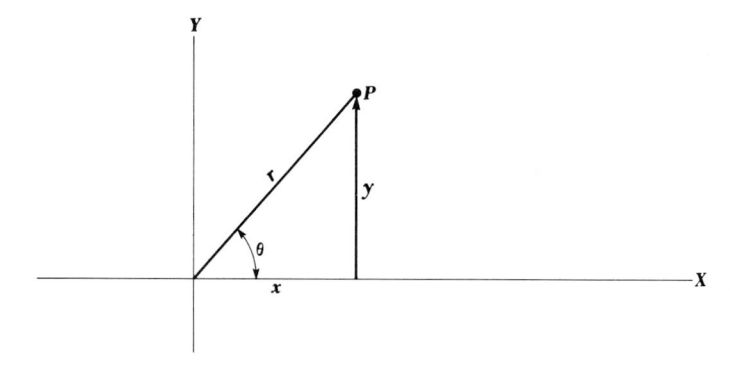

Fig. 1-7

34 ANALYTIC GEOMETRY IN THE PLANE

The definitions of the trigonometric functions are:*

$$\sin \theta = \frac{\text{ordinate}}{\text{distance}} = \frac{y}{r} \qquad \csc \theta = \frac{\text{distance}}{\text{ordinate}} = \frac{r}{y}$$

$$\cos \theta = \frac{\text{abscissa}}{\text{distance}} = \frac{x}{r} \qquad \sec \theta = \frac{\text{distance}}{\text{abscissa}} = \frac{r}{x}$$

$$\tan \theta = \frac{\text{ordinate}}{\text{abscissa}} = \frac{y}{x} \qquad \cot \theta = \frac{\text{abscissa}}{\text{ordinate}} = \frac{x}{y}$$

These six functions are positive or negative, zero, or undefined, the results depending on the signs of x and y and the values of x, y, and r.

The following chart gives the signs of the six functions for the respective quadrants:

Quadrant	$\sin \theta$	$\cos \theta$	$\tan \theta$	$\cot \theta$	$\sec \theta$	$\csc \theta$
I	+	+	+	+	+	+
II	+	−	−	−	−	+
III	−	−	+	+	−	−
IV	−	+	−	−	+	−

1–17. The functions of $0°$, $30°$, $45°$, $60°$, $90°$ and related measures

The values of the functions of these angles are extremely important and are listed in the table below for convenience.

Angle Degree	Radian	sin	cos	tan	cot	sec	csc
0°	0	0	1	0	undef.	1	undef.
30°	$\frac{\pi}{6}$	$\frac{1}{2}$	$\frac{\sqrt{3}}{2}$	$\frac{\sqrt{3}}{3}$	$\sqrt{3}$	$\frac{2\sqrt{3}}{3}$	2
45°	$\frac{\pi}{4}$	$\frac{\sqrt{2}}{2}$	$\frac{\sqrt{2}}{2}$	1	1	$\sqrt{2}$	$\sqrt{2}$
60°	$\frac{\pi}{3}$	$\frac{\sqrt{3}}{2}$	$\frac{1}{2}$	$\sqrt{3}$	$\frac{\sqrt{3}}{3}$	2	$\frac{2\sqrt{3}}{3}$
90°	$\frac{\pi}{2}$	1	0	undef.	0	undef.	1
120°	$\frac{2\pi}{3}$	$\frac{\sqrt{3}}{2}$	$-\frac{1}{2}$	$-\sqrt{3}$	$-\frac{\sqrt{3}}{3}$	−2	$\frac{2\sqrt{3}}{3}$
135°	$\frac{3\pi}{4}$	$\frac{\sqrt{2}}{2}$	$-\frac{\sqrt{2}}{2}$	−1	−1	$-\sqrt{2}$	$\sqrt{2}$
150°	$\frac{5\pi}{6}$	$\frac{1}{2}$	$-\frac{\sqrt{3}}{2}$	$-\frac{\sqrt{3}}{3}$	$-\sqrt{3}$	$-\frac{2\sqrt{3}}{3}$	2
180°	π	0	−1	0	undef.	−1	undef.

*Throughout the text, many items of particular importance are indicated by shading.

1–18. Important identities

The elementary relations

$$\sin x \csc x = 1 \quad \text{or} \quad \csc x = \frac{1}{\sin x}$$

$$\cos x \sec x = 1 \quad \text{or} \quad \sec x = \frac{1}{\cos x}$$

$$\tan x \cot x = 1 \quad \text{or} \quad \cot x = \frac{1}{\tan x}$$

$$\tan x = \frac{\sin x}{\cos x} \qquad \cot x = \frac{\cos x}{\sin x}$$

Pythagorean relations

$$\sin^2 x + \cos^2 x = 1$$
$$1 + \tan^2 x = \sec^2 x$$
$$1 + \cot^2 x = \csc^2 x$$

✶Sum and difference relations

$$\sin(x + y) = \sin x \cos y + \cos x \sin y$$
$$\sin(x - y) = \sin x \cos y - \cos x \sin y$$
$$\cos(x + y) = \cos x \sin y - \sin x \cos y$$
$$\cos(x - y) = \cos x \sin y + \sin x \cos y$$
$$\tan(x + y) = \frac{\tan x + \tan y}{1 - \tan x \tan y}$$
$$\tan(x - y) = \frac{\tan x - \tan y}{1 + \tan x \tan y}$$
$$\cot(x + y) = \frac{\cot x \cot y - 1}{\cot x + \cot y}$$
$$\cot(x - y) = \frac{\cot x \cot y + 1}{\cot y - \cot x}$$

✶Double angle relations

$$\sin 2x = 2 \sin x \cos x$$
$$\cos 2x = \cos^2 x - \sin^2 x = 2 \cos^2 x - 1 = 1 - 2 \sin^2 x$$
$$\tan 2x = \frac{2 \tan x}{1 - \tan^2 x} \qquad \cot 2x = \frac{\cot^2 x - 1}{2 \cot x}$$

✶Half-angle relations*

$$\sin \frac{x}{2} = \pm \sqrt{\frac{1 - \cos x}{2}}, \qquad \cos \frac{x}{2} = \pm \sqrt{\frac{1 + \cos x}{2}}$$

$$\tan \frac{x}{2} = \pm \sqrt{\frac{1 - \cos x}{1 + \cos x}} = \frac{1 - \cos x}{\sin x} = \frac{\sin x}{1 + \cos x}$$

$$\cot \frac{x}{2} = \pm \sqrt{\frac{1 + \cos x}{1 - \cos x}} = \frac{\sin x}{1 - \cos x} = \frac{1 + \cos x}{\sin x}$$

---

* When a double sign, $\pm$, appears, the choice of the sign depends upon the quadrant in which the terminal side of the angle $\frac{x}{2}$ is located.

1–19. Reduction formulas

	sin	cos	tan	cot	sec	csc
$-x$	$-\sin x$	$+\cos x$	$-\tan x$	$-\cot x$	$+\sec x$	$-\csc x$
$90° + x$	$+\cos x$	$-\sin x$	$-\cot x$	$-\tan x$	$-\csc x$	$+\sec x$
$90° - x$	$+\cos x$	$+\sin x$	$+\cot x$	$+\tan x$	$+\csc x$	$+\sec x$
$180° + x$	$-\sin x$	$-\cos x$	$+\tan x$	$+\cot x$	$-\sec x$	$-\csc x$
$180° - x$	$+\sin x$	$-\cos x$	$-\tan x$	$-\cot x$	$-\sec x$	$+\csc x$
$270° + x$	$-\cos x$	$+\sin x$	$-\cot x$	$-\tan x$	$+\csc x$	$-\sec x$
$270° - x$	$-\cos x$	$-\sin x$	$+\cot x$	$+\tan x$	$-\csc x$	$-\sec x$
$360° + x$	$+\sin x$	$+\cos x$	$+\tan x$	$+\cot x$	$+\sec x$	$+\csc x$
$360° - x$	$-\sin x$	$+\cos x$	$-\tan x$	$-\cot x$	$+\sec x$	$-\csc x$

The above table may be summarized and extended by the following easily remembered rule:

$$f(\pm x + n90°) = \pm g(x)$$

where n may be any integer, positive, negative, or zero,
  f is any one of the six trigonometric functions: sin, cos, tan, cot, sec, or csc,
  x may be any real angle measure.

If n is even, then g is the same function as f. If n is odd, then g is the *cofunction* of f.

(Sine and cosine, tangent and cotangent, secant and cosecant, are cofunctions of each other.)

The second $\pm$ sign is not necessarily the same as the first one, but is determined as follows: for a given function f, a given value of n, and a given choice of the first $\pm$ sign, the second $\pm$ sign will be the same for all values of x. Thus it is only necessary to check the sign for any one value of x, and the formula will be complete.

Examples

  $\tan(x + 270°) = \pm \cot x$. Since $n(=3)$ is odd, we use the cofunction. To determine the sign, assume a value of x in the first quadrant. Then $x + 270°$ is in the fourth quadrant, where the tangent is negative, so a minus sign is required. Thus the formula becomes

$$\tan(x + 270°) = -\cot x, \text{ valid for } all \text{ values of } x$$

  $\cos(x - 450°) = \pm \sin x$. Again assuming a value of x in the first quadrant, we find that $x - 450°$ is in the fourth quadrant. Thus

$\cos(x - 450°)$ would be positive, and no minus sign is needed. Hence the formula becomes

$$\cos(x - 450°) = \sin x$$

$\sec(180° - x) = \pm \sec x$. Here $n(=2)$ is even, so we use the same function. To determine the sign, again assume a value of x in the first quadrant. Then $180° - x$ is in the second quadrant, where the secant is negative. Thus the formula, valid for all values of x, is

$$\sec(180° - x) = -\sec x$$

1–20. Relations for triangles

In Fig. 1–8, A, B, C denote both the vertices and measure of the angles of any plane triangle and a, b, c the corresponding measures of the opposite sides.

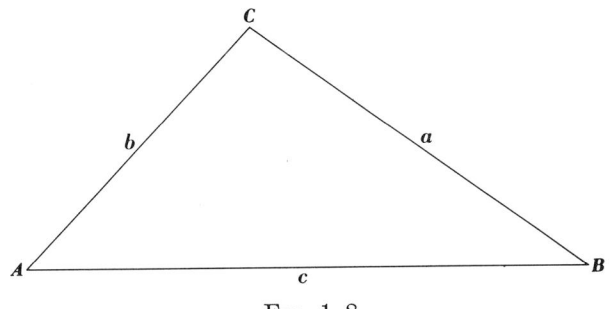

Fig. 1–8

Law of sines

$$\frac{a}{\sin A} = \frac{b}{\sin B} = \frac{c}{\sin C}$$

Law of cosines

$$c^2 = a^2 + b^2 - 2ab \cos C$$
$$b^2 = c^2 + a^2 - 2ca \cos B$$
$$a^2 = b^2 + c^2 - 2bc \cos A$$

Area of the triangle

$$S = \frac{1}{2} ab \sin C = \frac{1}{2} bc \sin A = \frac{1}{2} ca \sin B$$

or

$$S = \sqrt{s(s-a)(s-b)(s-c)}$$

where $s = \dfrac{a+b+c}{2}$.

1–21. Inverse trigonometric functions

The notation arcsin x (or $\sin^{-1} x$) is used to denote any angle whose sine is x; Arcsin x (or $\text{Sin}^{-1} x$) is usually used to denote the *principal value*. Similar notation is used for the other inverse trigonometric functions. The principal values of the inverse trigonometric functions are defined as follows:

$-\pi/2 \leq \text{Arcsin } x \leq \pi/2,$	$-1 \leq x \leq 1$
$0 \leq \text{Arccos } x \leq \pi,$	$-1 \leq x \leq 1$
$-\pi/2 < \text{Arctan } x < \pi/2,$	$-\infty < x < \infty$
$0 < \text{Arccsc } x \leq \pi/2,$ $-\pi < \text{Arccsc } x \leq -\pi/2,$	$x \geq 1$ $x \leq -1$
$0 \leq \text{Arcsec } x < \pi/2,$ $-\pi \leq \text{Arcsec } x < -\pi/2,$	$x \geq 1$ $x \leq -1$
$0 < \text{Arccot } x < \pi,$	$-\infty < x < \infty$

Note: There is no uniform agreement on the definitions of Arccsc x, Arcsec x, Arccot x for negative values of x. In this text we define these inverses in accord with the above table.

2

The Point and Plane Vectors

2–1. Introduction

Analytic geometry is a branch of mathematics that reduces the study of a geometric problem to the study of an algebraic problem. In this way many geometric problems are simplified and their results are more easily interpreted. With the aid of analytic geometry it is also possible to give a geometric interpretation to many algebraic equations. Réné Descartes, a Frenchman who lived from 1596 to 1650, was one of the first to introduce the theory of algebra into the study of geometry. His methods were introduced to the world through his book *La Geometrie* which appeared about 1637. Accordingly analytic geometry is sometimes called "Cartesian" geometry.

2–2. The rectangular coordinate system

Let $X'X$ and $Y'Y$ be two mutually perpendicular lines intersecting at O. These reference lines, called *axes*, divide the plane into four parts, or *quadrants*. The quadrants are numbered I, II, III, and IV, starting with the upper right-hand quadrant and reading counter-clockwise. The point of intersection of the axes is called the *origin*. The line $X'X$ is called the *axis of abscissas*, or *x-axis*, and is generally taken to be horizontal. The line $Y'Y$ is called the *axis of ordinates*, or *y-axis*, and is generally taken to be vertical. All distances measured horizontally to the right or vertically upward will be considered positive, and all distances measured horizontally to the left or vertically downward will be considered negative. Positive distances are represented by positive numbers, and negative distances are represented by negative numbers. A point is an undefined object which we shall represent pictorially by a *dot*. Any point in the plane is determined by two real numbers which represent the distances of the point from the axis of ordinates and the axis of abscissas, respectively. These two numbers are called the *coordinates* of the point. The first number is called the *abscissa* and the second number is called the *ordinate*. When written, the two numbers are enclosed in parentheses and are separated by a comma, as $(7, -5)$. Several points and their coordinates are shown in Fig. 2–1.

40   ANALYTIC GEOMETRY IN THE PLANE

For a point in quadrant I both coordinates are positive, as (3, 2). A point in quadrant II has a negative abscissa and a positive ordinate, as (−5, 1). In quadrant III a point has a negative abscissa and a negative ordinate, as (−2, −2). In quadrant IV a point has a positive abscissa and a negative ordinate, as (4, −2). If a point lies on one of the axes, the coordinate representing the distance from that axis to the point is zero. Thus, (−4, 0) is a point on the x-axis; (0, 5) is a point on the y-axis; and (0, 0) is the origin.

An arbitrary point in the plane may be represented by (x, y). The letters x and y represent either *positive* or *negative* distances. The abscissa, or x, is *always measured from* the axis $Y'Y$ along a line parallel to the axis $X'X$. The ordinate, or y, is *always measured from* the axis $X'X$ along a line parallel to the axis $Y'Y$. In locating a point whose coordinates are (x, y) it is customary to start at the origin and to measure the x-distance first, and then, from the point whose coordinates are $(x, 0)$, to measure the y-distance. *We shall use either the notation $P(x, y)$ or the notation $P = (x, y)$ to indicate that the point P has coordinates (x, y).* We shall sometimes speak of the coordinates of a point as the *name* of the point.

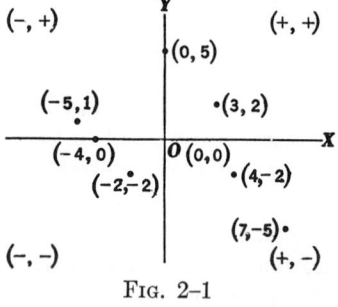

FIG. 2–1

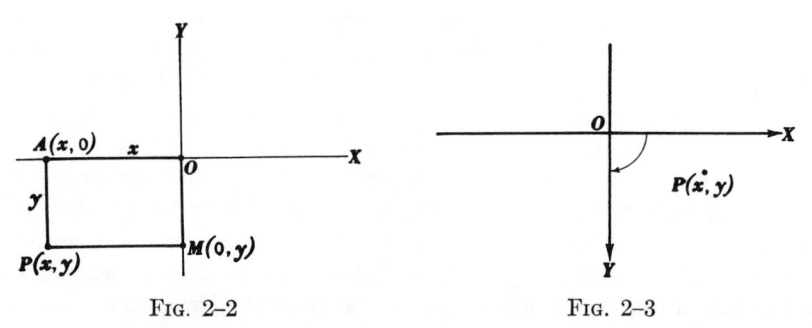

FIG. 2–2   FIG. 2–3

It is extremely important for the student to note that, when a point is named arbitrarily as (x, y), the symbols x and y represent *directed distances*. The symbol $\overline{AB}$ will represent the *directed distance* from A to B. For example, in Fig. 2–2, $x = \overline{OA} = \overline{MP}$ and is negative since it is measured to the left; $y = \overline{AP} = \overline{OM}$ and is negative since it is measured downward.

When we have named in this way the points of the plane by ordered pairs of real numbers, we say that we have attached a rectangular *reference frame* to the plane, and the numbers x and y for the point P are termed its coordinates *relative to this reference frame*.

Any rectangular reference frame in which the turn from the positive x-axis to the positive y-axis is counter-clockwise is a *right-handed* frame, and all such reference frames will be said to have the *same character*. If the turn from the positive x-axis to the positive y-axis is clockwise, the reference frame is said to be *left-handed*. The frame in Fig. 2-2 is a right-handed frame and that in Fig. 2-3 is a left-handed frame. In the discussions which follow we shall consider all points to be named relative to right-handed reference frames.

EXERCISES

1. Define carefully the following terms: abscissa; ordinate; coordinates of a point; name of a point.

2. Locate the following points: $(-1, 3)$, $(4, 0)$, $(2, 6)$, $(4, -3)$, $(-2, 0)$, $(0, -5)$, $(-4, -6)$, $(5, -3)$.

3. Locate $P(x, y)$ in the third quadrant. Determine the coordinates of the three other vertices of the rectangle whose sides are parallel to the axes and which has P as one vertex and the origin as its center.

4. Follow the instructions in Exercise 3 with $P(x, y)$ in the second quadrant.

5. One vertex of a rectangle is (a, b) in quadrant II. What are the coordinates of the other vertices if the sides are parallel, respectively, to the x-axis and the y-axis and if the origin is the center of the rectangle? Check your result by letting $a = -3$ and $b = 4$.

6. An equilateral triangle has two of its vertices at $(0, 0)$ and $(a, 0)$. Where is the third vertex? Is there more than one solution?

7. What is the locus* of a point whose abscissa is always 6? Draw it.

8. What is the locus of a point whose abscissa is always a constant? Draw it.

9. Describe the locus of a point whose ordinate is: (a) always -3; (b) always 4; (c) always a constant. Draw each locus.

10. What is the locus of a point whose abscissa is always equal to its ordinate? Draw it.

11. What is the locus of a point if its abscissa is always equal to the negative of its ordinate? Draw it.

12. Two vertices of a square are $(0, 0)$ and $(a, 0)$. What are the coordinates of two other possible vertices?

---

*By *locus* of a point having a specified property we mean the *set* or totality of all points having this property. Locus literally means path.

13. Two vertices of a square are (0, 0) and (0, b). What are the coordinates of the other vertices, and how many solutions are there?

14. A parallelogram has three of its vertices at (0, 0), (a, 0), and (b, c). Find a fourth vertex. Give all possible solutions.

15. A square has its center at the origin and its vertices on the axes. If the length of a side is a, what are the coordinates of the vertices?

16. What is the locus of a point whose distance from the origin is a given number?

17. What is the locus of a point whose ordinate is always twice the abscissa?

18. Describe the set of points for which $x \leq 6$.

19. Describe the set of points for which $y \geq -2$.

20. Describe the set of points for which $x < 6$ and $y > -2$.

21. Two vertices of a square are (0,0) and (a,a). Find the coordinates of the other vertices. How many solutions are there?

22. Describe the set of points for which $x \geq -3$ and $y \leq 0$.

23. Describe the set of points having the property that the distance of each point of the set from the origin is less than 4.

24. What set of points satisfies the condition that $|x| < 3$?

25. Describe the set of points for which (a) $|x| \leq 4$, $|y| \leq 2$; (b) $|x| < 3$, $|y| < 1$; (c) $|x| > 2$, $|y| > 3$; (d) $|x| < 1$, $|y| > 3$.

26. What is the locus of a point satisfying the condition that $x = y$; $x = -y$?

27. What set of points satisfies the condition that the distance of each point from the origin is greater than 5?

2-3. Symmetry

Two points A and B, Fig. 2-4, are said to be *symmetric with respect to a point* C provided C is the mid-point of the line segment joining A and B. The points A and B are also called *images* of each other with respect to C. The point C is called the *center of symmetry* of the two points.

Two points A and B, Fig. 2-5, are said to be *symmetric with respect to a line* l provided l is the perpendicular bisector of the line segment joining the two points. The points A and B are also said to be *images* of each other in the line. The line is called the *axis of symmetry* of the two points.

For example, the points $(-3, 4)$ and $(3, -4)$ are symmetric with respect

FIG. 2-4

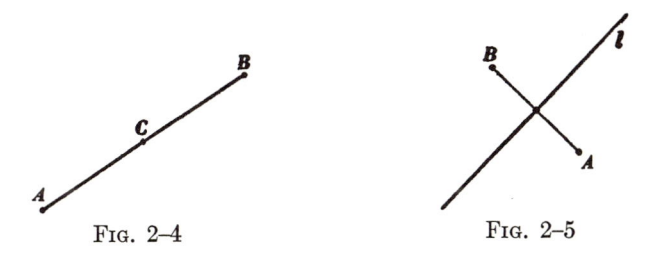

FIG. 2-5

to the origin; the points $(-3, 4)$ and $(3, 4)$ are images of each other in the y-axis; and the x-axis is an axis of symmetry of the points $(-3, 4)$ and $(-3, -4)$; and the line bisecting the first and third quadrants is an axis of symmetry of the points $(3,4)$ and $(4,3)$.

EXERCISES

1. Determine the coordinates of the point symmetric to $(5, 4)$ with respect to: (a) the origin; (b) the x-axis; (c) the y-axis.

2. If $P = (x, y)$ is a point in the third quadrant, what are the coordinates of the points symmetric to P with respect to the origin, the x-axis, and the y-axis, respectively?

3. What are the images of $P = (-1, 2)$ in the x-axis and in the y-axis?

4. What is the image of $P = (-2, -4)$ with respect to the origin?

5. In the set of points, $\{(-2,1), (2,3), (2,-1), (4,0), (0,-4)\}$, insert those that should be included if the symmetric condition to be satisfied is symmetry
   (a) with respect to the origin
   (b) with respect to the x-axis
   (c) with respect to the y-axis
   (d) with respect to the line bisecting the first and third quadrants.

6. Same as Exercise 5 with the following set of points:
$$\left\{(c,d), (-c,-d), \left(\frac{a+b}{2}, \frac{a-b}{2}\right), (c,-d)\right\}.$$

7. If symmetry with respect to both the x-axis and y-axis is satisfied, what additional points must be included in the following set?
$$\{(-2,1), (2,-3), (2,1), (-4,0)\}.$$

8. If symmetry with respect to the origin and the line bisecting the first and third quadrants is satisfied, what other points must be included for the following set?
$$\left\{(c,-d), (-c,d), \left(\frac{a+b}{2}, \frac{b-a}{2}\right), (c,d)\right\}.$$

2–4. Projections

The projection of a point on a given line is the foot of the perpendicular dropped from the point to the line. In Fig. 2–6 point A is the projection of P on line l.

If a segment has its end points at P_1 and P_2, then P_1P_2 will be understood to mean the *directed segment* from P_1 to P_2.

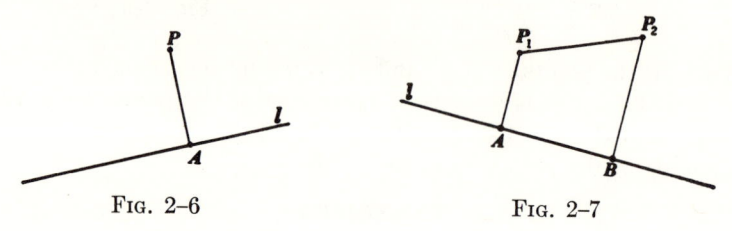

Fig. 2–6   Fig. 2–7

In Fig. 2–7 the projection of the directed segment P_1P_2 on the line l is the directed distance from A to B, where A is the projection of P_1 on l and B is the projection of P_2 on l.

We shall be particularly interested in the projections of a segment on the x-axis and the y-axis. In Fig. 2–8, the projection of P_1P_2 on the x-axis

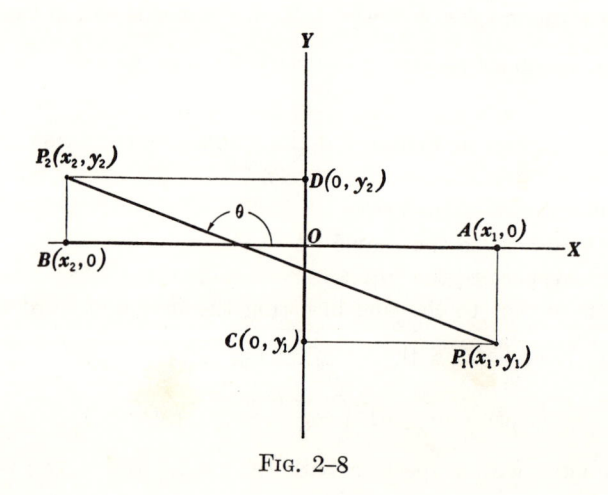

Fig. 2–8

is $\overline{AB}$, while the projection of P_1P_2 on the y-axis is $\overline{CD}$. If we represent the *length* of P_1P_2 by the absolute value symbol* $|P_1P_2|$, then from trigonometry it follows immediately that $\overline{AB} = |P_1P_2| \cos \theta$, where θ is the angle generated from the positive x-axis, or a half-line parallel to it through P_1,

---
* See Section 1–11.

to the segment P_1P_2. The projection $\overline{CD}$ of P_1P_2 on the y-axis is $|P_1P_2|$ sin θ.

When the coordinates of the end points of the segment are known, it is very easy to determine the projections of the segment on the axes. In Fig. 2-8, regardless of the position of A and B on the x-axis or C and D on the y-axis, we have $\overline{AB} = \overline{AO} + \overline{OB}$ and $\overline{CD} = \overline{CO} + \overline{OD}$.

Since $\overline{OA} = x_1$, and since $\overline{AO}$ is measured in the direction opposite to that of $\overline{OA}$, it follows that $\overline{AO} = -x_1$. Also, we know that $\overline{OB} = x_2$. Therefore,

$$\overline{AB} = -x_1 + x_2 = x_2 - x_1$$

In the same way, $\overline{OC} = y_1$ and, hence, $\overline{CO} = -y_1$. Therefore,

$$\overline{CD} = -y_1 + y_2 = y_2 - y_1$$

These results may be stated as follows: *The projection of a segment P_1P_2 on the x-axis is equal to the abscissa of the second point P_2 minus the abscissa of the first point P_1. The projection of a segment P_1P_2 on the y-axis is equal to the ordinate of the second point P_2 minus the ordinate of the first point P_1.*

EXAMPLE 2-1. What are the projections of the segment P_1P_2 on the axes, where $P_1 = (2, 3)$ and $P_2 = (-2, -1)$?

Solution. In Fig. 2-9, $\overline{AB} = x_2 - x_1 = (-2) - 2 = -4$; $\overline{CD} = (-1) - 3 = -4$.

The geometric significance of the signs must be considered. Since $\overline{AB} = -4$, we know that the direction from A to B is to the left. Since $\overline{CD} = -4$, we know that the direction from C to D is downward.

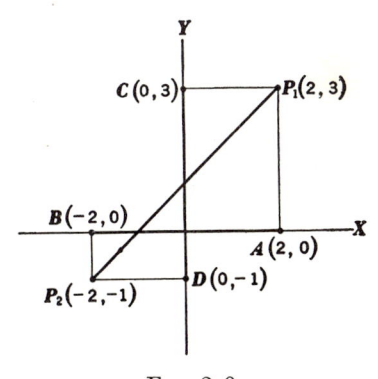

FIG. 2-9

EXERCISES

1. Refer to Fig. 2–10 and prove that the sum of the projections on the x-axis of the directed segments which are the sides of any quadrilateral is zero.

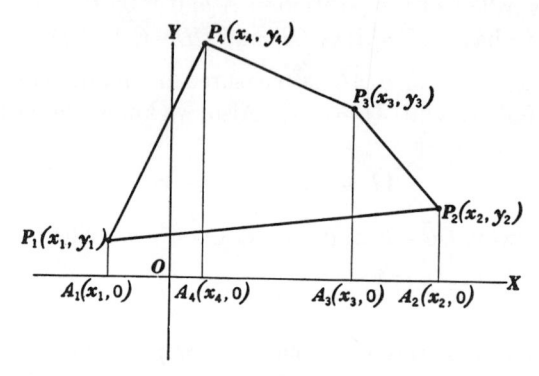

FIG. 2–10

2. Develop a proof like that in Exercise 1 for the projections on the y-axis.

3. Find the projections of each of the following segments on the x-axis and on the y-axis, respectively:

   (a) From $P_1 = (-3, -5)$ to $P_2 = (4, -6)$
   (b) From $L = (2, -3)$ to $M = (-2, 5)$
   (c) From $Q = (4, 5)$ to $R = (7, -3)$
   (d) From $A = (-1, 0)$ to $B = (0, -5)$
   (e) From $P_1 = (4, -3)$ to $P_2 = (0, 7)$
   (f) From $R = (-6, 0)$ to the origin

2–5. Scalar components of a segment

Let $P_1(x_1, y_1)$ and $P_2(x_2, y_2)$ represent any two points in the plane. We shall define the scalar* components of the segment P_1P_2 to be the projections of the segment on the x-axis and on the y-axis, respectively. So $x_2 - x_1$ and $y_2 - y_1$ are the scalar components of P_1P_2. We shall represent scalar components by Δx and Δy. So, for the segment P_1P_2, $\Delta x = x_2 - x_1$ and $\Delta y = y_2 - y_1$. We shall call Δx the *increment of x* and Δy the *increment of y*. It is extremely important for the student to realize that there is a difference between the scalar components of P_1P_2 and those of P_2P_1. The scalar components of P_2P_1 are $x_1 - x_2$ and $y_1 - y_2$ and are the negatives of the original pair.

---

* The adjective *scalar* is used to signify that the components as used here are *numbers* and not *vectors*.

THE POINT AND PLANE VECTORS   47

In order to distinguish numerical values of the scalar components of a segment from the coordinates of a point, we shall enclose scalar components in brackets whereas the coordinates of a point are enclosed in parentheses. So the notation $P_1P_2 = [\Delta x, \Delta y]$ means that the directed segment from P_1 to P_2 has the scalar components Δx and Δy.

EXAMPLE 2-2. Find the scalar components of the segment from $A(-2, 3)$ to $B(5, -1)$.

Solution. In Fig. 2-11, $\Delta x = \overline{EF}$ and $\Delta y = \overline{CD}$. Hence,

$$\Delta x = 5 - (-2) = 7 \text{ and } \Delta y = -1 - 3 = -4$$

Thus, the scalar components of AB are $[7, -4]$.

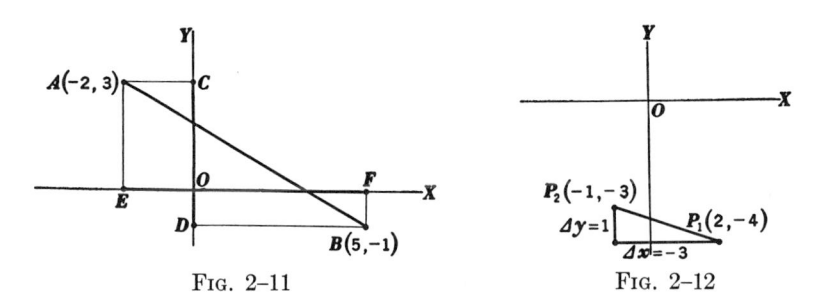

FIG. 2-11

FIG. 2-12

When the scalar components of a segment and the coordinates of one end point are known, the other end point is uniquely determined. Call the given point $P_1(x_1, y_1)$ and call the other end point of the segment $P_2(x_2, y_2)$. Then, since Δx and Δy are known, the coordinates of P_2 can be found from the relations $x_2 = x_1 + \Delta x$ and $y_2 = y_1 + \Delta y$.

EXAMPLE 2-3. Given $P_1 = (2, -4)$ and the scalar components $[-3, 1]$ of the segment P_1P_2, find the coordinates of P_2 and construct the segment.

Solution. Here,

$$x_2 = x_1 + \Delta x = 2 + (-3) = -1$$
$$y_2 = y_1 + \Delta y = -4 + 1 = -3$$

Hence, the other end point is $P_2 = (-1, -3)$. The segment is shown in Fig. 2-12.

EXERCISES

1. Find the scalar components of each of the following segments:
   (a) From $P_1 = (-3, 4)$ to $P_2 = (2, -5)$
   (b) From $P_1 = (5, 6)$ to $P_2 = (-3, 1)$

48 ANALYTIC GEOMETRY IN THE PLANE

    (c) From $P_1 = (5, 0)$ to $P_2 = (-2, -1)$
    (d) From $P_1 = (0, 6)$ to $P_2 = (5, 3)$
    (e) From $P_1 = (-1, -3)$ to $P_2 = (6, -1)$

2. In each case in Exercise 1 what are the scalar components of the segment from P_2 to P_1? How are these pairs of components related to the corresponding pairs found in Exercise 1?

3. Given the initial point and the scalar components of each of the following segments, find the other end point and construct the segment:

    (a) $(3, -4)$, $[2, 6]$ (d) $(0, 2)$, $[-1, -1]$
    (b) $(-1, -1)$, $[-3, 4]$ (e) $(-3, 0)$, $[0, 4]$
    (c) $(-4, 2)$, $[1, -6]$ (f) $(0, 0)$, $[4, -2]$

4. Given the terminal point and the scalar components, find the initial point of each of the following segments:

    (a) $(-2, 5)$, $[3, -1]$ (d) $(-4, -1)$, $[2, -3]$
    (b) $(6, 1)$, $[-2, -1]$ (e) $(2, -4)$, $[-5, 2]$
    (c) $(-4, -1)$, $[3, -2]$ (f) $(0, 0)$, $[3, -4]$

2–6. Distance between two points

Let $P_1(x_1, y_1)$ and $P_2(x_2, y_2)$ be any two points in the plane. Construct a right triangle having P_1P_2 for its hypotenuse and having its other two sides parallel, respectively, to the x-axis and the y-axis and intersecting at R (see Figs. 2–13 and 2–14). By the Pythagorean Theorem it follows that $|P_1P_2|^2 = |P_1R|^2 + |RP_2|^2 = |\Delta x|^2 + |\Delta y|^2$, or

$$d = |P_1P_2| = \sqrt{|\Delta x|^2 + |\Delta y|^2} \qquad (2\text{-}1)$$

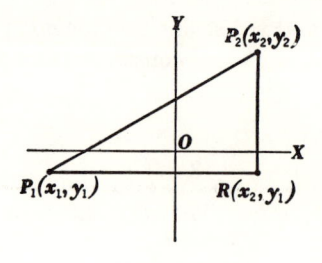

FIG. 2–13 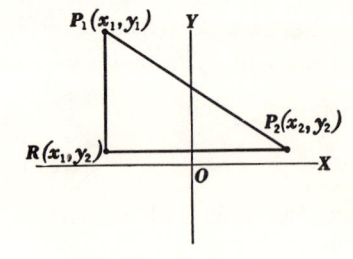   FIG. 2–14

To find the distance between two points, take the square root of the sum of the squares of the scalar components of the segment joining the two points. We are interested in the unsigned distance here, and hence the order of the points does not matter. Also, since the square of either a positive number or a negative number is a positive quantity, the absolute values of Δx and Δy may be used and the signs of those components may be disregarded.

Hereafter the word *distance* will be understood to mean the *unsigned distance* or the *absolute distance*.

EXAMPLE 2–4. Find the length of the segment P_1P_2 where P_1 has the coordinates $(-2, 5)$ and P_2 has the coordinates $(3, -1)$.

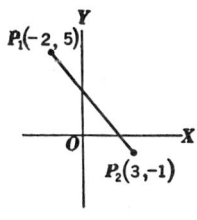

FIG. 2–15

Solution. The segment is shown in Fig. 2–15. If P_1 is considered the initial point of the segment,

$$\Delta x = 3 - (-2) = 5 \text{ and } \Delta y = -1 - 5 = -6$$

Hence, the scalar components of P_1P_2 are $[5, -6]$ and the length of the segment P_1P_2 is

$$d = \sqrt{5^2 + 6^2} = \sqrt{25 + 36} = \sqrt{61}$$

EXAMPLE 2–5. Prove that the diagonals of *any* rectangle are equal in length.

Solution. Since it is immaterial which two perpendicular lines of the plane are selected to be the x-axis and y-axis of our coordinate system, we find it convenient in this example to choose the axes to contain two of the sides of the rectangle, as in Fig. 2–16.

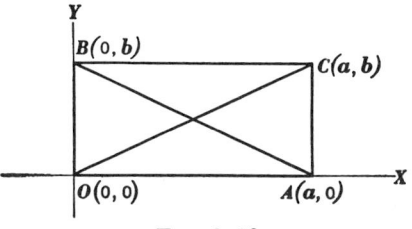

FIG. 2–16

Since the scalar components of OC are $[a, b]$ and those of BA are $[a, -b]$,

and
$$|BA| = \sqrt{a^2+b^2}$$
$$|OC| = \sqrt{a^2+b^2}$$

Hence,
$$|OC| = |BA|$$

In all problems like that just considered, the positions of the axes should be selected with care in order to simplify the calculations as much as possible. However, the figures used must not be *more special* than the problem permits. For example, if the problem concerns a rectangle, we must not use a square; if the problem deals with a general triangle, neither a right triangle nor an isosceles triangle should be used; and, if it concerns any quadrilateral, a general four-sided figure rather than a parallelogram must be used. For example, the following figures (Figs. 2-17, 2-18, and 2-19) may be used to represent general situations respectively for an isosceles trapezoid, a parallelogram, and a triangle.

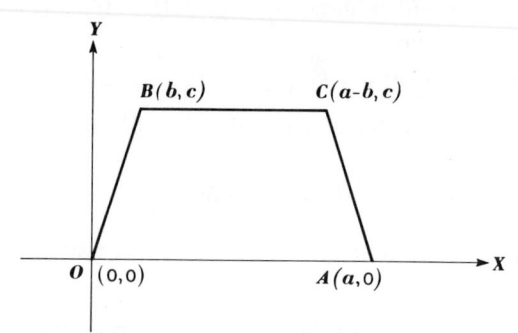

Fig. 2-17

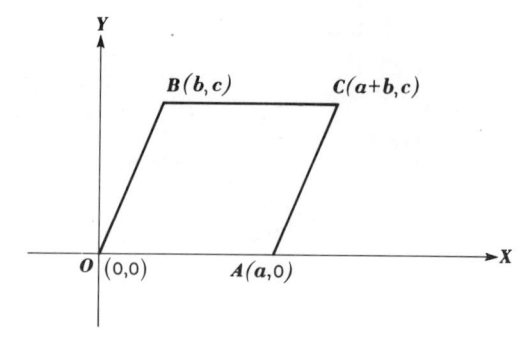

Fig. 2-18

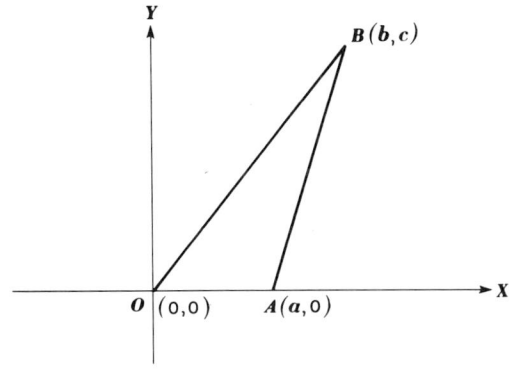

FIG. 2-19

See Exercise 6 in the next group.

EXERCISES

1. In each of the following find the distance between the two points whose coordinates are given.

    (a) $(-2, 3)$ and $(4, 6)$
    (b) $(5, 1)$ and $(4, -3)$
    (c) $(3, 0)$ and $(0, 4)$
    (d) $(1, 5)$ and $(-2, -1)$
    (e) $(6, -2)$ and $(0, -4)$
    (f) $(0, -2)$ and $(0, -10)$
    (g) $(-4, 3)$ and $(5, -2)$
    (h) $(5, -12)$ and $(0, 0)$
    (i) $(2, 3)$ and $(-7, 3)$
    (j) $(-4, 6)$ and $(-4, -5)$

2. In each of the following find the perimeter of the triangle whose vertices have the given coordinates.

    (a) $(-1, 2)$, $(2, 0)$, and $(3, -1)$
    (b) $(5, 3)$, $(-4, 3)$, and $(11, -5)$
    (c) $(-2, 4)$, $(2, -3)$, and $(1, 1)$
    (d) $(3, 5)$, $(8, 0)$, and $(-1, 4)$
    (e) $(4, -1)$, $(0, 0)$, and $(-2, -3)$
    (f) $(7, 1)$, $(-2, -4)$, and $(-3, 5)$

3. Prove that the points $(-1, 3)$, $(5, 1)$, and $(0, -4)$ are vertices of an isosceles triangle.

4. Two vertices of an equilateral triangle are $(0, 0)$ and $(8, 0)$. Find the coordinates of a third vertex.

5. Two vertices of an equilateral triangle are $(5, 5)$ and $(5\sqrt{3}, -5\sqrt{3})$. Find the coordinates of a third vertex.

6. Prove each of the following theorems analytically:

    (a) The diagonals of an isosceles trapezoid are equal in length.

52 ANALYTIC GEOMETRY IN THE PLANE

(b) The sum of the squares of the four sides of a parallelogram is equal to the sum of the squares of the diagonals.

(c) The square of the side opposite an acute angle of any triangle is equal to the sum of the squares of the other two sides decreased by twice the product of one of those sides and the projection of the other upon it.

7. What is the area of the quadrilateral which has its vertices at the points (5, 3), (0, 3), (0, −1), and (5, −1) in that order?

8. Find the coordinates of the point equidistant from the points (1, 0), (−1, −2), and (3, −2).

9. Prove that the sum of any two sides of a triangle is greater than the third side, and the difference between any two sides is less than the third side.

10. Two points P and Q are on opposite legs of an isosceles triangle, each equidistant from the vertex. Show analytically that the lengths of the segments from P and Q to the opposite vertices are equal.

11. Prove that any side of a quadrilateral is less than the sum of the other three sides.

12. Show analytically that the perimeter of a parallelogram is greater than the sum of its diagonals.

13. Prove analytically that if the diagonals of a parallelogram are not equal, the parallelogram is not a rectangle.

2–7. Direction cosines of a segment

Let $P_1(x_1, y_1)$ and $P_2(x_2, y_2)$ be any two distinct points in the plane. The direction cosines of P_1P_2 are defined to be the cosines of the angles between P_1P_2 and the rays, or half-lines, through P_1 parallel to the positive

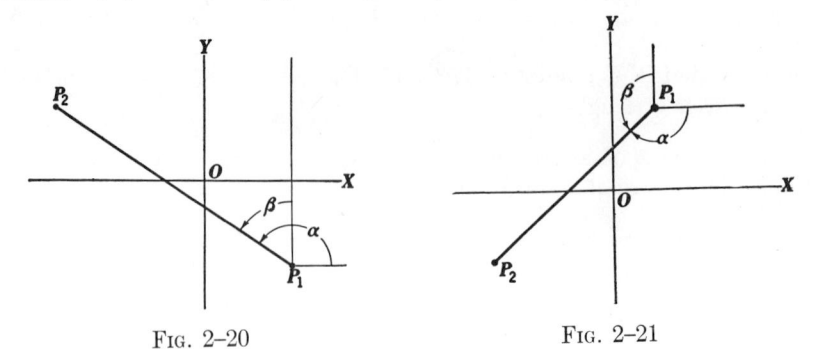

FIG. 2–20

FIG. 2–21

x-axis and the positive y-axis, respectively. Since $\cos(360° - \alpha) \equiv \cos \alpha$, it does not matter whether the angles are measured from the directed segment to the axes or from the axes to the segment. Moreover, since $\cos(-\alpha) \equiv \cos \alpha$, it does not matter whether we choose the positive angles between the directed segment and the axes or the negative angles. See Figs. 2–20 and 2–21. If l and m denote the direction cosines of P_1P_2, then $l = \cos \alpha$ and $m = \cos \beta$. Since $\cos(180° - \alpha) = -\cos \alpha$, the direction cosines of P_2P_1 are $-l$ and $-m$.

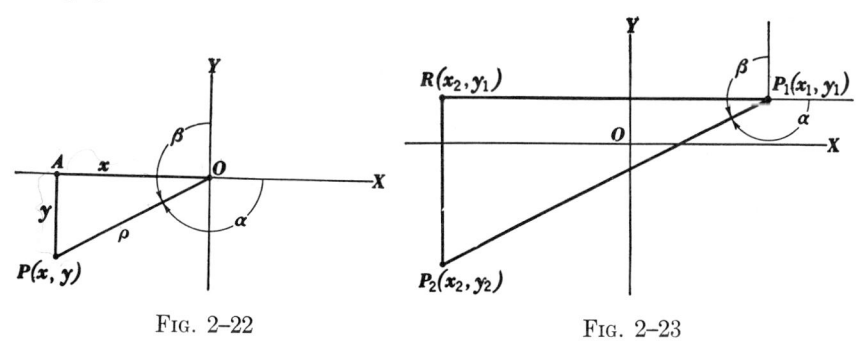

FIG. 2–22   FIG. 2–23

Consider any segment having its initial point at the origin, as in Fig. 2–22. Then, if $P(x, y)$ is the terminal point of the segment,

$$l = \frac{x}{\rho}$$
$$m = \frac{y}{\rho} \tag{2-2}$$

where $\rho = |OP| = \sqrt{x^2 + y^2}$.

For an arbitrary segment P_1P_2, Fig. 2–20,

$$l = \frac{x_2 - x_1}{d} = \frac{\Delta x}{d}$$
$$m = \frac{y_2 - y_1}{d} = \frac{\Delta y}{d} \tag{2-3}$$

where $d = |P_1P_2|$.

Squaring l and m and adding the squares, we have:

$$l^2 + m^2 = \frac{(\Delta x)^2 + (\Delta y)^2}{d^2} = 1 \tag{2-4}$$

since $d^2 = (\Delta x)^2 + (\Delta y)^2$.

Hence, *a fundamental property of the two direction cosines of a segment is that the sum of their squares is 1.*

EXAMPLE 2–6. What are the direction cosines of the segment OP, where O is the origin and P has the coordinates $(1, -3)$?

54   ANALYTIC GEOMETRY IN THE PLANE

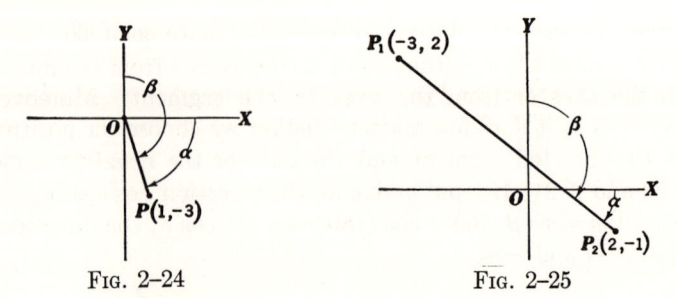

FIG. 2-24

FIG. 2-25

Solution. The segment is shown in Fig. 2–24. Since $\rho = |OP| = \sqrt{10}$,

$$l = \frac{1}{\sqrt{10}} \quad \text{and} \quad m = \frac{-3}{\sqrt{10}}$$

EXAMPLE 2–7. Find the direction cosines of the segment joining $P_1(-3, 2)$ to $P_2(2, -1)$.

Solution. The segment is plotted in Fig. 2–25. The scalar components of P_1P_2 are [5, −3]. Hence,

$$l = \frac{5}{\sqrt{34}} \quad \text{and} \quad m = \frac{-3}{\sqrt{34}}$$

EXERCISES

1. What do we mean by the direction cosines of a segment? What important property do they have?

2. Why do we use direction cosines instead of direction sines?

3. (a) Are $[\frac{2}{3}, \frac{1}{4}]$ direction cosines? Explain.
   (b) Are $[\frac{3}{5}, -\frac{4}{5}]$ direction cosines? Explain.

4. (a) Are direction cosines also scalar components? Explain.
   (b) Are scalar components also direction cosines? Explain.

5. Could $[\frac{3}{5}, -\frac{4}{5}]$ and $[-\frac{3}{5}, \frac{4}{5}]$ be direction cosines of the same directed segment? Explain.

6. What are the direction cosines of a segment OP if the coordinates of P are $(-1, 0)$?

7. Determine the direction cosines of: (a) a segment from the origin to $A = (x, 0)$; (b) a segment from the origin to $B = (0, y)$.

8. In each of the following find the direction cosines of the directed segment P_1P_2 whose end points are:
   (a) $P_1(3, 5)$ and $P_2(-1, 2)$
   (b) $P_1(-2, 7)$ and $P_2(1, 4)$
   (c) $P_1(-2, -1)$ and $P_2(5, 8)$
   (d) $P_1(4, 0)$ and $P_2(-5, 1)$
   (e) $P_1(0, 3)$ and $P_2(8, -4)$
   (f) $P_1(0, 0)$ and $P_2(-4, -1)$

2–8. Plane vectors

The "direction" of a segment is determined by its scalar components or by its direction cosines. If two segments AB and PQ have the *same* direction cosines, they have the same "direction." If AB and PQ have the same scalar components, they have the same magnitude as well as the same "direction." By the *magnitude* of a segment is meant the length of the segment.

We shall define a plane vector to be the *collection, considered as an entity, of all segments in the plane having the same direction and the same magnitude as a given one*. Any one of the segments may be used to *represent* the vector. Algebraically, a vector may be described by the scalar components of any segment representing it. We shall call these the *scalar components* of the vector and shall enclose them in brackets. Three geometric representations of the vector $[3, -4]$ are shown in Fig. 2–26. Here the scalar components of the vector are $[3, -4]$. Representing a vector by a small letter in heavy type, as **u**, we see that **u** is represented by CD, OQ, or MP. Any other segment equal in magnitude and direction to those used would have done just as well.

We shall also represent the vector having P_1P_2 as a representative segment by the symbol $\overrightarrow{P_1P_2}$. *The magnitude of a vector is the length of any one of the segments representing it.* We shall symbolize the magnitude of **u** by $|\mathbf{u}|$ or of $\overrightarrow{P_1P_2}$ by $|\overrightarrow{P_1P_2}|$. From the definition of a vector it is clear that *two vectors are equal if and only if the scalar components of each are equal*. For example, in order that $\mathbf{u} = [u_1, u_2]$ may be equal to $\mathbf{v} = [v_1, v_2]$, we must have $u_1 = v_1$ and $u_2 = v_2$. The student should note this carefully and understand clearly that the vector $[-1, 2]$ is *not* equal to the vector $[1, -2]$ or to the vector $[2, -1]$. Since the location of the segment representing the vector does not matter, it is often convenient to choose the origin as the initial point of the representative segment. In this case the coordinates of the end point $P(x, y)$ are also the scalar components of the vector $\overrightarrow{OP}$.

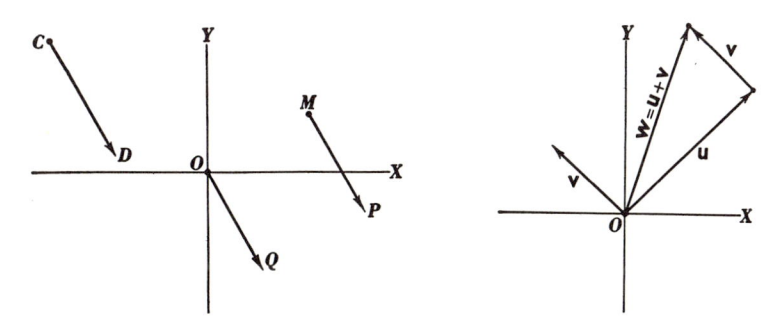

Fig. 2–26   Fig. 2–27

56   ANALYTIC GEOMETRY IN THE PLANE

We shall define the sum of two vectors to be that vector which has for its scalar components the sum of the scalar components of the two vectors. Thus, if $\mathbf{u} = [u_1, u_2]$ and $\mathbf{v} = [v_1, v_2]$, then $\mathbf{w} = \mathbf{u} + \mathbf{v} = [u_1 + v_1, u_2 + v_2]$.

Geometrically, we add two vectors in the manner shown in Fig. 2–27. Consider the segment representing one of the vectors to have its initial point at the origin. Place the initial point of the segment representing the other vector at the terminal point of the first segment. When this is done draw the segment from the initial point of the first segment to the terminal point of the second segment. This segment is defined to be a representative segment of the *sum vector*.

If we project $\mathbf{u}$, $\mathbf{v}$, and $\mathbf{w}$ on the x-axis and y-axis, we see that the scalar components of $\mathbf{w}$ are the sums of the scalar components of $\mathbf{u}$ and $\mathbf{v}$.

If a vector $\mathbf{u} = [u_1, u_2]$ is known, the vector $-\mathbf{u} = [-u_1, -u_2]$ is defined to be *the negative of* $\mathbf{u}$. This negative vector has the same magnitude as $\mathbf{u}$ but has the opposite direction. It is clear that $\mathbf{u} = -(-\mathbf{u})$. This relation suggests the following method for subtracting vectors: Change the sign of the vector being subtracted and add. Thus, in Fig. 2–28,

$$\mathbf{w} = \mathbf{u} - \mathbf{v} = \mathbf{u} + (-\mathbf{v})$$

By the notation $k[u_1, u_2]$, we shall mean the vector $[ku_1, ku_2]$. *If k is positive, the new vector has the same "direction" as the vector $[u_1, u_2]$; if k is negative, the new vector has the "direction" opposite to that of the vector $[u_1, u_2]$.*

We shall define the vector $\mathbf{0} = [0,0]$ to be the zero vector, even though it has no "direction." So the vector $\mathbf{0} = \mathbf{u} - \mathbf{u}$ is the zero vector. A segment P_1P_2 represents the zero vector exactly when $P_1 = P_2$.

A unit vector is defined as a vector whose magnitude is 1 unit.

The direction cosines of a vector are defined to be the direction cosines of any representative segment of the vector. Hence, the direction cosines of a unit vector are also the scalar components of the vector. The zero vector has no direction cosines.

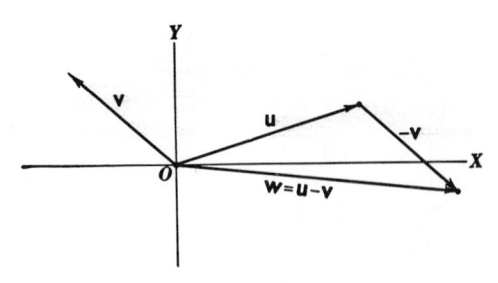

FIG. 2–28

Using the definition $k[u_1, u_2] = [ku_1, ku_2]$, we can express any vector as a constant times a unit vector. For example, if $\mathbf{u}$ is a non-zero vector,

$$\mathbf{u} = [u_1, u_2] = \sqrt{u_1^2 + u_2^2}\left[\frac{u_1}{\sqrt{u_1^2 + u_2^2}}, \frac{u_2}{\sqrt{u_1^2 + u_2^2}}\right] = |\mathbf{u}|\left[\frac{u_1}{|\mathbf{u}|}, \frac{u_2}{|\mathbf{u}|}\right] \quad (2\text{-}5)$$

where $\sqrt{u_1^2 + u_2^2}$ is the magnitude of $\mathbf{u}$; and $\frac{u_1}{\sqrt{u_1^2 + u_2^2}}$ and $\frac{u_2}{\sqrt{u_1^2 + u_2^2}}$ are the direction cosines of $\mathbf{u}$. The zero vector is equal to 0 times an arbitrary unit vector.

EXAMPLE 2–8. Find the sum and difference of the vectors $\mathbf{u} = [-3, 2]$ and $\mathbf{v} = [4, -1]$.

Solution. The scalar components of the required vectors are:

$$\mathbf{w} = \mathbf{u} + \mathbf{v} = [1, 1]$$
$$\mathbf{q} = \mathbf{u} - \mathbf{v} = [-7, 3]$$

EXAMPLE 2–9. Express $\mathbf{u} = [3, -2]$ in terms of a unit vector, and find the direction cosines of $\mathbf{u}$.

Solution. The expression representing the vector is

$$\mathbf{u} = [3, -2] = \sqrt{13}\left[\frac{3}{\sqrt{13}}, \frac{-2}{\sqrt{13}}\right]$$

The direction cosines of $\mathbf{u}$ are $\frac{3}{\sqrt{13}}$ and $\frac{-2}{\sqrt{13}}$.

EXERCISES

1. Give four geometric representations of the vector $[-2, 3]$. What is its magnitude?

2. In each of the following add the vectors and find the magnitude of the sum vector:
   (a) $[2, 5]$ and $[1, 4]$
   (b) $[-3, 2]$ and $[-1, -2]$
   (c) $[-3, -5]$ and $[2, 6]$
   (d) $[3, -6]$ and $[4, -1]$
   (e) $[4, 7]$ and $[2, -6]$
   (f) $[-5, 6]$ and $[4, -3]$

3. (a) Are all unit vectors equal? Explain.
   (b) Is the vector $[3, -1]$ equal to the vector $[-3, 1]$? Explain.

4. Express each of the following vectors in terms of a unit vector: (a) $[5, -3]$; (b) $[2, 1]$; (c) $[-1, 3]$; (d) $[-3, -4]$; (e) $[0, 5]$.

5. State which of the following vectors are equal: (a) $[3, -2]$, (b) $[-3, 2]$, (c) $[-3, 2]$, (d) $[3, 2]$, (e) $[-3, 2]$, and (f) $[-3, -2]$.

6. In Exercise 2 subtract the second vector from the first in each case and find the magnitudes of the difference vectors.

7. Show that, if P is the point (x, y), then $\overrightarrow{OP} = [x, y]$.

58   ANALYTIC GEOMETRY IN THE PLANE

8. If $\mathbf{i} = [1, 0]$ and $\mathbf{j} = [0, 1]$, express any vector $\mathbf{u} = [u_1, u_2]$ as a linear combination* of $\mathbf{i}$ and $\mathbf{j}$. Hint: Write $\mathbf{u} = c_1\mathbf{i} + c_2\mathbf{j}$ and solve for c_1 and c_2.

9. Show that the direction cosines of the non-zero vector $\mathbf{u} = [\Delta x, \Delta y]$ are
$$\frac{\Delta x}{\sqrt{|\Delta x|^2 + |\Delta y|^2}} \quad \text{and} \quad \frac{\Delta y}{\sqrt{|\Delta x|^2 + |\Delta y|^2}}$$

10. Show that the coordinates of the point where a vector $\overrightarrow{OP}$ intersects a circle of unit radius with the origin at the center are the direction cosines of the vector.

11. Given $\mathbf{u} = [-3, 2]$, $\mathbf{v} = [1, -4]$, $\mathbf{w} = [2, 3]$, construct geometrically and find:

   (a) $\mathbf{u} - 3\mathbf{v}$, $|\mathbf{u} - 3\mathbf{v}|$
   (b) $-3\mathbf{w}$, $|-3\mathbf{w}|$
   (c) $\mathbf{w} + 2\mathbf{u}$, $|\mathbf{w} + 2\mathbf{u}|$
   (d) $2\mathbf{v} - 4\mathbf{w}$, $|2\mathbf{v} - 4\mathbf{w}|$
   (e) $\frac{1}{3}\mathbf{w}$, $|\frac{1}{3}\mathbf{w}|$
   (f) $2\mathbf{w} - 3\mathbf{u} + 4\mathbf{v}$, $|2\mathbf{w} - 3\mathbf{u} + 4\mathbf{v}|$
   (g) $\mathbf{v} + 2\mathbf{u} - \mathbf{w}$, $|\mathbf{v} + 2\mathbf{u} - \mathbf{w}|$
   (h) $\mathbf{u} + \mathbf{v} + \mathbf{w}$, $|\mathbf{u} + \mathbf{v} + \mathbf{w}|$

12. Express the following vectors as linear combinations of $\mathbf{i}$ and $\mathbf{j}$: $[-1, 3]$; $[2, 2]$; $[4, -3]$; $[1, 1]$; $[-2, -3]$. (See Exercise 8.)

13. Find where the vector $\overrightarrow{OP}$ intersects the unit circle having its center at the origin, given the point P to be
   (a) $(-1, 3)$
   (b) $(3, -2)$
   (c) $(4, 0)$
   (d) $(0, -5)$
   (e) $(-3, -4)$
   (f) $(5, 3)$
   (g) $(-2, 0)$
   (h) $(-2, 5)$

14. Using the vectors $\mathbf{v}$, $\mathbf{w}$ in Exercise 11, draw representations for:
   (a) $3\mathbf{v}$, $2(3\mathbf{v})$, and $6\mathbf{v}$
   (b) $-3\mathbf{v}$, $2(-3\mathbf{v})$, and $-6\mathbf{v}$
   (c) $2\mathbf{v}$, $3\mathbf{v}$, $5\mathbf{v}$, and $2\mathbf{v} + 3\mathbf{v}$
   (d) $2\mathbf{v}$, $-3\mathbf{v}$, $-\mathbf{v}$, and $2\mathbf{v} - 3\mathbf{v}$
   (e) $3\mathbf{v}$, $3\mathbf{w}$, $3\mathbf{v} + 3\mathbf{w}$, $\mathbf{v} + \mathbf{w}$, and $3(\mathbf{v} + \mathbf{w})$
   (f) $-3\mathbf{v}$, $-3\mathbf{w}$, $(-3)\mathbf{v} + (-3)\mathbf{w}$, $\mathbf{v} + \mathbf{w}$, $-3(\mathbf{v} + \mathbf{w})$

2–9. Angle between two vectors

A circle with its center at the origin and a unit radius is called a *unit circle*. Consider $\mathbf{u}$ and $\mathbf{v}$ to be any two non-zero vectors. Let l_1 and m_1 be the direction cosines of $\mathbf{u}$, and let l_2 and m_2 be the direction cosines of $\mathbf{v}$; and place the initial points of representative segments of both vectors at the origin. Then, as shown in Fig. 2–29, $P_1(l_1, m_1)$ is the point where the vector $\mathbf{u}$ cuts the unit circle, and $P_2(l_2, m_2)$ is the point where the vector $\mathbf{v}$ cuts the unit circle.

---

*A *linear combination* of vectors $\mathbf{u}$ and $\mathbf{v}$ is a vector expression of the form $c_1\mathbf{u} + c_2\mathbf{v}$, where c_1 and c_2 are numbers.

THE POINT AND PLANE VECTORS   59

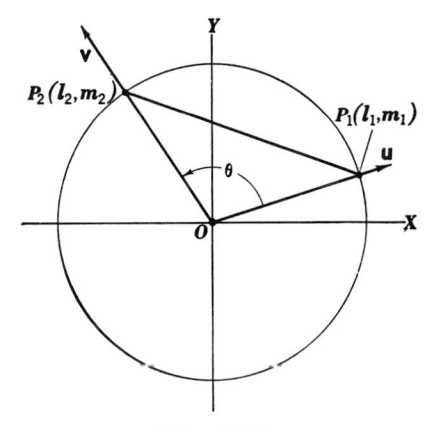

Fig. 2-29

From the law of cosines,

$$|P_1P_2|^2 = |OP_1|^2 + |OP_2|^2 - 2|OP_1|\,|OP_2|\cos\theta$$

Since $|OP_1| = |OP_2| = 1$,

$$|P_1P_2|^2 = 1 + 1 - 2\cos\theta = 2 - 2\cos\theta$$

But $|P_1P_2|^2 = (l_2 - l_1)^2 + (m_2 - m_1)^2$. Therefore,

$$2\cos\theta = 2 - [(l_2 - l_1)^2 + (m_2 - m_1)^2]$$
$$= 2 - [l_2^2 - 2l_1l_2 + l_1^2 + m_2^2 - 2m_2m_1 + m_1^2]$$

Since $l_1^2 + m_1^2 = 1$ and $l_2^2 + m_2^2 = 1$,

$$2\cos\theta = 2 - 2 + 2(l_1l_2 + m_1m_2)$$

or
$$\cos\theta = l_1l_2 + m_1m_2 \tag{2-6}$$

Hence, *the cosine of the angle between any two vectors is equal to the sum of the products of the respective direction cosines of the two vectors.*

If $\mathbf{u} = [u_1, u_2]$ and $\mathbf{v} = [v_1, v_2]$, then

$$l_1 = \frac{u_1}{\sqrt{u_1^2 + u_2^2}} \quad \text{and} \quad m_1 = \frac{u_2}{\sqrt{u_1^2 + u_2^2}}$$

and
$$l_2 = \frac{v_1}{\sqrt{v_1^2 + v_2^2}} \quad \text{and} \quad m_2 = \frac{v_2}{\sqrt{v_1^2 + v_2^2}}$$

Hence,
$$\cos\theta = \frac{u_1v_1 + u_2v_2}{\sqrt{u_1^2 + u_2^2}\sqrt{v_1^2 + v_2^2}} = \frac{u_1v_1 + u_2v_2}{|\mathbf{u}|\,|\mathbf{v}|} \tag{2-7}$$

since $|\mathbf{u}| = \sqrt{u_1^2 + u_2^2}$ and $|\mathbf{v}| = \sqrt{v_1^2 + v_2^2}$.

From equation (2–7) it follows that

$$u_1v_1 + u_2v_2 = |\mathbf{u}||\mathbf{v}| \cos \theta$$

The combination $u_1v_1 + u_2v_2$ is so important and useful that it is given a special title and a special notation. It is called the *symmetric product* or the *scalar product* and will be symbolized as $\mathbf{u} \cdot \mathbf{v}$. Because of this notation, we shall also call the symmetric product the *dot product*. So

$$\mathbf{u} \cdot \mathbf{v} = u_1v_1 + u_2v_2 \tag{2-8}$$

The dot product of two vectors is equal to the sum of the products of the corresponding scalar components of the two vectors.

We now have

$$\mathbf{u} \cdot \mathbf{v} = |\mathbf{u}||\mathbf{v}| \cos \theta$$

If θ is an *acute* angle, $\cos \theta$ is positive and, hence, $\mathbf{u} \cdot \mathbf{v}$ is positive. If θ is an *obtuse* angle, $\cos \theta$ is negative and, hence, $\mathbf{u} \cdot \mathbf{v}$ is negative.

Conversely, if $\mathbf{u} \cdot \mathbf{v}$ is positive, $\cos \theta$ is positive and θ is acute; if $\mathbf{u} \cdot \mathbf{v}$ is negative, $\cos \theta$ is negative and θ is obtuse.

In Section 2–10, we shall discuss the situation when two vectors are parallel or perpendicular to each other.

EXAMPLE 2–10. Find the cosines of the angles of the triangle whose vertices are $P_1(-3, 4)$, $P_2(2, 5)$, and $P_3(5, -1)$.

Solution. The triangle is shown in Fig. 2–30. Vectors represented by the directed segments that intersect at the vertex P_1 are

$$\overrightarrow{P_1P_2} = [5, 1] \quad \text{and} \quad \overrightarrow{P_1P_3} = [8, -5]$$

Therefore,

$$\cos \angle P_2P_1P_3 = \frac{40 - 5}{\sqrt{26}\sqrt{89}} = \frac{35}{\sqrt{26}\sqrt{89}}$$

For the segments that intersect at P_2,

$$\overrightarrow{P_2P_1} = [-5, -1] \quad \text{and} \quad \overrightarrow{P_2P_3} = [3, -6]$$

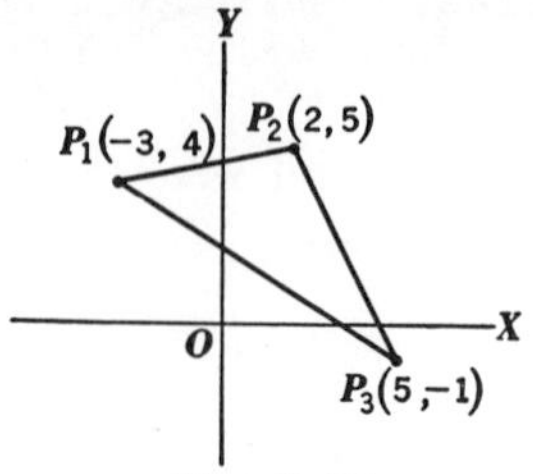

FIG. 2–30

Therefore,
$$\cos \angle P_1P_2P_3 = \frac{-15+6}{\sqrt{26}\ \sqrt{45}} = \frac{-9}{\sqrt{26}\ \sqrt{45}} = \frac{-3}{\sqrt{26}\ \sqrt{5}}$$

Finally,
$$\overrightarrow{P_3P_1} = [-8, 5] \quad \text{and} \quad \overrightarrow{P_3P_2} = [-3, 6]$$

Therefore,
$$\cos \angle P_1P_3P_2 = \frac{24+30}{\sqrt{89}\ \sqrt{45}} = \frac{54}{\sqrt{89}\ \sqrt{45}} = \frac{18}{\sqrt{5}\ \sqrt{89}}$$

The student should explain which angles of the triangle are acute and which one is obtuse.

The scalar product can also be used to determine the projection of one line segment on another line. Considering the definition of the projection of a directed line segment on a line (Section 2–4) and the definition of a unit vector (Section 2–8), we can prove that the projection of **u** on a line in the direction of **v** is equal to $\frac{\mathbf{u} \cdot \mathbf{v}}{|\mathbf{v}|}$. To do this consider Fig. 2–31.

FIG. 2–31

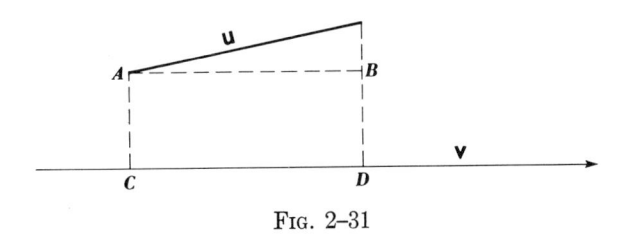

Through the endpoints of vector **u** pass lines perpendicular to **v**. Suppose these lines intersect **v** at C and D. The projection of **u** on **v** is then CD which equals AB.

But $AB = |\mathbf{u}|\cos\theta$

$\qquad = \dfrac{|\mathbf{u}|\,|\mathbf{v}|\cos\theta}{|\mathbf{v}|}$, when numerator and denominator are multiplied by $|\mathbf{v}|$

$\qquad = \dfrac{\mathbf{u}\cdot\mathbf{v}}{|\mathbf{v}|}$, since $|\mathbf{u}|\,|\mathbf{v}|\cos\theta = \mathbf{u}\cdot\mathbf{v}$

$\therefore AB = \dfrac{\mathbf{u}\cdot\mathbf{v}}{|\mathbf{v}|}.$

2–10. Parallel and perpendicular vectors

Two non-zero vectors will be said to be *parallel in the same sense* when the angle between them is 0°, and they will be said to be *parallel in the opposite sense* when the angle is 180°. If the vectors are either parallel in the same sense or parallel in the opposite sense, they are said to be *parallel*. When two vectors are parallel in the same sense, their direction cosines are equal. So, if **u** and **v** are two vectors that are parallel in the same sense, and if l_1 and m_1 and l_2 and m_2 are the direction cosines of **u** and **v**, respectively, then $l_1 = l_2$ and $m_1 = m_2$. In this case, $l_1 l_2 + m_1 m_2 = \cos 0° = 1$. When two vectors are parallel in the opposite sense, $l_1 = -l_2$ and $m_1 = -m_2$. Now $l_1 l_2 + m_1 m_2 = \cos 180° = -1$.

Since the direction cosines of a vector are determined when its scalar components are known, it follows that *two vectors are parallel if, and only if, their corresponding scalar components are proportional*. If the ratio of either pair of components is positive, the vectors are parallel in the same sense; and, if the ratio is negative, the vectors are parallel in the opposite sense. Hence, if $\mathbf{u} = [u_1, u_2]$ and $\mathbf{v} = [v_1, v_2]$, then **u** is parallel to **v** only when

$$\frac{u_1}{v_1} = \frac{u_2}{v_2} \qquad (2\text{–}9)$$

Here it is assumed that both v_1 and v_2 are not zero. If $v_1 = 0$, then $v_2 \neq 0$ and equation (2–9) is interpreted to mean that $u_1 = 0$ and $u_2 \neq 0$; if $v_1 = v_2 = 0$, then $\mathbf{v} = [0, 0]$ and parallelism is undefined. Note that identical vectors are parallel.

If two non-zero vectors are perpendicular, the angle θ between them is 90°. Hence, $\cos \theta = 0$ and so $l_1 l_2 + m_1 m_2 = 0$. Since $\mathbf{u} \cdot \mathbf{v} = |\mathbf{u}| \, |\mathbf{v}| \cos \theta$, it follows that the two vectors will be perpendicular only when

$$\mathbf{u} \cdot \mathbf{v} = 0 \qquad (2\text{–}10)$$

Hence, *two vectors are perpendicular if, and only if, the sum of the products of their scalar components is zero.*

Since equation (2–10) is satisfied when either **u** or **v** is the zero vector, we shall *define* the zero vector to be perpendicular to every vector.

EXAMPLE 2–11. Show that the points $P_1(1, 5)$, $P_2(3, 3)$, and $P_3(-4, 0)$ are vertices of a right triangle.

Solution. If we draw the triangle, as in Fig. 2–32, we notice that the right angle appears to be at P_1. Hence, we obtain $\overrightarrow{P_1 P_3} = [-5, -5]$ and $\overrightarrow{P_1 P_2} = [2, -2]$ and we compute their dot product. Since $\overrightarrow{P_1 P_3} \cdot \overrightarrow{P_1 P_2} = -10 + 10 = 0$, the triangle $P_1 P_2 P_3$ is a right triangle.

EXAMPLE 2–12. Find the fourth vertex of the parallelogram whose other vertices are $P_1(1, 4)$, $P_2(2, 1)$, and $P_3(8, 2)$ in that order.

FIG. 2-32

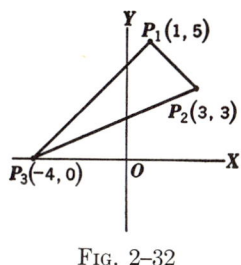

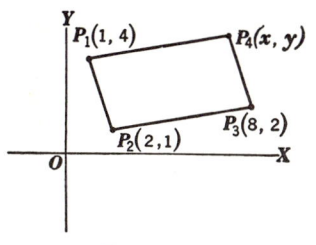

FIG. 2-33

Solution. Since the figure is to be a parallelogram whose vertices in order are P_1, P_2, P_3, and P_4, as shown in Fig. 2-33, we know that the directed segments $\overrightarrow{P_1P_4}$ and $\overrightarrow{P_2P_3}$ are parallel in the same sense. Hence, $\overrightarrow{P_1P_4} = \overrightarrow{P_2P_3}$. Since $\overrightarrow{P_1P_4} = [x-1, y-4]$ and $\overrightarrow{P_2P_3} = [6, 1]$, we then have $[x-1, y-4] = [6, 1]$. It follows that $x-1 = 6$, or $x = 7$; and $y-4 = 1$, or $y = 5$. Therefore, the coordinates of the fourth vertex P_4 are $(7, 5)$.

EXERCISES

1. In each of the following triangles for which the coordinates of the vertices are given, find the cosines of the angles:
   (a) $P_1(3, -2)$, $P_2(-1, 4)$, and $P_3(-6, -1)$
   (b) $P_1(-3, 5)$, $P_2(1, 1)$, and $P_3(2, -4)$
   (c) $P_1(0, 0)$, $P_2(-3, 1)$, and $P_3(2, -3)$
   (d) $P_1(1, 3)$, $P_2(-5, 2)$, and $P_3(-3, -4)$
   (e) $P_1(2, 4)$, $P_2(0, 6)$, and $P_3(-1, 0)$
   (f) $P_1(-1, -4)$, $P_2(-5, 2)$, and $P_3(1, 1)$

2. Find the value of k if the segment whose end points are $(1, 3)$ and $(2, k)$ is perpendicular to the segment whose end points are $(-3, 1)$ and $(5, 6)$.

3. Show that the points $(0, -4)$, $(10, 0)$, $(6, 10)$, and $(-4, 6)$ are vertices of a square.

4. Three vertices of a rectangle are $(0, 0)$, $(6, 3)$, and $(-1, 2)$. Find the coordinates of the fourth vertex.

5. Prove analytically that the diagonals of any square are perpendicular to each other.

6. Find the value of a if the points $(5, 0)$, $(a, 4)$, and $(-4, 3)$ are vertices of a right triangle.

7. For what value of k is the vector $\overrightarrow{P_1P_2}$, from $P_1 = (2, -1)$ to $P_2 = (4, k)$, perpendicular to the vector $\overrightarrow{P_1P_3}$, from P_1 to $P_3 = (6, 1)$.

64   Analytic Geometry in the Plane

8. Show that $\mathbf{u} \cdot \mathbf{v} = \mathbf{v} \cdot \mathbf{u}$.

9. If two vectors $\mathbf{u}$ and $\mathbf{v}$ are parallel, show that there exists a number k such that $\mathbf{u} = k\mathbf{v}$.

10. Prove analytically that the base angles of an isosceles triangle are equal.

11. Show analytically that if the diagonals of a rectangle are perpendicular, the rectangle is a square.

12. If two medians of a triangle are equal, show analytically that the triangle is isosceles.

13. Let $ABCD$ be the vertices of a parallelogram with A and C opposite vertices. From B and D, drop perpendiculars on the diagonal AC at P and Q, respectively. Prove that the figure $BPDQ$ is a parallelogram.

14. Determine which of the following sets of points are collinear:
    (a) (1, 2), (3, 8), (−1, −4)
    (b) (4, 1), (−1, −9), (2, −3)
    (c) (0, 1), (1/3, 0), (−1, 4)
    (d) (4, −1), (−1, 2), (6, −2)
    (e) (2, 2), (6, 4), (0, 1)
    (f) (−1, −1), (2, 1), (5, 4)

15. (a) Determine x so that $(x, 3)$, $(-1, 2)$, $(3, -5)$ are collinear.
    (b) Determine y so that $(2, 5)$, $(8, 2)$, $(6, y)$ are collinear.

16. Prove that $\mathbf{u} \cdot (\mathbf{v} + \mathbf{w}) = \mathbf{u} \cdot \mathbf{v} + \mathbf{u} \cdot \mathbf{w}$.

17. Prove that $(\mathbf{u} + \mathbf{v}) \cdot (\mathbf{w} + \mathbf{q}) = \mathbf{u} \cdot \mathbf{w} + \mathbf{u} \cdot \mathbf{q} + \mathbf{v} \cdot \mathbf{w} + \mathbf{v} \cdot \mathbf{q}$.

18. Show that $|\mathbf{u} \cdot \mathbf{v}| \leq |\mathbf{u}| \, |\mathbf{v}|$. When is $|\mathbf{u} \cdot \mathbf{v}| = |\mathbf{u}| \, |\mathbf{v}|$?

19. In finding $k\mathbf{u}$, explain carefully what effect the scalar k has on the vector $\mathbf{u}$ if $0 < k < 1$; if $k > 1$; if $k < 0$. Demonstrate your answer for $\mathbf{u} = [-2, 3]$ and k equal to the following values: $\frac{1}{2}$; -2; $-\frac{2}{3}$; 3; 2; $-\frac{3}{2}$; $\frac{1}{3}$; $\frac{5}{4}$; -1; 0.

20. Exercises 11 and 14 in Section 2–8 and the preceding exercise (19) suggest the following properties for vectors $\mathbf{v}$ and $\mathbf{w}$, and scalars r and s:

    (a) $r(s\mathbf{v}) = (rs)\mathbf{v}$
    (b) $(r + s)\mathbf{v} = r\mathbf{v} + s\mathbf{v}$
    (c) $r(\mathbf{v} + \mathbf{w}) = r\mathbf{v} + r\mathbf{w}$

    Verify (a), (b), (c), if $r = 2$, $s = 3$, $\mathbf{v} = [2, -1]$, $\mathbf{w} = [-3, 4]$; if $r = -\frac{1}{2}$, $s = -2$, $\mathbf{v} = [\frac{1}{2}, 1]$, $\mathbf{w} = [-2, 3]$.

2–11. The coordinates of a point that divides a segment in a given ratio

In Fig. 2–34 let $P(x, y)$ be a point on the line containing the point $P_1(x_1, y_1)$ and $P_2(x_2, y_2)$ and having the property that $\overrightarrow{P_1P} = k\,\overrightarrow{P_1P_2}$,

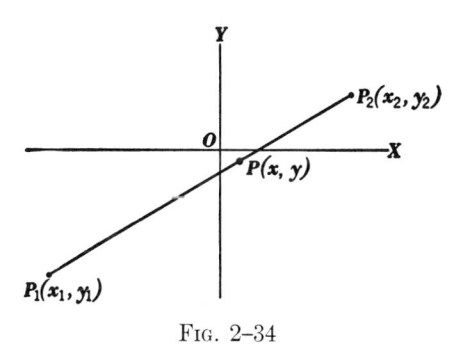

FIG. 2–34

where k is a given number. When the coordinates of P_1 and P_2 are known, the coordinates of P are easily determined for any given value of k. We find them as follows. We know that $\overrightarrow{P_1P} = [x - x_1, y - y_1]$ and that $\overrightarrow{P_1P_2} = [\Delta x, \Delta y]$. Hence, we have $[x - x_1, y - y_1] = k[\Delta x, \Delta y] = [k\,\Delta x, k\,\Delta y]$. From these relations, we obtain $x - x_1 = k\,\Delta x$ and $y - y_1 = k\,\Delta y$; or

$$x = x_1 + k\,\Delta x$$
$$y = y_1 + k\,\Delta y \qquad (2\text{--}11)$$

From the properties of vectors the following conclusions can be drawn: When $k = 0$, $P = P_1$; when $k = 1$, $P = P_2$; when $0 < k < 1$, P lies between P_1 and P_2; when $k > 1$, P lies beyond P_2; when k is negative, P lies on the half-line through P_1 not containing P_2. The last conclusion follows because, if k is negative, $\overrightarrow{P_1P}$ and $\overrightarrow{P_1P_2}$ must extend in opposite directions.

EXAMPLE 2–13. Find the coordinates of P if $\overrightarrow{P_1P} = \frac{2}{3}\overrightarrow{P_1P_2}$, where $P_1 = (-3, 4)$ and $P_2 = (2, -1)$.

Solution. The conditions are shown in Fig. 2–35. Since $\overrightarrow{P_1P} = [x + 3, y - 4]$ and $\overrightarrow{P_1P_2} = [5, -5]$, we have:

$$x + 3 = \tfrac{2}{3}(5)$$
$$y - 4 = \tfrac{2}{3}(-5)$$

Hence, $\qquad x = \tfrac{1}{3}$ and $y = \tfrac{2}{3}$

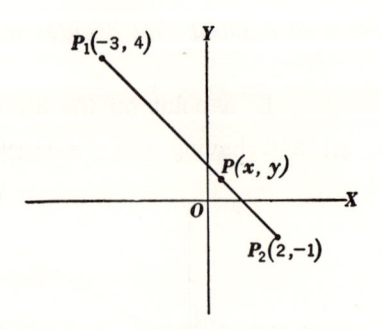

FIG. 2-35

Sometimes we are given the ratio of $\overline{P_1P}$ to $\overline{PP_2}$. This ratio is, by definition, *the ratio in which P divides the segment* P_1P_2. When this is the case we base our equation on the relation $\overrightarrow{P_1P} = r\,\overrightarrow{PP_2}$ and proceed as before. We have:

$$\overrightarrow{P_1P} = [x - x_1,\ y - y_1] \quad \text{and} \quad \overrightarrow{PP_2} = [x_2 - x,\ y_2 - y]$$

Then, $[x - x_1,\ y - y_1] = r[x_2 - x,\ y_2 - y] = [r(x_2 - x),\ r(y_2 - y)]$

Hence, $x - x_1 = r(x_2 - x) \quad \text{and} \quad y - y_1 = r(y_2 - y)$

Solving for x and y, we obtain:

$$x = \frac{x_1 + rx_2}{1 + r}$$

$$y = \frac{y_1 + ry_2}{1 + r}$$

(2-12)

When $r = 0$, $P = P_1$; when r is positive, P lies on the segment; and, when r is negative, P lies outside of the segment.

It is not advisable to memorize either equations (2-11) or equations (2-12). Instead the student should be sure to understand the method of approach and to rebuild each problem accordingly.

EXAMPLE 2-14. Find the coordinates of the point P which divides the segment from $P_1 = (1, -4)$ to $P_2 = (5, 6)$ in the ratio $2:1$.

Solution. Here $\overrightarrow{P_1P} = 2\overrightarrow{PP_2}$. In Fig. 2-36,

$$\overrightarrow{P_1P} = [x - 1,\ y + 4] \quad \text{and} \quad \overrightarrow{PP_2} = [5 - x,\ 6 - y]$$

Hence,

$$[x - 1,\ y + 4] = 2[5 - x,\ 6 - y] = [10 - 2x,\ 12 - 2y]$$

It follows that

$$x - 1 = 10 - 2x$$
$$y + 4 = 12 - 2y$$

Solving for x and y, we obtain:

$$x = \tfrac{11}{3} \text{ and } y = \tfrac{8}{3}$$

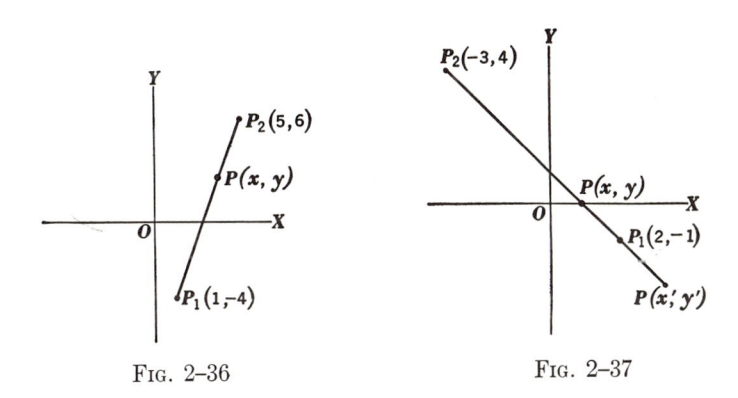

Fig. 2-36

Fig. 2-37

EXAMPLE 2–15. Find two points on the line containing $P_1 = (2, -1)$ and $P_2 = (-3, 4)$, each of which is four times as far from P_2 as it is from P_1.

Solution. We could state this problem symbolically as $4|P_1P| = |PP_2|$, where the absolute-value symbols imply that direction has not been considered. As indicated in Fig. 2–37, one position P of the required point is between P_1 and P_2, while the other position P' is beyond P_1.

To find the internal point P, where $\overrightarrow{P_1P}$ and $\overrightarrow{PP_2}$ have the same direction, we have $4\ \overrightarrow{P_1P} = \overrightarrow{PP_2}$. Since $\overrightarrow{P_1P} = [x-2, y+1]$ and $\overrightarrow{PP_2} = [-3-x, 4-y]$, it follows that

$$4[x-2, y+1] = [-3-x, 4-y]$$

Hence,
$$4x - 8 = -3 - x$$
$$4y + 4 = 4 - y$$

and we obtain $x = 1$ and $y = 0$.

For the external point P', we have $4\ \overrightarrow{P_1P'} = -\overrightarrow{P'P_2}$. Hence,

$$4[x'-2, y'+1] = [3+x', -4+y']$$

From this equation,
$$4x' - 8 = 3 + x'$$
$$4y' + 4 = -4 + y'$$

and we obtain $x' = \tfrac{11}{3}$ and $y' = -\tfrac{8}{3}$.

2–12. The mid-point of a segment

If P is the mid-point of a segment P_1P_2, it follows that $\overrightarrow{P_1P} = \overrightarrow{PP_2}$. From this we obtain $[x-x_1, y-y_1] = [x_2-x, y_2-y]$. Solving for x and y, we have:

$$x = \frac{x_1+x_2}{2}$$
$$y = \frac{y_1+y_2}{2}$$
(2–13)

Hence, *the coordinates of the mid-point of a segment are equal to one-half the sum of the abscissas and one-half the sum of the ordinates of the end points of the segment.*

EXAMPLE 2–16. Find the mid-point of the segment joining $P_1(2, -1)$ and $P_2(-3, 4)$.

Solution. The segment is shown in Fig. 2–38. Using equation (2–13), we have:

$$x = \frac{2+(-3)}{2} = -\tfrac{1}{2}$$
$$y = \frac{-1+4}{2} = \tfrac{3}{2}$$

EXAMPLE 2–17. Prove that the diagonals of any parallelogram bisect each other.

Solution. Consider the parallelogram in Fig. 2–39. The mid-point

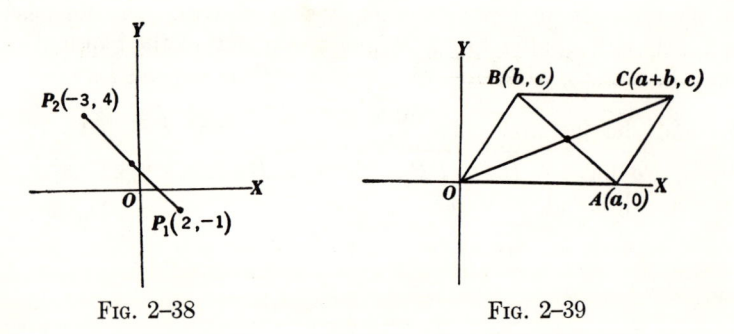

FIG. 2–38

FIG. 2–39

of diagonal AB is $\left(\frac{a+b}{2}, \frac{c}{2}\right)$. The mid-point of diagonal OC is $\left(\frac{a+b}{2}, \frac{c}{2}\right)$. Since the two mid-points are the same, the diagonals bisect each other.

EXERCISES

1. Find the coordinates of the point that divides the segment from $P_1(3, 4)$ to $P_2(-2, 7)$ in the ratio $3:5$.

2. What are the coordinates of the point P such that $\overrightarrow{P_1P} = \tfrac{3}{5}\overrightarrow{P_1P_2}$, where P_1 has the coordinates $(-2, -1)$ and P_2 has the coordinates $(5, 6)$?

3. Find the coordinates of all the points on the line containing $P_1(-2, 4)$ and $P_2(6, -3)$ each of which is 4 times as far from P_1 as it is from P_2.

4. Find the coordinates of all the points on the line containing $A(4, -6)$ and $B(-3, -2)$ each of which is 3 times as far from one given point as it is from the other.

5. What is the perimeter of the triangle obtained by joining the mid-points of the sides of the triangle whose vertices are $P_1(-1, 2)$, $P_2(5, -3)$, and $P_3(4, 7)$?

6. Two points of a given segment are $P_1(-3, 2)$ and $P_2(5, -4)$. Find the coordinates of the points P and P' on the line containing P_1 and P_2 such that: (a) $|P_1P| = \frac{7}{3}|P_1P_2|$; (b) $|P_1P'| = \frac{7}{3}|P'P_2|$.

7. One end point of a segment is $(3, -4)$, and the mid-point of the segment is $(-2, 1)$. What are the coordinates of the other end point of the segment?

8. Verify the conclusions immediately preceding Example 2–13 by applying equations (2–11).

9. (a) Show that the segment joining the mid-points of two sides of any triangle is equal in length to half the third side and is parallel to the third side.
(b) Show that the mid-point of the hypotenuse of a right triangle is equidistant from the three vertices.

10. Find the coordinates of the vertices of a triangle if the points $(-1, 1)$, $(0, 4)$, and $(5, 3)$ are the mid-points of the sides.

11. Prove that the lines joining the mid-points of adjacent pairs of sides of any quadrilateral form a parallelogram.

12. Show that the medians of the triangle with its vertices at $P_1(1, 3)$, $P_2(-2, 4)$, and $P_3(-4, -2)$ intersect in a point. (Hint: Find on each median the point two-thirds of the distance from a vertex to the mid-point of the opposite side.

13. Prove analytically each of the following theorems:
(a) The medians from the ends of the base of an isosceles triangle are equal in length.
(b) In any triangle the sum of the squares of the medians is equal to three-fourths the sum of the squares of the three sides.
(c) The sum of the squares of the four sides of any quadrilateral is equal to the sum of the squares of the diagonals increased by four times the square of the segment joining the mid-points of the diagonals.

14. Show analytically that: (a) the segments joining the mid-points of the opposite sides of any quadrilateral bisect each other; (b) the segment joining the mid-points of two opposite sides of any quadrilateral and the segment joining the mid-points of the diagonals of the quadrilateral bisect each other.

15. In what ratio does the point $(0, 2)$ divide the segment whose end points are $(-3, 0)$ and $(3, 4)$?

16. (a) Find y if the point $(2, y)$ lies on the line joining the points $(-3, 4)$ and $(6, -3)$.
(b) Find x if the point $(x, -3)$ lies on the line joining the points $(4, 2)$ and $(-1, 5)$.

17. Find the coordinates of the point where the segment from $P_1 = (2, -3)$ to $P_2 = (3, 1)$ cuts the x-axis and those of the point where the segment extended cuts the y-axis.

18. If P_1 has the coordinates $(-4, -3)$ and P_2 has the coordinates $(7, 2)$, find the coordinates of the points where the segment P_1P_2 cuts the axes.

19. In what ratio does the point $(\frac{22}{3}, -\frac{13}{3})$ divide the segment having end points $P_1(-2, 1)$ and $P_2(5, -3)$?

2–13. Complementary vectors

A vector which is 90° "ahead of" a given vector and has the same magnitude as the given vector is called the *complement* of the given vector. If the line segment representations of vectors **u** and **v** have their initial points in common and if the turn from **v** to **u** is 90° in a counter-clockwise sense, we say that **u** is *90° ahead of* **v**. If $\mathbf{u} = [x_1, y_1]$* is the given vector, then co-**u** is the complementary vector and the scalar components of co-**u** are $[-y_1, x_1]$. To see this we proceed as follows:

In Fig. 2–40 let l_1 and m_1 be the direction cosines of $\mathbf{u} = [x_1, y_1]$. Then $l_1 = \cos \alpha$ and $m_1 = \sin \alpha$. If l_2 and m_2 are the direction cosines of co-**u**, then $l_2 = \cos(90° + \alpha) = -\sin \alpha$ and $m_2 = \sin(90° + \alpha) = \cos \alpha$. Hence,

$$l_2 = -m_1 = \frac{-y_1}{|\text{co-}\mathbf{u}|} = \frac{-y_1}{|\mathbf{u}|}$$

$$m_2 = l_1 = \frac{x_1}{|\text{co-}\mathbf{u}|} = \frac{x_1}{|\mathbf{u}|}$$

Since the direction cosines of co-**u** are $\frac{-y_1}{|\mathbf{u}|}$ and $\frac{x_1}{|\mathbf{u}|}$, it follows that the scalar components of co-**u** are $[-y_1, x_1]$.

---

*The vector symbol $\mathbf{u} = [x_1, y_1]$ is used in Sections 2–13 and 2–14 to emphasize that the initial point of the vector has been selected to be the origin.

Thus, to obtain the scalar components of a vector complementary to a given vector, we interchange the scalar components of the given vector and, after the interchange, change the sign of the first scalar component. For example, if **u** = [3, 4], we know that co-**u** = [−4, 3].

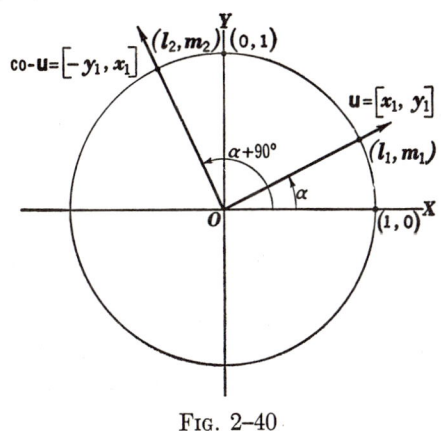

FIG. 2–40

EXERCISES

Find the vector complementary to each of the following vectors:
(a) [3, −2]; (b) [−1, 3]; (c) [−4, −1]; (d) [7, −1]; (e) [$\frac{3}{5}$, $\frac{4}{5}$]; (f) [u_1, u_2];
(g) [Δx, Δy]; (h) [a, b].

2–14. Area of a triangle and the bar product of two vectors

Let θ be the angle between two given vectors **u** = [x_1, y_1] and **v** = [x_2, y_2], as shown in Fig. 2–41. From Section (2–9) we know that $|\mathbf{u}|\,|\mathbf{v}|\cos\theta = x_1 x_2 + y_1 y_2$. We also know that co-**v** = [−y_2, x_2] and the angle between **u** and co-**v** is $90° + \theta$. Then,

$$\mathbf{u} \cdot \text{co-}\mathbf{v} = |\mathbf{u}|\,|\mathbf{v}|\cos(90°+\theta) = -x_1 y_2 + x_2 y_1$$

Since $\cos(90° + \theta) = -\sin\theta$,

$$-|\mathbf{u}|\,|\mathbf{v}|\sin\theta = -x_1 y_2 + x_2 y_1$$

or
$$|\mathbf{u}|\,|\mathbf{v}|\sin\theta = x_1 y_2 - x_2 y_1 \tag{2–14}$$

This result is either positive or negative, the sign depending on the measure and direction of the angle θ.

From trigonometry, we know that the area of a triangle is equal to $\frac{1}{2} d_1 d_2 \sin\theta$, where d_1 and d_2 are the lengths of two of the sides of the triangle and θ is the positive angle between these two sides. Hence, if two of the sides of a triangle are taken to be the segment representations of two vectors **u** and **v**, and if θ is the angle between them, the area A of the triangle is

$$A = \pm \frac{|\mathbf{u}| \, |\mathbf{v}| \sin \theta}{2} = \pm \frac{x_1 y_2 - x_2 y_1}{2} \tag{2-15}$$

where the minus sign must be used if $\sin \theta$ is negative.

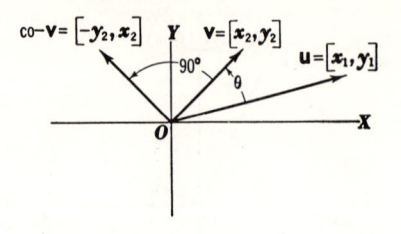

FIG. 2-41

The number $x_1 y_2 - x_2 y_1$ is a second important product of two plane vectors and we shall call it the *bar product* and write it $\mathbf{u}|\mathbf{v}$. This quantity is also called the *alternating product* of the vector $\mathbf{u}$ and the vector $\mathbf{v}$. So, if $\mathbf{u} = [x_1, y_1]$ and $\mathbf{v} = [x_2, y_2]$, we have

$$\mathbf{u}|\mathbf{v} = x_1 y_2 - x_2 y_1 \tag{2-16}$$

The geometric significance of the bar product is that its absolute value is equal to twice the area of the triangle having $|\mathbf{u}|$ and $|\mathbf{v}|$ as the lengths of two of its sides.

Since $\mathbf{u}|\mathbf{v} = |\mathbf{u}| \, |\mathbf{v}| \sin \theta$ involves the sine of the angle between $\mathbf{u}$ and $\mathbf{v}$, the bar product will be negative if the turn from $\mathbf{u}$ to $\mathbf{v}$ is clockwise and $0° < \theta < 180°$. If $\mathbf{u}|\mathbf{v} = 0$ and neither $\mathbf{u}$ nor $\mathbf{v}$ is the zero vector, it follows that $\sin \theta = 0$. Therefore, $\theta = 0°$ or $180°$, and the two vectors are parallel.

EXAMPLE 2-18. Find the area of the triangle whose vertices are $P_1(1, -4)$, $P_2(2, 3)$, and $P_3(-3, 5)$.

Solution. The triangle is shown in Fig. 2-42. We shall consider the two sides passing through P_3. We have:

$$\overrightarrow{P_3 P_2} = [5, -2] \quad \text{and} \quad \overrightarrow{P_3 P_1} = [4, -9]$$

Then,
$$\overrightarrow{P_3 P_2} | \overrightarrow{P_3 P_1} = -45 + 8 = -37$$

The sign is found to be negative since the turn from $\overrightarrow{P_3 P_2}$ to $\overrightarrow{P_3 P_1}$ is clockwise. The area is

$$A = \frac{|-37|}{2} = 18\tfrac{1}{2} \text{ square units}$$

EXAMPLE 2-19. Find k if the vectors $\mathbf{u} = [-3, 5]$ and $\mathbf{v} = [k, 2]$ are parallel.

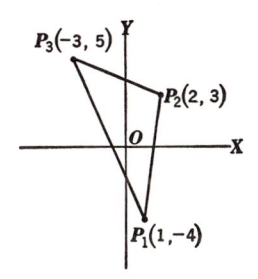

FIG. 2-42

Solution. Since the bar product of the vectors must be zero,
$$u|v = -6 - 5k = 0$$
and
$$k = -\tfrac{6}{5}$$
Hence, if $k = -\tfrac{6}{5}$, the vector **v** is parallel to the vector **u**.

EXERCISES

1. Show that $u|v$ is the negative of $v|u$.
2. If $u = [a_1, b_1]$ and $v = [a_2, b_2]$, determine: (a) their bar product and (b) their dot product.
3. (a) Which product of two vectors is used to test for their perpendicularity? Explain.
   (b) Which product is used to test for parallelism? Explain.
   (c) Which product is used to find *area*? Explain.
4. Find the area of each of the following triangles the vertices of which have the given coordinates:
   (a) $P_1(0, 0)$, $P_2(6, 1)$, and $P_3(4, 2)$
   (b) $P_1(8, 2)$, $P_2(-5, 1)$, and $P_3(-1, -1)$
   (c) $P_1(7, 1)$, $P_2(3, 6)$, and $P_3(-2, 1)$
5. Find the area of each of the following quadrilaterals the vertices of which have the given coordinates:
   (a) $P_1(-3, 2)$, $P_2(8, 1)$, $P_3(4, -1)$, and $P_4(-2, 0)$
   (b) $P_1(0, 0)$, $P_2(6, 1)$, $P_3(3, 7)$, and $P_4(-5, 2)$
   (c) $P_1(-4, -2)$, $P_2(-2, 3)$, $P_3(3, 1)$, and $P_4(6, 0)$
6. Determine which of the following vectors are perpendicular and which are parallel to one another:
$$[2, -3], [3, -2], [-6, 9], [-3, 2], [3, 2], [\tfrac{1}{3}, \tfrac{1}{2}]$$
7. If $u = [2, k]$ and $v = [-3, 5]$, find k: (a) when **u** is perpendicular to **v**; (b) when **u** is parallel to **v**.

8. Prove that the area of the triangle whose vertices are $P_1(x_1, y_1)$, $P_2(x_2, y_2)$, and $P_3(x_3, y_3)$ is the absolute value of
$$\tfrac{1}{2}(x_1y_2 + x_2y_3 + x_3y_1 - x_2y_1 - x_3y_2 - x_1y_3)$$

9. Prove that the area of the triangle in Exercise 8 is the absolute value of
$$\tfrac{1}{2}\begin{vmatrix} x_1 & y_1 & 1 \\ x_2 & y_2 & 1 \\ x_3 & y_3 & 1 \end{vmatrix}$$

10. If P is any point on the line containing P_1P_2, prove that $\overrightarrow{P_1P} = k\,\overrightarrow{P_1P_2}$, where $k=0$ if $P=P_1$; $k=1$ if $P=P_2$, k is positive if $\overrightarrow{P_1P}$ has the same direction as $\overrightarrow{P_1P_2}$, and k is negative if $\overrightarrow{P_1P}$ has the direction opposite to that of $\overrightarrow{P_1P_2}$.

11. Find the area of each of the following polygons the vertices of which have the given coordinates:

   (a) $P_1(1, -1)$, $P_2(5, 1)$, $P_3(4, 2)$, $P_4(0, 6)$, $P_5(-2, 1)$
   (b) $A(0, -3)$, $B(4, -1)$, $C(5, 3)$, $D(1, 5)$, $E(-2, 0)$
   (c) $P_1(0, 0)$, $P_2(3, 0)$, $P_3(5, 3)$, $P_4(4, 6)$, $P_5(1, 5)$, $P_6(-2, 4)$

12. Derive the formula for finding the area of a trapezoid.

REVIEW EXERCISES

1. (a) How far is the point $(3, -4)$ from the origin?
   (b) Find the coordinates of another point at the same distance from the origin.
   (c) How many points are there at this distance from the origin, and how would one find their coordinates?

2. (a) Find the coordinates of the point halfway between the origin and the point $(-6, -8)$.
   (b) Find the coordinates of the point such that the point $(-6, -8)$ is halfway between the desired point and the origin.

3. (a) Where are the points in the plane for which $x < 7$?
   (b) Where are the points in the plane for which $x^2 + y^2 > 9$?

4. (a) Where are the points for which $-1 < x \leq 0$ and $-1 < y \leq 0$?
   (b) Where are the points for which $-1 < x < y < 0$?

5. (a) Where are the points for which $xy = 0$?
   (b) Where are the points for which $x < 6$ and $y > -2$?
   (c) Where are the points for which $|x| < 6$ and $|y| > 2$?

6. Find the coordinates of the image of the point $P_1 = (3, -4)$: (a) in the x-axis; (b) in the y-axis; (c) with respect to the origin; (d) with respect to the point $Q = (-2, 1)$.

7. Determine the coordinates of the terminal point of the line segment with scalar components [4, 5] and initial point (1, 1).

8. Determine the coordinates of the initial point of the line segment with scalar components [−3, −2] and terminal point (0, 2).

9. The extremities of a segment are $P_1 = (-1, 2)$ and $P_2 = (5, 0)$; and a point on another segment is $P_3 = (0, -3)$. Find the coordinates of the point $P_4 = (x_4, y_4)$ such that $\overrightarrow{P_3P_4}$ has the same direction as $\overrightarrow{P_1P_2}$ and $|P_3P_4| = 2|P_1P_2|$.

10. In Exercise 9 find the coordinates of the point $P_4 = (x_4, y_4)$ if the direction of $\overrightarrow{P_3P_4}$ is opposite to the direction of $\overrightarrow{P_1P_2}$ and $|P_3P_4| = 2|P_1P_2|$.

11. Select y so that the triangle with vertices $(6, y)$, $(2, -4)$ and $(5, -1)$ will be isosceles.

12. Find the coordinates of the point on the x-axis which is equidistant from the points $(2, 4)$ and $(6, 8)$.

13. Find the coordinates of the point on the y-axis which is equidistant from the points $(2, 4)$ and $(6, 8)$.

14. Find the coordinates of the third vertex of an equilateral triangle if two of its vertices are at the points $(0, 0)$ and $(0, 1)$. Is there a unique solution?

15. (a) Find two representative segments of the vector $\mathbf{v} = [3, -4]$. (b) Find $|\mathbf{v}|$. (c) Find $-3\mathbf{v}$ and $|-3\mathbf{v}|$.

16. Given the vectors $\mathbf{u} = [2, 3]$, $\mathbf{v} = [4, 0]$, and $\mathbf{w} = [5, -3]$, find each of the following vectors both algebraically and geometrically:
    (a) $\mathbf{u} + \mathbf{v}$
    (b) $\mathbf{u} - \mathbf{v}$
    (c) $\mathbf{u} + \mathbf{v} + \mathbf{w}$
    (d) $\mathbf{u} - \mathbf{v} - \mathbf{w}$
    (e) $\mathbf{u} + \mathbf{v} - \mathbf{w}$
    (f) $\mathbf{u} - \mathbf{v} + \mathbf{w}$
    (g) $\mathbf{u} - 2\mathbf{v} + 3\mathbf{w}$
    (h) $2\mathbf{u} - \mathbf{v} + \mathbf{w}$

17. Two points of a segment are $P_1 = (1, 2)$ and $P_2 = (4, -3)$. (a) Find the point P_3 such that $\overrightarrow{P_1P_3} = -3\overrightarrow{P_1P_2}$. (b) In what ratio does P_3 divide P_1P_2?

18. Find the coordinates of the mid-point and the points of trisection of the segment P_1P_2 between $P_1 = (1, 2)$ and $P_2 = (3, 4)$.

19. Find the coordinates of the point which divides the line segment from $P_1 = (-6, 3)$ to $P_2 = (3, 9)$ in the ratio $-\frac{1}{2}$.

20. Show that the points $(1, 2)$, $(3, 8)$, and $(-1, -4)$ are collinear. Three or more points are said to be *collinear* when they all lie on the same straight line.

21. Find the ratio in which the point $(2, 0)$ divides the line segment from $P_1 = (-1, -2)$ to $P_2 = (5, 2)$.

22. Show that the medians of the triangle with vertices $P_1 = (-4, -3)$, $P_2 = (3, -1)$ and $P_3 = (2, 5)$ intersect in a point, by finding a point on each median which is twice as far from the corresponding vertex as it is from the opposite side.

76 ANALYTIC GEOMETRY IN THE PLANE

23. A point C lies on the line containing points $A = (2, 4)$ and $B = (-3, 2)$; and $|AC| = 3|AB|$. Find the coordinates of C. How many solutions are there?

24. Show that the points $(0, 0)$, $(6, 3)$, and $(-2, 4)$ are vertices of a right triangle.

25. What kind of quadrilateral has its vertices at the points $(5, 3)$, $(0, 3)$, $(0, -1)$, and $(5, -1)$ in that order?

26. Show analytically that a segment which bisects one non-parallel side of a trapezoid and is parallel to one base also bisects the other non-parallel side.

27. Show analytically that the length of the segment joining the mid-points of two adjacent sides of a quadrilateral is equal to one-half the length of *one* of the diagonals of the quadrilateral.

28. If the base of an isosceles triangle is trisected, prove analytically that the lengths of the segments drawn from the vertex to the points of trisection are equal.

29. The mid-point of one side of a parallelogram is joined to a vertex not adjacent to the side, and the mid-point of the opposite side is joined to the vertex opposite the first. Prove analytically that the two segments thus determined are equal in length and parallel to each other.

30. A segment is drawn from the intersection of the diagonals of a parallelogram to the opposite sides. Prove analytically that this segment is bisected by the point of intersection of the diagonals.

31. Show that the vectors $\mathbf{u} = [3, -1]$ and $\mathbf{v} = [2, 5]$ are linearly independent. Two plane vectors are said to be linearly independent provided it is *impossible* to find two constants c_1 and c_2 not both zero which will make the vector $c_1\mathbf{u} + c_2\mathbf{v} = [0, 0]$. Hint: Form $c_1[3, -1] + c_2[2, 5] = [0, 0]$ and solve for c_1 and c_2. This will show that in order for $c_1\mathbf{u} + c_2\mathbf{v} = \mathbf{0}$, c_1 and c_2 *must* be zero. Hence, $\mathbf{u}$ and $\mathbf{v}$ are linearly independent.

32. Show that if $\mathbf{u} = [u_1, u_2]$ and $\mathbf{v} = [v_1, v_2]$ are linearly independent, then *any* other non-zero vector $\mathbf{w} = [w_1, w_2]$ may be expressed as a linear combination of $\mathbf{u}$ and $\mathbf{v}$. Hint: Write $\mathbf{w} = c_1\mathbf{u} + c_2\mathbf{v}$ and solve for c_1 and c_2.

33. Show that the vectors $\mathbf{u} = [2, -1]$ and $\mathbf{v} = [-6, 3]$ are linearly dependent. Two vectors are said to be linearly dependent when they are not linearly independent. This means that it *is possible* to find two constants c_1 and c_2 not both zero such that $c_1\mathbf{u} + c_2\mathbf{v} = [0, 0]$.

34. Prove that two linearly dependent vectors are parallel.

35. Show that the vectors $\mathbf{i} = [1, 0]$ and $\mathbf{j} = [0, 1]$ are linearly independent.

36. Is it possible to have three linear independent *plane* vectors? What this means is: Is it possible to have three vectors no *one* of which can be expressed as a linear combination of the other two? Hint: See Exercise 32.

THE POINT AND PLANE VECTORS 77

37. Express the vectors **i** and **j** as linear combinations of $\mathbf{u} = [-3, 2]$ and $\mathbf{v} = [-1, -1]$.

38. Two plane vectors that are linearly independent are said to form a *basis* for *all* plane vectors. Show that **i** and **j** form a basis for all plane vectors.

39. Show that if $P_1, P_2, \cdots, P_n$ are the vertices of any plane polygon, $\overrightarrow{P_1P_2} + \overrightarrow{P_2P_3} + \cdots + \overrightarrow{P_{n-1}P_n} + \overrightarrow{P_nP_1} = [0, 0]$.

40. Prove that $\mathbf{u}|\mathbf{v}$ can be written as a second order determinant in which the elements of the first row are the scalar components of **u** while the elements of the second row are the scalar components of **v**.

41. With the information contained in Exercises 31 to 38, direct methods using vectors may be used to prove various theorems in Euclidean geometry. This is illustrated in the following:

EXAMPLE: Prove that the diagonals of a parallelogram bisect each other.

Proof: Let $ABCD$ (Fig. 2–43) be any parallelogram. G is the intersection of the diagonals.

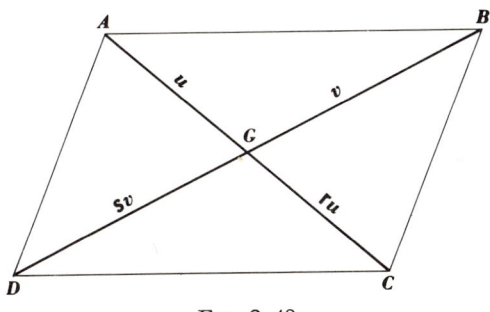

FIG. 2–43

Let $\overrightarrow{AG} = \mathbf{u}$ and $\overrightarrow{GB} = \mathbf{v}$. Then $\overrightarrow{GC} = r\mathbf{u}$ and $\overrightarrow{DG} = s\mathbf{v}$.
Thus $\overrightarrow{DG} + \overrightarrow{GC} = \overrightarrow{DC} = s\mathbf{v} + r\mathbf{u}$
But $\overrightarrow{DC} = \overrightarrow{AB}$, and $\overrightarrow{AB} = \overrightarrow{AG} + \overrightarrow{GB} = \mathbf{u} + \mathbf{v}$

It follows that $s\mathbf{v} + r\mathbf{u} = \mathbf{u} + \mathbf{v}$ and since **u** and **v** are linearly independent $(s - 1)\mathbf{u} + (r - 1)\mathbf{v} = 0$ from which we obtain $s - 1 = 0$ and $r - 1 = 0$ or $s = 1$ and $r = 1$.

The theorem now follows.

Use vector methods to prove the following:

(a) The line segment joining the midpoints of two sides of any triangle is parallel to the third side and one-half its length.
(b) The medians of a triangle meet in a point G (called its centroid) that trisects each median.
(c) The line segment joining the midpoints of the nonparallel sides of a trapezoid is parallel to the parallel sides of the trapezoid and equal to half their sum in magnitude.

3

The Straight Line

3-1. Direction numbers and direction cosines of a line*

We shall define a pair of *direction numbers of a line* to be the scalar components of any vector, excluding the zero vector, on the line. A vector is *on* a line if one of its representative segments is on the line. Since every line contains an infinite number of vectors, an infinite number of pairs of direction numbers is associated with every line. Hence, for the line in Fig. 3-1 which is determined by the two points $P_1(x_1, y_1)$ and $P_2(x_2, y_2)$, the direction numbers are $[\Delta x, \Delta y]$, $[-\Delta x, -\Delta y]$, or $[k\,\Delta x, k\,\Delta y]$, where $k \neq 0$.

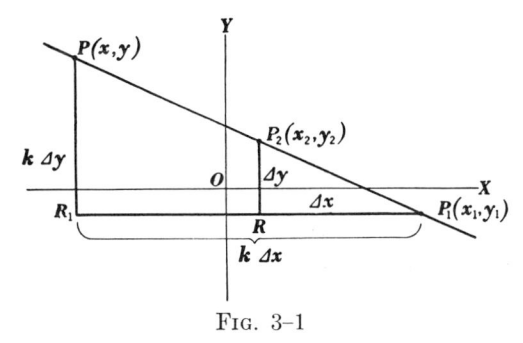

Fig. 3-1

EXAMPLE 3-1. Find a pair of direction numbers of the line through the points $P_1(-4, 2)$ and $P_2(2, -6)$.

Solution. Since $\mathbf{u} = \overrightarrow{P_1P_2} = [6, -8]$ is a vector on the line, it follows that a pair of direction numbers of the line is $[6, -8]$. Another pair of direction numbers of the line is $[3, -4]$ because these numbers can be obtained by multiplying 6 and -8 by the same factor, $\frac{1}{2}$.

Every line contains two unit vectors, whose scalar components differ from each other in sign only. *We shall define the direction cosines of a line to be the scalar components of a unit vector on the line.* Associated

---
* Throughout this chapter, the term *line* will be understood to mean a "straight" line.

with every line there are *two* pairs of direction cosines; for, if $\mathbf{u} = [l, m]$ is a unit vector on a line, $\mathbf{v} = [-l, -m]$ is also on the line and both $[l, m]$ and $[-l, -m]$ are pairs of direction cosines of the line. If $[\Delta x, \Delta y]$ is a pair of direction numbers of a line, then a pair of direction cosines of the line is

$$l = \frac{\Delta x}{\sqrt{(\Delta x)^2 + (\Delta y)^2}}$$
$$m = \frac{\Delta y}{\sqrt{(\Delta x)^2 + (\Delta y)^2}} \qquad (3\text{-}1)$$

If $\mathbf{u} = [u_1, u_2]$ is any non-zero vector on a line, a pair of direction cosines is

$$l = \frac{u_1}{|\mathbf{u}|}$$
$$m = \frac{u_2}{|\mathbf{u}|} \qquad (3\text{-}2)$$

Another pair of direction cosines of the line is

$$-l = \frac{-u_1}{|\mathbf{u}|}$$
$$-m = \frac{-u_2}{|\mathbf{u}|} \qquad (3\text{-}3)$$

A pair of direction cosines of any horizontal line is $[1, 0]$.
A pair of direction cosines of any vertical line is $[0, 1]$.

EXAMPLE 3-2. Find a pair of direction cosines of the line through the points $P_1 = (-1, 2)$ and $P_2 = (2, -5)$.

Solution. We know that $\mathbf{u} = \overrightarrow{P_1 P_2} = [3, -7]$. Since this is a pair of direction numbers for the line, $l = \frac{3}{\sqrt{58}}$ and $m = \frac{-7}{\sqrt{58}}$.

EXAMPLE 3-3. Determine a pair of direction cosines of the line through the points $P_1 = (-1, 2)$ and $P_2 = (-1, 5)$.

Solution. Since a pair of direction numbers is $[0, 3]$, the line is vertical and a pair of direction cosines is $[0, 1]$.

EXAMPLE 3-4. Determine a pair of direction cosines of a line that contains the vector $[3, -1]$.

Solution. By equations (3-2), a pair of direction cosines is $l = \frac{3}{\sqrt{10}}$ and $m = \frac{-1}{\sqrt{10}}$.

3-2. The slope of a line

The *inclination* of a line is defined to be the *smallest* counter-clockwise angle through which the positively directed x-axis must turn in order to coincide with the line. This angle is always measured from the x-axis to the line. In both Fig. 3-2 and Fig. 3-3, angle α is the inclination of the line. The inclination of any horizontal line is defined to be 0°. The inclination is always less than 180°.

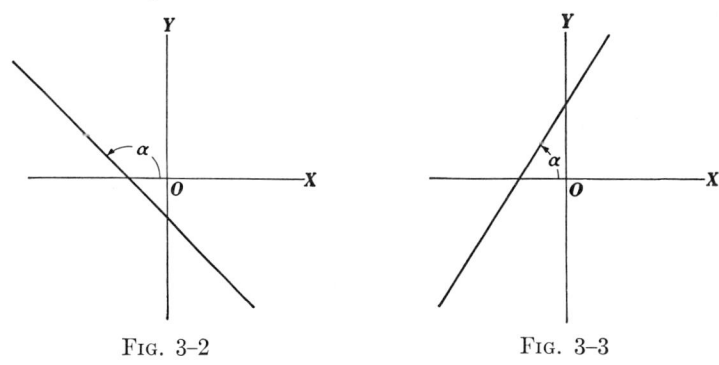

FIG. 3-2   FIG. 3-3

The tangent of the inclination of a line is called the slope of the line. We shall represent slope by s. If the line makes an acute angle with the positive x-axis, the slope is positive; if the line makes an obtuse angle with the positive x-axis, the slope is negative; if the line is parallel to the x-axis,

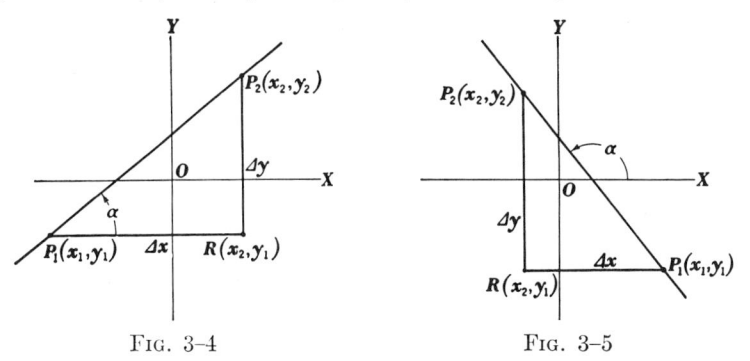

FIG. 3-4   FIG. 3-5

the slope is 0; and, if the inclination is 90°, the slope is not defined. Conversely, if s is positive, the inclination is an acute angle; if s is negative, the inclination is an obtuse angle; if $s = 0$, the inclination is 0°; if s is undefined, the inclination is 90°.

In either Fig. 3-4 or Fig. 3-5,

$$s = \tan \alpha = \frac{\Delta y}{\Delta x} = \frac{y_2 - y_1}{x_2 - x_1} \tag{3-4}$$

Hence, the slope of a line is equal to the quotient obtained by dividing the vertical direction number by the horizontal direction number, unless the horizontal direction number is zero, in which case the slope is not defined. Since $\dfrac{y_2 - y_1}{x_2 - x_1} = \dfrac{y_1 - y_2}{x_1 - x_2}$, it follows that the order in which the points are selected is immaterial in determining the slope of a line.

EXAMPLE 3-5. Find the slope of a line that passes through the points $P_1 = (-3, 4)$ and $P_2 = (5, -1)$.

Solution. We know that $\overrightarrow{P_1P_2} = [8, -5]$; therefore, $s = -\tfrac{5}{8}$. The student should notice that, if we had taken $\overrightarrow{P_2P_1} = [-8, 5]$, we would also have obtained $s = -\tfrac{5}{8}$.

EXERCISES

1. Find a pair of direction numbers of each of the lines containing the following points:

    (a) $P_1(-1, 2)$ and $P_2(3, 4)$
    (b) $P_1(0, 3)$ and $P_2(1, -1)$
    (c) $P_1(4, 0)$ and $P_2(3, 2)$
    (d) $P_1(1, -3)$ and $P_2(-4, -2)$
    (e) $P_1(4, -2)$ and $P_2(6, -2)$
    (f) $P_1(2, 0)$ and $P_2(0, -3)$

2. Find the direction cosines of each of the lines in Exercise 1.

3. Find the slope of each of the lines in Exercise 1.

4. Show that a pair of direction numbers for a line of slope s is $[1, s]$.

5. Find direction numbers of each of the lines whose slopes are: (a) $\tfrac{3}{5}$; (b) $-\tfrac{1}{3}$; (c) 4; (d) -7; (e) 0; (f) not defined.

6. Construct the line that satisfies the following geometric conditions:

    (a) It contains the point $(-1, 2)$ and has direction numbers $[2, 3]$.
    (b) It contains the point $(3, 4)$ and has direction numbers $[5, 6]$.
    (c) It contains the point $(-4, -1)$ and has direction numbers $[-1, 0]$.
    (d) It contains the point $(7, -2)$ and its slope is $\tfrac{1}{3}$.
    (e) It contains the point $(-2, 0)$ and its slope is -4.
    (f) It contains the point $(2, 6)$ and its slope is 0.
    (g) It contains the point $(7, 3)$ and has the direction numbers $[0, 2]$.
    (h) It contains the point $(-3, 5)$ and the vector $[6, -4]$.

7. Determine the direction cosines of the lines that are parallel to the following vectors: (a) $[-4, 10]$; (b) $[6, -2]$; (c) $[10, -5]$; (d) $[3, -9]$.

8. What are the slopes of the lines in Exercise 7?

3–3. Parametric equations of a line

The *set* of all points $P(x, y)$ having the property that the vectors $\overrightarrow{P_1P}$ are parallel, where P_1 is a fixed point, lie on a line. In this section we shall obtain a pair of algebraic equations which will be satisfied by the coordinates of every point on a certain line and will not be satisfied by the coordinates of any point not on that line. These equations are called *parametric equations* of the line.

Let two distinct points $P_1(x_1, y_1)$ and $P_2(x_2, y_2)$ be given; that is, either $x_1 \neq x_2$ or $y_1 \neq y_2$. For example, consider the line containing the segment P_1P_2 in Fig. 3–6. If $P(x, y)$ is any arbitrary point on this line, then there is a number t such that

(1) $$\overrightarrow{P_1P} = t\,\overrightarrow{P_1P_2}$$

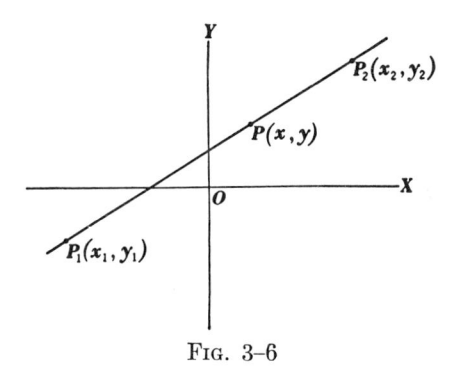

FIG. 3–6

If $t = 0$, P coincides with P_1; if $t = 1$, P coincides with P_2; if $t > 0$, P lies on the half-line through P_1 containing P_2; if $t < 0$, P lies on the half-line through P_1 extended in the direction opposite to that of $\overrightarrow{P_1P_2}$.

Since $\overrightarrow{P_1P} = [x - x_1, y - y_1]$ and $\overrightarrow{P_1P_2} = [x_2 - x_1, y_2 - y_1]$, it follows from equation (1) that $[x - x_1, y - y_1] = [t(x_2 - x_1),\ t(y_2 - y_1)]$. Hence, $x - x_1 = t(x_2 - x_1)$ and $y - y_1 = t(y_2 - y_1)$. As t varies, the corresponding values of x and y are the coordinates of points lying on the line containing $\overrightarrow{P_1P_2}$. We call t a *parameter* for the line; and the *parametric equations* of the line are:

$$x = x_1 + t(x_2 - x_1)$$
$$y = y_1 + t(y_2 - y_1)$$
(3–5)

Since $\Delta x = x_2 - x_1$ and $\Delta y = y_2 - y_1$,

$$x = x_1 + t\,\Delta x$$
$$y = y_1 + t\,\Delta y$$
(3–6)

Note the following characteristics: (1) The coordinates x and y are expressed as linear* functions of t. (2) The coefficients of the parameter are direction numbers of the line. (3) The constant terms are the coordinates of a point on the line. The two parametric equations of a pair must be taken together. They are often written with a comma between them, as $x = x_1 + t\,\Delta x,\ y = y_1 + t\,\Delta y$.

For a vertical line, $x = x_1 + 0t$ and $y = y_1 + (\Delta y)t$. Hence,

$$x = x_1 \quad \text{and} \quad y = y_1 + t\,\Delta y \tag{3-7}$$

For a horizontal line, $x = x_1 + (\Delta x)t$ and $y = y_1 + 0t$. Therefore,

$$x = x_1 + t\,\Delta x \quad \text{and} \quad y = y_1 \tag{3-8}$$

EXAMPLE 3–6. Find parametric equations of the line through $P_1 = (-3, 2)$ and $P_2 = (4, 3)$.

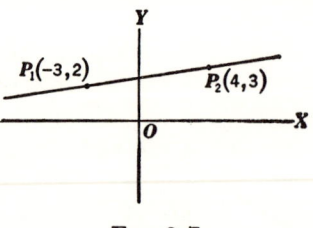

FIG. 3–7

Solution. The line is shown in Fig. 3–7. We find that $\overrightarrow{P_1P_2} = [7, 1]$. Hence, by equations (3–6),

$$x = -3 + 7t \quad \text{and} \quad y = 2 + t$$

If $a + bt$ and $c + dt$ are any linear functions (provided both b and d are not zero), then the parametric equations of a line through the point (a, c) and having the direction numbers $[b, d]$ are:

$$\begin{aligned} x &= a + bt \\ y &= c + dt \end{aligned} \tag{3-9}$$

To prove this, consider the distinct points $P_1 = (a, c)$ and $P_2 = (a + b, c + d)$. A pair of direction numbers of the line through P_1 and P_2 is $[b, d]$. Hence, parametric equations of the line containing $\overrightarrow{P_1P_2}$ are $x = a + bt$ and $y = c + dt$, which are the given equations.

When the equations of a line appear in the parametric form shown by equations (3–9), we know by inspection that a and c are the coordinates of a point on the line and that b and d are direction numbers of the line.

---

* A *linear function* of t is of the form $a + bt$, where a and b are numbers.

EXAMPLE 3-7. Draw the line whose equations are $x = -3 - 2t$ and $y = 1 + 3t$.

Solution. We know that the coordinates of a point on this line are $(-3, 1)$. Also a pair of direction numbers of the line is $[-2, 3]$. Hence, we locate the point $(-3, 1)$, and through this point we draw a line containing the vector $[-2, 3]$, as shown in Fig. 3-8. This is the required line.

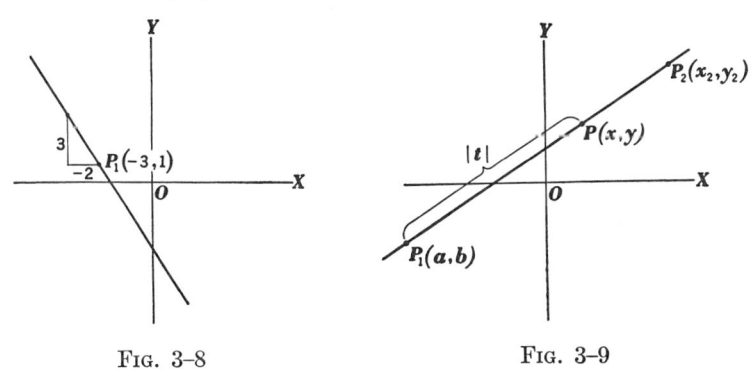

FIG. 3-8

FIG. 3-9

EXERCISES

1. Find parametric equations of the line through each of the following pairs of points and draw the lines:

   (a) $P_1(3, 5)$ and $P_2(-4, 1)$
   (b) $P_1(-2, 6)$ and $P_2(5, -1)$
   (c) $P_1(3, 0)$ and $P_2(0, 4)$
   (d) $P_1(-1, 2)$ and $P_2(-1, 7)$
   (e) $P_1(4, -2)$ and $P_2(-3, -2)$
   (f) $P_1(0, 0)$ and $P_2(-1, -3)$
   (g) $P_1(3, 4)$ and $P_2(7, 4)$
   (h) $P_1(0, 2)$ and $P_2(0, -1)$
   (i) $P_1(1, 1)$ and $P_2(5, 2)$
   (j) $P_1(7, -3)$ and $P_2(-1, -3)$

2. Find direction numbers of the line, find the coordinates of a point on the line, and draw the graph of each line, the equations of which follow:

   (a) $x = -1 + 2t$ and $y = 2 - 3t$
   (b) $x = 5 - 3t$ and $y = -1 + t$
   (c) $x = 2$ and $y = 3t$
   (d) $x = -2 - 5t$ and $y = 3$
   (e) $x = -3 - t$ and $y = 2 - 3t$
   (f) $x = 0$ and $y = t$
   (g) $x = t$ and $y = 0$
   (h) $x = \frac{1}{2} + \frac{3}{5}t$ and $y = 1 + \frac{4}{5}t$
   (i) $x = -5 + 2t$ and $y = 3 + t$
   (j) $x = 4 - 3t$ and $y = 2t$

3. What are the slopes (if defined) of the lines of Exercise 2?

4. Given $x = a + lt$ and $y = b + mt$, where l and m are direction cosines of the line. Show that the absolute value of the parameter t is the distance from the fixed point $P_1(a, b)$ to the variable point $P(x, y)$. See Fig. 3-9.

86   Analytic Geometry in the Plane

5. Determine parametric equations: (a) of the x-axis; (b) of the y-axis.

6. Determine which of the following lines are parallel and which are perpendicular:

   (a) $x = -3 + 2t,\ y = 4 - t$ (e) $x - 2 = -6t,\ y + 8 = -3t$
   (b) $x = 5 - 6t,\ y = -1 + 3t$ (f) $x = -1,\ y = -t$
   (c) $x = 2 + 5t,\ y = -7 + 10t$ (g) $x = t,\ y = 3$
   (d) $x + 1 = t,\ y - 3 = -2t$ (h) $x + 3t + 5 = 0,\ 2y + 3t + 1 = 0$

7. Show that two vectors on the same line are linearly dependent.

8. Given the lines $x = a_1 t + b_1,\ y = c_1 t + d_1$ and $x = a_2 t + b_2,\ y = c_2 t + d_2$, what is the condition on the constants that the lines be parallel; be perpendicular?

9. Find parametric equations of lines satisfying the following conditions:

   (a) Contains $P = (-1, 3)$ and is parallel to $\mathbf{u} = [2, 5]$.
   (b) Contains $Q = (3, -2)$ and is parallel to $\mathbf{v} = [-1, -2]$.
   (c) Contains $P = (-3, -2)$ and is perpendicular to $\mathbf{u} = [-4, 1]$.
   (d) Contains $A = (4, 5)$ and is perpendicular to $\mathbf{w} = [3, -4]$.
   (e) Perpendicular to the segment from $A = (-1, 3)$ to $B = (2, -3)$, and contains the point P where $\overrightarrow{AP} = 2\overrightarrow{PB}$.
   (f) Perpendicular to the segment from $C = (4, -5)$ to $D = (1, 0)$, and contains the point P where $\overrightarrow{CP} = -2\overrightarrow{DC}$.
   (g) Contains the point $(3, 2)$ and is perpendicular to the x-axis.
   (h) Contains the point $(-1, 4)$ and is parallel to the x-axis.

3–4. Equation of a line in direction number form

From the relation $\overrightarrow{P_1P} = t\ \overrightarrow{P_1P_2}$ for the line in Fig. 3–10, we obtained the parametric equations of the line, which are

$$x = x_1 + t(\Delta x) \text{ and } y = y_1 + t(\Delta y)$$

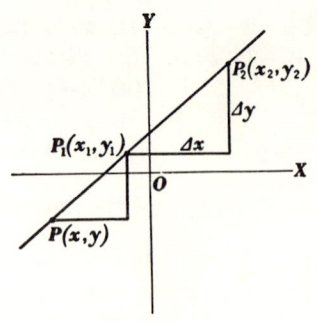

Fig. 3–10

The Straight Line

It is implied that both Δx and Δy are not equal to 0. If we eliminate t from the two equations, we obtain

$$\frac{x - x_1}{\Delta x} = \frac{y - y_1}{\Delta y} \tag{3-10}$$

Since $[\Delta x, \Delta y]$ is a pair of direction numbers of the line, we call equation (3-10) a *direction number* equation of the line. It is also called a *symmetric* equation of the line.

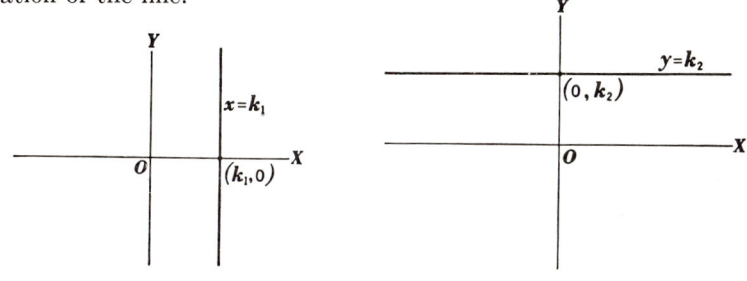

FIG. 3-11   FIG. 3-12

If $\Delta x = 0$, the line is vertical, as shown in Fig. 3-11, and an equation for it is

$$x = k_1 \tag{3-11}$$

where k_1 is the abscissa of every point on the line.

If $\Delta y = 0$, the line is horizontal, as shown in Fig. 3-12, and an equation for it is

$$y = k_2 \tag{3-12}$$

where k_2 is the ordinate of every point on the line.

Since equations (3-10), (3-11), and (3-12) are first-degree equations in x and y, it follows that *an equation of a straight line is of the first degree in x and y*.

If we expand equation (3-10), we have

$$(x - x_1)\Delta y - (y - y_1)\Delta x = 0$$

or
$$x\,\Delta y - y\,\Delta x = k = x_1\,\Delta y - y_1\,\Delta x \tag{3-13}$$

This suggests the following method of writing an equation of a line when a point on the line and a pair of direction numbers are given:

Rule: *To obtain an equation of a line when a point on it and its direction numbers are known, multiply x by the second direction number and add this to y times the negative of the first direction number. Set this sum equal to k and determine k by substituting for x and y the coordinates of the known point on the line.*

EXAMPLE 3-8. Find an equation of the line which passes through the point $(-1, 2)$ and has direction numbers $[3, -2]$.

88   ANALYTIC GEOMETRY IN THE PLANE

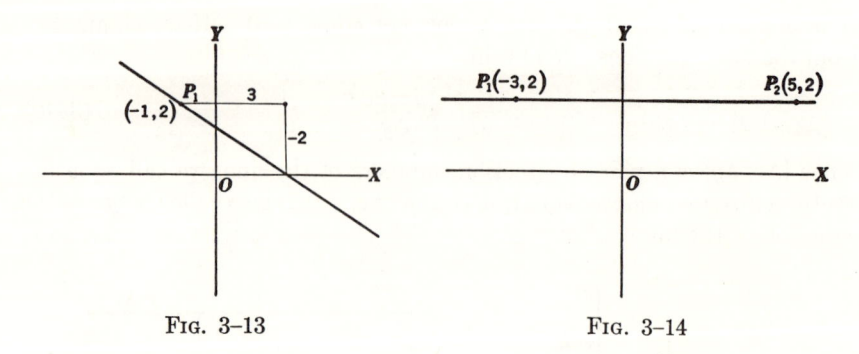

FIG. 3-13

FIG. 3-14

Solution. The conditions are shown in Fig. 3-13. In accordance with the foregoing rule, we have:

$$-2x - 3y = k = -2(-1) - 3(2) = -4$$

Hence, the required equation is

$$2x + 3y = 4$$

Although the foregoing rule applies to horizontal or vertical lines, it is easier to use equation (3-11) or equation (3-12) when Δy or Δx is zero.

EXAMPLE 3-9. Find an equation of the line through the points $P_1 = (-3, 2)$ and $P_2 = (5, 2)$.

Solution. The line is shown in Fig. 3-14. Since $\Delta y = 0$, the line is horizontal and an equation of the line is $y = 2$.

EXERCISES

1. By using similar triangles derive an equation of the line passing through the point $P_1 = (x_1, y_1)$, which lies in quadrant IV, and $P_2 = (x_2, y_2)$, which lies in quadrant II. Do the actual positions of the points have any effect on the result?

2. Derive an equation of the line through each of the following pairs of points:

   (a) $P_1(-3, 2)$ and $P_2(5, 4)$
   (b) $P_1(2, 5)$ and $P_2(-3, 6)$
   (c) $P_1(3, -1)$ and $P_2(-2, -7)$
   (d) $P_1(2, -3)$ and $P_2(7, -3)$
   (e) $P_1(1, 4)$ and $P_2(-2, 4)$
   (f) $P_1(5, 2)$ and $P_2(-1, -1)$
   (g) $P_1(3, -1)$ and $P_2(3, 5)$
   (h) $P_1(7, 3)$ and $P_2(4, 0)$
   (i) $P_1(-2, -4)$ and $P_2(0, 0)$
   (j) $P_1(0, -4)$ and $P_2(5, 0)$

3. Given a point on the line and a pair of direction numbers of the line, find an equation of each of the following lines:

(a) (1, 2) and [−1, 3]
(b) (0, 0) and [2, 5]
(c) (−6, 3) and [−4, −2]
(d) (5, −3) and [4, 0]
(e) (−1, −2) and [−3, 2]
(f) (7, 2) and [0, −3]
(g) (8, 0) and [2, 1]
(h) (0, 6) and [−3, −4]
(i) (4, 2) and [1, −1]
(j) (0, 1) and [−3, 4]

4. Show that $[\Delta x, \Delta y] \mid [x - x_1, y - y_1] = 0$, where the first member is the bar product of two vectors, is an equation of a line through the point (x_1, y_1) with direction numbers Δx and Δy.

5. Prove that $\begin{vmatrix} x & y & 1 \\ x_1 & y_1 & 1 \\ x_2 & y_2 & 1 \end{vmatrix} = 0$ is an equation of a line through the points whose coordinates are (x_1, y_1) and (x_2, y_2).

6. Find an equation of each of the lines satisfying the following conditions:

   (a) Contains the point $(-1, 3)$ and is parallel to the vector $[2, -1]$.
   (b) Is the perpendicular bisector of the segment from $A(-3, 2)$ to $B(5, -6)$.
   (c) Contains the point $(4, 3)$ and is perpendicular to the vector $\mathbf{v} = [2, 1]$.

3–5. The general linear equation $ax + by + c = 0$

Let $P_1 = (x_1, y_1)$ be the initial point of a representative segment of the non-zero vector $\mathbf{u} = [a, b]$, where a and b are constants. The set of all points $P(x, y)$ having the property that $\overrightarrow{P_1P}$ is *perpendicular* to $\mathbf{u}$ will lie on a line. See Fig. 3–15. Hence we have $\mathbf{u} \cdot \overrightarrow{P_1P} = 0$ or $[a, b] \cdot \overrightarrow{P_1P} = 0$ and

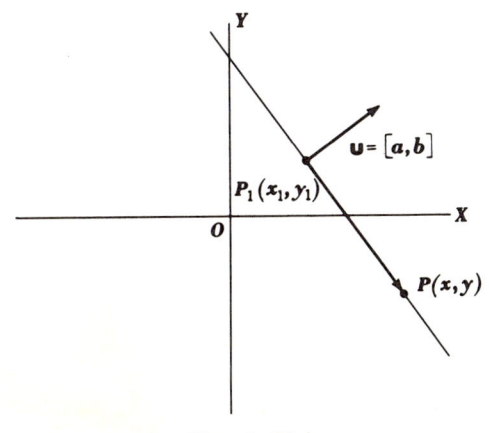

Fig. 3–15

$a(x-x_1)+b(y-y_1)=0$. If we let $-ax_1-by_1=c$, an equation of the line becomes $ax+by+c=0$. When the equation of a line is written in this form it is said to be in *general* form. Now we shall show that the equation $ax+by+c=0$ is an equation of a straight line, provided that both a and b are not 0. Let us assume that (x_1, y_1) is a pair of numbers which satisfies the equation

(1) $$ax+by+c=0$$

Then, it follows that

(2) $$ax_1+by_1+c=0$$

Subtracting equation (2) from equation (1), we obtain:

(3) $$a(x-x_1)+b(y-y_1)=0$$

But the first member of equation (3) is the dot product of vectors $[a, b]$ and $[x-x_1, y-y_1]$. Hence, $[a, b]\cdot[x-x_1, y-y_1]=0$, and the constant vector $[a, b]$ is perpendicular to the variable vector $[x-x_1, y-y_1]$ for any pair of values (x, y) that satisfies equation (1). This means that any such point (x, y) lies on a line which passes through the point (x_1, y_1) and is perpendicular to the vector $[a, b]$. It is always possible to find the coordinates x_1 and y_1 of some point on a given line. Hence, the equation $ax+by+c=0$ is an equation of a straight line perpendicular to the vector $[a, b]$.

Any straight line can be represented by an equation of the form

$$ax+by+c=0 \qquad (3\text{--}14)$$

the *general form* of the equation of a straight line.

A line l will be said to be perpendicular to a vector v provided every vector on the line is perpendicular to v. In the same way, a line will be said to be parallel to a vector v provided every vector on the line is parallel to v. Since the geometric representation of the first-degree equation is a line, it is called a linear equation. We shall call the vector $[a, b]$ the *coefficient vector* of the equation $ax+by+c=0$.

If $b=0$ in equation (1), equation (3) becomes $a(x-x_1)+0(y-y_1)=0$. In this case, the variable vector is perpendicular to a horizontal vector and the line is vertical. So $ax+c=0$ is the equation of a vertical line; or the general form of the equation of a vertical line is

$$ax+c=0 \qquad (3\text{--}15)$$

In the same way, $by+c=0$ is the equation of a horizontal line; or the general form of the equation of a horizontal line is

$$by+c=0 \qquad (3\text{--}16)$$

In order to draw the graph of a straight line when its equation is known, we obtain two points on it. If the line does not pass through the origin and it is not vertical or horizontal, two points easy to obtain are the points where the line crosses the x-axis and the y-axis, respectively. These points are found by first letting $y = 0$ and solving for x and then letting $x = 0$ and solving for y. If the line passes through the origin, as indicated by the fact that $c = 0$, only one other point need be located; its coordinates can be found by assuming any arbitrary value for x or y and determining the corresponding value of the other coordinate from the given equation. If the line is horizontal or vertical, only one intercept suffices.

The x-coordinate or abscissa of the point where the line crosses the x-axis is called the *x-intercept* of the line.

The y-coordinate or ordinate of the point where the line crosses the y-axis is called the *y-intercept* of the line.

EXAMPLE 3–10. Draw the graph of the line whose equation is $2x - 3y + 6 = 0$.

Solution. When $x = 0$, then $y = 2$. Hence, the y-intercept of the line is 2. When $y = 0$, then $x = -3$. So -3 is the x-intercept of the line. Hence, two points on the line are $(-3, 0)$ and $(0, 2)$; and the line is drawn as shown in Fig. 3–16.

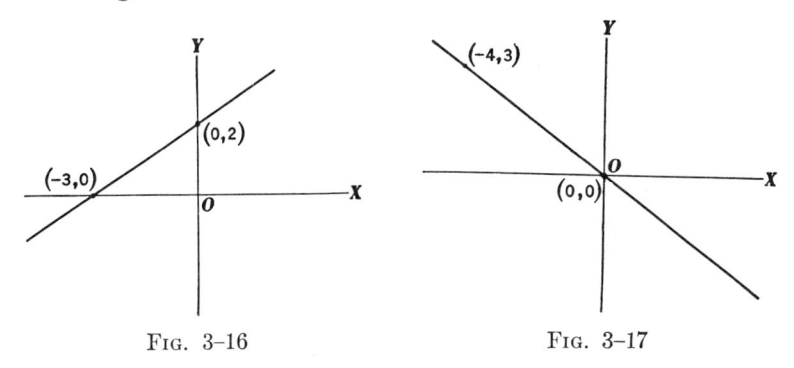

FIG. 3–16

FIG. 3–17

EXAMPLE 3–11. Draw the graph of the line $3x + 4y = 0$.

Solution. In this case the x-intercept and the y-intercept are each equal to zero. Hence, the line goes through the origin and one point on the line is $(0, 0)$. We find another point by letting $y = 3$ and obtaining $x = -4$. The line through $(0, 0)$ and $(-4, 3)$ is the graph of $3x + 4y = 0$. This line is shown in Fig. 3–17.

Since we know that the coefficient vector $\mathbf{u} = [a, b]$ is a vector perpendicular to the line $ax + by + c = 0$, it is easy to determine a pair of direction numbers of the line. We know also that a vector $\mathbf{v} = [-b, a]$ is perpendicular to the vector $[a, b]$. It therefore follows that the given line is parallel to $\mathbf{v}$ and $-b$ and a are direction numbers of that line.

If the line is vertical, as indicated by the fact that its equation is $ax + c = 0$, then $\mathbf{u} = [a, 0]$ is a vector perpendicular to the line. Since $\mathbf{v} = [0, a]$ is perpendicular to $\mathbf{u}$, a pair of direction numbers of the line is $[0, a]$.

If the line is horizontal, its equation being $by + c = 0$, it follows that the coefficient vector of the line is $[0, b]$ and a pair of direction numbers of the line is $[b, 0]$.

EXAMPLE 3–12. Find a pair of direction numbers of the line whose equation is $2x - 3y + 4 = 0$; and construct the graph.

Solution. Since the vector $\mathbf{u} = [2, -3]$ is perpendicular to the given line, it follows that a pair of direction numbers of the line is $[3, 2]$. To construct the line, we locate the point $(-2, 0)$ and through this point we draw the line having direction numbers $[3, 2]$. The line is shown in Fig. Fig. 3–18,

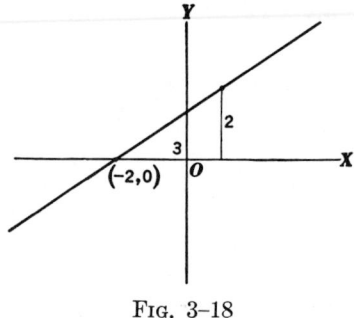

FIG. 3–18

EXERCISES

1.* Draw the graphs of the lines whose equations are:
   (a) $2x + 3y + 12 = 0$
   (b) $x + 3y = 0$
   (c) $3x - 4y + 6 = 0$
   (d) $4x - y = 0$
   (e) $2x + 5 = 0$
   (f) $5x + y - 7 = 0$
   (g) $3y + 2 = 0$
   (h) $7x - 2y = 5$

2. What are direction numbers of the lines of Exercise 1?
3. What are direction cosines of the lines of Exercise 1?
4. What is the relation between the lines whose equations are $3x - y + 6 = 0$ and $9x - 3y + 18 = 0$? Explain.

---
* Problems of this type may also be expressed by the use of set notation. For example, 1(a) may be stated as $f = \{(x,y) \mid 2x + 3y + 12 = 0\}$.

THE STRAIGHT LINE   93

5. What is the relation between the lines whose equations are $3x-y+6=0$ and $x+3y-2=0$? (Hint: Investigate direction numbers of the two lines.)

6. What are values of the coefficients a, b, and c in the equation $ax+by+c=0$, if the line contains the points $(-1, 3)$ and $(2, 5)$?

7. Show that the points $(-2, 0)$, $(-1, 3)$, and $(2, 12)$ are collinear.*

8. Show that the points $(0, 0)$, $(1, 4)$, and $(-\frac{1}{2}, -2)$ are collinear.*

9. (a) Where are the points for which $x-y>0$?
   (b) Where are the points for which $x-y<0$?

10. What are equations of the sides of a triangle ABC whose vertices are: $A(0, 0)$, $B(3, 5)$, and $C(-1, 2)$?

11. Find direction cosines of the lines whose equations are:
    (a) $2x-3y+6=0$ (e) $5x+12y-20=0$
    (b) $4x-5=0$ (f) $3y+7=0$
    (c) $x-y+6=0$ (g) $3x-2y-7=0$
    (d) $3x-4y+12=0$ (h) $x+3y=0$

12. What is the slope (if defined) of each of the lines of Exercise 11?

13. Given the line whose equation is $3x-4y+12=0$, find:
    (a) A vector perpendicular to the line.
    (b) A vector on the line.
    (c) The slope of the line.
    (d) A set of direction cosines of the line.
    (e) The x- and y-intercepts of the line.
    (f) Draw the line.

14. Find the information asked in parts (a) to (e) of Exercise 13 for the lines whose equations are:
    (a) $5x-12y+12=0$ (b) $2x-3y-6=0$ (c) $3x+4y+12=0$

15. Identify the sets of points that satisfy the following conditions:
    (a) $x-3y<0$ (f) $5x-3y+3\leq 2$
    (b) $2x+3y>0$ (g) $3x+4y-4\geq 0$
    (c) $3x+2y<0$ (h) $y-2\leq -1$
    (d) $x+y-4>0$ (i) $x+3\geq 2$
    (e) $2x-3y+6<-1$ (j) $4x-3y-12\geq -6$

---

* Three or more points are *collinear* if they lie on the same straight line.

16. Using the general form of a line, find an equation for each line having the following properties:
   (a) Contains the point $(-1, 4)$ and is perpendicular to $\mathbf{u} = [2, -3]$.
   (b) Contains the point $(2, 3)$ and is perpendicular to $\mathbf{u} = [-3, -1]$.
   (c) Contains the point $(-1, -4)$ and is parallel to $\mathbf{u} = [4, -2]$.
   (d) Contains the point $(3, -2)$ and is parallel to $\mathbf{v} = [2, 5]$.
   (e) Is the perpendicular bisector of the segment from $A = (2, -1)$ to $B = (-3, 4)$.
   (f) Is the perpendicular bisector of the segment from $C = (-1, -4)$ to $D = (3, -1)$.

3–6. Angle between two lines

Cosine formula: Let θ represent the angle between two lines whose direction numbers are $[u_1, u_2]$ and $[v_1, v_2]$, respectively, where $0° \leq \theta \leq 180°$. It follows from equation (2–7) that

$$\cos \theta = \frac{u_1 v_1 + u_2 v_2}{\sqrt{u_1^2 + u_2^2} \sqrt{v_1^2 + v_2^2}} = \frac{\mathbf{u} \cdot \mathbf{v}}{|\mathbf{u}| |\mathbf{v}|} \qquad (3\text{--}17)$$

where $\mathbf{u} = [u_1, u_2]$ and $\mathbf{v} = [v_1, v_2]$ are vectors on the lines.

The sign of $\cos \theta$ is the same as that of $\mathbf{u} \cdot \mathbf{v}$. When $\mathbf{u} \cdot \mathbf{v} > 0$, θ is an acute angle; when $\mathbf{u} \cdot \mathbf{v} < 0$, θ is an obtuse angle; and, when $\mathbf{u} \cdot \mathbf{v} = 0$, θ is 90°.

EXAMPLE 3–13. Given two lines with direction numbers $[2, -3]$ and $[3, 4]$, respectively. Find the cosine of the angle between them.

Solution. We have

$$\cos \theta = \frac{(2)(3) + (-3)(4)}{(\sqrt{13})(5)} = \frac{-6}{5\sqrt{13}}$$

Since the result is negative, θ is the obtuse angle between the lines.

If two lines are perpendicular to each other, the angle between them is 90°. Hence, from equation (3–17),

$$\cos 90° = \frac{\mathbf{u} \cdot \mathbf{v}}{|\mathbf{u}| |\mathbf{v}|} = 0$$

or

$$u_1 v_1 + u_2 v_2 = 0 \qquad (3\text{--}18)$$

If two lines are perpendicular, the sum of the products of their corresponding direction numbers is zero. Conversely, if $u_1 v_1 + u_2 v_2 = 0$, then $\cos \theta = 0$ and $\theta = 90°$, since $0° \leq \theta \leq 180°$.

Tangent formula: In both Fig. 3–19 and Fig. 3–20, we shall be interested in finding the tangent of the angle θ obtained by turning in a counterclockwise direction from the line whose slope is s_1 to the line whose slope is

s_2. Let α_1 and α_2 be the corresponding inclinations of the two lines. Then, in Fig. 3-19,
$$\theta = \alpha_2 - \alpha_1$$
Hence,
$$\tan \theta = \tan(\alpha_2 - \alpha_1) = \frac{\tan \alpha_2 - \tan \alpha_1}{1 + \tan \alpha_2 \tan \alpha_1}$$
or
$$\tan \theta = \frac{s_2 - s_1}{1 + s_1 s_2} \tag{3-19}$$

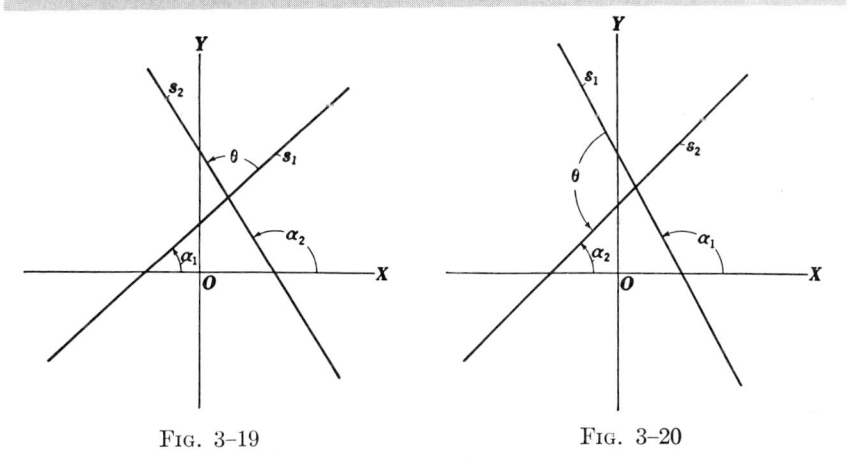

FIG. 3-19   FIG. 3-20

In Fig. 3-20, $\pi - \theta = \alpha_1 - \alpha_2$, or $\tan(\pi - \theta) = -\tan \theta = \tan(\alpha_1 - \alpha_2)$. Hence,
$$\tan \theta = -\frac{\tan \alpha_1 - \tan \alpha_2}{1 + \tan \alpha_1 \tan \alpha_2}$$
or
$$\tan \theta = \frac{\tan \alpha_2 - \tan \alpha_1}{1 + \tan \alpha_1 \tan \alpha_2} = \frac{s_2 - s_1}{1 + s_1 s_2}$$

Thus, the resulting formula is the same whether α_1 is greater or less than α_2.

If the two given lines are parallel, $\theta = 0°$, $\tan \theta = 0$, and $s_1 = s_2$. That is, *two lines are parallel when their slopes are equal.* Since the slope is the quotient of the direction numbers, $\frac{u_2}{u_1} = \frac{v_2}{v_1}$. From this it follows that $\frac{u_1}{v_1} = \frac{u_2}{v_2}$. That is, *two lines are parallel when their direction numbers are proportional.* This rule applies only to lines which are neither horizontal nor vertical.

If the two given lines are perpendicular, $\theta = 90°$, $\tan \theta$ is not defined, and $1 + s_1 s_2 = 0$. Hence, $s_1 s_2 = -1$, or $s_1 = -\frac{1}{s_2}$. That is, *two lines are perpendicular if the product of their slopes is -1.* This rule does not apply in the case where the lines are horizontal and vertical. We recognize such lines readily, however, since the slope of the horizontal line is 0 and the slope of the vertical line is not defined.

EXERCISES

1. Determine direction cosines of the lines through the following pairs of points:
   (a) $P_1(3, -2)$ and $P_2(-1, 4)$
   (b) $P_1(-4, 3)$ and $P_2(6, 1)$
   (c) $P_1(-1, -1)$ and $P_2(-1, 6)$
   (d) $P_1(3, 6)$ and $P_2(0, 4)$
   (e) $P_1(5, 8)$ and $P_2(-3, -2)$
   (f) $P_1(0, 0)$ and $P_2(-1, -6)$
   (g) $P_1(0, 3)$ and $P_2(4, 0)$
   (h) $P_1(1, 2)$ and $P_2(-6, 3)$

2. What is the slope (if defined) of each of the lines of Exercise 1?

3. Find direction cosines of the lines whose equations are:
   (a) $2x - 3y + 6 = 0$
   (b) $3x + 4y - 12 = 0$
   (c) $3x - 10 = 0$
   (d) $x + 5y = 0$
   (e) $x - y + 2 = 0$
   (f) $2y + 3 = 0$
   (g) $5x - 12y + 20 = 0$
   (h) $4x - 3y + 24 = 0$

4. What is the slope (if defined) of each of the lines of Exercise 3?

5. (a) Can $[\tfrac{1}{3}, \tfrac{2}{5}]$ be direction cosines of a line? Explain.
   (b) Can $[\tfrac{3}{5}, -\tfrac{4}{5}]$ be direction cosines of a line? Explain.
   (c) Can $[\tfrac{4}{5}, \tfrac{3}{5}]$ be direction cosines of a line? Explain.

6. Find the cosine of an angle between the following pairs of lines:
   (a) $2x - 3y - 6 = 0$ and $x + y - 4 = 0$
   (b) $x - 3y + 7 = 0$ and $5x - 12y + 20 = 0$
   (c) $x - 6 = 0$ and $y + 2 = 0$
   (d) $x + y - 1 = 0$ and $x - 2y + 3 = 0$
   (e) $2x - y + 7 = 0$ and $x + 2y - 3 = 0$

7. (a) Derive the formula for the cosine of an angle between two lines.
   (b) Derive the formula for the tangent of an angle between two lines.

8. The equations of the sides of a triangle are, respectively, $2x - y + 6 = 0$, $x + y - 1 = 0$, and $x + 3y - 3 = 0$. Find: (a) the tangents of the interior angles of the triangle; (b) the direction cosines of the sides; and (c) the cosines of the interior angles of the triangle.

9. Given $\mathbf{u} = [u_1, u_2]$ and $\mathbf{v} = [v_1, v_2]$ to be non-zero vectors, prove that $\tan \theta = \dfrac{\mathbf{u} | \mathbf{v}}{\mathbf{u} \cdot \mathbf{v}}$, where θ is the angle generated from $\mathbf{u}$ to $\mathbf{v}$.

10. A triangle has for its vertices the points $P_1(-2, 3)$, $P_2(4, -1)$, $P_3(0, 6)$. Find:
    (a) The cosine of angle $P_1P_2P_3$
    (b) The tangent of the exterior angle of the triangle at P_2
    (c) The area of the triangle

(d) Find the tangents of the interior angles of the triangle and show that their sum is equal to their product, i.e.

tan $P_1P_2P_3$ + tan $P_1P_3P_2$ + tan $P_2P_1P_3$ = (tan $P_1P_2P_3$) (tan $P_1P_3P_2$) (tan $P_2P_1P_3$)

11. The equations of the three sides of a triangle are $2x - 3y + 8 = 0$, $2x + y = 0$, and $x + 2y - 10 = 0$. Find:
    (a) The cosines of the angles of the triangle
    (b) The tangents of the angles of the triangle
    (c) The area of the triangle
    (d) Show that the sum of the tangents of the angles found in (b) is equal to their product.

12. Find the tangent of the obtuse angle between the lines whose equations are $2x - 3y + 6 = 0$ and $3x + 4y - 12 = 0$.

3–7. Point-slope form of equation of a line

Since the slope s of a non-vertical line is the quotient obtained by dividing the second direction number of the line by its first direction number, we know that a pair of direction numbers of the line is $[1, s]$. Hence, the direction number equation of a line having a slope s and passing through the point $P_1 = (x_1, y_1)$ is $\frac{x - x_1}{1} = \frac{y - y_1}{s}$. If we clear of fractions, we obtain

$$y - y_1 = s(x - x_1) \qquad (3\text{--}20)$$

This is called the *point-slope form* of the equation of a line.

EXAMPLE 3–14. Using the point-slope form, find an equation of the line whose slope is -3 and which passes through the point $(-2, 1)$

Solution. By equation (3–20),

$$y - 1 = -3(x + 2)$$

or
$$3x + y + 5 = 0$$

3–8. Intercept form of equation of a line

Let h and k represent the x-intercept and y-intercept of a line, neither h nor k being 0. Then, the coordinates of the points at which the line intersects the x-axis and the y-axis are $(h, 0)$ and $(0, k)$ and $s = \frac{k - 0}{0 - h} = -\frac{k}{h}$. Hence, a pair of direction numbers of the line is $[h, -k]$, and a vector perpendicular to the line is $\mathbf{u} = [k, h]$. So the equation of the line becomes

$$kx + hy = k(h) + h(0) = kh$$

98   ANALYTIC GEOMETRY IN THE PLANE

If we divide every term by hk, we obtain

$$\frac{x}{h}+\frac{y}{k}=1 \qquad (3\text{-}21)$$

This is called the *intercept form* of the equation of a line. In order for a linear equation to be in the intercept form, the right-hand side of the equation must be $+1$, and the *numerators* of the *coefficients* of x and y also must be $+1$.

EXAMPLE 3-15. Write the equation $2x-3y+5=0$ in the intercept form.

Solution. We first write the equation as

$$2x-3y = -5$$

Dividing each term by -5, we have

$$\frac{2x}{-5}-\frac{3y}{-5}=1$$

Finally, dividing the numerator and the denominator of the x-term by 2, and dividing both parts of the y-term by -3, we obtain

$$\frac{x}{-\frac{5}{2}}+\frac{y}{\frac{5}{3}}=1$$

The x-intercept is $-\frac{5}{2}$ and the y-intercept is $\frac{5}{3}$. The line is shown in Fig. 3–21.

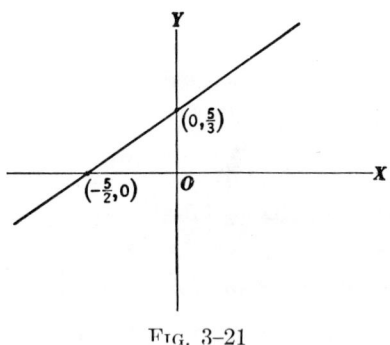

FIG. 3–21

EXAMPLE 3-16. What is an equation of the line whose x-intercept and y-intercept are -3 and 2, respectively?

Solution. By equation (3–21),

$$\frac{x}{-3}+\frac{y}{2}=1$$

or

$$2x-3y+6=0$$

3–9. Slope-intercept form of equation of a line

If the slope s of a line and the y-intercept k are known, we can easily write an equation of the line as follows:

$$sx - y = s(0) - 1(k) = -k$$

Solving for y, we obtain

$$y = sx + k \tag{3-22}$$

This is known as the *slope-intercept form* of the equation of a line. For an equation to be in this form, we must have y, by itself, on one side of the equality sign with the coefficient $+1$. Then the coefficient of x is the *slope*, and the constant term is the *y-intercept*.

EXAMPLE 3–17. Write the equation $2x - 3y + 5 = 0$ in the slope-intercept form.

Solution. First we write the equation in the form

$$-3y = -2x - 5$$

Then, dividing each term of the equation by -3, we obtain

$$y = \tfrac{2}{3}x + \tfrac{5}{3}$$

Hence, the slope is $\tfrac{2}{3}$ and the y-intercept is $\tfrac{5}{3}$.

EXAMPLE 3–18. The slope of a line is $-\tfrac{1}{4}$ and the y-intercept is 3. Find an equation of the line.

Solution. According to equation (3–22),

$$y = -\tfrac{1}{4}x + 3$$

or

$$x + 4y - 12 = 0$$

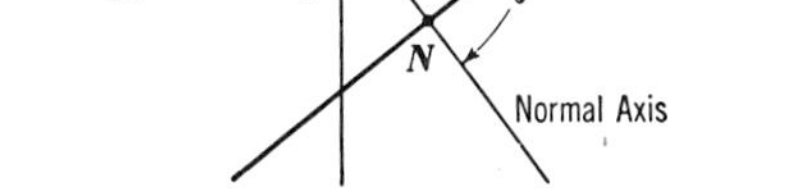

FIG. 3–22

FIG. 3–23

3–10. Normal form of equation of a line

We shall define the *normal axis* of a line l to be the line that passes through the origin and is perpendicular to l. If the line l does not go through the origin, as in Fig. 3–22, let N denote the point where the line

100   ANALYTIC GEOMETRY IN THE PLANE

intersects its normal axis. Also, let α denote the angle from the positive x-axis to $\overrightarrow{ON}$. We shall define *the normal vector* **n** of the line l to be the unit vector in the direction from O to N. In Fig. 3-22, $\mathbf{n} = \overrightarrow{OQ}$. The *normal intercept* of a line l is defined to be $|\overrightarrow{ON}|$ and will be denoted by p.

If the line l does go through the origin, as in Fig. 3-23, and O and N coincide, we shall define *the normal vector* to be *a unit vector* $\mathbf{n} = \overrightarrow{OQ}$ or $\mathbf{n}' = \overrightarrow{OQ}'$ that is perpendicular to l. Hence, a line passing through the origin O has two possible unit normal vectors.

Let $P(x, y)$ be any point on the line l, Fig. 3-22. Then, if θ is the angle between the vectors $\overrightarrow{OP} = [x, y]$ and $\mathbf{n} = [\cos \alpha, \sin \alpha]$, and $-90° < \theta < 90°$, we have

$$\cos \theta = \frac{[x, y] \cdot [\cos \alpha, \sin \alpha]}{\sqrt{x^2 + y^2}}$$

from which

$$[x, y] \cdot [\cos \alpha, \sin \alpha] = \sqrt{x^2 + y^2} \cos \theta = |\overrightarrow{OP}| \cos \theta = p$$

Hence,

$$x \cos \alpha + y \sin \alpha = p \qquad (3\text{-}23)$$

This is called the *normal form* of the equation of a line.

If the line l passes through the origin, $p = 0$; but equation (3-23) holds. In this case, the normal form becomes

$$x \cos \alpha + y \sin \alpha = 0 \qquad (3\text{-}24)$$

A linear equation is said to be in normal form when it meets the following conditions: (1) the coefficient vector is a unit vector; (2) the constant term is positive or zero when written by itself on one side of the equality sign.

In order to reduce the linear equation $ax + by + c = 0$ to the normal form, we divide each term by $e\sqrt{a^2 + b^2}$. The equation then becomes

$$\frac{a}{e\sqrt{a^2 + b^2}} x + \frac{b}{e\sqrt{a^2 + b^2}} y = \frac{-c}{e\sqrt{a^2 + b^2}}$$

If the line does not pass through the origin, the right-hand member must be positive. Therefore, if c is positive, $e = -1$; and, if c is negative, $e = +1$. In case the line passes through the origin, $c = 0$ and e may be either $+1$ or -1. Since $\left(\dfrac{a}{e\sqrt{a^2+b^2}}\right)^2 + \left(\dfrac{b}{e\sqrt{a^2+b^2}}\right)^2 = 1$, the vector $\left[\dfrac{a}{e\sqrt{a^2+b^2}}, \dfrac{b}{e\sqrt{a^2+b^2}}\right]$ is a unit vector. Also, $\dfrac{-c}{e\sqrt{a^2+b^2}} \geq 0$. The equation is therefore in the normal form.

It should be noted that a line not passing through the origin has only one normal form; whereas a line passing through the origin has two normal forms.

In a particular linear equation in the normal form, the signs of the coefficients of x and y show whether the angle α is clockwise or counter-clockwise and whether that angle is acute or obtuse. The angle α is measured counter-clockwise when the coefficient of y is positive and is measured clockwise when that coefficient is negative. The angle α is acute when the coefficient of x is a positive number and is obtuse when that coefficient is negative.

Since the constant term written by itself on one side of the equality sign in the normal form of the equation can never be a negative number, the normal intercept is always laid off from the origin along the terminal side of the angle α.

EXAMPLE 3–19. Write the equation $3x - 4y + 10 = 0$ in the normal form, and draw the graph of the line by using the normal vector.

Solution. Dividing each term of the given equation by $-\sqrt{9+16} = -5$, we have
$$-\tfrac{3}{5}x + \tfrac{4}{5}y = 2$$

In this example, the angle α is measured counter-clockwise because the coefficient of y is positive; and this angle is obtuse because the coefficient of x is negative. The magnitude of the acute angle $(180° - \alpha)$ may be determined by using either of the following relations: $180° - \alpha = \sin^{-1} \tfrac{4}{5}$*

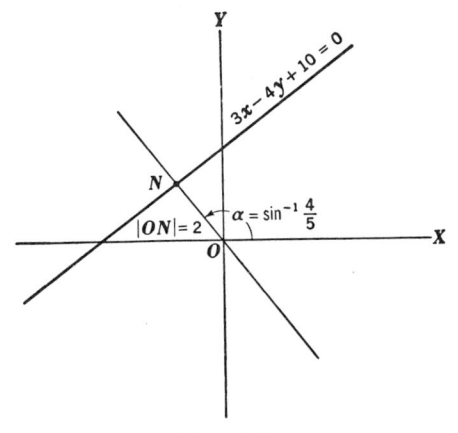

FIG. 3–24

or $180° - \alpha = \cos^{-1} \tfrac{3}{5}$. The normal intercept is 2 and is located as shown in Fig. 3–24. The required line is then drawn through the end of the normal intercept and perpendicular to it.

When the direction and magnitude of *the normal vector* of a line are known, the right-hand member in equation (3–23) is always made positive; and the signs of the coefficients of x and y are determined by the signs of $\cos \alpha$ and $\sin \alpha$ as established by the usual rules for the signs of $\cos A$, $\cos (180° - A)$, $\cos (-A)$, $\sin A$, $\sin (180° - A)$, and $\sin (-A)$, where A is

---
* $\sin^{-1} \tfrac{4}{5}$ is read "an angle whose sine is $\tfrac{4}{5}$."

an acute angle. The angle is considered positive when measured counter-clockwise and is considered negative when measured clockwise.

EXAMPLE 3–20. Find an equation of the line for which *the normal vector* makes a counter-clockwise angle of 120° with the positive x-axis and the normal intercept is 5.

Solution. In equation (3–23), $\alpha = 120°$ and $p = 5$. Since this angle is positive and obtuse, $\sin \alpha$ is positive and $\cos \alpha$ is negative. Hence,

$$x \cos 120° + y \sin 120° = 5$$

which becomes

$$-\frac{1}{2}x + \frac{\sqrt{3}}{2}y = 5$$

or

$$x - \sqrt{3}y + 10 = 0$$

EXERCISES

1. Find an equation of each line from the following data:
   (a) $s = \frac{1}{3}$ and $P = (2, 3)$
   (b) $s = -4$ and y-intercept $= 2$
   (c) $h = -1$ and $k = 3$
   (d) $s = \frac{1}{2}$ and x-intercept $= -5$
   (e) $h = 7$ and $k = -3$
   (f) $s = -\frac{3}{4}$ and $P = (3, -5)$
   (g) $s = \frac{2}{5}$ and $k = -2$
   (h) $s = 0$ and $k = 5$
   (i) x-intercept $= \frac{1}{2}$ and $s = -2$
   (j) $h = -3$ and $k = -5$

2. Write each of the following equations in both the *intercept form* and the *slope-intercept form*:
   (a) $2x - 3y + 6 = 0$ (b) $3x + 4y - 12 = 0$ (c) $5x - 7y + 10 = 0$

3. Determine *the normal vector* of each line in Exercise 2.

4. What is s if a line through the point $(3, 2)$ forms with the x-axis and the y-axis a triangle in the first quadrant having an area of 16 square units?

5. Write the equation $ax + by + c = 0$ in the slope-intercept form and also in the intercept form.

6. State in words how one can determine when an equation of a line is: (a) in the slope-intercept form; (b) in the intercept form; (c) in the normal form.

7. Show that $\begin{vmatrix} x & y & 1 \\ h & 0 & 1 \\ 0 & k & 1 \end{vmatrix} = 0$ is an equation of the line whose x-intercept and y-intercept are h and k, respectively. (Compare Exercise 5 in Section 3–4.)

THE STRAIGHT LINE   103

8. Given the equation $ax+by+c=0$. Show that the normal form of the equation of the line is $\dfrac{ax+by+c}{e\sqrt{a^2+b^2}}=0$, where e is $+1$ if c is negative, is -1 if c is positive, and is either $+1$ or -1 if $c=0$. Give reasons for the choices of e.

9. Given the equation $ax+c=0$, show that the normal form of the equation of the line is $ex=p$ where $p=\left|\dfrac{c}{a}\right|$ and $e=+1$ if ac is negative, $e=-1$ if ac is positive, and $e=\pm 1$ if $c=0$.

10. Write an equation of each line and construct the line from the following data:

   (a) $\alpha=30°$ and $p=10$
   (b) $\alpha=120°$ and $p=4$
   (c) $\alpha=-90°$ and $p=1$
   (d) $\alpha=0°$ and $p=6$
   (e) $\alpha=-30°$ and $p=4$
   (f) $\alpha=135°$ and $p=6$
   (g) $\alpha=-120°$ and $p=2$
   (h) $\alpha=-60°$ and $p=4$

11. Write each of the following equations in normal form:

   (a) $3x-4y+12=0$
   (b) $x+2y-3=0$
   (c) $3x+4y-12=0$
   (d) $x-y-2=0$
   (e) $5x-12y+65=0$
   (f) $2x-3y=0$
   (g) $2x-5=0$
   (h) $3y+2=0$
   (i) $x-3y=0$

12. Find an equation of a line containing the point $(5, 2)$ and having equal intercepts with the axes.

13. A line makes equal intercepts with the x- and y-axes and contains the point $(-1, -4)$. What is an equation of the line?

14. A line passing through the point $(-12, 3)$ makes with the axes a triangle of area 9. What is its equation?

15. Show that the lines $x+2y+1=0$, $6x-3y-5=0$, $y=2x-1$, and $4x+8y+7=0$ form a rectangle.

16. Show that lines drawn from any vertex of a parallelogram to the mid-points of the opposite sides will divide the diagonals which they intersect into three equal parts.

17. The line through the point $(2, -4)$ perpendicular to a given line meets it in the point $(-5, 3)$. What is an equation of the given line?

18. Find an equation of the line whose intercepts are twice those of the line $x-2y+6=0$.

19. Find an equation of the line that contains the point $(5, -1)$ and the point on the line $3x-y=7$ whose ordinate is 2.

20. The perpendicular from the origin intersects a line in the point $(4, 3)$. What is an equation of the line?

104   ANALYTIC GEOMETRY IN THE PLANE

21. A line has the same intercept on the x-axis as the line $\sqrt{3}x - y + 6 = 0$, but it makes with that axis half the angle. What is an equation of the line?

22. Two sides of a parallelogram lie along the lines $2x - 3y + 6 = 0$ and $x + 2y = 4$. A vertex is at $(1, -3)$. Find equations for the other two sides.

3-11. Lines parallel and perpendicular to a given line

Since the vector $\mathbf{u} = [a, b]$ is perpendicular to the line $ax + by + c = 0$, it follows that $\mathbf{u} = [a, b]$ is perpendicular to any line $ax + by = k$, for an arbitrary value of k. Hence, $ax + by + c = 0$ and $ax + by = k$ are equations of parallel* lines. If $k = -c$, the lines not only are parallel but they coincide.

To write an equation of a line that is *parallel* to a given line $ax + by + c = 0$ and passes through a given point (x_1, y_1), we first write the equation

(1) $$ax + by = k$$

where k is unknown. We determine k by substituting x_1 for x and y_1 for y and obtaining the relation

(2) $$ax_1 + by_1 = k$$

The required equation is

$$ax + by = ax_1 + by_1 \qquad (3\text{-}25)$$

To prove that this is correct, we observe the following facts: The equation represents a line, since it is a linear equation; the line is parallel to $ax + by + c = 0$, since it is perpendicular to the vector $\mathbf{u} = [a, b]$; and it passes through the point (x_1, y_1), since these coordinates satisfy the equation.

EXAMPLE 3-21. Find an equation of the line passing through the point $(5, -2)$ and parallel to the line $2x - 3y + 6 = 0$.

Solution. We know that the required equation may be written in the form $2x - 3y = k$. Substituting $x = 5$ and $y = -2$, we obtain $2(5) - 3(-2) = k = 16$. Therefore, the required equation is

$$2x - 3y = 16$$

The two lines are shown in Fig. 3-25.

To derive the equation of a line that is *perpendicular* to the line $ax + by + c = 0$ and passes through the point (x_1, y_1), we proceed as follows: We interchange the coefficients of x and y and we change one sign. We thus obtain

---
* In this book coincident lines are regarded as parallel.

(1) $$bx - ay = k$$
where k is unknown. Since (x_1, y_1) is a point on this line, we have
(2) $$bx_1 - ay_1 = k$$
Hence, the required equation is

$$bx - ay = bx_1 - ay_1 \qquad (3\text{-}26)$$

This is a linear equation, and it therefore represents a line. The line is perpendicular to $ax + by + c = 0$, since it is parallel to the vector $\mathbf{v} = [a, b]$; and it contains the point (x_1, y_1).

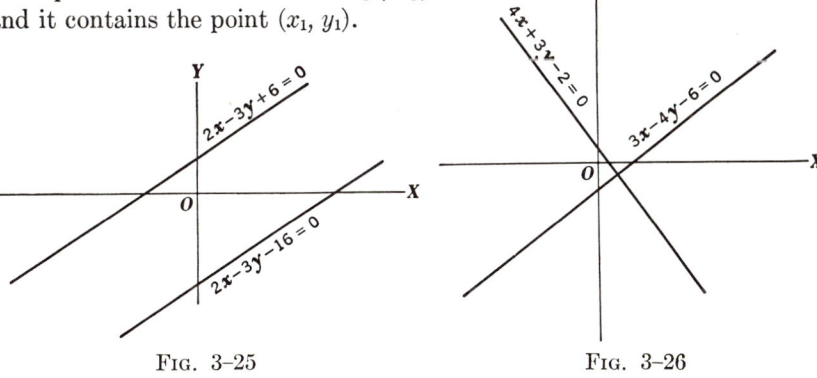

FIG. 3-25

FIG. 3-26

EXAMPLE 3-22. Find an equation of the line passing through the point $(-1, 2)$ and perpendicular to the line $3x - 4y - 6 = 0$.

Solution. Interchanging the coefficients and changing one sign, we have $4x + 3y = k$. Then $4(-1) + 3(2) = k = 2$. Therefore, the required equation is
$$4x + 3y = 2$$
The two lines are shown in Fig. 3-26.

EXERCISES

1. Find an equation of each of the following lines:
    (a) Perpendicular to the vector $\mathbf{u} = [-3, 2]$ and passing through the point $P = (1, 4)$
    (b) Perpendicular to $\mathbf{u} = [1, -4]$ and passing through $P = (-3, -2)$
    (c) Parallel to $\mathbf{u} = [-1, -2]$ and passing through $P = (2, -1)$
    (d) Parallel to $\mathbf{u} = [3, -5]$ and passing through $P = (-4, 2)$

2. Find a vector perpendicular to each of the following lines:
    (a) $2x - 3y + 6 = 0$
    (b) $x - 7y + 7 = 0$
    (c) $3x + 4y - 12 = 0$
    (d) $2x + y - 4 = 0$
    (e) $x - y + 3 = 0$
    (f) $5x + y + 5 = 0$

3. Find a vector parallel to each of the lines in Exercise 2.

4. Find equations of the lines that are parallel and perpendicular, respectively, to each of the following lines and contain the given points:
   (a) $2x - 3y + 6 = 0$ and $(2, 1)$
   (b) $3x + 4y + 12 = 0$ and $(-1, 4)$
   (c) $x - y + 2 = 0$ and $(-1, -2)$
   (d) $x - 4 = 0$ and $(3, 5)$
   (e) $y + 2 = 0$ and $(6, -1)$
   (f) $x - 5y + 6 = 0$ and $(-4, -1)$
   (g) $3x + 2y - 1 = 0$ and $(1, 1)$
   (h) $5x - 12y + 60 = 0$ and $(2, 0)$
   (i) $x - 3y = 0$ and $(4, -2)$
   (j) $x = 0$ and $(1, -6)$

5. Given the triangle ABC with vertices $A(1, 0)$, $B(3, 4)$, and $C(5, -2)$. Find: (a) an equation of each of the sides; (b) an equation of each of the altitudes; (c) an equation of the perpendicular bisector of each side; (d) an equation of a line passing through each vertex and parallel to the opposite side; (e) equations of lines that pass through the vertices and are parallel and perpendicular to the coordinate axes.

6. Find the information required in Exercise 5 for the triangle whose vertices are $A(0, 0)$, $B(6, 8)$, and $C(4, -2)$.

7. Find the information asked for in Exercise 5 for the triangle whose vertices are $P_1(-1, 2)$, $P_2(3, 1)$, and $P_3(5, -6)$.

8. Find equations of the lines which contain the perpendicular bisectors and the altitudes of the triangle whose vertices are $A(0,0)$, $B(a, 0)$, and $C(b, c)$.

9. The diagonals of a square lie along the coordinate axes and their length is 4 units. Find equations of the lines that contain the four sides.

10. What are equations of the lines parallel to $x + 2y = 6$ and forming with the coordinate axes triangles of area 16?

11. What are equations of the lines perpendicular to $x - 2y + 5 = 0$ and forming with the coordinate axes triangles of area 9?

12. The coordinates of the projection of the origin on a given line are (a, b). What is an equation of the line?

13. A line is parallel to the line $3x + 2y - 12 = 0$ and forms a triangle in the first quadrant with the lines $2x - y = 0$ and $x - 2y = 0$ whose area is 21. Find an equation of the line.

14. Find the coordinates of the center of a circle which passes through the points $P_1(-1, -1)$, $P_2(3, 1)$, and $P_3(5, -3)$.

15. Find the coordinates of the center of a circle which passes through the points $(-3, 1)$, $(1, -1)$, $(7, 3)$.

THE STRAIGHT LINE   107

16. The vertices of a triangle are $O(0, 0)$, $A(a, 0)$, and $B(b, c)$. Find an equation of the median line from O on side BA. Prove that the length of a median is *less* than half the sum of the two adjacent sides.

17. Factor the equation $4x^2 - 12xy + 9y^2 - 16 = 0$ and describe the set of points that satisfies the equation.

3–12. Distance from a line to a point

Let us consider the problem of determining the distance δ from a line whose equation is known to a point whose coordinates are known. Let $ax + by + c = 0$ be an equation of the given line, and let $P_1 = (x_1, y_1)$ be the coordinates of the given point. From any point on the line we draw a representative segment of the coefficient vector $\mathbf{u} = [a, b]$. The point P_1 and the vector $\mathbf{u}$ may lie either on the *same* side of the line or on opposite sides of the line.

We shall assume first that P_1 and $\mathbf{u}$ are on the same side of the line, as in Fig. 3–27. Since the point P_1 is not on the line, we know that $ax_1 + by_1 + c \neq 0$. Let $P_2 = (x_2, y_2)$ be the projection of P_1 on the line. Then it follows that $ax_2 + by_2 + c = 0$ or $c = -ax_2 - by_2$. Since $\overrightarrow{P_2P_1}$ and $\mathbf{u}$ are parallel in the same sense, the angle between $\mathbf{u}$ and $\overrightarrow{P_2P_1}$ is $0°$. By the cosine formula,

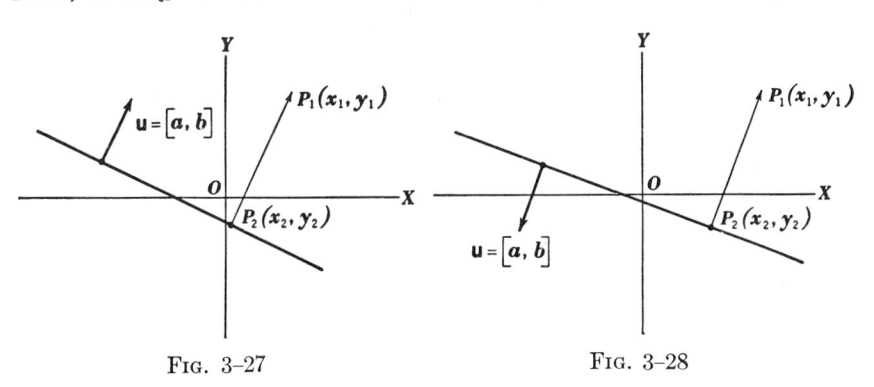

FIG. 3–27

FIG. 3–28

(1) $$\cos 0° = \frac{\mathbf{u} \cdot \overrightarrow{P_2P_1}}{|\mathbf{u}|\,|\overrightarrow{P_2P_1}|}$$

Since $\mathbf{u} = [a, b]$ and $\overrightarrow{P_2P_1} = [x_1 - x_2, y_1 - y_2]$, equation (1) becomes

(2) $$\cos 0° = 1 = \frac{[a, b] \cdot [x_1 - x_2, y_1 - y_2]}{\sqrt{a^2 + b^2}\,|\overrightarrow{P_2P_1}|} = \frac{ax_1 - ax_2 + by_1 - by_2}{\sqrt{a^2 + b^2}\,|\overrightarrow{P_2P_1}|}$$

But $-ax_2 - by_2 = c$. Hence, multiplying both sides by $|\overrightarrow{P_2P_1}|$, we have

$$\delta = |\overrightarrow{P_2P_1}| = \frac{ax_1+by_1+c}{\sqrt{a^2+b^2}} \qquad (3\text{-}27)$$

Now let us assume that P_1 and $\mathbf{u}$ lie on opposite sides of the line, as in Fig. 3-28; in this case, the angle between $\mathbf{u}$ and $\overrightarrow{P_2P_1}$ is 180°. Hence,

(3) $$\cos 180° = \frac{\mathbf{u} \cdot \overrightarrow{P_2P_1}}{|\mathbf{u}|\,|\overrightarrow{P_2P_1}|}$$

or

(4) $$-1 = \frac{[a,\,b]\cdot[x_1-x_2,\,y_1-y_2]}{\sqrt{a^2+b^2}\,|\overrightarrow{P_2P_1}|} = \frac{ax_1-ax_2+by_1-by_2}{\sqrt{a^2+b^2}\,|\overrightarrow{P_2P_1}|}$$

Multiplying both sides by $|\overrightarrow{P_2P_1}|$ and replacing $-ax_2-by_2$ by c, we have

$$-|\overrightarrow{P_2P_1}| = \frac{ax_1+by_1+c}{\sqrt{a^2+b^2}} \qquad (3\text{-}28)$$

Regardless of the position of the point with respect to the line, the *distance* δ from the line whose equation is $ax+by+c=0$ to the point $P_1 = (x_1,\,y_1)$ is

$$\delta = \left|\frac{ax_1+by_1+c}{\sqrt{a^2+b^2}}\right| \qquad (3\text{-}29)$$

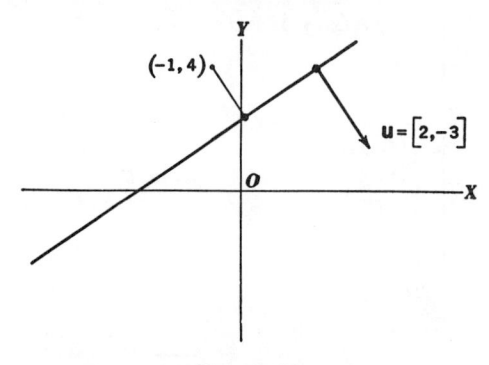

Fig. 3-29

When the number $\dfrac{ax_1+by_1+c}{\sqrt{a^2+b^2}}$ is positive, the point P_1 lies on the side of the line toward which the coefficient vector points; and, when the number is negative, P_1 lies on the side of the line opposite to that toward which the coefficient vector points.

EXAMPLE 3-23. Find the distance from the line $2x-3y+6=0$ to the point $(-1,\,4)$.

Solution. The conditions are represented in Fig. 3–29. By equation (3–29),

$$\delta = \left|\frac{2(-1) - 3(4) + 6}{\sqrt{13}}\right| = \frac{8}{\sqrt{13}}$$

EXERCISES

1. Prove that $\delta = |lx_1 + my_1 + c|$, if l and m are direction cosines of the normal vector to the line whose equation is $lx + my + c = 0$.

2. Prove that the distance from the line $ax + by + c = 0$ to the origin is $\left|\dfrac{c}{\sqrt{a^2 + b^2}}\right|$.

3. Find the distance from each of the given lines to the given points:
   (a) $3x - 4y + 12 = 0$ and $(1, -2)$
   (b) $x - y - 3 = 0$ and $(2, -3)$
   (c) $2x + y - 6 = 0$ and $(-4, -5)$
   (d) $x + 2y - 6 = 0$ and $(-2, 1)$
   (e) $3x + 4y - 24 = 0$ and $(0, 0)$
   (f) $5x - y + 10 = 0$ and $(-3, 4)$

4. Find the distance between the following pairs of parallel lines:
   (a) $\begin{array}{l}2x - 3y + 6 = 0 \\ 2x - 3y + 4 = 0\end{array}$
   (b) $\begin{array}{l}x - 3y + 4 = 0 \\ 2x - 6y + 15 = 0\end{array}$
   (c) $\begin{array}{l}ax + by + c_1 = 0 \\ ax + by + c_2 = 0\end{array}$

5. Find the lengths of the altitudes of each of the following triangles:
   (a) $P_1(-1, 3)$, $P_2(3, -4)$, and $P_3(-4, 0)$
   (b) $P_1(2, 1)$, $P_2(3, 6)$, and $P_3(-1, 3)$
   (c) $P_1(6, 1)$, $P_2(-2, 1)$, and $P_3(0, 0)$
   (d) $P_1(4, -3)$, $P_2(1, 5)$, and $P_3(-3, 1)$

6. What is the area of each of the triangles in Exercise 5?

7. Prove that $\delta = |x_1 \cos \alpha + y_1 \sin \alpha - p|$, where $x \cos \alpha + y \sin \alpha = p$ is the normal form of the equation of the line and δ is the distance from the line to the point $P = (x_1, y_1)$.

8. There are two points on the y-axis which are a distance of 4 units from the line $3x + 4y - 6 = 0$. What are their coordinates?

9. Find equations of the lines that are parallel to $3x + 4y + 25 = 0$ and a distance of 1 unit from the origin.

10. Find equations of the lines that are a distance of 2 units from the origin and are also parallel to the line $x - 2y + 6 = 0$.

11. Two lines are parallel to the line $2x+3y-6=0$ and 3 units from it. What are equations of these lines?

12. Find the locus of the centers of all circles of radius 6 which touch the line $3x+4y-12=0$.

13. What are equations of the lines which are parallel to the line $x-3y-6=0$ and at a distance $\sqrt{10}$ units from the point $(-1, 2)$?

14. Find equations of the lines which pass through the point $(9, -6)$ and are at a distance of 1 unit from the point $(4, -1)$.

15. Find equations of the lines that are perpendicular to $3x-4y+12=0$ and a distance of 2 units from the point $(4, -1)$.

16. The base of a triangle joins the points $(-2, 4)$ and $(-1, 3)$. What is the locus of the third vertex if the area of the triangle is $\frac{9}{2}$?

17. The base of a triangle joins the points $(2, -5)$ and $(-6, 3)$. What is the locus of the third vertex if the triangle is isosceles?

18. Describe the set of points (x_1, y_1) for which $2x_1-3y_1+3<-3$. Hint: Adding 3 to both sides of the inequality, we obtain $2x_1-3y_1+6<0$. From the distance formula, it follows that $\frac{2x_1-3y_1+6}{\sqrt{13}}<0$ (and, hence, $2x_1-3y_1+6<0$) for all points (x_1, y_1) which lie on the opposite side of the line $2x-3y+6=0$ from which the coefficient vector $\mathbf{u}=[2, -3]$ points. Hence, this is the set (x_1, y_1) for which $2x_1-3y_1+3<-3$.

19. Describe the set of points that satisfies each of the following inequalities:
 (a) $x-y<0$
 (b) $2x+3y<0$
 (c) $x+2y>0$
 (d) $x-y<4$
 (e) $2x+3y<2$
 (f) $x+2y>3$
 (g) $3x+4y\leq -2$
 (h) $2x-3y+8\leq -3$
 (i) $3x+4y-4\geq 8$

20. Describe the set of points satisfying each of the following conditions:
 (a) $x+y<0$
  $x-y>0$
 (b) $2x-y>4$
  $3x+4y>12$
 (c) $x-2y+3<-3$
  $2x+3y-2>4$

3–13. Intersecting lines

The geometric problem of finding in what point or points (if any) two lines intersect is the same as the algebraic problem of solving simultaneously two linear equations in two unknowns. Consider any two linear equations $a_1x+b_1y+c_1=0$ and $a_2x+b_2y+c_2=0$, each of which represents a line in the plane. These two lines may be parallel, they may coincide, or they may intersect in a single point. These possible conditions are shown in Fig. 3–30.

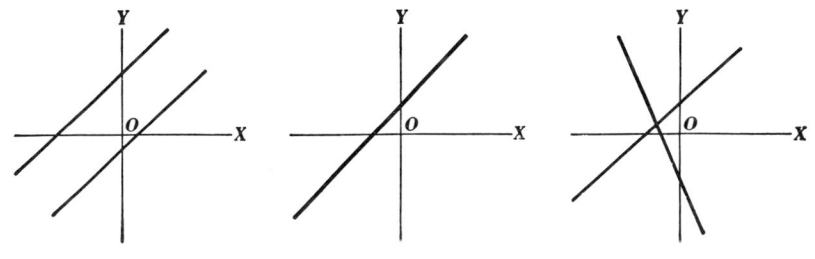

FIG. 3–30

If $\begin{vmatrix} a_1 & b_1 \\ a_2 & b_2 \end{vmatrix} = 0$ and $\begin{vmatrix} a_1 & c_1 \\ a_2 & c_2 \end{vmatrix}$ or $\begin{vmatrix} c_1 & b_1 \\ c_2 & b_2 \end{vmatrix} \neq 0$, the two lines are parallel.

If $\begin{vmatrix} a_1 & b_1 \\ a_2 & b_2 \end{vmatrix} = \begin{vmatrix} a_1 & c_1 \\ a_2 & c_2 \end{vmatrix} = \begin{vmatrix} c_1 & b_1 \\ c_2 & b_2 \end{vmatrix} = 0$, the two lines coincide.

If $\begin{vmatrix} a_1 & b_1 \\ a_2 & b_2 \end{vmatrix} \neq 0$, the two lines intersect in a point. In this case, we can solve simultaneously the equations $a_1 x + b_1 y + c_1 = 0$ and $a_2 x + b_2 y + c_2 = 0$, and we obtain the following results:

$$x = \frac{\begin{vmatrix} -c_1 & b_1 \\ -c_2 & b_2 \end{vmatrix}}{\begin{vmatrix} a_1 & b_1 \\ a_2 & b_2 \end{vmatrix}}$$

$$y = \frac{\begin{vmatrix} a_1 & -c_1 \\ a_2 & -c_2 \end{vmatrix}}{\begin{vmatrix} a_1 & b_1 \\ a_2 & b_2 \end{vmatrix}}$$

(3–30)

These are the coordinates of the point of intersection of the two lines.

EXAMPLE 3–24. Find the point of intersection of the lines whose equations are $2x - y + 3 = 0$ and $x + 3y - 4 = 0$.

Solution. By equations (3–30),

$$x = \frac{\begin{vmatrix} -3 & -1 \\ 4 & 3 \end{vmatrix}}{\begin{vmatrix} 2 & -1 \\ 1 & 3 \end{vmatrix}} = -\frac{5}{7}$$

and
$$y = -\frac{\begin{vmatrix} 2 & -3 \\ 1 & 4 \end{vmatrix}}{\begin{vmatrix} 2 & -1 \\ 1 & 3 \end{vmatrix}} = \frac{11}{7}$$

We check the results by substituting these values back in either of the original equations. Thus,
(1) $\qquad 2(-\frac{5}{7}) - \frac{11}{7} + 3 = 0$
(2) $\qquad -\frac{5}{7} + 3(\frac{11}{7}) - 4 = 0$

3–14. Three concurrent lines

If three lines intersect in a point, as in Fig. 3–31, they are said to be *concurrent*. Let us assume that the equations of the lines are: $l_1 \equiv a_1x + b_1y + c_1 = 0$, $l_2 \equiv a_2x + b_2y + c_2 = 0$, and $l_3 \equiv a_3x + b_3y + c_3 = 0$; and let $P(x_1, y_1)$ be a point on all three lines. Hence, its coordinates satisfy the three equations, and we have:

(1)
$$a_1x_1 + b_1y_1 + c_1 1 = 0$$
$$a_2x_1 + b_2y_1 + c_2 1 = 0$$
$$a_3x_1 + b_3y_1 + c_3 1 = 0$$

Let us use the following notation:

(2)
$$\Delta = \begin{vmatrix} a_1 & b_1 & c_1 \\ a_2 & b_2 & c_2 \\ a_3 & b_3 & c_3 \end{vmatrix}$$

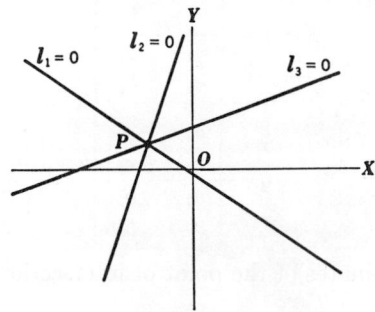

FIG. 3–31

It will be assumed that Δ is not equal to zero. Equations (1) may be regarded as a system of three linear equations for the numbers x_1, y_1, and 1. Solving by determinants for the third number, we obtain:

$$1 = \frac{\begin{vmatrix} a_1 & b_1 & 0 \\ a_2 & b_2 & 0 \\ a_3 & b_3 & 0 \end{vmatrix}}{\Delta} = \frac{0}{\Delta} = 0$$

Hence our assumption that $\Delta \neq 0$ has led us to the false statement $1 = 0$. Therefore, $\Delta = 0$.

However, the converse of this theorem is not true. In fact, we shall now prove that $\Delta = 0$ if any *two* of the lines coincide or if the three lines are parallel.

First, assume that the first two lines in equation (1) coincide; hence, there exists a number k such that $a_2 = ka_1$, $b_2 = kb_1$, and $c_2 = kc_1$. It follows that the second row in the determinant for Δ is proportional to the first row, and $\Delta = 0$.

If the lines are parallel, it follows that

$$\begin{vmatrix} a_2 & b_2 \\ a_3 & b_3 \end{vmatrix} = 0, \quad \begin{vmatrix} a_1 & b_1 \\ a_3 & b_3 \end{vmatrix} = 0, \text{ and } \begin{vmatrix} a_1 & b_1 \\ a_2 & b_2 \end{vmatrix} = 0$$

Therefore, expanding Δ by the elements of the last column, we obtain:

$$\Delta = c_1 \begin{vmatrix} a_2 & b_2 \\ a_3 & b_3 \end{vmatrix} - c_2 \begin{vmatrix} a_1 & b_1 \\ a_3 & b_3 \end{vmatrix} + c_3 \begin{vmatrix} a_1 & b_1 \\ a_2 & b_2 \end{vmatrix} = 0$$

Note that if it is desired to ascertain whether three lines are concurrent it is useful to compute Δ, since the lines cannot be concurrent unless $\Delta = 0$. But this in itself is insufficient; one may be certain of concurrency by finding a point which lies on all three lines.

EXAMPLE 3-25. Determine k in order that the lines $2x - 3y + 6 = 0$, $x + ky - 1 = 0$, and $x - y + 4 = 0$ will be concurrent.

Solution. In order that the lines may be concurrent, we must have

$$\begin{vmatrix} 2 & -3 & 6 \\ 1 & k & -1 \\ 1 & -1 & 4 \end{vmatrix} = 0$$

Adding the elements of the first column to the elements of the second column, and adding minus four times the elements of the first column to the elements of the third column, we obtain:

$$\begin{vmatrix} 2 & -1 & -2 \\ 1 & k+1 & -5 \\ 1 & 0 & 0 \end{vmatrix} = 0$$

Expanding by the elements of the third row, we obtain

$$5 + 2k + 2 = 0$$

or

$$k = -\tfrac{7}{2}$$

The student should check the result by finding the coordinates of the common point of intersection of the three lines.

EXERCISES

1. Find k if the following lines are concurrent: $2x-3ky+4=0$, $x+y-2=0$, and $x+2y-1=0$.

2. Find k if the following lines are concurrent: $x-3y-6=0$, $kx-y-3=0$, and $3x-y-1=0$.

3. Find the distance from the point of intersection of the lines whose equations are $x-y+2=0$ and $2x+3y-6=0$ to the line whose equation is $3x+4y-12=0$.

4. Find the coordinates of the point of intersection of the altitudes of the triangle whose vertices are $P_1(8, 0)$, $P_2(-1, 4)$, and $P_3(0, 3)$.

5. Find equations of the lines that are parallel and perpendicular, respectively, to the line whose equation is $3x-4y+6=0$ and that pass through the intersection of the lines whose equations are $2x+y-6=0$ and $x-y+3=0$.

6. Given the triangle whose vertices are $A(1, 0)$, $B(3, 4)$, and $C(5, -2)$. Prove: (a) that the altitudes meet in a point; (b) that the medians meet in a point; (c) that the perpendicular bisectors of the sides meet in a point; (d) that the three points determined in (a), (b), and (c) lie on a line.

7. Given the triangle $A(0, 0)$, $B(a, 0)$, and $C(b, c)$. Find: (a) where the medians intersect; (b) where the altitudes meet; (c) where the perpendicular bisectors of the sides meet.

8. Prove that the three points found in Exercise 7 are collinear.

9. Given a triangle with vertices $(a, 0)$, $(b, 0)$, and $(0, c)$, where $a<b$, show that the altitudes meet in a point, the medians meet in a point, and the perpendicular bisectors meet in a point. Prove the three points so determined lie on a line.

10. Show that the lines whose equations are $ax+by-1=0$, $bx+ay-1=0$, and $y-x=0$ are concurrent.

11. Show that the line drawn through the mid-points of the parallel sides of a trapezoid passes through the point of intersection of the non-parallel sides.

12. Prove that the diagonals and the line drawn through the mid-points of the parallel sides of a trapezoid meet in a point.

3-15. Bisectors of angles

Consider the two lines whose equations are $a_1x+b_1y+c_1=0$ and $a_2x+b_2y+c_2=0$, where $\begin{vmatrix} a_1 & b_1 \\ a_2 & b_2 \end{vmatrix} \neq 0$. Our problem will be to determine the equations of the two bisectors of the angles formed by the two intersecting lines. Let $P=(x_1, y_1)$ be an arbitrary point on the bisector of one of the angles, as shown in Fig. 3–32. Let R be the projection of P on the line

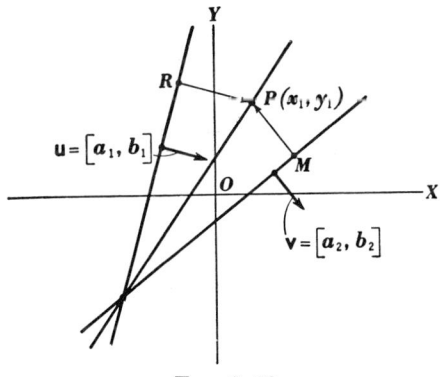

FIG. 3–32

$a_1x+b_1y+c_1=0$ and let M be the projection of P on the line $a_2x+b_2y+c_2=0$. Then $|\overrightarrow{MP}|=|\overrightarrow{RP}|$. In Section 3–12, we saw that, if $\overrightarrow{RP}$ is parallel in the same sense to the coefficient vector $\mathbf{u}=[a_1, b_1]$, then the number $\dfrac{a_1x_1+b_1y_1+c_1}{\sqrt{a_1^2+b_1^2}}$ is *positive*; while, if $\overrightarrow{RP}$ is parallel in the opposite sense to $\mathbf{u}$, the number is *negative*. Hence, if $\overrightarrow{RP}$ has the same direction as $\mathbf{u}$ and $\overrightarrow{MP}$ has the same direction as $\mathbf{v}$ (or if $\overrightarrow{RP}$ and $\overrightarrow{MP}$ have opposite directions to $\mathbf{u}$ and $\mathbf{v}$, respectively), we have:

$$\frac{a_1x_1+b_1y_1+c_1}{\sqrt{a_1^2+b_1^2}} = \frac{a_2x_1+b_2y_1+c_2}{\sqrt{a_2^2+b_2^2}} \qquad (3\text{-}31)$$

On the other hand, if $\overrightarrow{RP}$ is directed in the same sense as $\mathbf{u}$ and $\overrightarrow{MP}$ is directed in the opposite sense to $\mathbf{v}$, as in Fig. 3–32 (or vice versa), we have:

$$\frac{a_1x_1+b_1y_1+c_1}{\sqrt{a_1^2+b_1^2}} = -\frac{a_2x_1+b_2y_1+c_2}{\sqrt{a_2^2+b_2^2}} \qquad (3\text{-}32)$$

Since (x_1, y_1) is an arbitrary point on one of the bisectors, we drop the subscripts from x_1 and y_1 and obtain equations of the two bisectors to be

$$\frac{a_1x+b_1y+c_1}{\sqrt{a_1^2+b_1^2}} = \pm\frac{a_2x+b_2y+c_2}{\sqrt{a_2^2+b_2^2}} \qquad (3\text{-}33)$$

116 ANALYTIC GEOMETRY IN THE PLANE

We can prove that this equation represents a line. If it does not, both the coefficient of x and the coefficient of y are zero. Thus, using the negative sign in equation (3–33), we have:

(1) $$\frac{a_1}{\sqrt{a_1^2+b_1^2}} + \frac{a_2}{\sqrt{a_2^2+b_2^2}} = 0$$

(2) $$\frac{b_1}{\sqrt{a_1^2+b_1^2}} + \frac{b_2}{\sqrt{a_2^2+b_2^2}} = 0$$

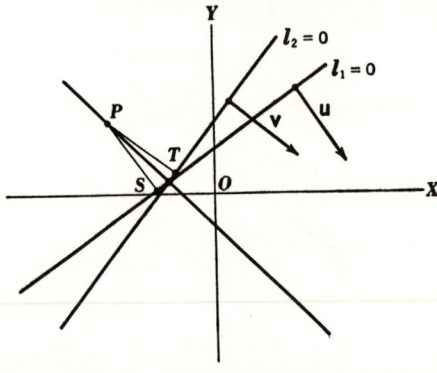

FIG. 3–33

Multiplying equation (1) by b_1 and multiplying equation (2) by a_1, and then subtracting the second result from the first, we obtain:

$$\frac{a_2 b_1}{\sqrt{a_2^2+b_2^2}} - \frac{a_1 b_2}{\sqrt{a_2^2+b_2^2}} = 0$$

or $$a_2 b_1 - a_1 b_2 \equiv \begin{vmatrix} a_1 & b_1 \\ a_2 & b_2 \end{vmatrix} = 0$$

Since this is contrary to our hypothesis, the bisector must exist.

The student should verify that, if we had chosen P in Fig. 3–32 on the other bisector, the equation would have been

$$\frac{a_1 x + b_1 y + c_1}{\sqrt{a_1^2+b_1^2}} = \frac{a_2 x + b_2 y + c_2}{\sqrt{a_2^2+b_2^2}}$$

EXAMPLE 3–26. Find an equation of the bisector of the angle that is included between the lines $l_1 \equiv 3x - 4y + 10 = 0$ and $l_2 \equiv 4x - 3y + 12 = 0$ and contains the origin.

Solution. The two lines are shown in Fig. 3–33. It will be found that the *numbers* $\dfrac{3x_1-4y_1+10}{5}$ and $\dfrac{4x_1-3y_1+12}{5}$ are both positive when (x_1, y_1) is the point $(0, 0)$. If P is a point on the bisector of the angle that contains the origin, either both vectors $\overrightarrow{SP}$ and $\overrightarrow{TP}$ have the same directions as **u** and **v**, respectively, or both vectors have the opposite directions. Hence, an equation of the bisector desired is

$$\frac{3x-4y+10}{5} = \frac{4x-3y+12}{5}$$

or
$$3x-4y+10 = 4x-3y+12$$

This reduces to
$$x+y+2=0$$

EXERCISES

1. Find equations of the bisectors of the angles formed by the lines whose equations are $x-2y+6=0$ and $3x+y+9=0$.

2. The origin lies in one of the regions formed by the lines $x-3y+3=0$ and $2x+4y-1=0$. Find an equation for the bisector of this angle.

3. Find an equation of the bisector of the obtuse angle between the lines $x-3=0$ and $3x-4y=0$.

4. Find the point of intersection of the bisectors of the interior angles of the triangle whose sides have the equations $3x-4y-4=0$, $4x-3y+5=0$, and $y=3$.

5. Find equations of bisectors of the angles formed by the lines $x+3y+6=0$ and $3x+y-4=0$.

6. Find an equation of the bisector of the acute angle between the lines $3x-4y+12=0$ and $3y-5=0$.

7. Given the lines $6x-3y+8=0$ and $2x-4y-11=0$, what is an equation of the bisector of the region between the lines which contain the origin?

8. Find an equation of the bisector of the obtuse angle between the lines $x+3y-9=0$ and $6x+2y-11=0$. What is the tangent of this angle?

9. Given the lines $x-2y+8=0$ and $6x+3y-10=0$, find an equation of the bisector of that region between the two lines that contain the point $(1, -6)$.

10. Find the coordinates of the point of intersection of the bisectors of the interior angles of the triangle whose sides are $3x+4y-3=0$, $3x-4y-3=0$, and $12x-5y+15=0$.

11. Prove that the bisectors of two adjacent angles of a parallelogram are perpendicular to each other.

12. Prove that the line passing through the vertex and the mid-point of the base of an isosceles triangle bisects the vertex angle.

13. Given the triangle with vertices at $P_1(1, 0)$, $P_2(-2, 4)$, and $P_3(-5, -8)$, prove that the bisector of the interior angle at P_1 divides the side P_2P_3 into segments whose lengths are proportional to $|P_1P_2|$ and $|P_1P_3|$.

14. Given the triangle with vertices $O(0, 0)$, $A(0, 3)$, $B(-4, 0)$, prove that the bisectors of the exterior angles of the triangle at A and B and the bisector of the interior angle at O intersect in a point.

15. Prove that the bisector of an exterior angle of an isosceles triangle, formed by producing one of the equal sides through the vertex, is parallel to the base.

3–16. Sets of lines

A single linear equation in x and y has been seen to represent a line. If it is desired to discuss analytically a *set* of lines, which set may consist of many individual lines, it is natural to represent each member of the set by its equation, thus obtaining a *set* of *linear equations*. When the set consists of but a few lines, the corresponding individual equations may be written and studied. This is what was done in Sections 3–13 and 3–14. However, some other device must be used when the set is large. In some cases, the totality of equations of the lines of the set may be written as a single equation containing an *arbitrary constant* or *parameter*.

Consider, for example, the equation $y = 3x + k$. If we interpret k to be arbitrary, capable of being assigned *any* real value, the single equation represents actually an entire set of equations, such as $y = 3x + 1$, $y = 3x - 3$,

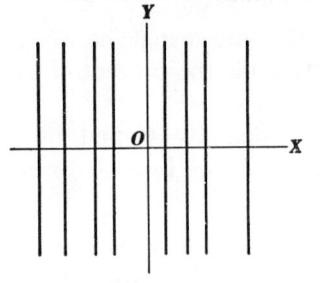

Fig. 3-34

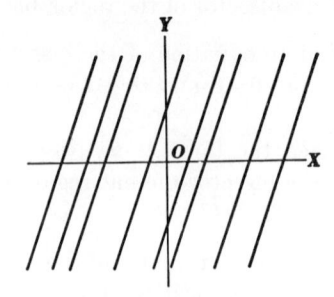

Fig. 3-35

etc. Since any line of slope 3 is represented by one of these equations, and since any such equation represents a line of slope 3, we say that $y = 3x + k$ represents the set of *all* lines of slope 3. See Fig. 3–35.

In general, any linear equation that contains one, and only one, arbitrary constant represents the *set* of lines having in common some specific geometric property. In the case of $y = 3x + k$, the common property is that of having slope 3. Similarly, $x = h$ represents the *set* of all vertical lines in the plane, as shown in Fig. 3–34, and $y = sx + 3$ represents the *set* of all non-vertical lines passing through (0, 3).

EXERCISES

1. Write an algebraic equation of the lines constituting each of the following sets:
   (a) Their slopes are 3.
   (b) They are non-horizontal and their x-intercepts are 4.
   (c) They are non-vertical and contain the point (2, −1).
   (d) They are parallel to the line $2x − 3y + 6 = 0$.
   (e) They are perpendicular to the line $x + 2y − 6 = 0$.
   (f) They are non-vertical and their y-intercepts are 2.
   (g) They are non-vertical and contain the origin.
   (h) They are parallel to the x-axis.
   (i) They have an inclination of 135°.
   (j) They are 6 units from the origin.

2. Identify the geometric property that characterizes each of the following systems of lines:
   (a) $y = sx − 2$
   (b) $y + 2 = s(x − 3)$
   (c) $y = k$
   (d) $kx + 2y = 2k$
   (e) $x = c$
   (f) $7x − by = 7b$
   (g) $kx + \sqrt{1 − k^2}\, y = 4$
   (h) $2x + by − 6 = 0$
   (i) $3x + 2y − k = 0$
   (j) $ax + by = 0$
   (k) $ax − 3y + 9 = 0$
   (l) $ax + ay = −4$

3. Write an algebraic equation of the lines constituting each of the following sets:
   (a) They are non-vertical and their y-intercepts are −1.
   (b) They are parallel to the line $3x − 4y + 6 = 0$.
   (c) They are perpendicular to the vector $\mathbf{u} = [2, −1]$.
   (d) They contain the vector $\mathbf{u} = [3, 2]$.
   (e) They are non-horizontal and their x-intercepts are −5.
   (f) They are 4 units from the origin.
   (g) They are non-vertical and contain the point (−2, 5).
   (h) Their slopes are −5.
   (i) They are perpendicular to the line $5x − 12y + 12 = 0$.

4. Identify the geometric property that characterizes each of the following sets of lines:

(a) $y = c$
(b) $2x + by - 4 = 0$
(c) $kx + ky + 5 = 0$
(d) $x = h$
(e) $2x - 3y + c = 0$
(f) $ax - 3y + 6 = 0$
(g) $y - sx + 5 = 0$
(h) $y + 3 = s(x + 1)$
(i) $ax - b = 0$
(j) $by + c = 0$
(k) $3x - 3y + h = 0$
(l) $ax - y = a$

3–17. Pencils of lines

The collection of *all* lines through a point is called a *pencil of lines*. In particular the set of lines through the intersection of two given lines is a pencil of lines. If we are given $l_1 \equiv a_1 x + b_1 y + c_1 = 0$ and $l_2 \equiv a_2 x + b_2 y + c_2 = 0$, where $\begin{vmatrix} a_1 & b_1 \\ a_2 & b_2 \end{vmatrix} \neq 0$, we can form the equation

(1) $$t_1(a_1 x + b_1 y + c_1) + t_2(a_2 x + b_2 y + c_2) = 0$$

where t_1 and t_2 are arbitrary constants and both are not zero. Since this equation is linear in x and y, it is an equation of a line. It follows that the coefficients $t_1 a_1 + t_2 a_2$ and $t_1 b_1 + t_2 b_2$ of x and y are not both zero; for, if $a_1 t_1 + a_2 t_2 = 0$ and $b_1 t_1 + b_2 t_2 = 0$, we have, after solving for t_1 and t_2 by determinants,

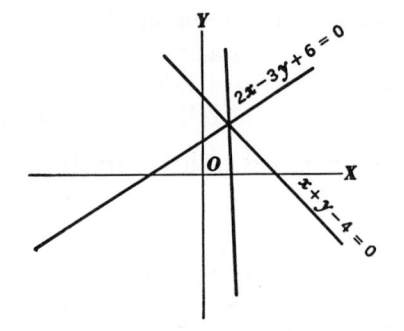

FIG. 3–36

(2) $$t_1 = -\frac{\begin{vmatrix} 0 & a_2 \\ 0 & b_2 \end{vmatrix}}{\begin{vmatrix} a_1 & a_2 \\ b_1 & b_2 \end{vmatrix}} = 0 \text{ and } t_2 = -\frac{\begin{vmatrix} a_1 & 0 \\ b_1 & 0 \end{vmatrix}}{\begin{vmatrix} a_1 & a_2 \\ b_1 & b_2 \end{vmatrix}} = 0$$

This is contrary to the assumption.

Furthermore, if x_1 and y_1 are the coordinates of the point of intersection of the lines $l_1 = 0$ and $l_2 = 0$, we know that $a_1 x_1 + b_1 y_1 + c_1 = 0$ and $a_2 x_1 + b_2 y_1 + c_2 = 0$. Hence, $t_1(a_1 x_1 + b_1 y_1 + c_1) + t_2(a_2 x_1 + b_2 y_1 + c_2) = 0$, regardless of the values of t_1 and t_2. To see that equation (1) includes *all* the lines through the point (x_1, y_1), let $l \equiv Ax + By + C = 0$ represent *any* line through (x_1, y_1).

Since the lines $a_1x+b_1y+c_1=0$, $a_2x+b_2y+c_2=0$, and $Ax+By+C=0$ are concurrent, it follows that

$$\begin{vmatrix} a_1 & b_1 & c_1 \\ a_2 & b_2 & c_2 \\ A & B & C \end{vmatrix} = 0$$

Now the three equations $a_1t_1+a_2t_2 = Ak$, $b_1t_1+b_2t_2 = Bk$, and $c_1t_1+c_2t_2 = Ck$ have solutions t_1, t_2, and k, provided all three are not zero. Moreover, if $k=0$, then the first two equations become $a_1t_1+a_2t_2 = 0$ and $b_1t_1+b_2t_2 = 0$; and $t_1 = t_2 = 0$, as we have seen. This contradiction shows that $k \neq 0$. Hence, $A = \frac{a_1t_1+a_2t_2}{k}$, $B = \frac{b_1t_1+b_2t_2}{k}$, and $C = \frac{c_1t_1+c_2t_2}{k}$; and equation (1) represents the line $l=0$, since the coefficients A, B, and C are proportional to those in equation (1). So equation (1) is an equation of the pencil of lines passing through the intersection of the two given lines.

EXAMPLE 3-27. Find an equation of the line passing through the intersection of the lines whose equations are $2x-3y+6=0$ and $x+y-4=0$ and containing the point $(2, -3)$.

Solution. The two given lines are shown in Fig. 3-36. An equation of the pencil of lines is

$$t_1(2x-3y+6) + t_2(x+y-4) = 0$$

Since the required line must contain the point $(2, -3)$, we have

$$t_1(4+9+6) + t_2(2-3-4) = 0$$

or

$$19t_1 - 5t_2 = 0$$

Hence, $\frac{t_1}{t_2} = \frac{5}{19}$. If we *choose* $t_1 = 5$ and $t_2 = 19$, our equation becomes

$$10x - 15y + 30 + 19x + 19y - 76 = 0$$

or

$$29x + 4y - 46 = 0$$

The student should check that $(2, -3)$ satisfies this equation.

EXAMPLE 3-28. Find equations of three lines passing through the intersection of the lines whose equations are $3x-4y+12=0$ and $x+3y-6=0$. The required lines are: (a) parallel to $2x+y-1=0$; (b) parallel to the x-axis; (c) parallel to the y-axis.

Solution. An equation of the pencil is

$$t_1(3x-4y+12) + t_2(x+3y-6) = 0$$

or

$$(3t_1+t_2)x + (3t_2-4t_1)y + 12t_1 - 6t_2 = 0$$

(a) Since we wish to find an equation of the line in the pencil that is

parallel to $2x+y-1=0$, we have
$$\frac{3t_1+t_2}{2} = \frac{3t_2-4t_1}{1}$$
or
$$3t_1+t_2 = 6t_2 - 8t_1$$
Hence,
$$11t_1 = 5t_2 \quad \text{or} \quad \frac{t_1}{t_2} = \frac{5}{11}$$

Choosing $t_1 = 5$ and $t_2 = 11$, we have
$$26x + 13y - 6 = 0$$
The student should check this result.

(b) In order that the line in the pencil may be parallel to the x-axis, we must have $3t_1 + t_2 = 0$. This gives $\frac{t_1}{t_2} = \frac{1}{-3}$. Choosing $t_1 = 1$ and $t_2 = -3$, we obtain
$$0x - 13y + 30 = 0$$
or
$$13y - 30 = 0$$
This is an equation of a line parallel to the x-axis.

(c) If the line in the pencil is parallel to the y-axis, we must have $3t_2 - 4t_1 = 0$. Solving for $\frac{t_1}{t_2}$, we obtain $\frac{t_1}{t_2} = \frac{3}{4}$. Choosing $t_1 = 3$ and $t_2 = 4$, we have
$$13x + 0y + 12 = 0$$
or
$$13x + 12 = 0$$
This is an equation of a line parallel to the y-axis.

EXAMPLE 3–29. Find an equation of the line that is in the pencil defined by $3x + 4y - 12 = 0$ and $x - 2y + 4 = 0$ and is perpendicular to the line $5x - 3y = 0$.

Solution. An equation of the pencil of lines is
$$t_1(3x + 4y - 12) + t_2(x - 2y + 4) = 0$$
We can rewrite this as
(1) $$(3t_1 + t_2)x + (4t_1 - 2t_2)y - 12t_1 + 4t_2 = 0$$
Since we want the line of the pencil that is perpendicular to $5x - 3y = 0$, we have
$$5(3t_1 + t_2) - 3(4t_1 - 2t_2) = 0$$
Expanding, we obtain
$$15t_1 + 5t_2 - 12t_1 + 6t_2 = 0$$

Hence, $3t_1 + 11t_2 = 0$ or $\dfrac{t_1}{t_2} = \dfrac{-11}{3}$

Choosing $t_1 = -11$ and $t_2 = 3$, and replacing these values in equation (1), we have
$$-30x - 50y + 144 = 0$$
or
$$15x + 25y - 72 = 0$$

The student should check that this line is perpendicular to $5x - 3y = 0$.

EXERCISES

1. Find an equation of the pencil of lines passing through the intersection of $3x - 4y + 12 = 0$ and $2x + y - 3 = 0$. Which line of the set contains the origin?

2. Find an equation of the pencil of lines passing through the intersection of the lines $x + y - 1 = 0$ and $2x - 3y + 6 = 0$. Determine the line of the pencil which passes through: (a) $(1, -1)$; (b) $(2, 3)$; (c) $(4, 0)$; (d) $(-2, 6)$.

3. Find an equation of each of the following lines:
   (a) Contains $P(1, -2)$ and the intersection of $x + y = 10$ and $3x - 2y = 12$.
   (b) Contains $P(5, 2)$ and the intersection of $2x - y = 10$ and $x + 3y = 0$.
   (c) Contains $P(0, 2)$ and the intersection of $x + 3 = 0$ and $2x + y = 12$.
   (d) Contains $P(-4, -2)$ and the intersection of $x + y - 1 = 0$ and $4x - 5y = 20$.

4. Find an equation of the line that passes through the intersection of the lines $2x - 3y + 6 = 0$ and $x + y - 4 = 0$ and also: (a) is parallel to $3x - 4y + 12 = 0$; (b) is perpendicular to $x - 3y + 6 = 0$; (c) is parallel to the x-axis; (d) is parallel to the y-axis.

5. Find an equation of the line that passes through the intersection of $3x - 4y + 8 = 0$ and $x - y + 6 = 0$ and also: (a) is parallel to $2x + 3y - 6 = 0$; (b) is perpendicular to $3x - 5y + 15 = 0$; (c) is parallel to the x-axis; (d) is parallel to the y-axis.

6. Find an equation of the line passing through the intersection of $2x - y + 6 = 0$ and $3x + 4y - 12 = 0$ and also: (a) having an x-intercept of -2; (b) having a y-intercept of 4; (c) containing the origin.

7. Show that an equation of the pencil of lines through the point (x_1, y_1) is $t_1(x - x_1) + t_2(y - y_1) = 0$.

8. Use the method of pencil of lines to find the point of intersection of the lines $a_1x + b_1y + c_1 = 0$ and $a_2x + b_2y + c_2 = 0$.

9. Find the locus of the centers of all circles of radius 4 which touch the line $5x - 12y + 12 = 0$.

10. A pencil of lines is determined by the lines $2x+y-3=0$ and $3x+4y-12=0$. Find which lines of the pencil satisfy the following conditions:

   (a) Contains the point $(2, -5)$.
   (b) Passes through $(0, 4)$.
   (c) Has an x-intercept 7.
   (d) Is parallel to $x-y+6=0$.
   (e) Is perpendicular to $5x-12y+12=0$.
   (f) Passes through the point $(-1, 1)$.
   (g) Is parallel to the x-axis.
   (h) Is perpendicular to the x-axis.

REVIEW EXERCISES

1. Verify that $x = x_1 + ld$ and $y = y_1 + md$ are parametric equations of a line containing the point (x_1, y_1) and having direction cosines l and m, where $|d|$ is the distance from the point (x_1, y_1) to any arbitrary point (x, y) on the line.

2. Find parametric equations of the line containing the point $(-1, 2)$ and having direction numbers 3 and -4.

3. Find parametric equations of the line containing the points $(2, -3)$ and $(-1, 4)$.

4. By eliminating the parameter, write an equation for each of the following lines in the form $ax + by + c = 0$:

   (a) $x = 2 - 3t$ and $y = 1 + t$ (c) $x = t$ and $y = 4 - 7t$
   (b) $x = 1 + 3t$ and $y = 2 + t$ (d) $x = t - 1$ and $y = 3t - 5$

5. Plot each of the lines in Exercise 4 directly from its parametric equations.

6. Find an equation for each of the lines through the following pairs of points:

   (a) $P_1(-1, 3)$ and $P_2(2, -4)$ (b) $P_1(5, -2)$ and $P_2(-1, -3)$

7. Determine an equation of each of the following lines:

   (a) Passes through $P_1(-3, -2)$ and has direction numbers $[4, -2]$
   (b) Passes through $P_1(4, -1)$ and has direction numbers $[-3, 2]$
   (c) Passes through $P_1(3, 5)$ and has direction numbers $[1, 0]$
   (d) Passes through $P_1(\frac{1}{2}, 4)$ and has direction numbers $[0, 1]$

8. Find direction cosines of each of the following lines:

   (a) $2x - 3y + 6 = 0$ (b) $x + 2y - 5 = 0$

9. Determine the slope of each of the following lines:

   (a) $3x + 4y - 6 = 0$ (b) $x - 3y + 4 = 0$

10. Write each of the following equations in the slope-intercept form, the intercept form, and the normal form:

    (a) $3x - 4y + 7 = 0$ (b) $5x + 12y - 60 = 0$

11. Find the scalar components of a vector perpendicular to each of the following lines:

    (a) $3x + 4y - 5 = 0$ (b) $2x - 3y + 6 = 0$ (c) $x - y - 2 = 0$

12. What is the slope of each line in Exercise 11?
13. What is the tangent of each of the interior angles of the triangle whose vertices are $P_1(1, -3)$, $P_2(4, 1)$ and $P_3(-1, 6)$?
14. Given the triangle with vertices $A(6, 4)$, $B(-1, 3)$, and $C(8, 0)$. Find: (a) an equation of AB; (b) an equation of the perpendicular bisector of AB; (c) an equation of the altitude through C on AB; (d) an equation of the line through C and parallel to AB; (e) a point P on BA extended, such that the ratio of $\overline{AP}$ to $\overline{PB}$ is $-1:3$; (f) an equation of the bisector of angle BAC; (g) the length of the altitude on AB; (h) the area of the triangle; (i) an equation of the median from A; (j) the tangent of angle CAB.
15. Given the triangle with vertices at $A(-1, -2)$, $B(1, 3)$, and $C(6, -2)$. Find the information asked for in Exercise 14.
16. Find the locus of centers of circles that touch the lines $3x - 4y + 20 = 0$ and $x = 0$. (Hint: The centers lie on the bisectors of the angles between the two lines.)
17. Find the locus of centers of circles that touch the lines $x - 4y + 6 = 0$ and $x + 4y - 12 = 0$. (Hint: See Exercise 16.)
18. Find the locus of a point equidistant from the points $P_1(-1, 4)$ and $P_2(3, 2)$.
19. Find an equation of the set of all lines which are 6 units from the origin.
20. (a) Find an equation of the pencil of lines passing through the intersection of the lines $x - 2y + 6 = 0$ and $3x - y + 2 = 0$.
    (b) Which line of the pencil contains the point $(-1, 2)$?
    (c) Which line of the pencil is parallel to the line $3x - 4y + 6 = 0$?
    (d) Which line of the pencil is perpendicular to the x-axis?
    (e) Which line of the pencil makes an angle of $45°$ with the x-axis?
    (f) Which line of the pencil is perpendicular to the line $5x - 2y + 10 = 0$?
21. After factoring, draw the graph of $x^2 - 4 = 0$.
22. After factoring, draw the graph of $x^2 - 2xy + y^2 - 9 = 0$.
23. Where are the points for which $x > 0$, $y > 0$, and $x + y < 2$?
24. In what ratio is the line segment with end points $(1, 4)$ and $(5, -2)$ divided by the point of intersection of that line segment and the line $6x + 7y = 0$?
25. Plot the line whose equations are $x = 2t - 3$ and $y = 4t + 4$.
26. Find equations of the two lines that pass through the point $(4, 5)$ and make with the x-axis as the base an isosceles triangle in which the base angles are $60°$.
27. Find the distance between the parallel lines whose equations are $2x - 3y + 6 = 0$ and $2x - 3y - 13 = 0$.
28. Find equations of the two lines that are parallel to the line whose equation is $3x + 4y - 12 = 0$ and are at a distance of 3 units from the given line.
29. Given the triangle whose sides have the equations $2x - y = 0$, $5x + 3y - 22 = 0$, and $x + 5y = 0$. Find: (a) the vertices; (b) the area of the triangle; (c) the cosines of the interior angles of the triangle; (d) equations of lines passing through the vertices and parallel and perpendicular to the opposite sides; (e)

equations of the perpendicular bisectors of the sides; (f) equations of the bisectors of the angles.

30. Find k if the lines $2x - 3y + k = 0$, $x - y + 4 = 0$, and $3x + 4y - 12 = 0$ are concurrent.

31. Find equations of the bisectors of the angles between the lines $3x + 4y - 12 = 0$ and $5x - 12y - 60 = 0$.

32. What are the center and the radius of the inscribed circle of the triangle whose sides have the equations $3x - 4y + 12 = 0$, $4x - 3y - 12 = 0$, and $5x + 12y - 60 = 0$? (Hint: Determine where the bisectors of the interior angles of the triangle intersect.)

33. Find equations of the lines which pass through the point (1, 4) and form with the axes triangles with an area of 9 square units.

34. Prove that, if (x_1, y_1), (x_2, y_2), and (x_3, y_3) are vertices of a triangle, the area of the triangle is the absolute value of
$$\frac{1}{2} \begin{vmatrix} x_1 & y_1 & 1 \\ x_2 & y_2 & 1 \\ x_3 & y_3 & 1 \end{vmatrix}$$

35. Represent by an equation the set of lines whose members are characterized by each of the following properties:
    (a) They are non-horizontal, and their x-intercepts are -2.
    (b) Their slopes are $\frac{1}{3}$.
    (c) They pass through the origin and are non-vertical.
    (d) They are 2 units from the origin.
    (e) They are non-vertical, and their y-intercepts are 5.
    (f) They are parallel to the line $x - 3y + 6 = 0$.
    (g) They are perpendicular to the line $2x + y - 4 = 0$.
    (h) They are non-vertical and pass through $(-1, 4)$.

36. In the pencil defined by the lines $2x - 3y - 1 = 0$ and $x + y - 3 = 0$, find an equation of the line that forms with the axes in the first quadrant a triangle having an area of 4 square units.

4

The Circle

4-1. Standard equation

A circle is the locus of all points and only those points which are equidistant from a fixed point. The constant distance is called the *radius*, and the fixed point is the *center* of the circle. If, in Fig. 4-1, h and k are

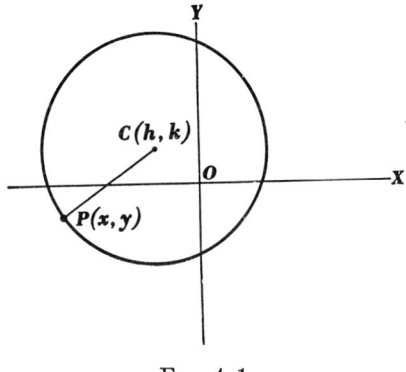

FIG. 4-1

the coordinates of the center C, and x and y are the coordinates of any point P at distance r from the fixed point, where r is the radius or $|\overrightarrow{CP}| = r$, we have

$$\sqrt{(x-h)^2 + (y-k)^2} = r$$

from which
$$(x-h)^2 + (y-k)^2 = r^2 \tag{4-1}$$

Conversely, if x_1 and y_1 are a pair of numbers which satisfy equation (4-1), then $P_1 = (x_1, y_1)$ is a point on the circle. This may be proved as follows: Since (x_1, y_1) satisfies the equation,

$$(x_1 - h)^2 + (y_1 - k)^2 = r^2$$

But this tells us that the magnitude of the vector $\mathbf{u} = \overrightarrow{CP_1}$ from $C = (h, k)$ to $P_1 = (x_1, y_1)$ is equal to r. Hence, we have $|\overrightarrow{CP_1}| = r$, and this is the condition that P_1 lie on a circle having its center at C and its radius equal to r.

127

128   ANALYTIC GEOMETRY IN THE PLANE

When equation (4–1) is given for a particular circle, the center and radius of the circle are known; and, when the center and radius of a circle are given, the equation of the circle is easily determined. We shall call equation (4–1) the *standard* equation of a circle.

When $h = k = 0$, the center is at the origin and our equation becomes

$$x^2 + y^2 = r^2 \qquad (4\text{–}2)$$

We shall call equation (4–2) the *simple* equation of a circle.

EXAMPLE 4–1. What is an equation of the circle having the point $(-2, 3)$ for its center and a radius of 2?

Solution. The circle is shown in Fig. 4–2. Using the standard equation of a circle, we have

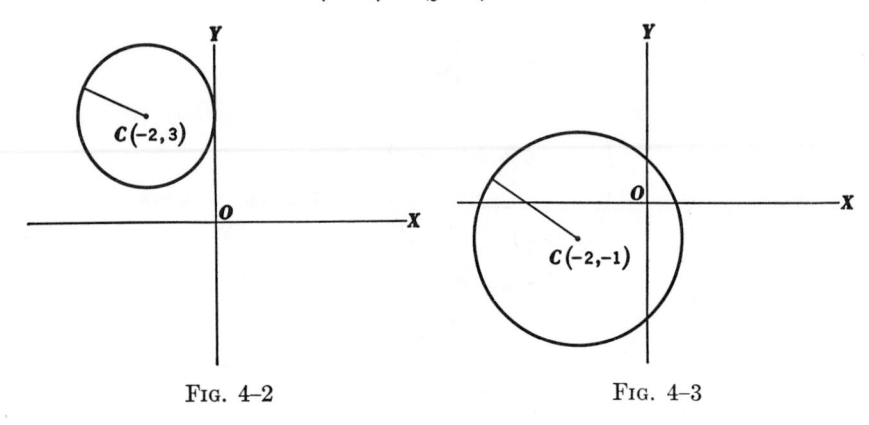

$$(x+2)^2 + (y-3)^2 = 4$$

FIG. 4–2   FIG. 4–3

EXAMPLE 4–2. What are the center and the radius of the circle whose equation is $(x+2)^2 + (y+1)^2 = 9$?

Solution. Since the standard equation is $(x-h)^2 + (y-k)^2 = r^2$, it follows that $h = -2$, $k = -1$, and $r = 3$. Hence, as shown in Fig. 4–3, the center is the point $C = (-2, -1)$ and the radius is 3.

EXAMPLE 4–3. What is an equation of the circle which has its center at $C = (-2, 1)$ and passes through $P = (1, 3)$?

Solution. Since $\overrightarrow{CP} = [3, 2]$, we have $r = |\overrightarrow{CP}| = \sqrt{13}$. Hence, by equation (4–1), an equation of the circle is

$$(x+2)^2 + (y-1)^2 = 13$$

EXAMPLE 4–4. Find an equation of the circle which has its center at the point $(2, -1)$ and which is tangent to the line $3x - 4y + 5 = 0$.

Solution. A *tangent* to a circle is perpendicular to the radius drawn to the point of tangency (see Section 4–5). Hence, the distance from the tangent to the given point is the length of the radius of the circle, and

$$r = \left|\frac{3(2)-4(-1)+5}{5}\right| = 3$$

The required equation of the circle is $(x-2)^2+(y+1)^2=9$.

EXERCISES

1. Find an equation of each of the following circles and draw the graph:
   (a) Center at $(2, -1)$ and $r=5$
   (b) Center at $(-3, 2)$ and $r=7$
   (c) Center at $(2, 4)$ and $r=1$
   (d) Center at $(-4, -1)$ and $r=1\frac{1}{2}$
   (e) Center at $(3, -2)$ and $r=6$
   (f) Center at $(-5, -6)$ and $r=7$

2. Find an equation of each of the following circles and draw the graph:
   (a) Center at $(-1, 3)$ and passing through $(2, -4)$
   (b) Center at $(0, 2)$ and passing through $(-1, -4)$
   (c) Center at $(3, 0)$ and passing through $(7, 3)$
   (d) Center at $(3, -3)$ and touching the axes

3. What is the equation of a circle if the end points of a diameter are $(2, 3)$ and $(-1, 5)$? Draw the circle.

4. Find an equation of each of the following circles and draw its graph:
   (a) Radius 6 and touching both axes
   (b) Radius 4 and touching both axes
   (c) Radius r and touching both axes

5. Find an equation of each of the following circles and draw its graph:
   (a) Center at $(6, 4)$ and touching the y-axis
   (b) Center at $(6, 4)$ and touching the x-axis
   (c) Center at $(-1, 3)$ and touching the line $3x-4y+12=0$

6. Find the center and radius of each of the following circles and draw the graph:
   (a) $(x-3)^2+(y-4)^2=16$
   (b) $(x+2)^2+(y+1)^2=25$
   (c) $(x-1)^2+(y+6)^2=8$
   (d) $x^2+(y+3)^2=9$
   (e) $(x+4)^2+y^2=25$
   (f) $x^2+y^2=7$

7. Find an equation of each of the following circles and draw its graph:
   (a) Radius 5 and touches both axes
   (b) Center at $(3, -1)$ and touching the x-axis
   (c) Center at $(-4, -2)$ and touching the y-axis
   (d) Center at $(-1, -4)$ and touching the line $4x+3y-6=0$

130 ANALYTIC GEOMETRY IN THE PLANE

8. Find an equation of the circle with the line segment joining the points $(-3, 0)$ and $(1, -4)$ as a diameter.

9. A circle has its center at $(2, -3)$ and is tangent to the line $3x + 4y - 4 = 0$. What is an equation of the circle?

10. Find equations of the circles tangent to the axes and passing through the point $(1, 2)$.

4–2. The general equation of a circle

If we expand the standard equation $(x-h)^2 + (y-k)^2 = r^2$, we obtain
$$x^2 + y^2 - 2hx - 2ky + h^2 + k^2 - r^2 = 0$$

This may be written as follows:

$$x^2 + y^2 + Dx + Ey + F = 0 \tag{4-3}$$

where $D = -2h$, $E = -2k$, and $F = h^2 + k^2 - r^2$. So we see that a circle is represented algebraically by an equation of the second degree having the form (4–3).

Now let us consider the equation

(1) $$A'x^2 + A'y^2 + D'x + E'y + F' = 0$$

where $A' \neq 0$. Dividing through by A', we have

(2) $$x^2 + y^2 + \frac{D'}{A'}x + \frac{E'}{A'}y + \frac{F'}{A'} = 0$$

If we let $\frac{D'}{A'} = D$, $\frac{E'}{A'} = E$, and $\frac{F'}{A'} = F$, we obtain

$$x^2 + y^2 + Dx + Ey + F = 0$$

Grouping the x-terms together and the y-terms together and completing their squares, we have

$$x^2 + Dx + \frac{D^2}{4} + y^2 + Ey + \frac{E^2}{4} = \frac{D^2}{4} + \frac{E^2}{4} - F$$

or

$$\left(x + \frac{D}{2}\right)^2 + \left(y + \frac{E}{2}\right)^2 = \frac{D^2 + E^2 - 4F}{4} \tag{4-4}$$

If $D^2 + E^2 - 4F > 0$, equation (4–4) is the standard equation of a circle, where $h = -\frac{D}{2}$, $k = -\frac{E}{2}$, and $r = \frac{\sqrt{D^2 + E^2 - 4F}}{2}$. There is, of course, no locus if $D^2 + E^2 - 4F < 0$; and the locus is only a point, namely, $\left(-\frac{D}{2}, -\frac{E}{2}\right)$, if $D^2 + E^2 - 4F = 0$. We agree to say, in all cases, that equation (4–3) represents a circle. Hence, it follows that a second-degree equation in which the coefficients of x^2 and y^2 are the same and in which there is no xy term represents a circle. We shall call equation (4–3) the *general* equation of a circle. The student must note particularly that the *general* equation of a circle is written so that the coefficients of x^2 and y^2 both are $+1$. Hence, we recognize a circle algebraically if the following conditions exist: The equation is of the second degree; the coefficients of x^2 and y^2 are equal; there is no xy term.

Furthermore, if $D^2+E^2-4F>0$, the graph of the circle exists; if $D^2+E^2-4F=0$, the circle is just a point, since the radius is zero (we shall call this a *point-circle*); and, if $D^2+E^2-4F<0$, the circle is said to be *imaginary*, since in this case r^2 would be equal to a negative number and no locus exists.

If $D=0$, equation (4–3) becomes

$$x^2+y^2+Ey+F=0 \qquad (4\text{–}5)$$

This is an equation of a circle having its center on the y-axis.

If $E=0$, equation (4–3) becomes

$$x^2+y^2+Dx+F=0 \qquad (4\text{–}6)$$

This is an equation of a circle having its center on the x-axis.

If $F=0$, equation (4–3) becomes

$$x^2+y^2+Dx+Ey=0 \qquad (4\text{–}7)$$

This is an equation of a circle that passes through the origin.

EXAMPLE 4–5. Find the center and the radius of the circle whose equation is $x^2+y^2-4x+6y-3=0$.

Solution. Completing the squares, we have

$$x^2-4x+4+y^2+6y+9=9+4+3$$

This can be written

$$(x-2)^2+(y+3)^2=16^*$$

Therefore, the center of the circle is at $(2, -3)$, and its radius is 4.

EXAMPLE 4–6. Find the center and the radius of the circle whose equation is $3x^2+3y^2-x+2y-4=0$.

Solution. Dividing through by 3, we get

$$x^2+y^2-\frac{x}{3}+\frac{2y}{3}=\frac{4}{3}$$

---

* If it becomes necessary to express y explicitly in terms of x, it can be done very conveniently from this form. Here, $y=-3+\sqrt{16-(x-2)^2}$ the upper half of the circle, and $y=-3-\sqrt{16-(x-2)^2}$ the lower half of the circle. Correspondingly x can be expressed explicitly in terms of y.

Completing the squares, we have

$$x^2 - \frac{x}{3} + \frac{1}{36} + y^2 + \frac{2}{3}y + \frac{1}{9} = \frac{4}{3} + \frac{1}{9} + \frac{1}{36}$$

or

$$\left(x - \frac{1}{6}\right)^2 + \left(y + \frac{1}{3}\right)^2 = \frac{48 + 4 + 1}{36} = \frac{53}{36}$$

Hence, the center of the circle is at $\left(\frac{1}{6}, -\frac{1}{3}\right)$ and the radius is $\frac{\sqrt{53}}{6}$.

EXERCISES

Find the center and the radius of each of the following circles:

1. $x^2 + y^2 - 2x + 4y + 1 = 0$
2. $x^2 + y^2 + 4x - 6y - 12 = 0$
3. $2x^2 + 2y^2 - 3x + 4y - 1 = 0$
4. $3x^2 + 3y^2 + 2x - 5y - 4 = 0$
5. $3x^2 + 3y^2 - 10 = 0$
6. $x^2 + y^2 + 4x + 16y + 12 = 0$
7. $x^2 + y^2 - 4x + 6y - 3 = 0$
8. $4x^2 + 4y^2 - 2x + 3y - 16 = 0$
9. $x^2 + y^2 - 2x + 3y = 0$
10. $x^2 + y^2 - 4x + 6y + 25 = 0$
11. $x^2 + y^2 - 2x + 4y + 5 = 0$
12. $x^2 + y^2 + x - y + 1 = 0$
13. $x^2 + y^2 + 6x - 2y + 15 = 0$
14. $2x^2 + 2y^2 + 3x - 5y + 17 = 0$

15. The point $(8, 5)$ bisects a chord of the circle whose equation is $x^2 + y^2 - 4x + 8y - 124 = 0$. What is an equation of this chord?

16. Find equations of circles having their centers on the line $3x - 5y = 8$ and tangent to the coordinate axes.

17. Find equations of the circles tangent to the axes and passing through the point $(2, 4)$.

18. In Exercises 1 to 7, express y in terms of x and interpret the resulting equation in relation to its graphical representation.

19. In Exercises 1 to 7, express x in terms of y and interpret the resulting equation in relation to its graphical representation.

4–3. Circle determined by three conditions

The standard equation of a circle contains the three arbitrary constants h, k, and r, and the general equation contains the arbitrary constants D, E, and F. To determine these constants, it is necessary to have three equations in the three unknowns D, E, and F or h, k, and r. Each equation will be obtained by some geometric condition that is imposed upon the circle. Hence, it is necessary to know *three* geometric conditions regarding a circle before its equation can be determined. The following examples illustrate the procedure.

EXAMPLE 4–7. Find an equation of the circle which passes through the three points $(-1, 1)$, $(2, 3)$, and $(1, -2)$.

Solution. Using the general form $x^2+y^2+Dx+Ey+F=0$, and applying the principle that the coordinates of any point on the circle satisfy the equation, we obtain:

(1) $\qquad\qquad 1+1-D+E+F=0$
(2) $\qquad\qquad 4+9+2D+3E+F=0$
(3) $\qquad\qquad 1+4+D-2E+F=0$

Using determinants, we find the unknowns as follows:

$$D = \frac{\begin{vmatrix} -2 & 1 & 1 \\ -13 & 3 & 1 \\ -5 & -2 & 1 \end{vmatrix}}{\begin{vmatrix} -1 & 1 & 1 \\ 2 & 3 & 1 \\ 1 & -2 & 1 \end{vmatrix}} = \frac{\begin{vmatrix} 2 & 1 & 1 \\ -11 & 2 & 0 \\ -3 & -3 & 0 \end{vmatrix}}{\begin{vmatrix} -1 & 0 & 0 \\ 2 & 5 & 3 \\ 1 & -1 & 2 \end{vmatrix}} = \frac{39}{-13} = -3$$

$$E = \frac{\begin{vmatrix} -1 & -2 & 1 \\ 2 & -13 & 1 \\ 1 & -5 & 1 \end{vmatrix}}{-13} = \frac{\begin{vmatrix} -1 & -2 & 1 \\ 3 & -11 & 0 \\ 2 & -3 & 0 \end{vmatrix}}{-13} = \frac{-9+22}{-13} = -1$$

$$F = \frac{\begin{vmatrix} -1 & 1 & -2 \\ 2 & 3 & -13 \\ 1 & -2 & -5 \end{vmatrix}}{-13} = \frac{\begin{vmatrix} -1 & 0 & 0 \\ 2 & 5 & -17 \\ 1 & -1 & -7 \end{vmatrix}}{-13} = \frac{-(-35-17)}{-13} = -4$$

Hence, the required equation is $x^2+y^2-3x-y-4=0$. The circle is shown in Fig. 4–4.

The student should check this result by substituting the coordinates of the given points in the equation of the circle.

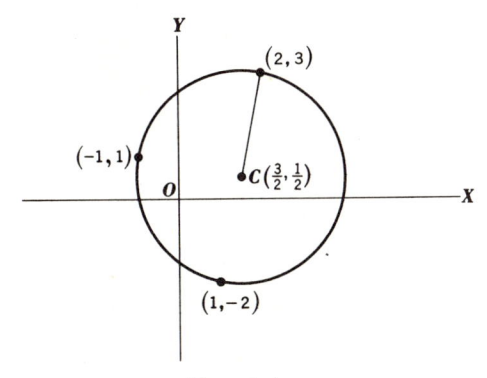

FIG. 4–4

EXAMPLE 4-8. Find an equation of the circle that contains the points $P_1(-1, 2)$ and $P_2(3, 0)$ and has its center on the line $x + 3y - 6 = 0$.

Solution. Using the standard equation of the circle, we have:

(1) $\quad\quad\quad\quad\quad (-1-h)^2 + (2-k)^2 = r^2$

(2) $\quad\quad\quad\quad\quad (3-h)^2 + (0-k)^2 = r^2$

(3) $\quad\quad\quad\quad\quad h + 3k - 6 = 0$

Expanding equations (1) and (2), we get:

(1') $\quad\quad\quad\quad 1 + 2h + h^2 + 4 - 4k + k^2 = r^2$

(2') $\quad\quad\quad\quad 9 - 6h + h^2 \quad\quad + k^2 = r^2$

Subtracting equation (2') from equation (1'), we obtain:

$$-8 + 8h + 4 - 4k = 0$$

or

(4) $\quad\quad\quad\quad\quad 2h - k = 1$

Solving equations (3) and (4) simultaneously, we find:

$$h = \frac{\begin{vmatrix} 6 & 3 \\ 1 & -1 \end{vmatrix}}{\begin{vmatrix} 1 & 3 \\ 2 & -1 \end{vmatrix}} = \frac{9}{7} \text{ and } k = \frac{\begin{vmatrix} 1 & 6 \\ 2 & 1 \end{vmatrix}}{\begin{vmatrix} 1 & 3 \\ 2 & -1 \end{vmatrix}} = \frac{11}{7}$$

So the coordinates of the center of the required circle are $\frac{9}{7}$ and $\frac{11}{7}$.

To find r, we substitute $h = \frac{9}{7}$ and $k = \frac{11}{7}$ in equation (2) and obtain

$r^2 = \left(3 - \frac{9}{7}\right)^2 + \left(0 - \frac{11}{7}\right)^2$, or $r = \frac{\sqrt{265}}{7}$. Hence, the standard equation of the circle is

$$\left(x - \frac{9}{7}\right)^2 + \left(y - \frac{11}{7}\right)^2 = \frac{265}{49}$$

The circle is shown in Fig. 4-5.

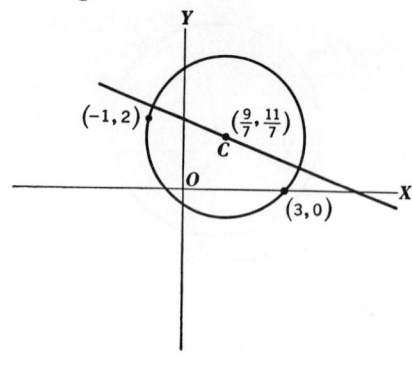

FIG. 4-5

EXAMPLE 4–9. Write in determinant form an equation of the circle through the three non-collinear points (x_1, y_1), (x_2, y_2), and (x_3, y_3).

Solution. Using determinant notation, we have

$$\begin{vmatrix} x^2+y^2 & x & y & 1 \\ x_1^2+y_1^2 & x_1 & y_1 & 1 \\ x_2^2+y_2^2 & x_2 & y_2 & 1 \\ x_3^2+y_3^2 & x_3 & y_3 & 1 \end{vmatrix} = 0$$

The student should explain how we know that this equation represents a circle which contains each of the three points. In particular, he should show that the common coefficient of x^2 and y^2 is not zero.

EXERCISES

1. Find an equation of the circle which passes through each of the following sets of three points:

   (a) $(2, 0)$, $(1, -3)$, and $(3, 1)$
   (b) $(1, -1)$, $(2, 1)$, and $(-3, 2)$
   (c) $(3, 2)$, $(-1, 1)$, and $(1, -1)$
   (d) $(3, 4)$, $(2, -1)$, and $(1, 2)$
   (e) $(0, 0)$, $(2, 0)$, and $(0, -4)$
   (f) $(1, 6)$, $(2, 5)$, and $(-6, 0)$
   (g) $(3, 2)$, $(2, -3)$, and $(-2, 4)$
   (h) $(1, -1)$, $(-1, 1)$, and $(0, 5)$

2. Find an equation of the circle that circumscribes the triangle whose vertices are the point $P_1(0, 0)$, $P_2(-1, 4)$, and $P_3(2, 6)$.

3. Find an equation of the circle that circumscribes the triangle whose sides are the lines $y = 0$, $x - y = 0$, and $x + 4y - 5 = 0$.

4. In each of the following problems, find an equation of each circle satisfying the given conditions:

   (a) The center is on the y-axis and the circle passes through the points $(1, 2)$ and $(-3, 1)$.
   (b) Its center is on the line $x + 2y - 6 = 0$ and it passes through the points $(5, 4)$ and $(1, -2)$.
   (c) Its center is on the line $x - y + 4 = 0$ and it touches both axes.
   (d) It is tangent to both axes and it passes through the point $(2, 1)$.
   (e) It is tangent to the line $3x - 4y - 31 = 0$ at the point $(1, -7)$ and its radius is 5.
   (f) It passes though the point $(1, -1)$ and its center is on the lines $x + 2y - 4 = 0$ and $2x - 3y + 13 = 0$.
   (g) It is concentric with the circle $x^2 + y^2 - 4x + 6y + 4 = 0$ and is tangent to the line $3x - 4y + 13 = 0$.
   (h) It is tangent to the line $x + 4y - 7 = 0$ at the point $(3, 1)$ and it passes through the point $(6, -2)$.

5. A circle touches the line $2x+y-7=0$ at the point $(4, -1)$ and passes through the point $(7, 2)$. Find its equation.

6. A circle touches the line $x+4y-5=0$ at the point $(1, 1)$ and passes through the point $(4, -4)$. Find its equation.

7. Find equations of the circles which touch the line $x+y=6$ at the point $(2, 4)$ and have a radius $\sqrt{2}$.

8. Find equations of the circles that touch the line $x-3y+5=0$ at the point $(-2, 1)$ and have a radius $3\sqrt{10}$.

9. A circle, having its center in the first quadrant, contains the points $(0, 9)$ and $(3, 10)$ and touches the x-axis. What is its equation?

10. Two circles pass through the origin, are tangent to the line $x+y=8$, and have their centers on the line $x=2$. What are equations of the circles?

11. Find an equation of the circle that touches the line $2x+3y=12$ and passes through the points $(-7, 0)$ and $(5, -8)$.

12. What are equations of the circles that are tangent to the line $2x-y+6=0$ at the point $(-1, 4)$ and have a radius $3\sqrt{5}$?

13. A circle is inscribed in the triangle formed by the coordinate axes and the line $3x-4y+12=0$. Find its equation.

14. Prove analytically that the perpendicular dropped from a point of a circle on a diameter is a mean proportional between the segments in which it divides the diameter.

15. Prove that the set of all mid-points of parallel chords of a circle lie on a diameter of the circle.

16. Describe the set of points for which $x^2+y^2=25$ and $x+y>7$. Are the points $(3, 4)$ and $(4, 3)$ elements of the set?

4-4. Symmetry

A curve (locus of points) is said to be *symmetric with respect to a point* C if, for every point P on the curve, the image Q of P with respect to C is also on the curve. The point C is the mid-point of the chord PQ and is called a *center of symmetry* of the curve. As shown in Fig. 4-6, the center C of a circle is a center of symmetry of the circle. It is the mid-point of any diameter, as PQ or $P'Q'$.

A curve is said to be *symmetric with respect to a line* l if, for every point P on the curve, its image Q with respect to the line l is also on the curve. The line is the perpendicular bisector of the chord PQ and is called an *axis of symmetry* of the curve. In Fig. 4-7, the line l is an axis of symmetry of the curve $P'PQQ'$. Any diameter of a circle is an axis of symmetry of the circle. In Fig. 4–6, the diameters PQ and $P'Q'$ are axes of symmetry.

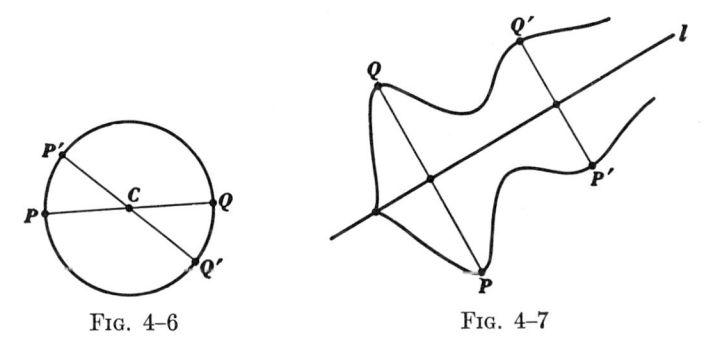

FIG. 4–6   FIG. 4–7

The symmetry of a curve with respect to the origin and to the coordinate axes may be ascertained by investigating the equation of the curve. If we represent the curve by the equation $f(x, y) = 0$, we may arrive at the following conclusions:

(a) A curve is symmetric with respect to the origin if, when we replace x by $-x$ and y by $-y$ in the original equation, we obtain an equation that is *equivalent** to the original equation. We know that $P = (x,y)$ and $Q = (-x, -y)$ are images of each other with respect to the origin. Hence, if $f(x, y) = 0$, we must have $f(-x, -y) = 0$. Conversely, if $f(-x, -y) = 0$, then $f(x, y) = f\{-(-x), -(-y)\} = 0$, and the equations $f(x, y) = 0$ and $f(-x, -y) = 0$ are equivalent.

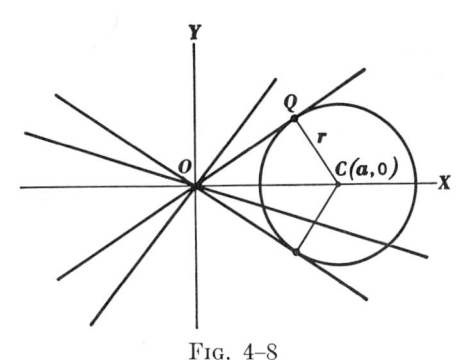

FIG. 4–8

(b) A curve is symmetric with respect to the x-axis provided $f(x, -y) = 0$ is equivalent to $f(x, y) = 0$. This follows since $P = (x, y)$ and $Q = (x, -y)$ are images with respect to the x-axis; and, if the coordinates of P satisfy the equation of the curve, then the coordinates of Q satisfy it also.

(c) A curve is symmetric with respect to the y-axis provided $f(-x, y) = 0$ is equivalent to $f(x, y) = 0$. This follows since $P = (x, y)$ and $Q = (-x,$

---
* Two equations are equivalent if they have the same solutions.

y) are images with respect to the y-axis; and, if the coordinates of P satisfy the equation of the curve, the coordinates of Q satisfy it also.

As an exercise the student should show that the circle $x^2+y^2=a^2$ is symmetric with respect to the x-axis, the y-axis, and the origin.

4–5. Tangents to a circle from an external point

Let $S = x^2+y^2+Dx+Ey+F = 0$ be an equation of a circle; and, as shown in Fig. 4–8, let the origin be a point outside the circle, that is, $|OC|>r$, where C is the center of the circle and r is its radius. A line through O may intersect the circle in two points, it may pass through one point on the circle, or it may have no contact with the circle. All three types of lines are shown in Fig. 4–8. When the line passes through only one point on the circle, or *touches* the circle in just one point, we say that *the line is tangent to the circle at this point*. Thus, the line OQ is tangent to the circle at the point Q.

We shall prove that *two* tangent lines may be drawn to a circle from an external point. Select the coordinate axes in such positions that the external point is the origin O and the center of the circle is on the x-axis at point $(a, 0)$, where $r<a$. Then an equation of the circle is

(1) $$(x-a)^2+y^2=r^2$$

Also, an equation of a line through the origin is

(2) $$y = sx$$

Combining equations (1) and (2), we obtain $(x-a)^2+s^2x^2=r^2$, or $x^2(1+s^2) - 2ax+a^2-r^2 = 0$. Solving for x, we have:

$$x = \frac{2a \pm \sqrt{4a^2-4(1+s^2)(a^2-r^2)}}{2(1+s^2)}$$

or
$$x = \frac{a \pm \sqrt{s^2(r^2-a^2)+r^2}}{1+s^2} \qquad (4\text{--}8)$$

If $s^2(r^2-a^2)+r^2>0$, the line intersects the circle in two distinct real points; if $s^2(r^2-a^2)+r^2<0$, there are no real points of intersection; and, if $s^2(r^2-a^2)+r^2=0$, the line touches the circle at one point. By definition, when $s^2(r^2-a^2)+r^2=0$, the line $y=sx$ is tangent to the circle. Hence, for tangency, the slope s of the line is $\pm \dfrac{r}{\sqrt{a^2-r^2}}$; and the lines whose equations are $y = \dfrac{rx}{\sqrt{a^2-r^2}}$ and $y = -\dfrac{rx}{\sqrt{a^2-r^2}}$ are tangent to the circle. It therefore follows that, from a point external to a given circle, two distinct tangents may be drawn to the circle.

If we substitute $\pm \dfrac{r}{\sqrt{a^2 - r^2}}$ for s in equation (4–8), we obtain $x = \dfrac{a^2 - r^2}{a}$. Substituting this expression for x in the equation $y = \pm \dfrac{rx}{\sqrt{a^2 - r^2}}$, we find that $y = \pm r \dfrac{\sqrt{a^2 - r^2}}{a}$. Hence, the coordinates of the point of tangency Q in Fig. 4–8 are

$$x = \frac{a^2 - r^2}{a}$$
$$y = \frac{r\sqrt{a^2 - r^2}}{a}$$
(4–9)

Also, $\overrightarrow{OQ} = \left[\dfrac{a^2 - r^2}{a}, \dfrac{r\sqrt{a^2 - r^2}}{a} \right]$ and $\overrightarrow{CQ} = \left[-\dfrac{r^2}{a}, \dfrac{r\sqrt{a^2 - r^2}}{a} \right]$. Hence, $\overrightarrow{OQ} \cdot \overrightarrow{CQ} = 0$, and we see that *a line tangent to a circle is perpendicular to the line containing the center of the circle and the point of tangency.*

4–6. Length of the tangent from a point outside a circle to the circle

If P is a point lying outside a given circle, as in Fig. 4–9, we know from the last section that it is possible to draw two tangents to the circle from this point. Our problem is to find the distance from P to the circle measured along either one of these tangents, as both tangents have the same length.

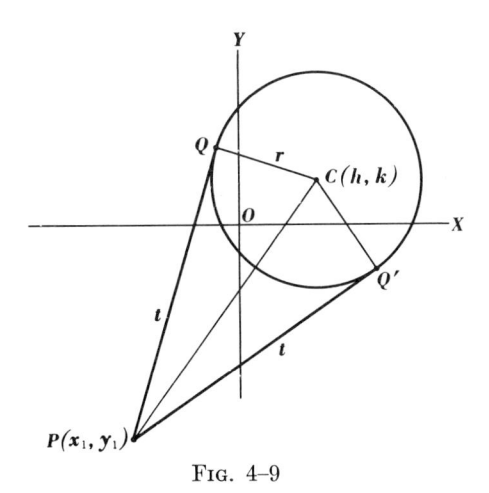

Fig. 4–9

Let $t = |QP|$, where Q is the point at which one of the tangents through P touches the circle. Using the right triangle CQP, we have

$$|\overrightarrow{QP}|^2 = |\overrightarrow{CP}|^2 - |\overrightarrow{CQ}|^2$$

or

$$t^2 = (x_1 - h)^2 + (y_1 - k)^2 - r^2 \qquad (4\text{–}10)$$

140 ANALYTIC GEOMETRY IN THE PLANE

Investigating this result carefully, we observe that, if the coordinates of P are substituted in the left side of the standard equation of the circle after *all* terms are placed on the left, we obtain the square of the distance measured from P along a tangent to the circle at Q. Since the standard equation becomes the *general* equation when expanded, we may write:

$$t^2 = x_1^2 + y_1^2 + Dx_1 + Ey_1 + F \qquad (4\text{--}11)$$

Hence, to find the square of the length of a tangent from a point to a circle, we substitute the coordinates of the point for x and y in the left side of either the standard equation or the general equation of the circle. If the result is positive, the point lies outside the circle; if t^2 is found to be negative, the point lies inside the circle and there are no tangents; and, if t^2 is found to be zero, the point is on the circle.

EXAMPLE 4–10. Find the length of a tangent from the point $(-4, 2)$ to the circle $(x-3)^2 + (y+2)^2 = 25$.

Solution. The conditions are shown in Fig. 4–10. We have:

$$t^2 = (-4-3)^2 + (2+2)^2 - 25 = 49 + 16 - 25 = 40$$

Therefore, $t = 2\sqrt{10}$.

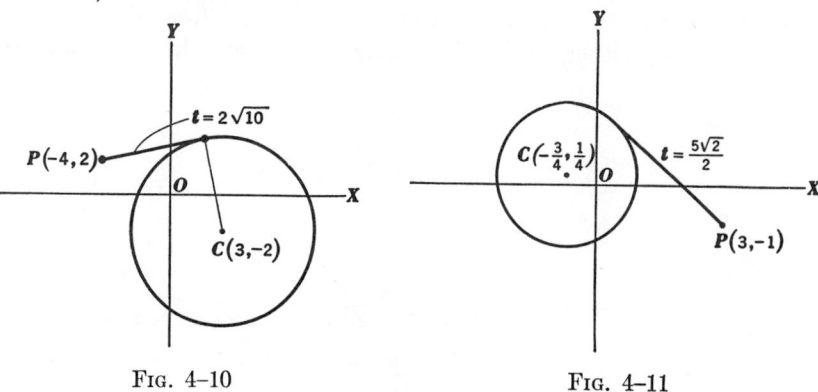

FIG. 4–10   FIG. 4–11

EXAMPLE 4–11. Find the length of a tangent from the point $(3, -1)$ to the circle $2x^2 + 2y^2 + 3x - y - 5 = 0$.

Solution. First, we divide every term by 2 in order to write the equation in the general form. Then we have

$$x^2 + y^2 + \frac{3}{2}x - \frac{1}{2}y - \frac{5}{2} = 0$$

The circle and one tangent are shown in Fig. 4–11. In this case,

$$t^2 = 9 + 1 + \frac{9}{2} + \frac{1}{2} - \frac{5}{2} = \frac{25}{2}$$

So
$$t = \frac{5}{\sqrt{2}} = \frac{5\sqrt{2}}{2}$$

EXERCISES

In each of the following, test to determine if the point lies on, inside, or outside the corresponding circle; and, if it lies outside, find the length of a tangent to the circle.

1. $(5, 2)$ and $x^2 + y^2 = 16$
2. $(-1, 3)$ and $x^2 + y^2 - x + y - 1 = 0$
3. $(3, -2)$ and $2x^2 + 2y^2 - 3x + y - 5 = 0$
4. $(1, -5)$ and $(x-3)^2 + y^2 = 16$
5. $(0, 0)$ and $(x-1)^2 + (y+1)^2 = 4$
6. $(3, 1)$ and $(x-2)^2 + (y+3)^2 = 17$

4-7. Radical axis

In order to solve simultaneously the equations of two circles which are not concentric, we proceed as follows: (1) Write the equations in the general form; (2) eliminate the second-degree terms; (3) solve the resulting linear equation with either of the second-degree equations.

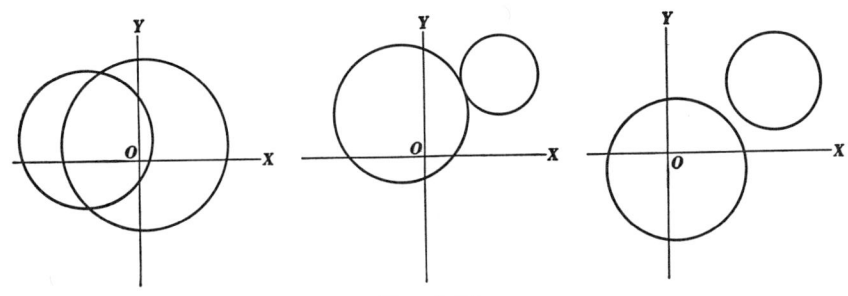

Fig. 4-12

If the solutions of the two equations are real and distinct, the circles intersect in two distinct points; if the solutions are real and equal, the circles have just one point in common or are tangent to each other at that point; and, if the solutions are imaginary, the circles have no points in common. The three conditions are illustrated in Fig. 4-12. However, in every case the linear equation resulting from eliminating the second-degree terms represents a line that is called the *radical axis* of the two circles. If the general equations of two circles are given as $x^2 + y^2 + D_1x + E_1y + F_1 = 0$ and $x^2 + y^2 + D_2x + E_2y + F_2 = 0$, the result obtained by subtracting the second equation from the first one is

$$(D_1 - D_2)x + (E_1 - E_2)y + F_1 - F_2 = 0$$

This is an equation of the radical axis of the two circles.

The radical axis has the important property that it contains every point from which tangents of equal length may be drawn to the two circles. To prove this, let $P_1 = (x_1, y_1)$, Fig. 4–13, be any point such that the tangents from it to any two circles $x^2 + y^2 + D_1 x + E_1 y + F_1 = 0$ and $x^2 + y^2 + D_2 x + E_2 y + F_2 = 0$ are equal in length. Then

$$t_1^2 = x_1^2 + y_1^2 + D_1 x_1 + E_1 y_1 + F_1$$

and

$$t_2^2 = x_1^2 + y_1^2 + D_2 x_1 + E_2 y_1 + F_2$$

Hence,

$$t_1^2 - t_2^2 = (D_1 - D_2)x_1 + (E_1 - E_2)y_1 + F_1 - F_2$$

Since $t_1 = t_2$, we have

$$(D_1 - D_2)x_1 + (E_1 - E_2)y_1 + F_1 - F_2 = 0$$

This tells us that the point (x_1, y_1) is on the radical axis.

Conversely, if $P_1 = (x_1, y_1)$ in Fig. 4–13 is a point on the radical axis which is outside both circles, then the distances t_1 and t_2 along the tangents from P_1 to the two circles are equal. Since $P_1 = (x_1, y_1)$ is on the radical

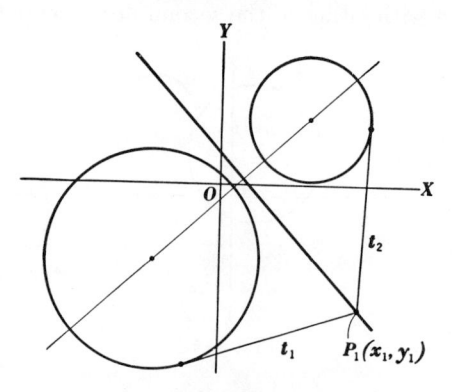

FIG. 4–13

axis, it follows that $(D_1 - D_2)x_1 + (E_1 - E_2)y_1 + F_1 - F_2 = 0$. Adding $x_1^2 + y_1^2$ to both sides of the equation, we have

$$x_1^2 + y_1^2 + (D_1 - D_2)x_1 + (E_1 - E_2)y_1 + F_1 - F_2 = x_1^2 + y_1^2$$

or

$$x_1^2 + y_1^2 + D_1 x_1 + E_1 y_1 + F_1 - (x_1^2 + y_1^2 + D_2 x_1 + E_2 y_1 + F_2) = 0$$

This tells us that $t_1^2 - t_2^2 = 0$. Hence, $t_1 = t_2$; and from *any* point on the radical axis tangents of equal length can be drawn to the two circles.

Another property of the radical axis of any two circles is that it is perpendicular to the line of centers of the two circles. The proof follows: Let the centers of the two circles be $\left(-\frac{D_1}{2}, -\frac{E_1}{2}\right)$ and $\left(-\frac{D_2}{2}, -\frac{E_2}{2}\right)$. Then, a pair of direction numbers of the line of centers is $\left[-\frac{D_2}{2}+\frac{D_1}{2}, -\frac{E_2}{2}+\frac{E_1}{2}\right]$. Also, scalar components of a vector perpendicular to the radical axis are $[D_1 - D_2, E_1 - E_2]$. So the line of centers is *parallel* to the vector perpendicular to the radical axis; and the radical axis is perpendicular to the line of centers.

If two circles intersect in two distinct points, the radical axis contains the common chord of the two circles; if the circles are tangent, the axis is the common tangent. If two circles are concentric, they have no radical axis.

EXAMPLE 4-12. Find an equation of the radical axis and the points of intersection of the two circles whose equations are:

(1) $$x^2 + y^2 + 8x + 12y - 28 = 0$$
(2) $$x^2 + y^2 - 4x - 12y + 20 = 0$$

Solution. Subtracting equation (2) from equation (1), we obtain

$$12x + 24y - 48 = 0$$

or

(3) $$x + 2y - 4 = 0$$

This is an equation of the radical axis.

To find where the circles intersect, we solve equation (3) with equation (1) and obtain:

$$(4 - 2y)^2 + y^2 + 8(4 - 2y) + 12y - 28 = 0$$

or $$16 - 16y + 4y^2 + y^2 + 32 - 16y + 12y - 28 = 0$$

From this, we obtain

(4) $$5y^2 - 20y + 20 = 0 \quad \text{or} \quad y^2 - 4y + 4 = 0$$

from which $$(y - 2)^2 = 0$$

Hence, both roots of equation (4) are $y = 2$.

When $y = 2$ in equation (3), $x = 0$. Since $(0, 2)$ is the only point on both the radical axis and the circle, it follows that the line is tangent to the circle at this point. Hence, in this case, the radical axis is the common tangent and the two circles touch each other at the point $(0, 2)$.

144   ANALYTIC GEOMETRY IN THE PLANE

EXAMPLE 4–13. Prove that the radical axes of three circles whose centers are not collinear intersect in a point.

Solution. Let the equations of the circles be:

(1) $$x^2+y^2+D_1x+E_1y+F_1=0$$
(2) $$x^2+y^2+D_2x+E_2y+F_2=0$$
(3) $$x^2+y^2+D_3x+E_3y+F_3=0$$

The radical axis of the circles designated by equations (1) and (2) is

(4) $$(D_1-D_2)x+(E_1-E_2)y+F_1-F_2=0$$

The radical axis of the circles given by equations (1) and (3) is

(5) $$(D_1-D_3)x+(E_1-E_3)y+F_1-F_3=0$$

and that for equations (2) and (3) is

(6) $$(D_2-D_3)x+(E_2-E_3)y+F_2-F_3=0$$

Since the centers are not collinear, the three radical axes are not parallel. If we form the pencil of lines defined by equations (4) and (5), we have:

$$k_1[(D_1-D_2)x+(E_1-E_2)y+F_1-F_2]+k_2[(D_1-D_3)x+(E_1-E_3)y+F_1-F_3]=0$$

If we select $k_1=-1$ and $k_2=1$, we obtain:

$$(D_2-D_3)x+(E_2-E_3)y+F_2-F_3=0$$

But this is the equation of the radical axis of the circles described by equations (2) and (3). Hence, the radical axes of three circles intersect in a point. Furthermore, if this point is outside the circles, it has the property that from it equal tangents can be drawn to the three circles.

EXAMPLE 4–14. Find the coordinates of the point from which equal tangents can be drawn to the three circles whose equations are:

(1) $$x^2+y^2+8y-20=0$$
(2) $$x^2+y^2-8x-48=0$$
(3) $$x^2+y^2-24x+128=0$$

Solution. The radical axis of the first two circles is

$$2x+2y=-7$$

and the radical axis of the first and third circles is

$$6x+2y=37$$

Solving these equations simultaneously, we obtain:

$$x = \frac{\begin{vmatrix} -7 & 2 \\ 37 & 2 \\ \hline 2 & 2 \\ 6 & 2 \end{vmatrix}}{} = \frac{-14-74}{-8} = 11$$

$$y = \frac{\begin{vmatrix} 2 & -7 \\ 6 & 37 \\ \hline 2 & 2 \\ 6 & 2 \end{vmatrix}}{} = \frac{74+42}{-8} = -\frac{29}{2}$$

In order to check this solution, we find the lengths of the tangents measured from the point $P = \left(11, -\frac{29}{2}\right)$ to each of the three circles. We have:

$$t_1^2 = 11^2 + \left(-\frac{29}{2}\right)^2 + 8\left(-\frac{29}{2}\right) - 20 = \frac{821}{4}$$

$$t_2^2 = 11^2 + \left(-\frac{29}{2}\right)^2 - 8(11) - 48 = \frac{821}{4}$$

$$t_3^2 = 11^2 + \left(-\frac{29}{2}\right)^2 - 24(11) + 128 = \frac{821}{4}$$

Since these three values are equal, $P = \left(11, -\frac{29}{2}\right)$ is the point from which equal tangents can be drawn to the three circles. Obviously, P lies outside the circles.

EXAMPLE 4–15. Show that two concentric circles do not have a radical axis.

Solution. Let (h, k) be the common center of two circles whose radii are r_1 and r_2, respectively, where $r_1 \neq r_2$. Then equations of the two circles are $(x-h)^2 + (y-k)^2 = r_1^2$ and $(x-h)^2 + (y-k)^2 = r_2^2$. To obtain the radical axis, we must eliminate the second-degree terms. However, in this case, the elimination of the square terms also eliminates the linear terms; and we are left with the condition $r_1^2 - r_2^2 = 0$, which does not represent a line. Hence, two concentric circles do not possess a radical axis.

EXERCISES

1. Find the radical axis of each of the following pairs of circles:
   (a) $x^2 + y^2 - 2x + 3y - 4 = 0$
   $x^2 + y^2 + x - 2y - 3 = 0$
   (b) $3x^2 + 3y^2 = 4$
   $x^2 + y^2 - x + 2y - 2 = 0$
   (c) $2x^2 + 2y^2 - x + 3y - 1 = 0$
   $x^2 + y^2 - 2x + 16y + 3 = 0$
   (d) $(x-2)^2 + (y+3)^2 = 4$
   $x^2 + y^2 = 25$

2. Find the points of intersection of each of the following pairs of circles:

(a) $(x-3)^2 + (y+2)^2 = 4$
$x^2 + y^2 = 25$
(b) $x^2 + y^2 + 6x - 16 = 0$
$x^2 + y^2 - 10y + 16 = 0$
(c) $x^2 + y^2 + 2x - 4y - 20 = 0$
$x^2 + y^2 - 6x + 2y - 40 = 0$
(d) $x^2 + y^2 + 6x - 91 = 0$
$x^2 + y^2 + 4x - 2y - 93 = 0$
(e) $x^2 + y^2 + 10x + 8y - 39 = 0$
$x^2 + y^2 - 2x - 16y = 45$
(f) $x^2 + y^2 + 10x + 4y + 9 = 0$
$x^2 + y^2 - 2x + 8y - 35 = 0$

3. Find a point from which tangents of equal length can be drawn to each of the pairs of circles of Exercise 1.

4. Find the point from which tangents of equal length can be drawn to the three circles in each of the following groups:

(a) $x^2 + y^2 + 12x + 24 = 0$
$x^2 + y^2 - 16x - 12y - 64 = 0$
$x^2 + y^2 - 64 = 0$
(b) $x^2 + y^2 + 3x - 6y - 1 = 0$
$2x^2 + 2y^2 - x + 2y + 2 = 0$
$x^2 + y^2 - 3y = 0$

5. Prove that the radical axis of two circles contains any common points of the two circles.

6. Prove that, if two circles are tangent to each other, the radical axis is the common tangent line.

4–8. Equation of the tangent to the circle $x^2 + y^2 = r^2$ at the point $P_1(x_1, y_1)$ on the circle

Vector $\overrightarrow{OP_1}$, Fig. 4–14, is perpendicular to vector $\overrightarrow{P_1P}$, where P is any point on the tangent to the circle through P_1. Forming the dot product of $\overrightarrow{OP_1}$ and $\overrightarrow{P_1P}$, we have:

$$\overrightarrow{OP_1} \cdot \overrightarrow{P_1P} = x_1(x - x_1) + y_1(y - y_1) = 0$$

or
$$x_1 x + y_1 y = x_1^2 + y_1^2$$

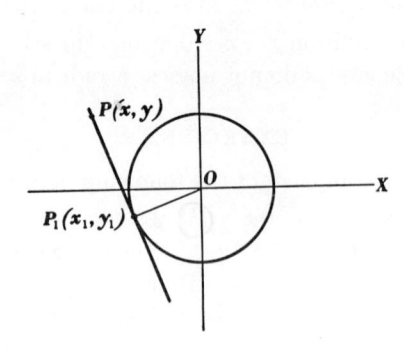

Fig. 4–14

Since the point (x_1, y_1) is on the circle, it follows that $x_1^2 + y_1^2 = r^2$. Also, since $\overrightarrow{P_1P}$ lies along the tangent to the circle at P_1, an equation of the tangent is

$$x_1 x + y_1 y = r^2 \qquad (4\text{-}13)$$

EXAMPLE 4-16. Find an equation of the tangent to the circle $x^2 + y^2 = 25$ at the point $(-3, 4)$.

Solution. By equation (4-13), we have for the required equation of the tangent:

$$-3x + 4y = 25$$

or

$$3x - 4y + 25 = 0$$

The *normal* to a circle at a point $P_1(x_1, y_1)$ on the circle is defined to be the line through P_1 perpendicular to the tangent at the point. Hence, an equation of the normal to the circle $x^2 + y^2 = r^2$ at the point (x_1, y_1) is

$$y_1 x - x_1 y = y_1 x_1 - x_1 y_1 = 0 \qquad (4\text{-}14)$$

Since the coordinates of the center of the circle satisfy this equation, it follows that all normals to a circle *pass through the center of the circle.*

In Example 4-16, an equation of the tangent to the circle $x^2 + y^2 = 25$ at the point $(-3, 4)$ was found to be $3x - 4y + 25 = 0$. Hence, an equation of the normal of the circle at that point is

$$4x + 3y = 0$$

As a check, $4(-3) + 3(4) = 0$.

EXERCISES

1. Find equations of the tangent and the normal to each of the following circles at the point indicated:

   (a) $x^2 + y^2 = 16$ and $(3, \sqrt{7})$
   (b) $x^2 + y^2 = 5$ and $(-1, 2)$
   (c) $x^2 + y^2 = 25$ and $(0, -5)$
   (d) $x^2 + y^2 = 13$ and $(2, -3)$
   (e) $x^2 + y^2 = 41$ and $(-4, -5)$
   (f) $x^2 + y^2 = 9$ and $(3, 0)$

2. Prove that an equation of the tangent to the circle $(x-h)^2 + (y-k)^2 = r^2$ at the point (x_1, y_1) is

   $$(x_1 - h)(x - h) + (y_1 - k)(y - k) = r^2$$

3. Find equations of the tangent and the normal to each of the following circles at the point indicated:

   (a) $(x-3)^2 + (y+4)^2 = 25$ and $(3, 1)$
   (b) $(x+1)^2 + (y-2)^2 = 34$ and $(2, -3)$

(c) $(x+2)^2+(y-1)^2=10$ and $(-3, -2)$
(d) $(x+2)^2+(y-1)^2=25$ and $(1, 5)$
(e) $(x-2)^2+y^2=2$ and $(3, -1)$
(f) $(x-5)^2+(y+7)^2=97$ and $(1, 2)$

4. Prove that an equation of the tangent to the circle $x^2+y^2+Dx+Ey+F=0$ at the point (x_1, y_1) is

$$x_1x+y_1y+\frac{D(x_1+x)}{2}+\frac{E(y_1+y)}{2}+F=0$$

5. Find equations of the tangent and the normal to each of the following circles at the point indicated:

(a) $x^2+y^2+2x-6y-10=0$ and $(1, -1)$
(b) $2x^2+2y^2-3x+4y-97=0$ and $(-3, -7)$
(c) $x^2+y^2-2x+y-6=0$ and $(2, -3)$
(d) $x^2+y^2+3x-4y+5=0$ and $(-1, 1)$
(e) $x^2+y^2-5x+6y-41=0$ and $(-2, 3)$
(f) $x^2+y^2-25=0$ and $(0, -5)$

4–9. Equation of the tangent to the circle $x^2+y^2=r^2$ when the slope of the tangent is known

Consider the line $y=sx+k$ to be tangent to the circle $x^2+y^2=r^2$. Since a circle is symmetric with respect to its center, there must be two such lines parallel to each other that are tangent to the circle. The conditions are represented in Fig. 4–15.

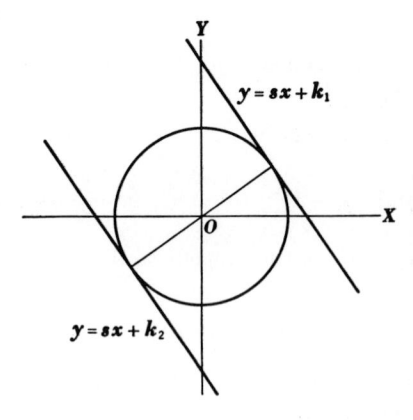

FIG. 4–15

Our problem is to determine k in terms of s. Solving the equation of the line simultaneously with that of the circle, we obtain:

$$x^2 + (sx+k)^2 = r^2$$

or

(1) $$x^2(1+s^2) + 2ksx + k^2 - r^2 = 0$$

from which $$x = \frac{-2ks \pm \sqrt{4k^2s^2 - 4(1+s^2)(k^2-r^2)}}{2(1+s^2)}$$

or $$x = \frac{-2ks \pm \sqrt{-4k^2 + 4(1+s^2)r^2}}{2(1+s^2)} \tag{4-15}$$

In order that the line $y = sx+k$ may be tangent to the circle $x^2+y^2 = r^2$, there must be only one point common to both the line and the circle. Hence, the quadratic expression in equation (1) must be a perfect square. In order for this to be so, the expression under the radical sign in equation (4-15) must be zero. So we have:

$$4k^2 = 4(1+s^2)r^2$$

or $$k = \pm r\sqrt{1+s^2} \tag{4-16}$$

It therefore follows that equations of the two lines with slope s that are tangent to the circle $x^2+y^2 = r^2$ are:

$$y = sx + r\sqrt{1+s^2}$$
$$y = sx - r\sqrt{1+s^2} \tag{4-17}$$

As s varies, these equations represent two sets of lines that are tangent to the circle. There are two tangents to the circle which are not included in the sets designated by equations (4-17). They are the vertical tangents at the points $(r, 0)$ and $(-r, 0)$; they are not included since their slopes are not defined. However, equations of these tangents are readily found to be $x = r$ and $x = -r$.

EXAMPLE 4-17. What are equations of the lines that are tangent to the circle $x^2+y^2 = 25$ and have a slope equal to $-\frac{4}{3}$?

Solution. By equations (4-17), equations of the tangents are

$$y = -\frac{4}{3}x + 5\sqrt{1+\frac{16}{9}} \quad \text{and} \quad y = -\frac{4}{3}x - 5\sqrt{1+\frac{16}{9}}$$

When simplified these become:

$$4x + 3y - 25 = 0 \quad \text{and} \quad 4x + 3y + 25 = 0$$

The circle and the tangents are shown in Fig. 4-16. To draw the tangents, construct the diametral line having direction numbers [4, 3];

150   ANALYTIC GEOMETRY IN THE PLANE

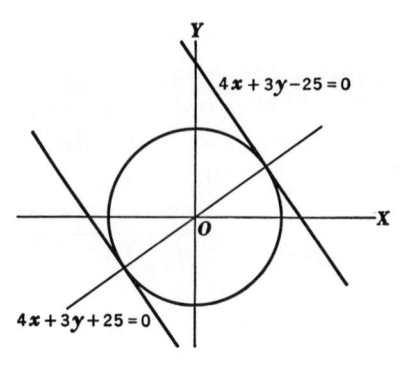

FIG. 4-16

and, at the points where this line cuts the circle, draw the tangents perpendicular to the diametral line.

EXERCISES

1. Find equations of the tangents to each of the following circles, the tangents having the slopes indicated:

   (a) $x^2 + y^2 = 25$ and $s = \dfrac{3}{4}$
   (b) $x^2 + y^2 = 16$ and $s = 2$
   (c) $x^2 + y^2 = 4$ and $s = \sqrt{15}$
   (d) $x^2 + y^2 = 64$ and $s = \dfrac{4}{3}$
   (e) $x^2 + y^2 = 5$ and $s = -2$
   (f) $x^2 + y^2 = 10$ and $s = 3$

2. Find equations of the lines that are tangent to each of the following circles and are parallel to the given line:

   (a) $x^2 + y^2 = 25$ and $3x - 4y + 12 = 0$
   (b) $x^2 + y^2 = 9$ and $5x + 12y = 0$
   (c) $x^2 + y^2 = 16$ and $8x - 15y - 20 = 0$
   (d) $x^2 + y^2 = 1$ and $x = 5$
   (e) $x^2 + y^2 = 64$ and $y = 5$
   (f) $x^2 + y^2 = 49$ and $2x - 3y + 6 = 0$

3. Find equations of the lines that are tangent to each of the following circles and are perpendicular to the given line:

   (a) $x^2 + y^2 = 64$ and $4x - 3y - 12 = 0$
   (b) $x^2 + y^2 = 289$ and $15x + 8y = 0$
   (c) $x^2 + y^2 = 225$ and $5x + 12y - 60 = 0$
   (d) $x^2 + y^2 = 25$ and $x = 10$
   (e) $x^2 + y^2 = 81$ and $y = 24$
   (f) $x^2 + y^2 = 9$ and $2x + 3y - 6 = 0$

THE CIRCLE   151

4. Prove that equations of the lines that are tangent to the circle $(x-h)^2 + (y-k)^2 = r^2$ and have slope s are
$$y - k = s(x - h) \pm r\sqrt{1+s^2}$$

5. Find equations of the tangents to each of the following circles, the tangents having the slope indicated:
   (a) $x^2 + y^2 - 6x - 36 = 0$ and $s = \frac{1}{2}$.
   (b) $x^2 + y^2 + 8y - 42 = 0$ and $s = -\frac{3}{7}$.
   (c) $x^2 + y^2 + 10x - 4y - 87 = 0$ and $s = -\frac{5}{2}$.

4–10. Pencil of circles

We have already seen that three geometric conditions usually determine a circle (or a *finite* number of circles). When only two conditions are specified, a *set* of circles, usually infinite, is determined. Such a set of circles may often be represented by an equation containing one or more arbitrary constants. For example, the equation $x^2 + y^2 = r^2$ with arbitrary r represents the set of all circles with center at the origin, as shown in Fig. 4–17. The equation $(x+3)^2 + (y-4)^2 = r^2$ with arbitrary r represents the set of all circles with center at the point $(-3, 4)$, as shown in Fig. 4–18; and the equation $(x-h)^2 + (y-3)^2 = 4$ with arbitrary h represents the set of all circles with center on the line $y = 3$ and having a radius of 2, as shown in Fig. 4–19.

Consider now the equation

(1) $$t_1(x^2 + y^2 + D_1x + E_1y + F_1) + t_2(x^2 + y^2 + D_2x + E_2y + F_2) = 0$$

where $x^2 + y^2 + D_1x + E_1y + F_1 = 0$ and $x^2 + y^2 + D_2x + E_2y + F_2 = 0$ are the equations of two distinct circles and both t_1 and t_2 are not zero.

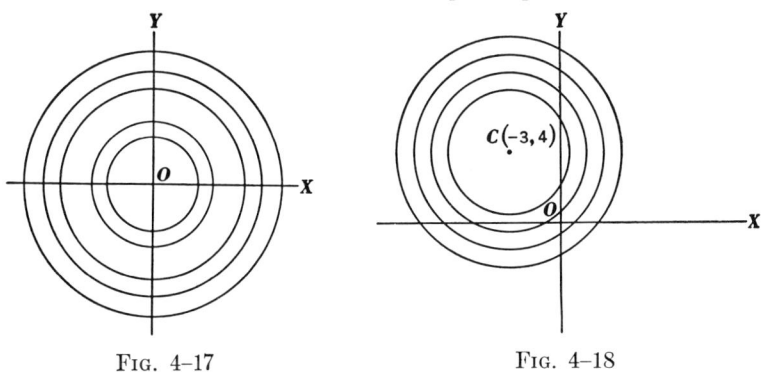

FIG. 4–17   FIG. 4–18

Except for the case when $t_1 = -t_2$, equation (1) represents a circle, since it is an equation of the second degree, it contains no xy term, and the

152   ANALYTIC GEOMETRY IN THE PLANE

coefficients of x^2 and y^2 are equal. Since both t_1 and t_2 are not zero, only *one* arbitrary constant, which is either $\dfrac{t_1}{t_2}$ or $\dfrac{t_2}{t_1}$, appears in equation (1). Hence, that equation represents a set of circles. We shall call such a set a *pencil of circles*.

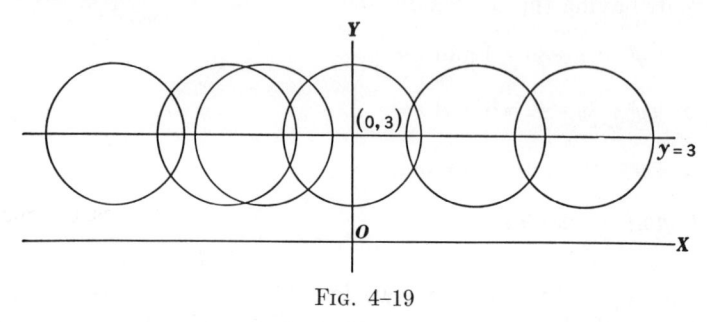

FIG. 4-19

When $t_1 = -t_2$, equation (1) reduces to $(D_2 - D_1)x + (E_2 - E_1)y + F_2 - F_1 = 0$, which is an equation of the radical axis of the two circles. In the discussion immediately following, we shall exclude this case since the radical axis has already been treated.

If the two given circles are concentric, the pencil represented by equation (1) is a set of concentric circles. To see this, we write the equations of the two given circles as $x^2 + y^2 + D_1x + E_1y + F_1 = 0$ and $x^2 + y^2 + D_1x + E_1y + F_2 = 0$. Then equation (1) becomes $t_1(x^2 + y^2 + D_1x + E_1y + F_1) + t_2(x^2 + y^2 + D_1x + E_1y + F_2) = 0$. After collecting like terms and dividing through by $t_1 + t_2$, we obtain:

$$x^2 + y^2 + D_1x + E_1y + \frac{t_1F_1 + t_2F_2}{t_1 + t_2} = 0 \qquad (4\text{--}18)$$

This is an equation of a pencil of concentric circles.

If the two given circles intersect, every circle in the pencil contains the points of intersection. To see this, let a point of intersection of the two circles be (x_1, y_1). Then we know that $x_1^2 + y_1^2 + D_1x_1 + E_1y_1 + F_1 = 0$ and $x_1^2 + y_1^2 + D_2x_1 + E_2y_1 + F_2 = 0$. Hence, when we substitute x_1 for x and y_1 for y in equation (1), we obtain $t_1(0) + t_2(0) = 0$ for every choice of t_1 and t_2. Thus, every circle of the pencil passes through any point of intersection of two of the circles of the pencil.

If the two given circles are tangent to each other at a point P, all the circles of the pencil are tangent at P.

If the two given circles are not concentric, every two distinct circles in the pencil have a common radical axis, regardless of whether the given circles intersect or not. To prove this last statement, we shall write an equation for the radical axis of any pair of circles of the pencil and show that it is the same as the radical axis for the two given circles. Let the equations of any two circles of the pencil be:

(2) $\quad t_1'(x^2+y^2+D_1x+E_1y+F_1)+t_2'(x^2+y^2+D_2x+E_2y+F_2)=0$

and

(3) $\quad t_1''(x^2+y^2+D_1x+E_1y+F_1)+t_2''(x^2+y^2+D_2x+E_2y+F_2)=0$

where t_1'' and t_2'' represent a *second* pair of numbers such that

$$\begin{vmatrix} t_1' & t_2' \\ t_1'' & t_2'' \end{vmatrix} \neq 0$$

This last condition is necessary, since we want equations (2) and (3) to represent distinct circles. If $\begin{vmatrix} t_1' & t_2' \\ t_1'' & t_2'' \end{vmatrix} = 0$, it follows that $t_1't_2''-t_1''t_2'=0$. If $t_1'=0$, then $t_1''=0$; or, if $t_2'=0$, then $t_2''=0$. In either case, equations (2) and (3) represent the same circle. If t_1', t_2', t_1'', and t_2'' are not 0, we can write $\dfrac{t_2'}{t_1'}=\dfrac{t_2''}{t_1''}$; and again equations (2) and (3) represent the same circle. So, in order that equations (2) and (3) may represent distinct circles, we must have $\begin{vmatrix} t_1' & t_2' \\ t_1'' & t_2'' \end{vmatrix} \neq 0$.

Expanding equations (2) and (3) and collecting the terms containing x^2, y^2, x, and y in each equation, we obtain:

(2′) $\quad (t_1'+t_2')(x^2+y^2)+(t_1'D_1+t_2'D_2)x+(t_1'E_1+t_2'E_2)y+t_1'F_1+t_2'F_2=0$

and

(3′) $\quad (t_1''+t_2'')(x^2+y^2)+(t_1''D_1+t_2''D_2)x+(t_1''E_1+t_2''E_2)y+t_1''F_1+t_2''F_2=0$

Dividing equation (2′) by $t_1'+t_2'$ and dividing equation (3′) by $t_1''+t_2''$, and subtracting the second result from the first, we obtain an equation of the radical axis for any pair of circles:

(4) $\quad \left(\dfrac{t_1'D_1+t_2'D_2}{t_1'+t_2'}-\dfrac{t_1''D_1+t_2''D_2}{t_1''+t_2''}\right)x+\left(\dfrac{t_1'E_1+t_2'E_2}{t_1'+t_2'}-\dfrac{t_1''E_1+t_2''E_2}{t_1''+t_2''}\right)y$

$\quad +\dfrac{t_1'F_1+t_2'F_2}{t_1'+t_2'}-\dfrac{t_1''F_1+t_2''F_2}{t_1''+t_2''}=0$

We can save considerable algebra, if we notice that the coefficients of x and y and the constant term are alike in form but involve different letters. Hence, if we simplify the coefficient of x, we can write at once the corresponding coefficient of y and the constant term. We have:

$$\dfrac{t_1'D_1+t_2'D_2}{t_1'+t_2'}-\dfrac{t_1''D_1+t_2''D_2}{t_1''+t_2''}=\dfrac{(t_1't_2''-t_1''t_2')(D_1-D_2)}{(t_1'+t_2')(t_1''+t_2'')}$$

Hence, equation (4) becomes

$$\frac{t'_1 t''_2 - t''_1 t'_2}{(t'_1+t'_2)(t''_1+t''_2)}[(D_1-D_2)x+(E_1-E_2)y+(F_1-F_2)]=0$$

Since $t'_1 t''_2 - t''_1 t'_2 \neq 0$, it follows that

$$(D_1-D_2)x+(E_1-E_2)y+(F_1-F_2)=0$$

Since this is an equation of the radical axis of the two given circles, it follows that every two distinct circles of the pencil defined by the given circles have the same radical axis.

EXAMPLE 4–18. Find an equation of the pencil of circles defined by the circles $x^2+y^2-2x+4y-4=0$ and $x^2+y^2=16$. Which circle of the pencil passes through the point (3, 1)?

Solution. An equation of the pencil is:

(1) $$t_1(x^2+y^2-2x+4y-4)+t_2(x^2+y^2-16)=0$$

Since the point (3, 1) must satisfy this equation,

$$t_1(9+1-6+4-4)+t_2(9+1-16)=0$$

Hence, $$4t_1-6t_2=0$$

or $$\frac{t_1}{t_2}=\frac{3}{2}$$

If we let $t_1=3$ and $t_2=2$, equation (1) becomes

$$3(x^2+y^2-2x+4y-4)+2(x^2+y^2-16)=0$$

or $$5x^2+5y^2-6x+12y-44=0$$

The three circles are shown in Fig. 4–20.

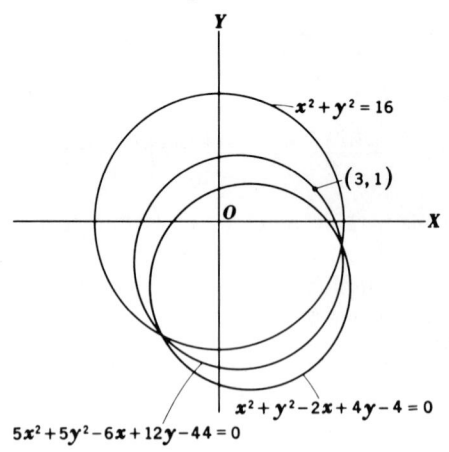

FIG. 4–20

EXAMPLE 4-19. Find an equation of the circle of the pencil defined by the circles $x^2+y^2+2x-4y+1=0$ and $x^2+y^2-6x+4y+12=0$ and having its center at a point whose abscissa is -3.

Solution. We have:
$$t_1(x^2+y^2-6x+4y+12)+t_2(x^2+y^2+2x-4y+1)=0$$
Expanding this equation and collecting the terms, we obtain:
$$(t_1+t_2)(x^2+y^2)+(-6t_1+2t_2)x+(4t_1-4t_2)y+12t_1+t_2=0$$
Dividing through by t_1+t_2, we get:
$$x^2+y^2+\left(\frac{-6t_1+2t_2}{t_1+t_2}\right)x+\left(\frac{4t_1-4t_2}{t_1+t_2}\right)y+\frac{12t_1+t_2}{t_1+t_2}=0$$

Since this result is in the form of the general equation of a circle, it follows from Section 4-2 that the coordinates of the center of any circle of the pencil are
$$\frac{6t_1-2t_2}{2(t_1+t_2)} \quad \text{and} \quad \frac{-4t_1+4t_2}{2(t_1+t_2)}$$

In our problem the abscissa of the center must be -3. Hence,
$$\frac{3t_1-t_2}{t_1+t_2}=-3$$
or
$$3t_1-t_2=-3t_1-3t_2$$

From this it follows that $6t_1=-2t_2$, or $\dfrac{t_1}{t_2}=-\dfrac{1}{3}$. If we let $t_1=-1$ and $t_2=3$, the equation of the circle in the pencil becomes
$$2x^2+2y^2+12x-16y-9=0$$
We check by finding the center of this circle to be $(-3, 4)$.

EXERCISES

1. Write an equation of each of the following families of circles:

    (a) All circles with center at $(2, -3)$
    (b) All circles with center at $(-1, 4)$
    (c) All circles having radius 4 and abscissa of centers 2
    (d) All circles having radius 5 and abscissa of centers -3
    (e) All circles having radius 3 and ordinate of centers -4
    (f) All circles having radius 5 and ordinate of centers 0.

2. A pencil of circles is defined by the circles $x^2+y^2=16$ and $x^2+y^2-4x+6y+9=0$. Find an equation of the circle of the pencil having each of the following properties:

(a) It passes through the point (4, 1).
(b) It passes through the point (−3, 5).
(c) It passes through the point (0, 0).
(d) It has its center on the y-axis.
(e) It has its center on the x-axis.
(f) It has its center on the line $x = 5$.
(g) It has its center on the line $y = 3$.
(h) It has its center on the line $y = -3$.

3. Work Exercise 2 for the pencil defined by the circles $x^2 + y^2 - 2x + 3y - 2 = 0$ and $x^2 + y^2 + 2x - 4y - 4 = 0$.

4–11. Parametric equations of the circle

Consider a circle with its center at $(0, 0)$ and radius r, as in Fig. 4–21. If $P = (x, y)$ is any point on the circle, and if θ is the angle which vector $\mathbf{u} = \overrightarrow{OP}$ makes with the positive x-axis, we have:

$$\mathbf{u} = \overrightarrow{OP} = [r \cos \theta, r \sin \theta]$$

Since $\mathbf{u} = [x, y] = [r \cos \theta, r \sin \theta]$, it follows that

$$x = r \cos \theta \quad \text{and} \quad y = r \sin \theta \qquad (4\text{–}19)$$

Equations (4–19) are called *parametric equations* of a circle. The *parameter* is the angle θ. Two such parametric equations are often written with a comma between them, as $x = r \cos \theta$, $y = r \sin \theta$.

If we square equations (4–19) and add the results, we obtain:

$$x^2 + y^2 = r^2 \cos^2 \theta + r^2 \sin^2 \theta = r^2$$

which is the simple equation of a circle.

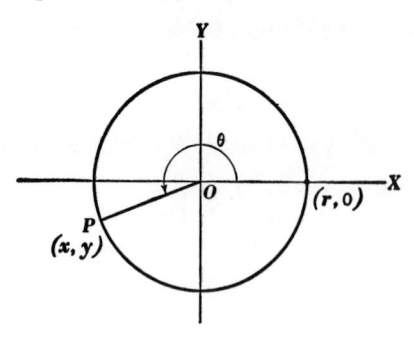

FIG. 4–21

If the center is at $C = (h, k)$ instead of at $(0, 0)$, and θ is the angle that $\overrightarrow{CP}$ makes with the ray through C parallel to the positive x-axis, we have:

$$\mathbf{u} = \overrightarrow{CP} = [x-h, y-k] = [r\cos\theta, r\sin\theta]$$

or
$$\begin{aligned} x - h &= r\cos\theta \\ y - k &= r\sin\theta \end{aligned} \qquad (4\text{-}20)$$

Squaring equations (4–20) and adding the results, we obtain
$$(x-h)^2 + (y-k)^2 = r^2$$
which is the standard equation of a circle.

Another pair of parametric equations of the circle with center at (0, 0), which is found useful in the calculus, is obtained as follows: Consider a circle whose equation is $x^2 + y^2 = r^2$. Combining this equation with $x = ty$ so as to eliminate x, we obtain $t^2 y^2 + y^2 = r^2$, or
$$y = \frac{r}{\pm\sqrt{1+t^2}}$$

Substituting this value of y in $x = ty$, we have:
$$x = \frac{tr}{\pm\sqrt{1+t^2}}$$

If we use the positive radical, we obtain a pair of parametric equations of the upper half of the circle to be:
$$\begin{aligned} x &= \frac{tr}{\sqrt{1+t^2}} \\ y &= \frac{r}{\sqrt{1+t^2}} \end{aligned} \qquad (4\text{-}21)$$

where the parameter t is the reciprocal of the slope of the line whose equation is $x = ty$.

If we use the negative radical, we obtain parametric equations of the lower half of the circle.

It should be noted that these parametric equations do not represent points on the x-axis. For example, in equations (4–21), the points $(+r, 0)$ and $(-r, 0)$ do not correspond to any values of t.

REVIEW EXERCISES

1. Find an equation of each circle satisfying each of the following conditions:
   (a) Its center is at $(-1, 2)$, and its radius is 5.
   (b) The end points of its diameter are $(-1, 4)$ and $(3, 2)$.
   (c) Its center is in quadrant III, it is tangent to both axes, and its radius is 3.
   (d) Its center is at $(3, 5)$ and it is tangent to the line $y = 2x + 1$.
   (e) Its center is on the y-axis and it passes through the points $(4, 2)$ and $(-6, -2)$.
   (f) It passes through the points $(1, -1)$ and $(5, 3)$ and its center is on the line $x - y + 7 = 0$.

158 ANALYTIC GEOMETRY IN THE PLANE

 (g) Its x-intercepts are -1 and 5, and its radius is 5.
 (h) It passes through the point (2, 1), it is tangent to the line $3x - 4y + 6 = 0$, and its radius is 4.
 (i) It contains the points (1, 1), (1, -1), and (3, 0).
 (j) It contains the points (2, 8), (5, 5), and (-3, 3).
 (k) It passes through the points (9, -4) and (5, -6) and its center is on the line $9x + 11y - 8 = 0$.

2. Find an equation of the inscribed circle of each of the following triangles:
 (a) $P_1(-1, 3)$, $P_2(7, -1)$, and $P_3(2, 9)$
 (b) $P_1(0, -1)$, $P_2(-7, 0)$, and $P_3(-1, 6)$

3. What is an equation of the radical axis of the circles $2x^2 + 2y^2 - 3x + y - 4 = 0$ and $x^2 + y^2 + 2x - 3y - 5 = 0$?

4. Find a point from which tangents of equal length can be drawn to the circles $x^2 + y^2 = 4$, $x^2 + y^2 + 6y - 2 = 0$, and $x^2 + y^2 - 4x + 1 = 0$; and determine the length of these tangents.

5. (a) What are the center and the radius of the circle whose equation is $3x^2 + 3y^2 - 2x - 4 = 0$?
 (b) What is the length of a tangent drawn from the point $(-3, 2)$ to the circle?

6. Consider the formula $t^2 = x_1^2 + y_1^2 + D_1 x_1 + E_1 y_1 + F_1$, and explain the geometric significance of the fact that: (a) $t^2 < 0$; (b) $t^2 > 0$; (c) $t^2 = 0$.

7. (a) In the pencil defined by $x^2 + y^2 + 2x - 4y - 8 = 0$ and $x^2 + y^2 - 3x - 3y - 2 = 0$, what is an equation of the circle which has its center on the y-axis?
 (b) Which circle of the pencil contains the point $(-3, 2)$?
 (c) Which circles of the pencil have a radius equal to 13?

8. Prove that the radical axis is perpendicular to the line of centers of the circles of the pencil defined by $x^2 + y^2 + D_1 x + E_1 y + F_1 = 0$ and $x^2 + y^2 + D_2 x + E_2 y + F_2 = 0$.

9. Write equations of the tangent and the normal to each of the given circles at the given point:
 (a) $x^2 + y^2 = 5$ and $(-1, 2)$
 (b) $(x + 5)^2 + (y - 1)^2 = 25$ and $(-2, 5)$
 (c) $2x^2 + 2y^2 - x + 5y - 8 = 0$ and the points whose abscissa is 1

10. Prove that an angle inscribed in any semicircle is a right angle.

11. Find the points of intersection of the line $2x + 3y - 5 = 0$ and the circle $x^2 + y^2 - 2x - 4y + 4 = 0$.

12. (a) Find the points of intersection of the circles $x^2 + y^2 + 2x - 4y - 8 = 0$ and $x^2 + y^2 - 3x - 3y - 2 = 0$.
 (b) What is the length of the common chord?

13. Given a pencil of circles defined by the equations $x^2 + y^2 - 2x + 4y - 6 = 0$ and $x^2 + y^2 - 4x + 2y - 4 = 0$. Find an equation of each of the following circles of the pencil:

(a) It passes through the point $(-4, 6)$.
(b) It passes through the point $(5, 1)$.
(c) Its center is on the x-axis.
(d) Its center is on the y-axis.

14. What is an equation of a circle that is tangent to $x^2 + y^2 = 16$ and has its center at $(-7, 0)$?

15. If $\mathbf{u} = \overrightarrow{OP} = [x, y]$ and $\mathbf{v} = \overrightarrow{OC} = [h, k]$, prove that an equation of the circle having its center at $C = (h, k)$, having a radius equal to r, and passing through $P = (x, y)$ may be written in either one of the following ways:
    (a) $(\mathbf{u} - \mathbf{v}) \cdot (\mathbf{u} - \mathbf{v}) = r^2$
    (b) $(\mathbf{u} \cdot \mathbf{u}) - 2(\mathbf{u} \cdot \mathbf{v}) + (\mathbf{v} \cdot \mathbf{v}) - r^2 = 0$

16. Find parametric equations of the following circles in terms of the parameter θ:
    (a) $x^2 + y^2 = 16$
    (b) $x^2 + y^2 = 25$
    (c) $x^2 + y^2 = 49$
    (d) $x^2 + y^2 - 2x + 4y - 11 = 0$
    (e) $x^2 + y^2 + 6x - 4y + 9 = 0$

17. What are parametric equations of the circles of Exercise 16 if the parameter is t, where t is the reciprocal of the slope of the line whose equation is $x = ty$?

18. Find the locus of a point having the property that the length of the tangent from this point to the circle $x^2 + y^2 + 4x + 6y = 27$ is 9 units.

19. Describe the set of points having the property that the length of the tangent from each point of the set to the circle $x^2 + y^2 - 2x + 4y + 1 = 0$ is 6 units.

20. Find equations of the tangents from the given points to each of the given circles:
    (a) $x^2 + y^2 - 6x - 36 = 0$ and $(6, 9)$
    (b) $x^2 + y^2 + 8y - 42 = 0$ and $(-4, 6)$
    (c) $x^2 + y^2 + 10x - 4y = 87$ and $(9, -4)$

21. The base of a triangle is fixed and the ratio of the lengths of the two sides is a positive constant. Show that the locus of the third vertex is a circle.

22. A point P moves so that its distance from a fixed line is proportional to the square of its distance to a fixed point not on the given line. If P remains always on the same side of the fixed line and the fixed point, describe its locus.

23. Find the locus of a point moving so that the sum of the squares of its distances from the sides of an equilateral triangle is constant. Discuss all cases.

24. The feet of the perpendiculars from the point $P = (x, y)$ on the sides of the triangle whose vertices are $(0, 0)$, $(4, 0)$, $(0, 2)$ lie on a line. Prove that the set of points P lies on the circumscribed circle of the triangle.

25. Work Exercise 24 if the vertices of the triangle are $(-2, 0)$, $(0, 6)$, $(3, 0)$.

26. Work Exercise 24 for the general triangle with vertices $(a, 0)$, $(c, 0)$, $(0, b)$.

SUPPLEMENTARY EXERCISES

1. Two intersecting circles are said to be *orthogonal* if the angle between their tangents at the points of intersection is 90°. Prove that, if the circles $x^2 + y^2 + D_1 x + E_1 y + F_1 = 0$ and $x^2 + y^2 + D_2 x + E_2 y + F_2 = 0$ are orthogonal, then $2(F_1 + F_2) = D_1 D_2 + E_1 E_2$.

2. If a circle $x^2 + y^2 + Dx + Ey + F = 0$ is orthogonal to two given circles $x^2 + y^2 + D_1 x + E_1 y + F_1 = 0$ and $x^2 + y^2 + D_2 x + E_2 y + F_2 = 0$, prove that its center lies on the radical axis of the two given circles. (Hint: Use the condition for orthogonality given in Exercise 1.)

3. Prove that, if a circle $x^2 + y^2 + Dx + Ey + F = 0$ is orthogonal to $x^2 + y^2 + D_1 x + E_1 y + F_1 = 0$ and also to $x^2 + y^2 + D_2 x + E_2 y + F_2 = 0$, it is orthogonal to *every* circle of the pencil defined by the two given circles.

4. (a) If we have any circle $x^2 + y^2 = a^2$ and a point $P_1(x_1, y_1)$ distinct from the center, the line $x_1 x + y_1 y = a^2$ is called the *polar* of the point with respect to the circle. If P_1 is on the circle, this line obviously is tangent to the circle at that point. Prove that, if the point P_1 is outside the circle, the polar is the line joining the points of contact of the tangents drawn from P_1 to the circle. (Hint: Write an equation of the circle having its center at the point $P_1(x_1, y_1)$ and having a radius equal to the length of the tangent from P_1 to the given circle. Combine the equations of the two circles so as to find an equation of the line through the two points of tangency. The point P_1 is called the *pole* of the line with respect to the circle.)
   (b) Prove that, if the polar of a point $P_2(x_2, y_2)$ passes through the point P_1, then the polar of P_1 passes through P_2.

5. Prove that the following circles are orthogonal: $x^2 + y^2 + 6x - 3y + 4 = 0$ and $x^2 + y^2 - 2x + 2y - 13 = 0$. Draw the circles and the tangents at their points of intersection.

6. Determine the constant D, E, or F, in order that the following pairs of circles may be orthogonal:
   (a) $x^2 + y^2 + 2x - 3y - 6 = 0$ and $x^2 + y^2 + x - y + F = 0$
   (b) $x^2 + y^2 + Dx + 2y - 1 = 0$ and $x^2 + y^2 - x - 8y - 3 = 0$
   (c) $x^2 + y^2 + 5x - 8y - 12 = 0$ and $x^2 + y^2 - 2x + Ey - 4 = 0$

7. Determine the constants D, E, and F in the following equations in order that each may represent a circle which is orthogonal to the given pair of circles $x^2 + y^2 + 2x - 3y - 6 = 0$ and $x^2 + y^2 - 3x + y - 1 = 0$:
   (a) $x^2 + y^2 + 2x + Ey + F = 0$
   (b) $x^2 + y^2 + Dx + 3y + F = 0$
   (c) $x^2 + y^2 + Dx + Ey - 5 = 0$

8. Find an equation of the circle that is orthogonal to all the following circles: $x^2 + y^2 + 2x + 2y - 14 = 0$, $x^2 + y^2 - 4x - 2y - 4 = 0$, and $x^2 + y^2 - 25 = 0$.

9. Find an equation of the polar of each of the following points with respect to the given circle:

   (a) $(2, -1)$ and $x^2 + y^2 = 1$
   (b) $(3, -5)$ and $x^2 + y^2 = 9$
   (c) $(5, -6)$ and $x^2 + y^2 = 16$
   (d) $(-1, 4)$ and $x^2 + y^2 = 4$
   (e) $(3, -4)$ and $x^2 + y^2 = 25$
   (f) $(0, 8)$ and $x^2 + y^2 = 16$

10. Select a point on each of the polars obtained in Exercise 9, find the polar of this point with respect to the same circle, and show that it contains the pole originally assigned.

11. Given the triangle whose vertices are $(18, 0)$, $(0, 24)$, and $(-10, 0)$. Prove each of the following theorems:

    (a) The medians of the triangle meet in a point M.
    (b) The altitudes of a triangle meet in a point H.
    (c) The perpendicular bisectors of the sides of the triangle meet in a point Q.
    (d) Points Q, M, and H are collinear; and M is a trisection point of line QH.
    (e) All the following points lie on the same circle (nine-point circle): the mid-points of the sides of the triangle, the feet of the altitudes, and the mid-points of the line segments joining point H to the vertices.
    (f) The center of the nine-point circle in (e) is the mid-point of the line segment QH.
    (g) The nine-point circle in (e) is tangent to all four circles that are tangent to the prolongations of the three sides of the given triangle. (This is known as the Feurbach Theorem.)

5

The Conics

5-1. The conic as a section of a cone

If a line $A'A$ making an angle θ with the y-axis is revolved about the y-axis so that θ is kept constant, a surface known as a right circular cone is generated. See Fig. 5-1. The line $A'A$ is called a *generator*. The point O is the *vertex* or *apex* of the cone, and the y-axis is the axis of the cone. The vertex divides the entire generated surface into two parts, each of which is called a *nappe*. The name *conic section*, or simply *conic*, is given to each of the plane sections of a right circular cone. The shape of a conic is dependent on the position of the cutting plane with reference to the generator. If the plane passes through the vertex O, we obtain a point, or one straight line, or two distinct straight lines, as shown in Fig. 5-2.

If the cutting plane is perpendicular to the axis and does not contain the vertex, we obtain a circle, as in Fig. 5-3.

If the cutting plane is parallel to a generator and does not contain the vertex, as indicated in Fig. 5-4, it intersects only one nappe of the cone and this conic section is called a *parabola*.

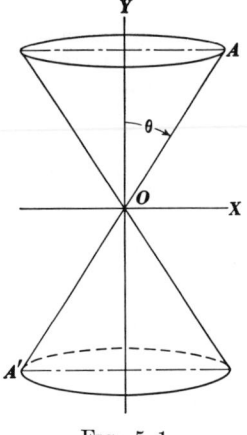

Fig. 5-1

If θ is the angle between a generator and the axis of a right circular cone and if, as in Fig. 5-5, the cutting plane intersects the axis in an angle greater than θ and does not contain the vertex, the conic section is called an *ellipse*.

Finally, if the plane cuts both nappes of the surface and does not contain the vertex, as in Fig. 5-6, the conic section is called a *hyperbola*. This curve has two open branches.

The early Greek mathematicians studied the conics from this point of view. In this chapter we shall first study the conics by analytic methods, and then at the end we shall show that the curves just defined geometrically are the same as the ones studied analytically.

FIG. 5–2

FIG. 5–3

FIG. 5–4

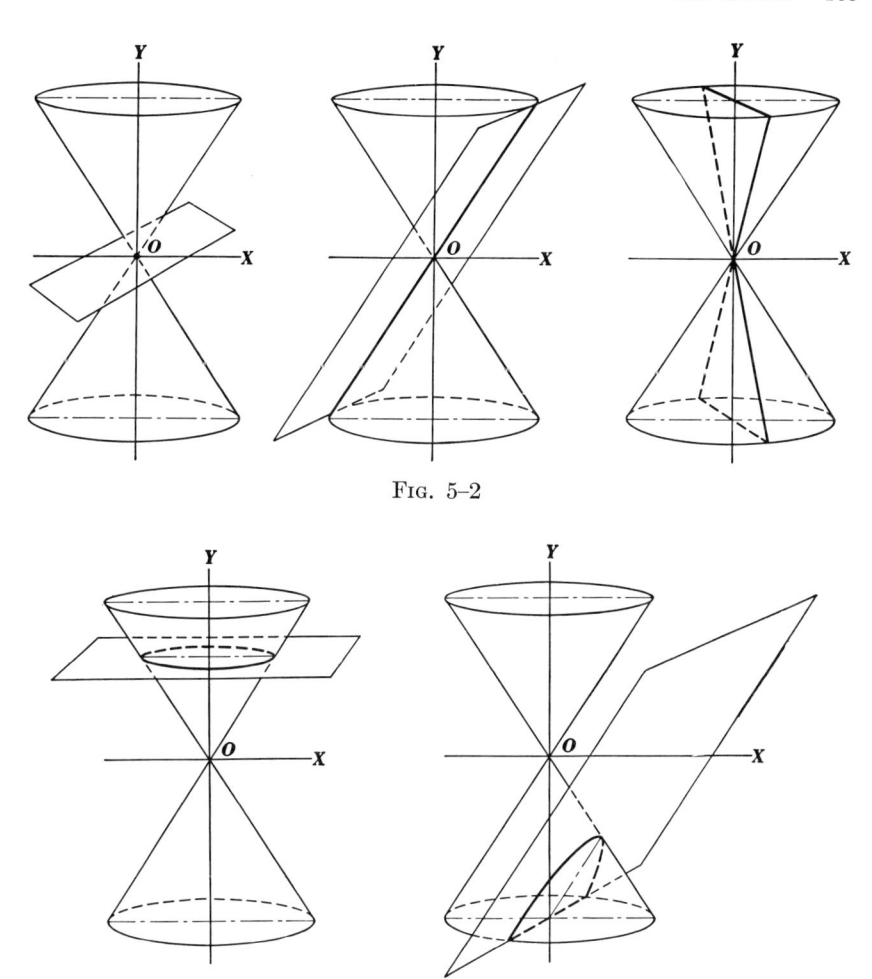

5–2. General definition of a conic

In the previous section we discussed the conic as a section of a cone. Now we shall consider it from an analytic viewpoint.

If each point of a given set satisfies the condition that the ratio of its distance from a fixed point to its distance from a fixed line is constant, the set is called a conic. The fixed point is called the focus; the fixed line is called the directrix; and the constant ratio is called the eccentricity of the conic and is denoted by e. If $e = 1$, the conic is a parabola; if $e < 1$, the conic is an ellipse; if $e > 1$, the conic is a hyperbola. A circle is a special kind of ellipse, for which $e = 0$.

164   ANALYTIC GEOMETRY IN THE PLANE

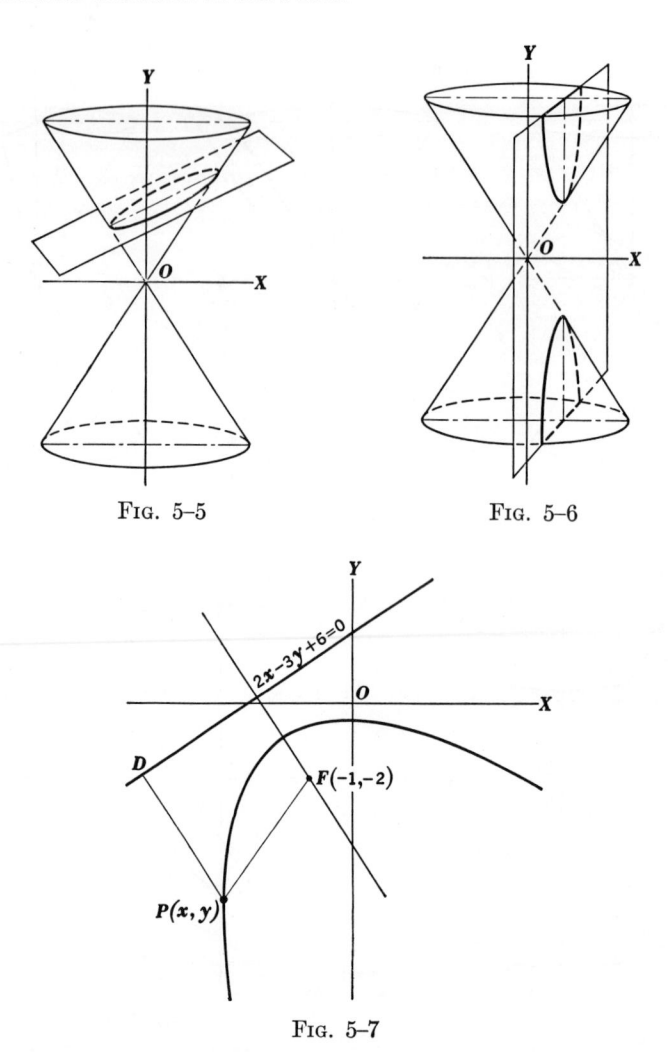

FIG. 5-5

FIG. 5-6

FIG. 5-7

EXAMPLE 5-1. Find an equation of the conic that has the line $2x - 3y + 6 = 0$ for its directrix, has the point $(-1, -2)$ for its focus, and has an eccentricity equal to 1.

Solution. Since $e = 1$, we know that the conic is a parabola. If, as shown in Fig. 5-7, we let $P(x, y)$ be an arbitrary point on the conic, $F(-1, -2)$ be the focus, and D be the projection of P on the directrix, it follows from the definition of a parabola that

$$\frac{|FP|}{|DP|} = 1 \text{ or } |FP| = |DP|$$

But $|FP| = \sqrt{(x+1)^2+(y+2)^2}$ and $|DP| = \left|\dfrac{2x-3y+6}{\sqrt{13}}\right|$. Hence, we have:

$$\sqrt{(x+1)^2+(y+2)^2} = \left|\dfrac{2x-3y+6}{\sqrt{13}}\right|$$

Squaring both sides and expanding, we obtain the equivalent equation

$$x^2+2x+1+y^2+4y+4 = \dfrac{4x^2+9y^2+36-12xy+24x-36y}{13}$$

This becomes

$$13x^2+26x+13+13y^2+52y+52 = 4x^2+9y^2+36-12xy+24x-36y$$

or
$$9x^2+12xy+4y^2+2x+88y+29 = 0$$

It is important to notice that this equation is of the second degree and that the second-degree terms form a perfect square. When the conic is a parabola, the second-degree terms always form a perfect square. See Section 6–8.

EXERCISES

Derive an equation of each of the following conics:
1. Focus is at $(-1, 2)$; directrix is the line $x = 3$; $e = 1$.
2. Focus is at $(0, 2)$; directrix is the line $y = 4$; $e = \frac{1}{2}$.
3. Focus is at $(0, 0)$; directrix is the line $x = 4$; $e = 2$.
4. Focus is at $(1, -2)$; directrix is the line $2x - y = 0$; $e = 1$.
5. Focus is at $(3, 0)$; directrix is the line $x = 5$; $e = \frac{1}{3}$.
6. Focus is at $(-5, 0)$; directrix is the line $x = 7$; $e = \frac{3}{5}$.
7. Focus is at $(0, -4)$; directrix is the line $y = -6$; $e = \frac{3}{2}$.
8. Focus is at $(-3, 2)$; directrix is the x-axis; $e = 1$.
9. Focus is at $(-5, -1)$; directrix is the line $2x + 3y - 6 = 0$; $e = \frac{1}{2}$.
10. Focus is at $(0, 0)$; directrix is the line $3x + 4y + 12 = 0$; $e = 3$.
11. Focus is at $(0, 0)$; directrix is the line $y = 3$; $e = 1$.
12. Focus is at $(2, -3)$; directrix is the line $4x - 3y + 24 = 0$; $e = 1$.

After working the foregoing exercises, the student will notice that all the equations are of the second degree and of the type $Ax^2 + Bxy + Cy^2 + Dx + Ey + F = 0$, where A, B, C, D, E, and F are numbers and A, B, C are not all zero. A discussion of this general equation will be given in the next chapter. In this chapter we shall content ourselves with a discussion of the *simple* equations of the parabola, ellipse, and hyperbola which arise from particular choices of the focus and the directrix.

THE PARABOLA

5–3. Explanation of terms

A *parabola* is a set of points, each of which satisfies the condition that its distance from a fixed point is equal to its distance from a fixed line. A typical parabola is shown in Fig. 5–8. From the definition, we know that, if P is any point on the parabola,

$$\frac{|FP|}{|DP|} = e = 1$$

or
$$|FP| = |DP| \tag{5–1}$$

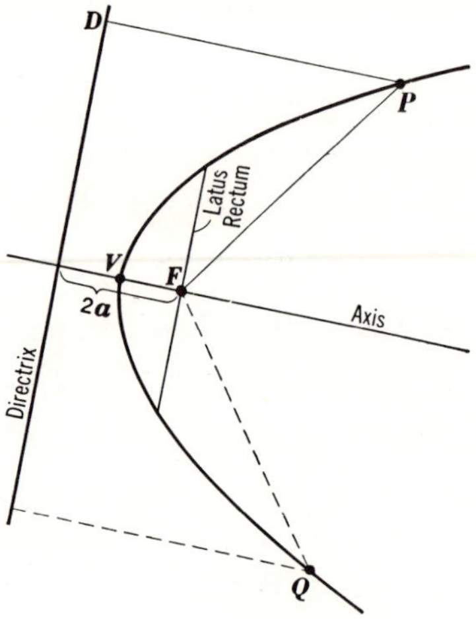

Fig. 5–8

The parabola is symmetric* with respect to the line through the focus perpendicular to the directrix, since the image Q of every point P on the parabola with respect to this line is also a point on the parabola. This line is called the *axis* of the parabola. The distance from the focus to the directrix is denoted by $2a$. The point, as V in Fig. 5–8, where the parabola intersects its axis is called the *vertex*. Hence, the distance from the focus to the vertex is a. A chord through the focus is called a *focal chord*. If the focal chord is perpendicular to the axis, it is called the *latus rectum* of the parabola. From the definition of the parabola and of a, it follows that the length of the latus rectum is $4a$. The student should verify this. Any line segment from the focus to a point on the parabola is called a *focal radius*. In Fig. 5–8, FP is a focal radius.

---
* See Section 4–4.

5–4. A geometric construction of the parabola

If several lines are drawn, all parallel to the directrix and on the same side of it as the focus, such that the distance from the directrix to each line is not less than a units, it is possible to construct many points of a parabola. In Fig. 5–9, for example, the several points on the curve are located in the following manner: With $|A_1D|$ as a radius and F as a center, swing arcs intersecting the perpendicular through A_1 at the points P_1 and P'_1, respectively. Perform similar operations with $|A_2D|$, $|A_3D|$, etc., as radii and F as a center; and obtain successively the points P_2 and P'_2, P_3 and P'_3, etc.

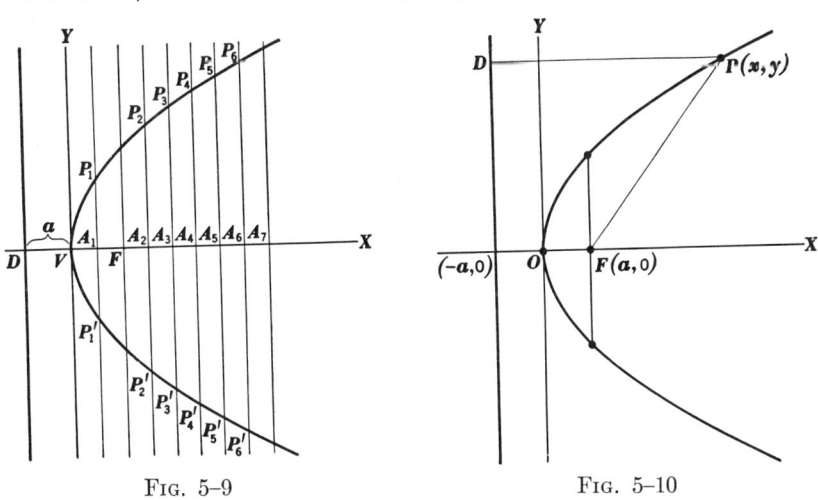

FIG. 5–9

FIG. 5–10

From the definition, we know that this set of points lies on a parabola. Joining them by a smooth curve, we have a geometric representation of the parabola.

5–5. Simple equations of the parabola

In order to obtain a simple form for an equation of a parabola, we choose the coordinate system so that the directrix is the line $x = -a$ and the focus is the point $F(a, 0)$, as shown in Fig. 5–10. Hence, the vertex is at the origin.

Let $P(x, y)$ be any point having the property that $|FP| = |DP|$, where D is the projection of P on the directrix. Then

$$\sqrt{(x-a)^2 + y^2} = |x+a|$$

Squaring both sides, we obtain

$$x^2 - 2ax + a^2 + y^2 = x^2 + 2ax + a^2$$

or

$$y^2 = 4ax \tag{5-2}$$

168   ANALYTIC GEOMETRY IN THE PLANE

Conversely, the equation $y^2 = 4ax$ represents the parabola with its vertex at the origin and its focus at the point $(a, 0)$ on the x-axis. The proof is as follows: Adding $x^2 - 2ax + a^2$ to both sides of equation (5-2), we obtain:
$$x^2 - 2ax + a^2 + y^2 = 4ax + x^2 - 2ax + a^2$$
or
$$(x - a)^2 + y^2 = (x + a)^2$$
Taking the non-negative square root of both sides, we find that
$$\sqrt{(x-a)^2 + y^2} = |x + a|$$
This relation states for $P(x, y)$ that $|FP| = |DP|$, where F is the point $(a, 0)$ and D is the point $(-a, y)$; hence, $y^2 = 4ax$ represents the parabola.

If we had chosen the focus to be the point $(-a, 0)$ and the directrix to be the line $x - a = 0$, an equation of the parabola would have been
$$y^2 = -4ax \tag{5-3}$$

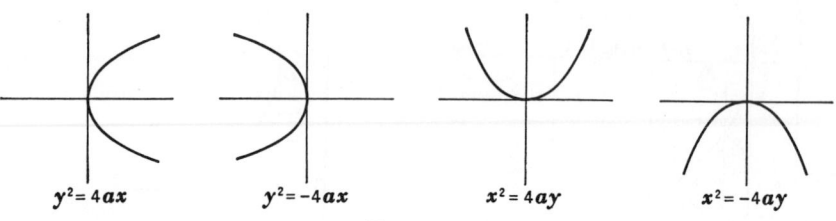

$y^2 = 4ax$ $y^2 = -4ax$ $x^2 = 4ay$ $x^2 = -4ay$

FIG. 5–11

When the focus is at the point $(0, a)$ on the y-axis and the directrix is the line $y + a = 0$, the equation becomes
$$x^2 = 4ay \tag{5-4}$$
If the focus is at the point $(0, -a)$ and the directrix is the line $y - a = 0$, the equation of the parabola becomes
$$x^2 = -4ay \tag{5-5}$$

As an exercise, the student should derive equations (5-3), (5-4), and (5-5). It is important for the student to understand clearly that the simple equations apply only when the focus is on one of the axes and the vertex is at the origin. It is advisable to associate the appearance of the curve with the equation. Thus, as indicated in Fig. 5–11,

$y^2 = 4ax$ opens to the right
$y^2 = -4ax$ opens to the left
$x^2 = 4ay$ opens upward
$x^2 = -4ay$ opens downward

Since the equations $Ax^2 + Ey = 0$ and $Cy^2 + Dx = 0$ are equivalent, respectively, to $x^2 = -\dfrac{E}{A} y$ and $y^2 = -\dfrac{D}{C} x$, provided $A \neq 0$ and $C \neq 0$, it follows that, if $D \neq 0$ and $E \neq 0$, each is an equation of a parabola with vertex at the origin. In the case of $x^2 = -\frac{E}{A} y$, we have $4a = \left|\frac{E}{A}\right|$; hence, the parabola opens either upward or downward, the direction depending on the sign of $-\frac{E}{A}$. In the case of $y^2 = -\frac{D}{C} x$, we have $4a = \left|\frac{D}{C}\right|$; hence, the parabola opens either to the right or to the left, the direction depending on the sign of $-\frac{D}{C}$. If $D = 0$ or $E = 0$, the equation becomes $y^2 = 0$ or $x^2 = 0$, which means that the parabola has degenerated into a line. This is what happens when the cutting plane like that in Fig. 5–4 passes through the vertex O of the cone. We shall say that the loci of the equations $x^2 = 0$ and $y^2 = 0$ are *degenerate parabolas*.

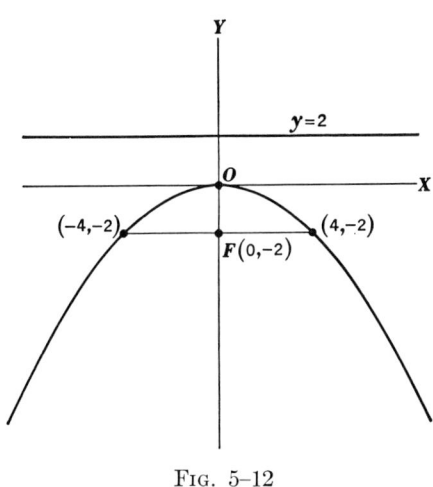

FIG. 5–12

EXAMPLE 5–2. Find the simple equation of the parabola that has its focus at $(0, -2)$.

Solution. Since the vertex is at $(0, 0)$, we have $a = 2$. As shown in Fig. 5–12, the parabola must open downward. Hence, the simple equation is $x^2 = -4ay$, or $x^2 = -8y$.

EXERCISES

1. Derive an equation of the parabola that has its focus at the point $(-a, 0)$ and has the line $x - a = 0$ for its directrix.

2. Derive an equation of the parabola that has its focus at the point $(0, a)$ and has the line $y + a = 0$ for its directrix.

3. Derive an equation of the parabola that has its vertex at the point $(0, 0)$ and has its focus at the point $(0, -a)$.

4. Find an equation for each of the following parabolas with its vertex at the origin and its focus at the point whose coordinates are given: (a) $(2, 0)$; (b) $(0, -3)$; (c) $(0, 3)$; (d) $(-4, 0)$; (e) $(0, 5)$; (f) $(6, 0)$.

5. Find an equation for each of the following parabolas having its vertex at the origin and having the given line as its directrix: (a) $x+2=0$; (b) $x-3=0$; (c) $y-4=0$; (d) $y+5=0$.

6. Find a *simple* equation of each of the following parabolas for which the coordinates of the focus and the equation of the directrix are given:
   (a) Focus is at $(1, 0)$ and directrix is $x = -1$.
   (b) Focus is at $(-3, 0)$ and directrix is $x - 3 = 0$.
   (c) Focus is at $(0, 5)$ and directrix is $y + 5 = 0$.
   (d) Focus is at $(0, -4)$ and directrix is $y - 4 = 0$.

7. Find a *simple* equation of each of the following parabolas:
   (a) Focus is on the y-axis, and curve passes through the point $(8, -3)$.
   (b) Focus is on the x-axis, and curve passes through the point $(-1, 5)$.

8. Find *simple* equations of the parabolas satisfying each of the following conditions:
   (a) The point $(2, -3)$ is on the parabola.
   (b) The point $(-4, -1)$ is on the parabola.

   Note: There are two answers to each part.

9. Prove that $x^2 = 4ay$ is an equation of a parabola that opens upward.

10. Prove that $y^2 = -4ax$ is an equation of a parabola that opens to the left.

11. Prove that $x^2 = -4ay$ is an equation of a parabola that opens downward.

5–6. The latus rectum

We have already defined the latus rectum as a focal chord perpendicular to the axis. In Fig. 5–13 is shown a parabola $y^2 = 4ax$ with its directrix and its latus rectum. An equation of the line that contains the latus rectum of such a parabola is $x = a$. Solving the equation of the line $x = a$ with that of the parabola $y^2 = 4ax$, we obtain $y^2 = 4a^2$, from which $y = \pm 2a$. Hence, as shown in Fig. 5–13, the coordinates of the points of intersection are $(a, 2a)$ and $(a, -2a)$; and the *length* of the latus rectum is $4a$.

The latus rectum is extremely useful in sketching the parabola since it enables us to locate two points on the curve. Notice that, when the equation is written in one of the simple forms $y^2 = \pm 4ax$ or $x^2 = \pm 4ay$, the absolute value of the coefficient of the linear term is the length of the latus rectum.

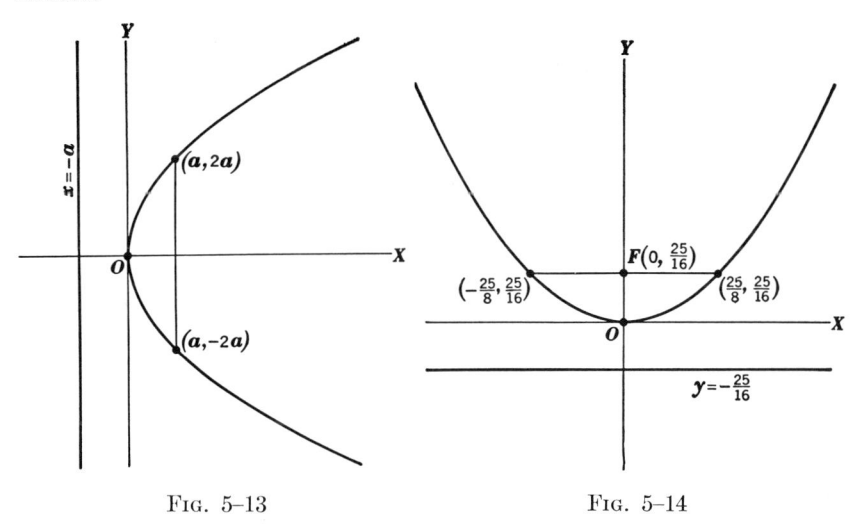

FIG. 5–13

FIG. 5–14

EXAMPLE 5–3. Discuss the equation $4x^2 - 25y = 0$ and sketch the curve. "Discuss" means to give the coordinates of the vertex, focus, and end points of the latus rectum, and an equation of the directrix.

Solution. First we write the equation as $x^2 = \frac{25}{4} y$. This is the simple form; hence, $4a = \frac{25}{4}$, or $a = \frac{25}{16}$. Since the vertex is at the origin, the parabola opens upward, as shown in Fig. 5–14. The focus is at $(0, \frac{25}{16})$. The end points of the latus rectum are $(\frac{25}{8}, \frac{25}{16})$ and $(-\frac{25}{8}, \frac{25}{16})$. An equation of the directrix is $y = -\frac{25}{16}$.

EXERCISES

1. Prove analytically that the length of the latus rectum of the parabola $x^2 = -4ay$ is $4a$.
2. Discuss the equation $x^2 + 16y = 0$ and sketch the parabola.
3. Discuss and sketch the curve represented by each of the following equations:
   (a) $y^2 - 36x = 0$
   (b) $y^2 + 4x = 0$
   (c) $2x^2 - 15y = 0$
   (d) $7y^2 - 36x = 0$
   (e) $3x^2 - 16y = 0$
   (f) $x^2 + 12y = 0$
   (g) $4y^2 - 25x = 0$
   (h) $x^2 - 8y = 0$

4. Using the definition of a parabola, find an equation of each of the parabolas having the following characteristics:
   (a) Focus is at (3, 4) and directrix is the x-axis.
   (b) Vertex is at (4, 3) and directrix is the y-axis.
   (c) Focus is at (0, 0) and directrix is the line $2x - 3y + 6 = 0$.
   (d) Focus is at (2, -1) and directrix is $x - 4 = 0$.
   (e) Vertex is at (-2, 4) and directrix is $y = 7$.
   (f) Focus is at (3, 1) and vertex is at (3, 5).
   (g) Focus is at (2, -4) and vertex is at (8, -4).
   (h) Focus is at (0, 0) and directrix is $3x + 4y - 12 = 0$.

5. Find an equation of the circle passing through the vertex and the extremities of the latus rectum of the parabola $x^2 = 4ay$.

6. An equilateral triangle having one vertex at the origin (0, 0) is inscribed in the parabola $x^2 = 4ay$. Determine the perimeter and the area of the triangle.

7. An isosceles right triangle is inscribed in the parabola $x^2 = 4ay$, the vertex of the right angle being at the origin (0, 0). Determine the area and the perimeter of the triangle.

8. What is an equation of the focal chord of the parabola $x^2 - 24y = 0$, one of whose end points has abscissa 6? What are the coordinates of the end points of the focal chord?

9. Prove that the abscissa of any point on the parabola $x^2 = 4ay$ is a mean proportional between the ordinate at that point and the latus rectum for the parabola.

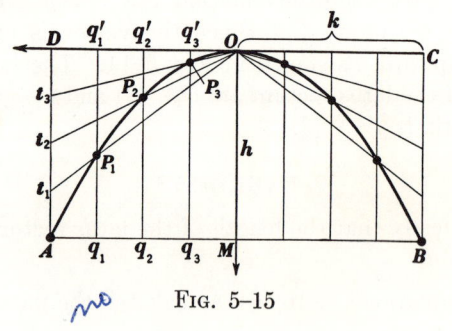

FIG. 5–15

5–7. Parabolic arch

Consider a rectangle $ABCD$, as in Fig. 5–15. Divide AD and the semi-base AM into the same number of equal parts. Designate the points of division on AM as q_1, q_2, q_3, etc.; and the points on AD as t_1, t_2, t_3, etc. Erect perpendiculars to AB from q_1, q_2, q_3, etc. Also, draw lines from O, which is the mid-point of DC, to the points t_1, t_2, t_3, etc. Locate points P_1, P_2, and P_3, respectively, at the intersections of the lines Ot_1 and q_1q_1', Ot_2 and q_2q_2' and Ot_3 and q_3q_3'. Locate similar points on the right-hand portion of the curve. These points lie on a parabola, and the arch AOB is called a parabolic arch.

To prove that the points do lie on a parabola, let O be the origin, OM the positive y-axis, and OD the positive x-axis. Denote OM by h, and AB by $2k$. The conditions to be considered in the following proof are shown in Fig. 5–16. If $P(x, y)$ is any point on arch AO other than O, we have:

(1) $$\frac{x}{y} = \frac{OD}{DL} = \frac{k}{DL}$$

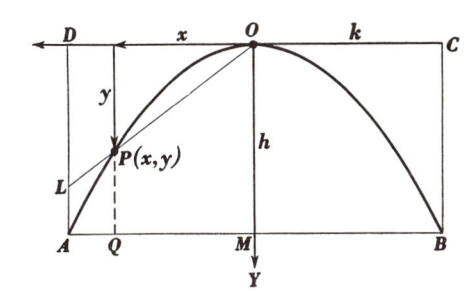

FIG. 5–16

Also, $\frac{DL}{DA} = \frac{MQ}{MA}$, or $\frac{DL}{h} = \frac{x}{k}$. Substituting $\frac{hx}{k}$ for DL in equation (1), we have:

$$\frac{x}{y} = \frac{k}{\frac{hx}{k}}$$

from which $$x^2 = \frac{k^2}{h} y$$

This is an equation of a parabola.

A distance like $|AB|$ in Fig. 5–16 is called the *span*; and a distance like $|MO|$ is called the *height* of the arch.

5–8. Parametric equations of the parabola

The simple forms of the parabola may be expressed in terms of a parameter as follows:

(1) $$x = k_1 t^2 \quad \text{and} \quad y = k_2 t; \quad \text{or} \quad x = k_1 t^2, \; y = k_2 t$$

where k_1 and k_2 are constants not equal to zero and t is the parameter.

174   ANALYTIC GEOMETRY IN THE PLANE

To see that these are equations of a parabola, we eliminate the parameter and obtain

$$x = k_1 \frac{y^2}{k_2^2} \quad \text{or} \quad y^2 = \frac{k_2^2}{k_1} x$$

If $\left|\frac{k_2^2}{k_1}\right|$ is denoted by $4a$, the equation becomes

$$y^2 = \pm 4ax$$

Let us consider now the equations

(2) $\qquad x = k_1 t \quad \text{and} \quad y = k_2 t^2$

Again eliminating the parameter, we obtain

$$y = k_2 \frac{x^2}{k_1^2} \quad \text{or} \quad x^2 = \frac{k_1^2 y}{k_2}$$

If $\left|\frac{k_1^2}{k_2}\right| = 4a$, the equation becomes

$$x^2 = \pm 4ay$$

In either case the coordinates of any point $P(x, y)$ on the parabola satisfy equation (1) or equation (2) with $t = \frac{y}{k_2}$ or $t = \frac{x}{k_1}$.

EXERCISES

1. By assigning convenient values to t and then plotting points, draw the graph of each of the following parabolas from its parametric equations:
   (a) $x = 2t^2$ and $y = t$
   (b) $y = \frac{t}{2}$ and $x = -t^2$
   (c) $x = 3t$ and $y = 2t^2$
   (d) $y = -t$ and $x = 5t^2$

2. Eliminate the parameter in each pair of equations in Exercise 1.

5–9. Applications of parabolic curves

We shall merely mention here a few of the applications of the parabola to the sciences:

(1) The path of the projectile from a gun, when it is acted upon by the force of gravity alone, is a parabola, as indicated in Fig. 5–17.

(2) If the weight of roadbed suspended from a cable and the weight of the cable are uniformly distributed horizontally, then the cable assumes the shape of a parabola.

(3) If a light is placed at the focus of a parabola, the rays that meet the

parabola will be reflected parallel to the axis of the parabola, as indicated in Fig. 5–18.

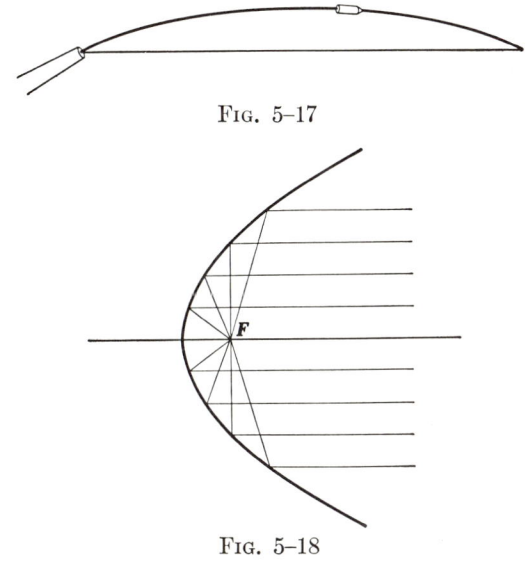

Fig. 5–17

Fig. 5–18

EXERCISES

1. Discuss and sketch the following parabolas:
   (a) $3y^2 + 10x = 0$
   (b) $5x^2 - 32y = 0$
   (c) $x^2 + 24y = 0$
   (d) $2y^2 = 37x$
   (e) $3x^2 - 16y = 0$
   (f) $y^2 + 32x = 0$

2. Find an equation of each of the following parabolas, for which the vertex and the focus are given:
   (a) Vertex at (2, 1) and focus at (4, 1)
   (b) Vertex at (−3, 4) and focus at (−3, −2)
   (c) Vertex at (−4, 7) and focus at (−4, 3)

3. Find an equation of each of the following parabolas, for which the data are:
   (a) Focus is at (1, 2) and directrix is $y = 0$.
   (b) Focus is at (2, 1) and directrix is $x = 0$.
   (c) Focus is at (−1, −3) and directrix is $2x + y − 1 = 0$.
   (d) Focus is at (3, −2) and directrix is $x − 2y = 0$.
   (e) Vertex is at (1, −2) and directrix is $x + 4 = 0$.

4. Find a simple equation of each of the following parabolas, for which the data are:
   (a) Focus is on the y-axis, and the length of the latus rectum is 16.

(b) Focus is on the x-axis, and the length of the latus rectum is 24.

5. Find an equation of each of the following parabolas, for which the data are:
   (a) Focus is at $(0, 0)$ and directrix is $ax + by + c = 0$.
   (b) Vertex is at (h, k), the axis is parallel to the x-axis, and the length of the latus rectum is $4a$.
   (c) Vertex is at (h, k), the axis is parallel to the y-axis, and the length of the latus rectum is $4a$.

6. Derive an equation of a parabolic arch when given: (a) span = 30 ft and height = 10 ft; (b) height = 12 ft and span = 48 ft.

7. Prove that the ordinate of any point on the parabola $y^2 = 4ax$ is a mean proportional between the abscissa at that point and the length of the latus rectum of the parabola.

8. Prove that the perpendicular drawn from any point other than the vertex on a parabola to its axis is a mean proportional between the latus rectum and the distance from the foot of the perpendicular to the vertex.

9. Prove that the segments drawn from the end points of the latus rectum of a parabola to the point where the axis of the parabola intersects its directrix are perpendicular to each other.

10. One end of a focal chord of the parabola $y^2 = 4ax$ is at $(4a, 4a)$. Find the coordinates of the other end.

11. Find the points of intersection of the parabolas $y^2 = 4ax$ and $x^2 = 4ay$.

12. The focal radius of a certain point on the parabola $y^2 = 4ax$ has the same length as the latus rectum. Find the coordinates of the point. How many answers are there?

13. If each point of a given set satisfies the condition that its distance from the point $(-3, 4)$ is 2 units greater than its distance from the line $x - 3 = 0$, find an equation of the set.

14. If each point of a given set satisfies the condition that its distance from the point $(0, -6)$ is 6 units less than its distance from the line $y = 12$, find an equation of the set.

15. Two vertices of a triangle are $P_1(2a, a)$ and $P_2(-2a, a)$. Find an equation of the locus of the third vertex $P(x, y)$, if the slope of P_1P is 1 unit greater than the slope of P_2P.

16. What is an equation of the set of points satisfying the condition that the distance of each point of the set from the origin is 3 units greater than its distance from the line $y - 2 = 0$?

17. For parts (a), (d), (e) of Exercise 1, express y explicitly in terms of x and associate the equations obtained with the appropriate portions of graph. If x is the independent variable, specify the domain and the range of each portion obtained. Hint: 17(a). $y = \dfrac{\sqrt{-30x}}{3}$. Here the vertex is at (0,0). It is a parabola opening to the left. This equation represents the upper half of the parabola. The domain is $\{x \mid -\infty < x \leq 0\}$ and the range is $\{y \mid 0 \leq y < \infty\}$. Do likewise for $y = -\dfrac{\sqrt{-30x}}{3}$.

18. The path of a projectile is given approximately by the parametric equations $x = (v_o \cos \alpha)t$, $y = (v_o \sin \alpha)t - 16t^2$ where v_o and α are constants and t is the time in seconds after the projectile is fired.

    The equations give the rectangular coordinates x and y (in feet) of the projectile in the vertical plane of motion and with the muzzle of the gun at the origin of the co-ordinate system, the x-axis horizontal, and the y-axis vertical; v_o is the muzzle velocity, i.e. the velocity of the projectile at the instant it leaves the gun; α is the angle of inclination of the projectile as it leaves the gun. If $v_o = 104$ ft/sec, $\cos \alpha = \dfrac{5}{13}$, $\sin \alpha = \dfrac{12}{13}$, what are the coordinates of the projectile at times $t = 1, 2,$ and 3?

    Find the time T when the projectile hits the ground, and find the distance D from the muzzle of the gun to the point at which the projectile strikes the ground.

19. Using the equations specified in Exercise 18, solve the following problem: a bullet is fired at an elevation of 30° with an initial velocity of 1600 ft/sec. Assuming that the bullet leaves the gun at ground level, find how far away the bullet strikes the earth. How far above the earth is the highest point of its path?

20. If in Exercise 19, the bullet strikes a target 784 feet above the ground, what distance is the target from the gun in a horizontal direction, i.e., what is the x co-ordinate of the point where the target is located?

21. An arch in the form of an arc of a parabola with its vertex at the center of the arch is 27 feet across at the base and its highest point is 12 feet above the base. What is the length of a beam which has its ends on the arch and which is parallel to the base and 9 feet above it?

22. The cable of a suspension bridge is in the shape of a parabola. The towers which support it are 60 feet high and 450 feet apart and the lowest point on the cable is 12 feet above the road. Find the length of a supporting beam 87 feet above the ground.

23. The circular front of a parabolic reflector is 18 inches in diameter and its depth is 6 inches. Find how far the focus is from the vertex.

THE ELLIPSE

5–10. Simple equations of the ellipse

If each point of a given set satisfies the condition that the ratio of its distance from a fixed point (*focus*) to its distance from a fixed line (*directrix*) is a constant e which is less than 1, the set is called an *ellipse*. By placing a system of axes in a convenient position relative to the focus and the directrix, it is possible to obtain a simple equation of the ellipse. To see how this may be done, we proceed as follows:

Denote the focus by F and the directrix by DD', as shown in Fig. 5–19. Draw the line through F perpendicular to the directrix, and let Q be the point of intersection of this line with DD'. If P is any point on the ellipse and L is the projection of P on $D'D$, we know that $|FP| = e|LP|$. In particular there exist two points, V and V', on the line containing F and Q such that $|FV| = e|VQ|$ and $|FV'| = e|V'Q|$, where V is contained within the segment FQ and V' is external to the segment FQ. This condition follows immediately from the ratio formula and from the fact that $e < 1$. Since both $|VQ|$ and $|V'Q|$ are distances from the directrix, it follows that V and V' are points on the ellipse.

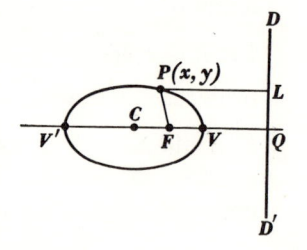

Fig. 5–19

Denote the mid-point of $V'V$ by C (this point C is called the *center* of the ellipse), and let $c = |CF|$ and $a = |CV|$. Then, since $a > c$, we have:

(1) $$|FV| = a - c$$

and

(1') $$|FV'| = a + c$$

We have already seen that

(2) $$|FV| = e|QV|$$

and

(2') $$|FV'| = e|QV'|$$

It is clear from Fig. 5-19 that

(3) $$|QV| = |CQ| - a$$

and

(3') $$|QV'| = |CQ| + a$$

By equations (1) and (3), equation (2) becomes

(4) $$a - c = e(|CQ| - a)$$

By equations (1') and (3'), equation (2') becomes

(5) $$a + c = e(|CQ| + a)$$

Adding equations (1) and (5), we obtain

$$2a = 2e|CQ| \quad \text{or} \quad |CQ| = \frac{a}{e}$$

This tells us that the distance from the center of an ellipse to the directrix is $\frac{a}{e}$. Putting $|CQ| = \frac{a}{e}$ back in equation (4), we find that the relation between a, c, and e is

$$c = ae \qquad (5\text{-}6)$$

In order to obtain a simple equation of the ellipse, we put the origin at the center C and we let the positive x-axis contain the focus F, as indicated in Fig. 5-20. Now the coordinates of F are $(ae, 0)$ or $(c, 0)$; the

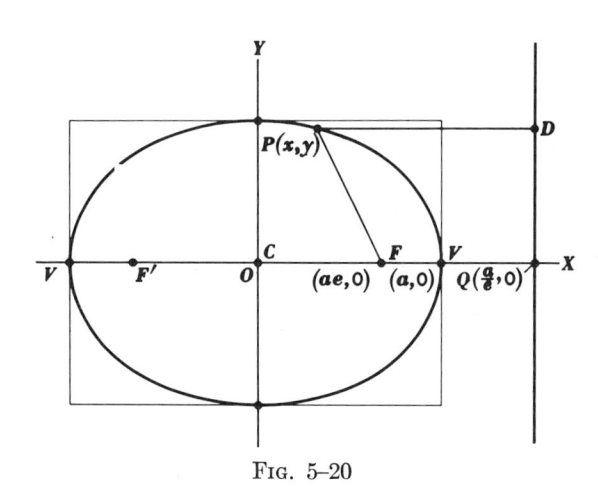

FIG. 5-20

coordinates of V are $(a, 0)$; and an equation of the directrix is $x = \frac{a}{e}$. Since $e < 1$, it follows that $ae < a < \frac{a}{e}$.

From the definition, $\dfrac{|FP|}{|PD|} = e$. Hence,

(6) $$\sqrt{(x-ae)^2+y^2} = e\left|\dfrac{a}{e} - x\right|$$

or

(7) $$\sqrt{(x-ae)^2+y^2} = |a-ex|$$

Squaring equation (6), we obtain
$$x^2 - 2aex + a^2e^2 + y^2 = a^2 - 2aex + e^2x^2$$

or
$$x^2(1-e^2) + y^2 = a^2(1-e^2)$$

and
$$\dfrac{x^2}{a^2} + \dfrac{y^2}{a^2(1-e^2)} = 1 \tag{5-7}$$

For simplification purposes, we let $b = a\sqrt{1-e^2}$. Then the equation becomes

$$\dfrac{x^2}{a^2} + \dfrac{y^2}{b^2} = 1 \tag{5-8}$$

Since $c = ae$, it follows that

$$b^2 = a^2 - c^2 \tag{5-9}$$

This is an important relation between a, b, and c. It tells us that $b < a$, and it may also be written

$$c^2 = a^2 - b^2 \tag{5-9a}$$

If we let $y = 0$ in equation (5-8), we obtain $x = \pm a$. Hence, the curve intersects the x-axis at the points ($\pm a$, 0). These are the points V and V'. If we let $x = 0$, we obtain $y = \pm b$. Hence, the curve intersects the y-axis at the points (0, $\pm b$). When drawing the graph of an ellipse, it is always advisable to plot these points.

Conversely, the equation $\dfrac{x^2}{a^2} + \dfrac{y^2}{b^2} = 1$ is an equation of an ellipse, if $b < a$. To prove this statement, let $e = \sqrt{1 - \dfrac{b^2}{a^2}}$, or $b^2 = a^2(1-e^2)$. Making this replacement for b^2 in equation (5-8) and clearing of fractions, we obtain:

$$(1-e^2)x^2 + y^2 = a^2(1-e^2)$$

Adding $-2aex$ to both sides of the equation, we obtain:

$$(1-e^2)x^2 + y^2 - 2aex = a^2(1-e^2) - 2aex$$

which becomes
$$x^2 - 2aex + a^2e^2 + y^2 = a^2 - 2aex + e^2x^2$$

or
$$(x-ae)^2 + y^2 = (a-ex)^2 = e^2\left(\dfrac{a}{e} - x\right)^2$$

Hence, $$\sqrt{(x-ae)^2+y^2} = e\left|\frac{a}{e}-x\right|$$

This is equation (6) and is equivalent to the following statement: If the focus of the curve is at $(ae, 0)$, then for any point $P(x, y)$ on the curve, $|FP| = e|PD|$, where $D\left(\frac{a}{e}, y\right)$ is the projection of P on the directrix. Hence, equation (5-8) represents an ellipse of which the focus is at $(ae, 0)$, the directrix is the line $x = \frac{a}{e}$, and the eccentricity is e.

If we had chosen the focus to be the point $(-ae, 0)$ and the directrix to be the line $x = -\frac{a}{e}$, we would have obtained exactly this same equation. We see from this that the ellipse has two foci and two directrices.

If we choose the foci to be the point $(0, ae)$ and $(0, -ae)$ and we choose the directrices to be the lines $y = \frac{a}{e}$ and $y = -\frac{a}{e}$, we obtain the simple form of the ellipse to be

$$\frac{y^2}{a^2} + \frac{x^2}{b^2} = 1 \tag{5-8a}$$

where again $b = a\sqrt{1-e^2}$.

It is important for the student to note that we obtain these simple equations because of our special choice of the system of axes with respect to the foci and the directrices; and that, if the origin is at the center and the foci are on one of the axes, the equation can be written when a and b are known.

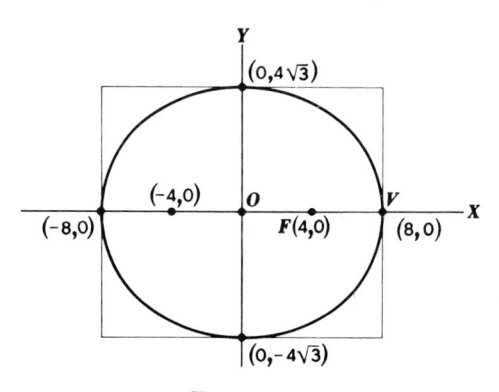

Fig. 5-21

EXAMPLE 5-4. Given an ellipse with its focus at $(4, 0)$, $e = \frac{1}{2}$, and the center at the origin; find an equation of the ellipse.

Solution. In Fig. 5–21, $|OF| = c = 4$. Since $e = \dfrac{c}{a} = \dfrac{1}{2}$, it follows that $\dfrac{4}{a} = \dfrac{1}{2}$, or $a = 8$. We know that $b^2 = a^2 - c^2 = 64 - 16 = 48$. The foci are on the x-axis, and the form of the equation is $\dfrac{x^2}{a^2} + \dfrac{y^2}{b^2} = 1$. So an equation of the ellipse is

$$\frac{x^2}{64} + \frac{y^2}{48} = 1$$

The x-intercepts are ± 8, and the y-intercepts are $\pm 4\sqrt{3}$.

EXERCISES

1. Given a focus $(-ae, 0)$ and a directrix $x = -\dfrac{a}{e}$, derive an equation of the ellipse.

2. Given a focus $(0, ae)$ and a directrix $y = \dfrac{a}{e}$, derive an equation of the ellipse.

3. Given a focus $(0, -ae)$ and a directrix $y = -\dfrac{a}{e}$, derive an equation of the ellipse.

4. Prove that the point $(-a, 0)$ is a point on the ellipse $\dfrac{x^2}{a^2} + \dfrac{y^2}{b^2} = 1$ by using the definition of the ellipse.

5. Find a simple equation of each of the following ellipses from the data:
   (a) Focus is at $(0, 2)$, directrix is $y = 4$, and center is at $(0, 0)$.
   (b) Focus is at $(3, 0)$, directrix is $x = 5$, and center is at $(0, 0)$.
   (c) Focus is at $(-5, 0)$, directrix is $x = -7$, and center is at $(0, 0)$.
   (d) Focus is at $(0, -4)$, directrix is $y = -6$, and center is at $(0, 0)$.
   (e) $a = 5$, $b = 4$, center is at $(0, 0)$, and foci are on the y-axis.
   (f) $a = 8$, $b = 6$, center is at $(0, 0)$, and foci are on the x-axis.

5–11. Explanation of terms for ellipse

In Fig. 5–22 the two points F_1 and F_2 are the *foci* of the ellipse, and the two lines through the points H_1 and H_2 are the *directrices*. The line through the foci of an ellipse is called the *major axis*. The mid-point of the segment joining the foci is the *center*. The line through the center perpendicular to the major axis is called the *minor axis*. The ellipse is symmetric with respect to its major axis, its minor axis, and its center. The points where the curve intersects its major axis are the *vertices*.

The distance from the center to either vertex is denoted by a. The distance from the center to either of the points where the ellipse cuts its minor axis is denoted by b. The distance from the center to either focus is denoted by c or ae. The distance from the center to either directrix is $\frac{a}{e}$. By the *length of the major axis* is meant the distance between the vertices. Hence, this length is equal to $2a$. Although we are using the term *major axis* in two senses, there is no need for confusion. By the *major axis* we shall mean the line that contains the foci; by the *length of the major axis* is meant the absolute distance between the vertices. In the same way the *minor axis* is the line through the center perpendicular to the major axis; and the *length of the minor axis*, which is $2b$, is the absolute distance between the points where the ellipse cuts its *minor axis*. A *focal chord* is a chord through either focus. A *focal radius* is a segment drawn

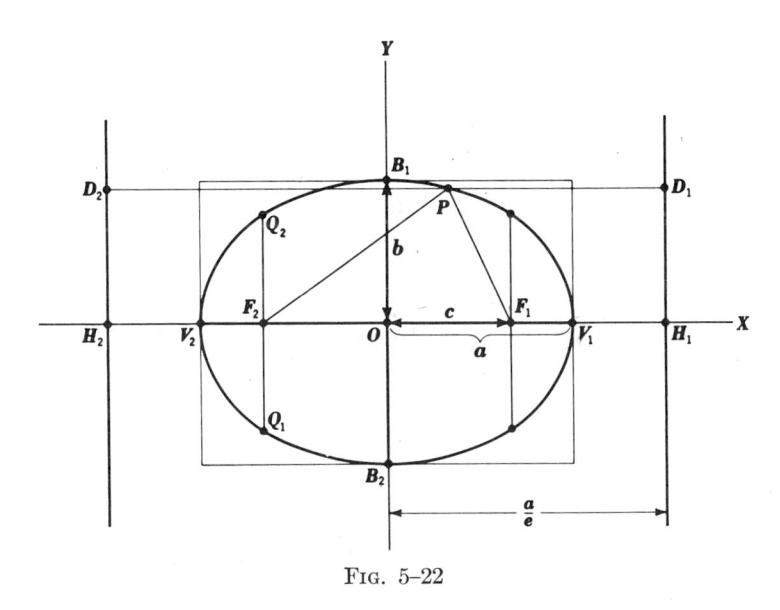

FIG. 5-22

from a focus to any point on the ellipse. A focal chord perpendicular to the major axis is called the *latus rectum*.

It is important to note that in any ellipse a, b, c, e, and $\frac{a}{e}$ are *positive* constants. Referring to Fig. 5-22, we see that

$$a = |OV_1| = |OV_2|, \ b = |OB_1| = |OB_2|, \text{ and } c = |OF_1| = |OF_2|$$

Also, $\quad \frac{a}{e} = |OH_1| = |OH_2|$ and $e = \frac{|F_1P|}{|D_1P|} = \frac{|F_2P|}{|D_2P|}$

We know that $c = ae < a < \frac{a}{e}$. Since $b = \sqrt{a^2 - c^2}$, it follows that $a > b$. These are important relations and should be known thoroughly.

EXERCISES

Find a simple equation of each of the following ellipses and draw its graph:

1. $a = 10$, $b = 6$, and foci are on the x-axis.
2. $a = 16$, $b = 12$, and foci are on the y-axis.
3. $e = \frac{2}{3}$, and focus is at $(6, 0)$.
4. $e = \frac{3}{4}$, and vertex is at $(8, 0)$.
5. $a = 5$, $c = 3$, and foci are on the y-axis.
6. A vertex is at $(6, 0)$, and focus is at $(4, 0)$.
7. $e = \frac{1}{2}$, $b = 5$, and foci are on the x-axis.
8. $b = 5$, $c = 3$, and foci are on the y-axis.
9. $e = \frac{1}{2}$, $c = 4$, and foci are on the y-axis.
10. Vertices are at $(0, \pm 5)$, and $b = 3$.

5–12. The focal radii

In Fig. 5–22, $|F_1P|$ and $|F_2P|$ are the lengths of the *focal radii* to the point P. If the foci are taken on the x-axis and the center is at the origin, it follows from the definition of an ellipse that

$$|F_1P| = a - ex \text{ and } |F_2P| = a + ex$$

Adding the focal radii, we obtain:

$$|F_1P| + |F_2P| = 2a \qquad (5\text{–}10)$$

This relation proves the following theorem: Every point of an ellipse has the property that $|F_1P| + |F_2P| = 2a$, where F_1 and F_2 are the foci and $2a$ is the length of the major axis.

Conversely, if we have an ellipse, every point P for which $|F_1P| + |F_2P| = 2a$ lies on the ellipse.

These two theorems imply that an ellipse is a set of points, each point of which satisfies the condition that the sum of its distances from two distinct fixed points is a constant. The two fixed points are the foci and the constant is 2a.

EXERCISES

1. Prove that every point P for which $|F_1P| + |F_2P| = 2a$ lies on a given ellipse. (Hint: Show that $\sqrt{(x-c)^2 + y^2} + \sqrt{(x+c)^2 + y^2} = 2a$ leads to the simple equation of the ellipse.)

2. Choose the foci to be the points $(0, \pm c)$ and the directrices to be the lines $y = \pm \dfrac{a}{e}$, and prove that the sum of the focal radii is $2a$.

3. Derive an equation of the ellipse if the foci are $(0, \pm c)$ and the sum of the focal radii is $2a$.

4. Find the locus of a point moving so that the sum of its distances from the fixed points $(4, 0)$ and $(-4, 0)$ is a constant 10.

5. What is the locus of a point moving so that the sum of its distances from the fixed points $(-3, 0)$ and $(3, 0)$ is a constant 8?

6. Each point of a given set has the property that the sum of its distances from the fixed points $(0, 5)$ and $(0, -5)$ is a constant 12. Find an equation of the set.

7. A set of points has the property that the sum of the distances of each point of the set from two fixed points $(0, -1)$ and $(0, 1)$ is 4. Find an equation of the set.

8. Prove that in an ellipse, the major axis is a mean proportional between the distance between the foci and the distance between the directrices.

5–13. Discussion of the equation $Ax^2 + Cy^2 + F = 0$, where A and C are of like sign and are not equal to zero

Assume that we have an equation $Ax^2 + Cy^2 + F = 0$, in which $A \neq C$, $F \neq 0$, and F is of different sign from A and C. If we divide the given equation through by $-F$, it becomes

$$\frac{Ax^2}{-F} + \frac{Cy^2}{-F} = 1$$

or

$$\frac{x^2}{-\dfrac{F}{A}} + \frac{y^2}{-\dfrac{F}{C}} = 1$$

If we let $-\dfrac{F}{A} = h^2$ and $-\dfrac{F}{C} = k^2$, where h and k are positive, we obtain

$$\frac{x^2}{h^2} + \frac{y^2}{k^2} = 1 \qquad (5\text{–}11)$$

This is the simple form of the equation of an ellipse having its foci on one of the axes and its center at the origin.

If $h > k$, then $h = a$ and $k = b$; and the x-axis is the major axis. If $h < k$, then $h = b$ and $k = a$; and the y-axis is the major axis. Letting $y = 0$, we find that the x-intercepts of the ellipse are h and $-h$. Letting $x=0$, we find that the y-intercepts of the ellipse are k and $-k$. Solving equation (5–11) for x, we obtain $x = \pm\frac{h}{k}\sqrt{k^2-y^2}$; hence, $|y|\leq k$. Solving equation (5–11) for y, we obtain $y = \pm\frac{k}{h}\sqrt{h^2-x^2}$; so $|x|\leq h$. It follows then that an ellipse is bounded by a rectangle the dimensions of which are $2h$ and $2k$, the sides of which are parallel to the axes of the ellipse, and the center of which is the center of the ellipse. It is helpful in drawing an ellipse to draw first this rectangle.

Now let us assume that $A = C$, $F \neq 0$, and F is of different sign from A and C. The equation $Ax^2 + Cy^2 + F = 0$ then becomes $Ax^2 + Ay^2 = -F$, or $x^2 + y^2 = -\frac{F}{A}$, which is the simple equation of a circle where $-\frac{F}{A} = r^2$. If we define $h = k = r$, our equation may be written in the form shown for equation (5–11). This condition suggests that a circle might be thought of as a special kind of ellipse. Such a view is consistent with the inclusion of a circle among the conic sections, as was done in Section 5–1. However, the definition of conic section adopted in Section 5–2, and in particular the definition of an ellipse adopted in Section 5–10, actually excludes the circle. The reason is as follows: If $h = k$, then $a = b$; hence, $c = \sqrt{a^2 - b^2} = 0$ and the foci coincide at the center. In other words, the eccentricity is zero, since $e = \frac{c}{a}$. But, from the definition of a conic, it follows that $|FP| = 0|DP|$ and $|FP| = 0$, which is impossible. However, for technical reasons, circles henceforth will be included among ellipses.

If $F = 0$, the equation $Ax^2 + Cy^2 + F = 0$ is satisfied by only one point, namely, the origin. When this condition exists, we shall speak of the locus as being a *point ellipse* or a *point circle*, as the case may be. Another term often used in this case is *degenerate ellipse* or *degenerate circle*.

If A, C, and F are of like sign, there is no real locus of the equation $Ax^2 + Cy^2 + F = 0$, and we shall speak of the equation as representing an *imaginary ellipse* or an *imaginary circle*.

EXAMPLE 5–5. Discuss the equation $9x^2 + 4y^2 = 36$ and sketch the curve. In the case of an ellipse, to *discuss* means to find a, b, c, e, $\frac{a}{e}$, the coordinates of the foci, the coordinates of the vertices, the coordinates of the points where the curve cuts its minor axis, equations of the directrices and the domain and range of the relation defined by the equation.

Solution. Writing the equation in the simple form, we have $\frac{x^2}{4} + \frac{y^2}{9} = 1$. The curve is an ellipse; and, since the larger number appears under y^2, the major axis is the y-axis and the minor axis is the x-axis. Hence,

$$a = 3, \ b = 2, \ c = \sqrt{5}, \ e = \frac{\sqrt{5}}{3}, \text{ and } \frac{a}{e} = \frac{9}{\sqrt{5}}.$$

The directrices are $y = \frac{9}{\sqrt{5}}$ and $y = -\frac{9}{\sqrt{5}}$, and the ellipse is shown in Fig. 5–23. As is easily seen from Fig. 5–23, the domain of the relation is the interval $[-2,2]$ and its range is the interval $[-3,3]$.

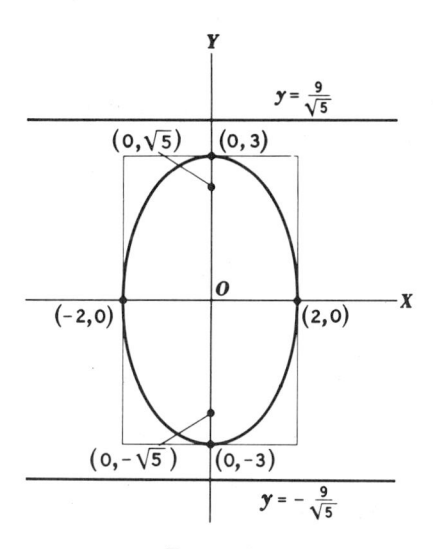

FIG. 5–23

EXAMPLE 5–6. Given one vertex at $(-8, 0)$, the center at $(0, 0)$, and $e = \frac{1}{2}$; find an equation of the conic.

Solution. Since $e = \frac{1}{2}$, we know that the conic is an ellipse. We also know that $a = 8$. So we have $c = ae = 8 \times \frac{1}{2} = 4$, and

$$b^2 = a^2 - c^2 = 64 - 16 = 48, \text{ or } b = 4\sqrt{3}$$

Since the vertices are on the x-axis, an equation of the ellipse is

$$\frac{x^2}{64} + \frac{y^2}{48} = 1$$

188   ANALYTIC GEOMETRY IN THE PLANE

EXERCISES

Discuss each of the following equations and sketch the curve.

1. $4x^2 + 9y^2 = 36$
2. $4x^2 + y^2 = 16$
3. $3x^2 + y^2 = 15$
4. $7x^2 + 3y^2 = 29$
5. $9x^2 + 4y^2 = 144$
6. $x^2 + 4y^2 = 64$
7. $9x^2 + 2y^2 = 18$
8. $4x^2 + 3y^2 = 100$
9. In Exercises 1 to 8, solve for y in terms of x and associate with each equation obtained the appropriate portion of the ellipse. State the domain and range of the relation defined by each equation.

LAST PART

5–14. The latus rectum of an ellipse

Each of the focal chords perpendicular to the major axis of an ellipse is called a *latus rectum* of the ellipse. To find the length of a latus rectum, we first find the end points by solving simultaneously the equation of the

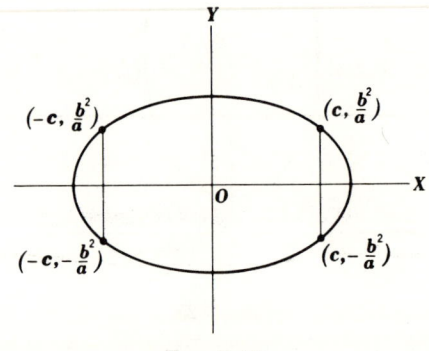

FIG. 5–24

ellipse and that of the line containing the latus rectum. If the major axis is the x-axis, then an equation of the line containing the latus rectum through $(c, 0)$ is $x = c$. Solving this equation with the equation $\dfrac{x^2}{a^2} + \dfrac{y^2}{b^2} = 1$, we have:

$$\frac{c^2}{a^2} + \frac{y^2}{b^2} = 1$$

from which

$$\frac{y^2}{b^2} = 1 - \frac{c^2}{a^2} = \frac{a^2 - c^2}{a^2} = \frac{b^2}{a^2}$$

Therefore,

$$y^2 = \frac{b^4}{a^2}, \text{ or } y = \pm \frac{b^2}{a}$$

Hence, the length of the latus rectum of an ellipse is $\dfrac{2b^2}{a}$. See Fig. 5–24. Note that the same result is obtained for the latus rectum through $(-c, 0)$. When no confusion can result, the length of the latus rectum is called simply the *latus rectum*.

EXERCISES

1. Discuss and sketch each of the following ellipses:
   (a) $x^2 + 2y^2 = 4$
   (b) $9x^2 + 4y^2 = 1$
   (c) $9x^2 + 4y^2 = 144$
   (d) $4x^2 + 9y^2 - 36 = 0$

2. Given the following conditions, write an equation of each ellipse having its center at $(0, 0)$, and sketch the curve:
   (a) $a = 6$, $b = 4$, and foci are on y-axis.
   (b) $a = 6$, $c = 4$, and foci are on x-axis.
   (c) $a = 6$, latus rectum = 3, and foci are on y-axis.
   (d) $b = 4$, latus rectum = 4, and foci are on x-axis.
   (e) $a = 9$, $b = 5$, and vertices are on x-axis.
   (f) $b = 3$, $c = 4$, and vertices are on y-axis.
   (g) Vertices are at $(\pm 5, 0)$, and foci are at $(\pm 4, 0)$.
   (h) Foci are at $(0, \pm 3)$, and $b = 4$.

3. Find a simple equation of each of the following ellipses and draw its graph.
   (a) $a = 16$, $e = \frac{3}{4}$, and foci are on x-axis.
   (b) $b = 16$, $e = \frac{4}{5}$, and foci are on y-axis.
   (c) $c = 8$, $e = \frac{2}{3}$, and foci are on y-axis.
   (d) $a = 4$, latus rectum = 2, and foci are on x-axis.
   (e) $b = 3$, latus rectum = 4, and foci are on y-axis.
   (f) $e = \frac{4}{5}$, latus rectum = 10, and foci are on y-axis.
   (g) A vertex is at $(8, 0)$, and length of minor axis is 6.
   (h) A focus is at $(6, 0)$, and length of major axis is 18.
   (i) A vertex is at $(0, 8)$, and a focus is at $(0, 6)$.
   (j) A vertex is at $(10, 0)$, and latus rectum = 5.
   (k) Latus rectum = 6, and an end of the minor axis is at $(0, 5)$.
   (l) An end of a latus rectum is at $(8, 3)$, and foci are on x-axis.
   (m) $a = 10$, $e = \frac{4}{5}$, and foci are on y-axis.
   (n) An end of a latus rectum is at $(10, 6)$, and foci are on y-axis.

4. Find a simple equation of the ellipse passing through each of the following pairs of points:
   (a) $(3, 6)$ and $(-4, 4)$
   (b) $(5, 1)$ and $(3, -2)$
   (c) $(5, 2)$ and $(1, 4)$
   (d) $(-1, 6)$ and $(2, 3)$

   Hint: Substitute the coordinates of the given points for x and y in equation (5–11).

5. Find the eccentricity of an ellipse when the length of the latus rectum is three-fourths of the length of the major axis.

6. Find the eccentricity of an ellipse when the length of the latus rectum is two-thirds of the length of the minor axis.

7. Show that the length of the latus rectum of an ellipse is equal to $2b\sqrt{1-e^2}$.

8. Show that the length of the latus rectum of an ellipse is equal to $2a(1-e^2)$.

9. Prove that the ratio of the squares of the semi-major axis and the semi-minor axis is equal to the ratio of the distances from the center of an ellipse to a directrix and from the directrix to the corresponding focus.

10. Deduce an equation representing the set of all ellipses having an eccentricity $\frac{1}{2}$, center at the origin, and foci on the x-axis.

11. If the segment joining the left-hand vertex with the upper end of the minor axis of a simple ellipse having its foci on the x-axis is parallel to the segment joining the center with the upper end of the right-hand latus rectum, what is the relationship between a, b, and c?

12. In an ellipse whose major axis is twice the minor axis, a segment of length equal to the minor axis has one end on the ellipse and the other end on the minor axis. The two ends are always on opposite sides of the major axis. Prove that the mid-point of the segment is always on the major axis.

13. The base of a triangle is fixed and the product of the *tangents* of the *base angles* is a positive constant. Find the locus of the vertex.

14. The orbit of the earth is an ellipse with the sun located at one focus. The length of its major axis is approximately 186,000 miles and its eccentricity $e = 0.0167$. Determine the greatest and least distances of the planet earth from the sun.

15. If the earth is placed at a focus and an orbit of a satellite is known to be elliptic such that its shortest distance (perigee) is known to be 238 miles and its furthest distance (apogee) is 1762 miles from the earth, find the eccentricity of the satellite's orbit.

16. An arch in the form of a semi-ellipse is 100 feet across the base and 30 feet high with the major axis horizontal. Find the distance from the level of the top to a point on the arch (a) 30 feet from the minor axis, (b) 40 feet from the minor axis.

5–15. Construction of an ellipse

Three methods of constructing an ellipse will be described here:

Method 1. In Section 5–12 we saw that an ellipse is the locus of a point moving in such a way that the sum of its distances from two fixed points is a constant. Using this definition, we have the following simple method for drawing the ellipse.

Place two thumb tacks in a piece of cardboard. As indicated in Fig. 5–25, denote their positions by F_1 and F_2, and let the distance between them be $2c$. Tie a piece of string in a loop such that the total length of string is equal to $2a + 2c$, where a is any quantity greater than c, and place

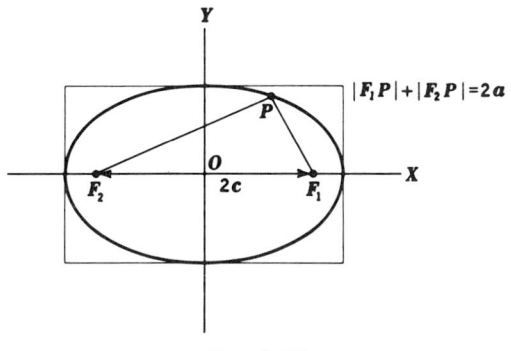

FIG. 5–25

the loop over the tacks. If a pencil is placed in the loop and is moved so as to keep the string taut, then $|F_1P| + |F_2P| = 2a$; hence, by definition, the pencil describes an ellipse. In case it is desired to construct an ellipse

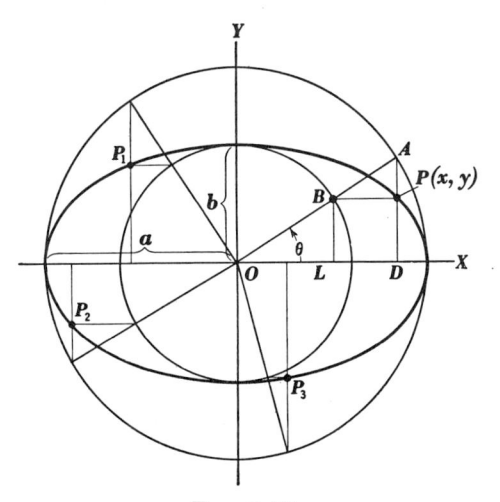

FIG. 5–26

with certain dimensions, c is selected as the required distance from the center to one focus and a is taken equal to the semi-major axis.

Method 2. As shown in Fig. 5-26, draw two concentric circles of radii a and b, respectively, where $a > b$. Draw any radius OA to the circumference of the large circle; and let B denote the point where OA cuts the small circle. Drop a perpendicular from A to the x-axis at D, and draw a line through B that is parallel to the x-axis and intersects AD at P. Every point, as P, obtained in this manner must lie on the ellipse. To prove this, we consider P to have the coordinates x and y and we let θ represent the *eccentric angle* XOA. Then

$$x = \overline{OD} = a \cos \theta$$

$$y = \overline{DP} = \overline{LB} = b \sin \theta$$

Hence, $\dfrac{x}{a} = \cos \theta$ and $\dfrac{y}{b} = \sin \theta$. If we square each of these two equations and add the results, we obtain

$$\frac{x^2}{a^2} + \frac{y^2}{b^2} = 1$$

This is the simple form of the equation of an ellipse. The two circles with radii a and b are sometimes called the *auxiliary circles* of the ellipse. When it is desired to construct an ellipse of a certain size, a is taken as the required semi-major axis and b as the semi-minor axis.

Method 3. Fix two points F_2 and F_1, as shown in Fig. 5-27, and let $|F_1 F_2| = 2c$, where c is any number greater than zero. Let a be any number such that $2a > 2c$. Then with a radius r that is less than $2a$ and greater than $(a-c)$, and with F_1 as a center, generate a circle. Also, with $(2a-r)$ as a radius and with F_2 as a center, generate a second circle. The points of

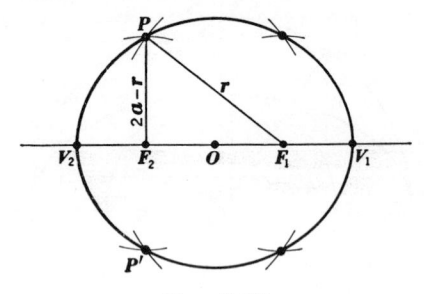

Fig. 5-27

intersection, P and P', of the two circles are points on an ellipse, since $|F_1 P| + |F_2 P| = r + 2a - r = 2a$. As in Method 1, an ellipse of a certain size can be constructed by taking c as the required distance from the center to one focus and taking a as the semi-major axis.

5–16. Parametric equations of the ellipse

In Section 5–15 the construction of an ellipse by means of the auxiliary circles shows that x and y may each be expressed in terms of a third variable θ. The equations are:

$$x = a \cos \theta$$
$$y = b \sin \theta \tag{5-12}$$

These are called *parametric equations* of the ellipse, where θ is the parameter. It is to be noted that for every point on the ellipse there is a corresponding angle θ; and for each value of θ a point on the ellipse is determined.

EXERCISES

1. Draw each of the following ellipses from its parametric equations, and find its simple equation by eliminating the parameter:

   (a) $x = 3 \cos \theta$ and $y = \sin \theta$
   (b) $x = 5 \cos \theta$ and $y = 4 \sin \theta$
   (c) $x = 8 \cos \theta$ and $y = 2 \sin \theta$
   (d) $x = 3 \sin \theta$ and $y = 5 \cos \theta$

2. What are the parametric equations of an ellipse if the radii of the auxiliary circles are 8 and 5, respectively?

3. What are the parametric equations of an ellipse if the radii of the auxiliary circles are 12 and 6, respectively.

4. Construct the ellipses given the radii of the auxiliary circles to be:

   (a) 9 and 4
   (b) 3 and 8
   (c) 3 and 2
   (d) 4 and 7

5–17. Applications of the ellipse

A few practical applications of the ellipse follow:

(1) In the construction of bridges, elliptic arches are used as well as parabolic arches.

(2) If the focal radii are drawn to any point on the ellipse, they make equal angles with the tangent to the ellipse at this point. Hence, if light rays or sound rays originate at either focus, they will be reflected from the surface so as to converge at the other focus.

(3) The paths or orbits of the planets with respect to the sun are approximate ellipses, the sun being at one focus.

194   ANALYTIC GEOMETRY IN THE PLANE

THE HYPERBOLA

5–18. *Simple equations of the hyperbola*

If each point of a given set satisfies the condition that the ratio of its distance from a fixed point (*focus*) to its distance from a fixed line (*directrix*) is a constant e, which is greater than 1, the set is called a *hyperbola*. If we denote the focus by F and the directrix by DD', as in Fig. 5–28, and we let L be the projection of F on the directrix, then we know by the ratio formula that there exist two points V and V' on the line containing FL such that $|FV| = e|VL|$ and $|FV'| = e|V'L|$. By definition, the points V and V' are on the hyperbola. The mid-point C of VV' is called the *center* of the hyperbola. If we denote $|CV|$ or $|CV'|$ by a and $|CF|$ by c, we find, by the reasoning of Section 5–10, that $|CL| = \dfrac{a}{e}$ and $c = ae$. The student should verify these results as an exercise.

To obtain a simple equation of a hyperbola, we choose the coordinate system so that the center is at $(0, 0)$. Then the focus is at $(ae, 0)$ and an equation of the directrix is $x = \dfrac{a}{e}$, as shown in Fig. 5–29. Since $e > 1$, it follows that $c = ae > a > \dfrac{a}{e}$.

By definition, $\dfrac{|F_1 P|}{|DP|} = e$, where D is the projection of P on the directrix. Hence,

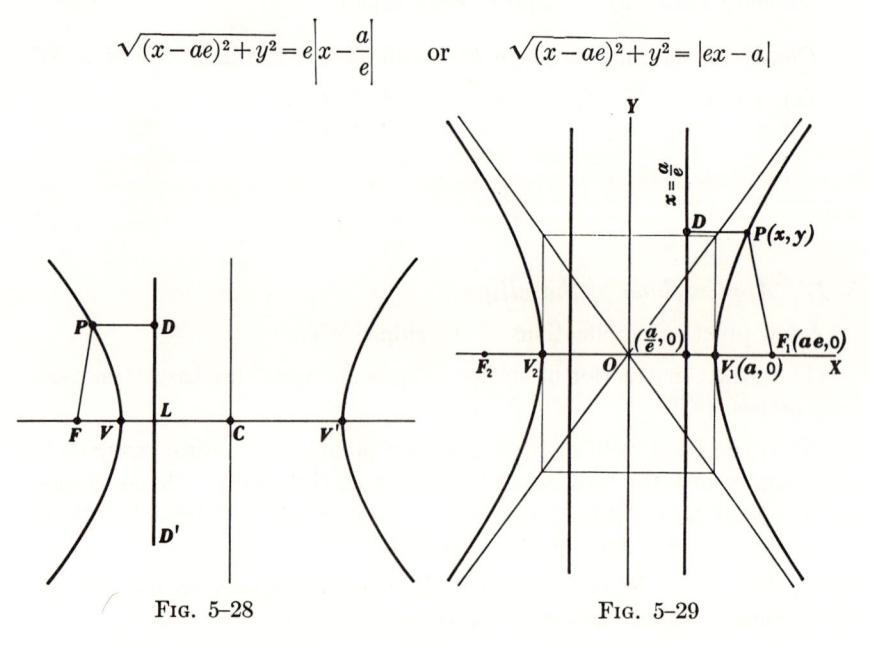

$$\sqrt{(x-ae)^2 + y^2} = e\left|x - \dfrac{a}{e}\right| \qquad \text{or} \qquad \sqrt{(x-ae)^2 + y^2} = |ex - a|$$

FIG. 5–28   FIG. 5–29

Squaring both sides, we obtain

$$x^2 - 2aex + a^2e^2 + y^2 = a^2 - 2aex + e^2x^2$$

or

$$x^2(1 - e^2) + y^2 = a^2(1 - e^2)$$

Dividing each term of this equation by $a^2(1 - e^2)$, we obtain

$$\frac{x^2}{a^2} + \frac{y^2}{a^2(1 - e^2)} = 1$$

Since $e > 1$, $a^2(1 - e^2)$ is a negative number. So we let $b = a\sqrt{e^2 - 1}$ or $a^2(1 - e^2) = -b^2$; and we have for the simple equation of the hyperbola:

$$\frac{x^2}{a^2} - \frac{y^2}{b^2} = 1 \qquad (5\text{-}13)$$

If we let $y = 0$ in equation (5-13), we obtain $x = \pm a$. Hence, the curve intersects the x-axis at the points $(a, 0)$ and $(-a, 0)$. If we let $x = 0$, we obtain $y^2 = -b^2$. This tells us that the locus of equation (5-13) does not intersect the y-axis. Since $c = ae$, we have $a^2 - a^2e^2 = -b^2$, or $a^2 - c^2 = -b^2$ and

$$c^2 = a^2 + b^2.$$

If we had taken the focus at the point $(-c, 0)$ and the directrix to be the line $x = -\frac{a}{e}$, we would have obtained the same equation. So the hyperbola, like the ellipse, has two foci, two directrices, and a center. Because ellipses and hyperbolas have centers, they are called the *central* conics.

Conversely, the equation $\frac{x^2}{a^2} - \frac{y^2}{b^2} = 1$ represents a hyperbola with its center at $(0, 0)$ and its foci on the x-axis. The proof of this follows the same reasoning as that given in Section 5-10 for the ellipse. Hence, this proof will be left as an exercise for the student.

When the coordinate system is so chosen that the foci are on the y-axis at $(0, ae)$ and $(0, -ae)$ and the directrices are the lines $y = \frac{a}{e}$ and $y = -\frac{a}{e}$, we obtain as an equation of the hyperbola

$$\frac{y^2}{a^2} - \frac{x^2}{b^2} = 1 \qquad (5\text{-}14)$$

In this case, the hyperbola intersects the y-axis but not the x-axis.

It is most important that the student understand that we obtain these simple equations only because of our special choice of the coordinate system with respect to the foci and the directrices.

EXERCISES

1. Derive equation (5–14).

2. Given a focus $(-ae, 0)$ and a directrix $x = -\dfrac{a}{e}$, derive an equation of the hyperbola.

3. Given a focus $(0, ae)$ and a directrix $y = \dfrac{a}{e}$, derive an equation of the hyperbola.

4. Given a focus $(0, -ae)$ and a directrix $y = -\dfrac{a}{e}$, derive an equation of the hyperbola.

5. Prove that the point $(a, 0)$ is a point on the hyperbola $\dfrac{x^2}{a^2} - \dfrac{y^2}{b^2} = 1$ by using the definition of the hyperbola.

6. Find a simple equation of each of the hyperbolas for the following conditions:

    (a) Focus is at $(0, 2)$, and directrix is $y = 1$.
    (b) Focus is at $(4, 0)$, and directrix is $x = 3$.
    (c) Focus is at $(0, -6)$, and directrix is $y = -4$.
    (d) Focus is at $(0, -3)$, and directrix is $y = -2$.

7. Prove that $\dfrac{x^2}{a^2} - \dfrac{y^2}{b^2} = 1$ is an equation of a hyperbola with center at $(0, 0)$ and foci on the x-axis. (Hint: See Section 5–10.)

5–19. Explanation of terms for hyperbola

The two fixed points F_1 and F_2, Fig. 5–30, are the *foci* and the two fixed lines through H_1 and H_2 are the *directrices*. The line through the foci is the *transverse axis*. The mid-point of the segment joining the foci is the *center*. The line through the center and perpendicular to the transverse

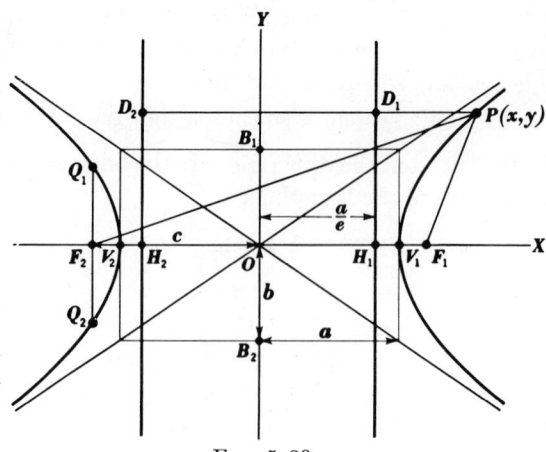

FIG. 5–30

axis is the *conjugate axis*. The hyperbola is symmetric with respect to its transverse axis, its conjugate axis, and its center. The points where the hyperbola intersects its transverse axis are its *vertices*. By the *length of the transverse axis* is meant the *distance between the vertices;* and this length is denoted by 2a, where a is the distance from the center to either vertex. It is important for the student to note that the *transverse axis* is a line and that the *length of the transverse axis* is the length of the segment of that line having the vertices of the hyperbola as its end points. Although the hyperbola does not intersect its conjugate axis, we represent the *length of the conjugate axis* by 2b, where b is the *absolute distance* from the center to either end of the segment that determines the *length of the conjugate axis*.

The *absolute distance from the center to either focus* is $c = ae$. The *absolute distance from the center to either directrix* is $\frac{a}{e}$. A *focal chord* is a chord through either focus. A *focal radius* is a segment drawn from a focus to any point on the hyperbola. A focal chord perpendicular to the transverse axis is called a *latus rectum*. It is important to note that a, b, c, e, and $\frac{a}{e}$ are positive constants. In Fig. 5-30 it is clear that $a = |OV_1| = |OV_2|$, $b = |OB_1| = |OB_2|$, $c = |OF_1| = |OF_2|$, $\frac{a}{e} = |OH_1| = |OH_2|$, and $e = \frac{|PF_1|}{|PD_1|} = \frac{|PF_2|}{|PD_2|}$.

From the definitions of a and e, we know that $ae > a > \frac{a}{e}$. Since $ae = c$, it follows that $c > a$. Also, since $b^2 = a^2(e^2 - 1)$, it follows that $b^2 = c^2 - a^2$ and $c > b$. There is no restriction in the relationship of a to b. That is, a may be greater than, less than, or equal to b. These are very important relations and the student must know them thoroughly. He should compare them with the corresponding relations for the ellipse.

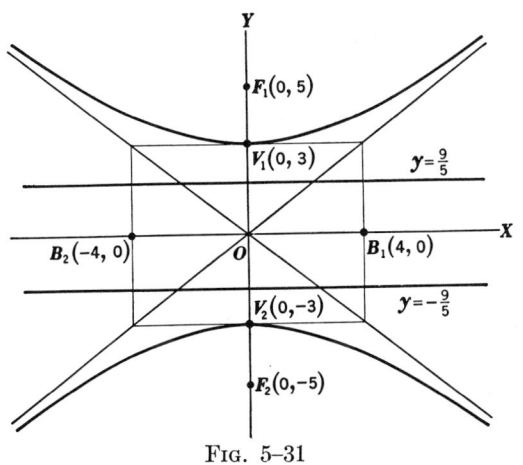

Fig. 5-31

EXAMPLE 5–7. Given the equation $\dfrac{y^2}{9}-\dfrac{x^2}{16}=1$. Discuss the curve by finding $a, b, c, e, \dfrac{a}{e}$, the coordinates of the foci, the coordinates of the vertices, and the equations of the directrices; and sketch the curve. Solve for y in terms of x and specify which parts of the curve the equations represent.

Solution. Since the right-hand side is $+1$, the equation is of the form $\dfrac{y^2}{a^2}-\dfrac{x^2}{b^2}=1$. Hence, $a^2=9$, $b^2=16$, and $c^2=a^2+b^2=25$; and so $a=3$, $b=4$, $c=5$, $e=\dfrac{5}{3}$, and $\dfrac{a}{e}=\dfrac{9}{5}$. Equations of the directrices are $y=\dfrac{9}{5}$ and $y=-\dfrac{9}{5}$, and the curve is shown in Fig. 5–31. Solving for y, we get $y=\dfrac{3}{4}\sqrt{16+x^2}$ and $y=-\dfrac{3}{4}\sqrt{16+x^2}$, the former is the equation of the upper part of the curve and the latter is the equation of the lower part.

EXAMPLE 5–8. Given $b=4$, $c=5$, the center at $(0, 0)$, and the foci on the x-axis; find an equation of the hyperbola.

Solution. Since the foci are on the x-axis, the simple equation is $\dfrac{x^2}{a^2}-\dfrac{y^2}{b^2}=1$. Since $b=4$ and $c=5$, we know that $a^2=25-16=9$. Hence, an equation of the hyperbola is

$$\dfrac{x^2}{9}-\dfrac{y^2}{16}=1$$

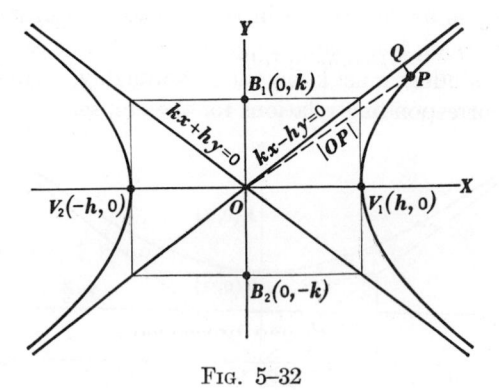

FIG. 5–32

5–20. Asymptotes of a hyperbola

If a line has the property that the distance between the line and a point P of a given curve approaches zero as the point P moves in such a way that the absolute distance $|OP|$ from a fixed point O increases indefinitely, the line is called an asymptote of the curve.

Consider the equation $\frac{x^2}{h^2} - \frac{y^2}{k^2} = n$. For positive values of n, this is an equation of a hyperbola with the x-axis the transverse axis; and if n is negative, the y-axis is the transverse axis. To see this, write the equation in the simple form as follows:

$$\frac{x^2}{h^2 n} - \frac{y^2}{k^2 n} = 1$$

When n is 0, the equation $\frac{x^2}{h^2} - \frac{y^2}{k^2} = n$ becomes $\frac{x^2}{h^2} - \frac{y^2}{k^2} = 0$; and we now have a pair of lines whose equations are $\frac{x}{h} + \frac{y}{k} = 0$ and $\frac{x}{h} - \frac{y}{k} = 0$. Let us investigate how these lines are related to the hyperbola $\frac{x^2}{h^2} - \frac{y^2}{k^2} = 1$. As shown in Fig. 5-32, both lines go through the center of the hyperbola; and each line is a diagonal of the rectangle the dimensions of which are $2h$ and $2k$, the sides of which are parallel to the axes of the hyperbola, and the center of which is the center of the hyperbola.

Consider the distance d from the diagonal line $kx - hy = 0$ to a point $P_1(x_1, y_1)$ on the hyperbola. We have

$$d = |QP_1| = \left| \frac{kx_1 - hy_1}{\sqrt{h^2 + k^2}} \right|$$

If we multiply both the numerator and the denominator by $|kx_1 + hy_1|$, we obtain

$$d = \left| \frac{k^2 x_1^2 - h^2 y_1^2}{\sqrt{h^2 + k^2}} \cdot \frac{1}{kx_1 + hy_1} \right|$$

Since (x_1, y_1) is on the hyperbola, it follows that $k^2 x_1^2 - h^2 y_1^2 = h^2 k^2$. Therefore,

$$d = \left| \frac{h^2 k^2}{\sqrt{h^2 + k^2}} \cdot \frac{1}{kx_1 + hy_1} \right| \qquad (5\text{-}15)$$

As P proceeds along the curve in quadrant I, x_1 and y_1 increase indefinitely. Therefore, $kx_1 + hy_1$ increases indefinitely; and, hence, d approaches 0. But this tells us that the line $kx - hy = 0$ is an asymptote to the hyperbola $\frac{x^2}{h^2} - \frac{y^2}{k^2} = 1$. If P_1 had proceeded along the other branch in quadrant III, x_1 and y_1 would have decreased indefinitely; and, again, we would have d approaching 0.

In exactly the same way we can show that $kx + hy = 0$ is also an asymptote to the curve. The student should do this as an exercise.

From now on, *discussion* of an equation of a hyperbola should include determination of the asymptotes; plotting the graph is greatly facilitated by first drawing the asymptotes.

EXAMPLE 5–9. Find an equation of the hyperbola for which the center is at $(0, 0)$, the foci are on the x-axis, one vertex is at $(-2, 0)$, and an equation of one asymptote is $y = 4x$.

Solution. Since the foci are on the x-axis, an equation of the hyperbola is $\dfrac{x^2}{a^2} - \dfrac{y^2}{b^2} = 1$. Hence, the equations of the asymptotes are $y = \pm \dfrac{b}{a} x$. Since $y = 4x$ is an asymptote, $\dfrac{b}{a} = 4$. Also, since we know that $a = 2$, it follows that $b = 8$. An equation of the hyperbola is

$$\frac{x^2}{4} - \frac{y^2}{64} = 1$$

EXAMPLE 5–10. Find an equation of the hyperbola from the following data: The center is at $(0, 0)$, one focus is at $(0, 5)$, and an equation of a directrix is $y = -3$.

Solution. The form of the equation is $\dfrac{y^2}{a^2} - \dfrac{x^2}{b^2} = 1$. We know that $c = 5$ and $\dfrac{a}{e} = 3$. Since $e = \dfrac{c}{a}$, it follows that $\dfrac{a^2}{c} = 3$, or $a^2 = 3c$. Hence, $a^2 = 15$ and $b^2 = c^2 - a^2 = 25 - 15 = 10$. Therefore, an equation of the hyperbola is

$$\frac{y^2}{15} - \frac{x^2}{10} = 1$$

EXAMPLE 5–11. Discuss the equation $4x^2 - 9y^2 + 36 = 0$, and sketch the curve.

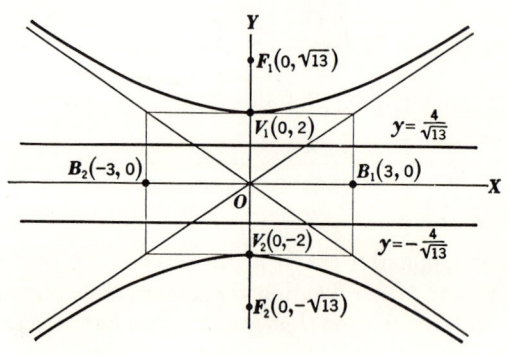

FIG. 5–33

Solution. We first write the equation as $\dfrac{y^2}{4} - \dfrac{x^2}{9} = 1$. Hence, $a = 2$, $b = 3$, $c = \sqrt{13}$, $e = \dfrac{\sqrt{13}}{2}$, and $\dfrac{a}{e} = \dfrac{4}{\sqrt{13}}$. The y-axis is the transverse axis and the x-axis is the conjugate axis. The curve is shown in Fig. 5–33. Equations of the directrices are: $y = \dfrac{4}{\sqrt{13}}$ and $y = -\dfrac{4}{\sqrt{13}}$. Equations of the asymptotes are $2x - 3y = 0$ and $2x + 3y = 0$. Here $y = \frac{2}{3}\sqrt{9 + x^2}$ represents the upper half of the hyperbola and $y = -\frac{2}{3}\sqrt{9 + x^2}$ represents the lower half.

EXERCISES

1. Discuss each of the following equations and sketch the curve:
    (a) $x^2 - y^2 = 4$
    (b) $9x^2 - 16y^2 = 144$
    (c) $x^2 - 4y^2 + 16 = 0$
    (d) $4x^2 - y^2 + 16 = 0$
    (e) $16x^2 - 9y^2 + 144 = 0$
    (f) $2x^2 - 3y^2 + 5 = 0$
    (g) $3y^2 - x^2 + 27 = 0$
    (h) $4y^2 - 9x^2 - 36 = 0$

2. Find a simple equation of each of the following hyperbolas:
    (a) $a = 10$, $b = 6$, and foci are on the x-axis.
    (b) $a = 6$, $b = 8$, and foci are on the y-axis.
    (c) $e = \frac{3}{2}$, and a focus is at $(0, 4)$.
    (d) $e = \frac{4}{3}$, and a vertex is at $(0, 6)$.
    (e) $a = 4$, $c = 5$, and foci are on the x-axis.
    (f) A vertex is at $(6, 0)$, and a focus is at $(8, 0)$.
    (g) $e = 2$, $b = 4$, and foci are on the y-axis.
    (h) $b = 5$, $c = 7$, and foci are on the x-axis.
    (i) $e = \frac{3}{2}$, $c = 6$, and foci are on the y-axis.
    (j) Vertices are at $(0, \pm 4)$, and $b = 3$.
    (k) Directrix is $y = 4$, and $a = 5$.
    (l) An asymptote is $2y - 3x = 0$, $a = 3$, and foci are on the x-axis.
    (m) An asymptote is $x + 3y = 0$, $b = 2$, and foci are on the y-axis.
    (n) An asymptote is $y = 4x$, $c = 17$, and foci are on the x-axis.

3. Find equations of the asymptotes of the hyperbola $\frac{x^2}{h^2} - \frac{y^2}{k^2} = -1$.

4. Prove that the product of the distances from a point on the hyperbola to its asymptotes is a constant. What is the value of this constant?

5. Prove that the length of a perpendicular drawn from a focus $(c, 0)$ to an asymptote of the hyperbola $\frac{x^2}{a^2} - \frac{y^2}{b^2} = 1$ is equal to b.

6. Find e in terms of a and b; find b in terms of a and e.

7. The asymptotes of a hyperbola have inclinations of $60°$ and $120°$, respectively. If the vertices are located at the points $(-1, 0)$ and $(1, 0)$, find an equation of the hyperbola.

8. The slope of one asymptote of a simple hyperbola is $\frac{2}{3}$. What is the eccentricity if the foci are on the x-axis?

9. The base of a triangle is fixed and the product of the tangents of the base angles is a negative constant. What is the locus of the vertex?

10. Find an equation of the hyperbola whose vertices are at the points $(0, \pm 4)$ and whose eccentricity is $\frac{3}{2}$.

11. The axes of a hyperbola are the coordinate axes. If the curve contains the points $(-1, -4)$ and $(2, -7)$, find an equation of the hyperbola.

12. What do the two conics $\dfrac{x^2}{24} + \dfrac{y^2}{8} = 1$ and $\dfrac{x^2}{4} - \dfrac{y^2}{12} = 1$ have in common?

13. What do the hyperbolas $9x^2 - y^2 = k$, where $k \neq 0$ is an arbitrary constant, have in common? Describe the conics if $k < 0$; if $k > 0$. What happens when $k = 0$?

14. In parts (a) to (h) of Exercise 1, solve for y in terms of x and sketch the graph of each y.

15. Sketch the graph of the equation $|\,x^2 - y^2\,| = 1$

16. Discuss and sketch the graph of the equation $x\,|\,x\,| - y\,|\,y\,| = 1$

[Hint: Consider $x^2 + y^2 = 1$; $x^2 - y^2 = 1$; $y^2 - x^2 = 1$; $-x^2 - y^2 = 1$; $x^2 = 1$; $y^2 = 1$.]

5–21. The focal radii

If P is any point on the hyperbola in Fig. 5–34, $|F_1P|$ and $|F_2P|$ are the lengths of the focal radii to point P. If the foci are at $(ae, 0)$ and $(-ae, 0)$ and P is on the right-hand branch of the hyperbola, it follows from the definition of a hyperbola that $|F_1P| = ex - a$ and $|F_2P| = ex + a$. See Section 5–18. Taking the difference between the lengths of the focal radii, we obtain

$$|F_2P| - |F_1P| = 2a$$

If P lies on the left-hand branch of the hyperbola, as indicated in Fig. 5–34, we have:

$$|F_1P| = e|D_1P| = e\left(-x + \frac{a}{e}\right) = -ex + a$$

and

$$|F_2P| = e|D_2P| = e\left(-x - \frac{a}{e}\right) = -ex - a$$

In this case we see that

$$|F_2P| - |F_1P| = -2a$$

Hence, we have proved the following theorem: Every point of a hyperbola has the property that $|F_1P| - |F_2P| = \pm 2a$, where F_1 and F_2 are the foci and $2a$ is the length of the transverse axis.

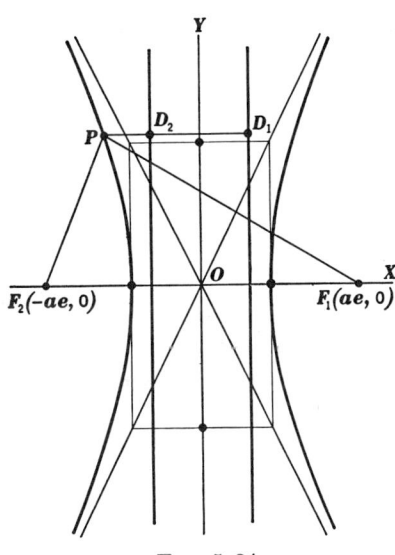

FIG. 5-34

Conversely, every point P for which $|F_1P| - |F_2P| = \pm 2a$ lies on the hyperbola.

These two theorems imply that a hyperbola is a set of points, each point of which satisfies the condition that the difference between its distances from two fixed points is a constant. It follows also that the two fixed points are the foci and the absolute value of the constant is 2a.

EXERCISES

1. Choose the foci to be $(\pm c, 0)$ and derive an equation for the hyperbola from the property that the constant difference between the focal radii is $\pm 2a$.

2. Choose the foci of a hyperbola to be $(0, \pm ae)$ and prove that the difference between the focal radii is $\pm 2a$.

3. Derive an equation for the hyperbola if the foci are $(0, \pm c)$ and the difference between the focal radii is equal to $\pm 2a$.

4. The foci of a hyperbola are at $(\pm 4, 0)$ and the difference between the focal radii is ± 6. What is an equation of the hyperbola?

5. The absolute value of the difference between the focal radii is 4. If the center of the hyperbola is at the origin and one focus is at $(0, 6)$, find an equation of the conic.

204   ANALYTIC GEOMETRY IN THE PLANE

6. Each point of a set has the property that the difference of its distances from the points (0, −6) and (0, 6) is 8. Find an equation of the set.

7. A set of points has the property that the difference of the distances from each point of the set to the points (−7, 0) and (7, 0) is 10. Find an equation of the set.

5–22. Discussion of the equation $Ax^2 + Cy^2 + F = 0$, where A and C are unlike in sign and different from zero

Assume that we have the equation $Ax^2 + Cy^2 + F = 0$, in which $F \neq 0$. If we divide through by $-F$, we can write the equation in the form

$$\frac{Ax^2}{-F} + \frac{Cy^2}{-F} = 1$$

or

$$\frac{x^2}{-\frac{F}{A}} + \frac{y^2}{-\frac{F}{C}} = 1$$

In this case, $-\frac{F}{A}$ and $-\frac{F}{C}$ are opposite in sign. If $-\frac{F}{A}$ is positive, the equation becomes

$$\frac{x^2}{h^2} - \frac{y^2}{k^2} = 1 \qquad (5\text{--}16)$$

If $-\frac{F}{C}$ is positive, the equation is

$$-\frac{x^2}{h^2} + \frac{y^2}{k^2} = 1 \qquad (5\text{--}17)$$

In either case $\left|-\frac{F}{A}\right| = h^2$ and $\left|-\frac{F}{C}\right| = k^2$, and h and k are greater than 0.

Either equation (5–16) or equation (5–17) is the simple equation of a hyperbola having its foci on one of the axes and its center at the origin. The curve is symmetric with respect to the x-axis, the y-axis, and the origin. When the right-hand side of the equation is $+1$, the plus sign on the left precedes the term involving the variable of the transverse axis, while the minus sign precedes the term involving the variable of the conjugate axis. We shall assume that the equation has been written in the form $\frac{x^2}{h^2} - \frac{y^2}{k^2} = 1$. Then $h = a$ and $k = b$; and, when $y = 0$, we have $x = \pm h$. When $x = 0$, we find that y is not defined. Therefore, the hyperbola does not cut its conjugate axis. Since $x = \pm h\sqrt{1 + \frac{y^2}{k^2}}$, we see that there is no restriction on the values that y can have. However, $y = \pm k\sqrt{\frac{x^2}{h^2} - 1}$, and this tells us that $|x| \geq h$. The asymptotes are the lines $\frac{x}{h} + \frac{y}{k} = 0$ and $\frac{x}{h} - \frac{y}{k} = 0$.

THE CONICS 205

Now let us assume that $F = 0$. The equation then becomes $Ax^2 + Cy^2 = 0$. Since A and C are unlike in sign, the equation can be factored. Thus, we have $x + \sqrt{-\frac{C}{A}}\, y = 0$ and $x - \sqrt{-\frac{C}{A}}\, y = 0$; and the locus is a pair of lines. We shall call the locus in this case a *degenerate* hyperbola.

EXAMPLE 5–12. Discuss the equation $9x^2 - y^2 = 81$.

Solution. Here $A = 9$, $C = -1$ and $F = -81$. Writing the equation in simple form, we have $\frac{x^2}{9} - \frac{y^2}{81} = 1$. Hence, we know that $a^2 = 9$ and $b^2 = 81$; and, therefore, $c^2 = 90$. So $a = 3$, $b = 9$, and $c = 3\sqrt{10}$. Hence, $e = \sqrt{10}$, and $\frac{a}{e} = \frac{3}{\sqrt{10}}$. The x-axis is the transverse axis and the y-axis is the conjugate axis. The coordinates of the vertices are $(\pm 3, 0)$, and the coordinates of the foci are $(\pm 3\sqrt{10}, 0)$. Equations of the directrices are $x = \frac{3}{\sqrt{10}}$ and $x = -\frac{3}{\sqrt{10}}$. Equations of the asymptotes are $3x - y = 0$ and $3x + y = 0$. The curve is shown in Fig. 5–35.

5–23. The latus rectum of the hyperbola

We have already defined a latus rectum to be a focal chord perpendicular to the transverse axis. Hence, for the hyperbola $\frac{x^2}{a^2} - \frac{y^2}{b^2} = 1$, each of the

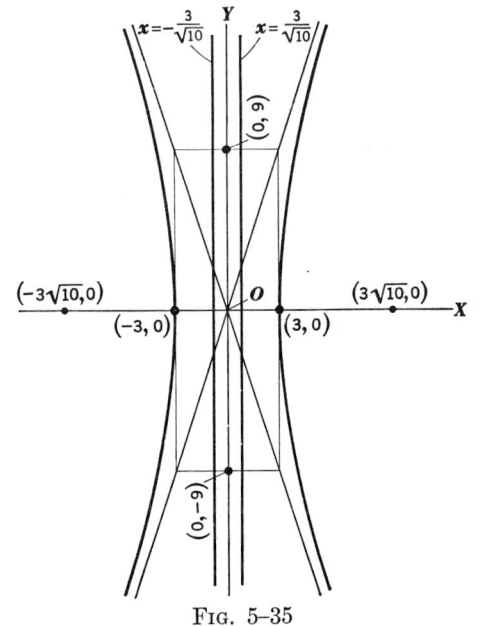

FIG. 5–35

lines whose equations are $x = c$ and $x = -c$ contains a latus rectum. Since the hyperbola is symmetric with respect to its conjugate axis, it is sufficient to find the length of one latus rectum.

Solving $x = c$ simultaneously with $\dfrac{x^2}{a^2} - \dfrac{y^2}{b^2} = 1$, we have $\dfrac{c^2}{a^2} - \dfrac{y^2}{b^2} = 1$. Hence, $\dfrac{y^2}{b^2} = \dfrac{c^2}{a^2} - 1 = \dfrac{c^2 - a^2}{a^2} = \dfrac{b^2}{a^2}$; and $y^2 = \dfrac{b^4}{a^2}$, or $y = \pm\dfrac{b^2}{a}$. Therefore, the coordinates of the points of intersection are $\left(c, \dfrac{b^2}{a}\right)$ and $\left(c, -\dfrac{b^2}{a}\right)$; and the length of the latus rectum is $\dfrac{2b^2}{a}$.

EXERCISES

1. Starting with $x = -c$ as the line which contains a latus rectum for the hyperbola $\dfrac{x^2}{a^2} - \dfrac{y^2}{b^2} = 1$, prove that its length is $\dfrac{2b^2}{a}$.

2. Starting with the equation $\dfrac{y^2}{a^2} - \dfrac{x^2}{b^2} = 1$, prove that the length of a latus rectum is $\dfrac{2b^2}{a}$.

3. If the length of the latus rectum of a hyperbola is 6, $a = 3$, the center is at $(0, 0)$, and the foci are on the y-axis, what is an equation of the curve? Discuss it completely.

5-24. Conjugate hyperbolas

The two hyperbolas $\dfrac{x^2}{h^2} - \dfrac{y^2}{k^2} = 1$ and $\dfrac{y^2}{k^2} - \dfrac{x^2}{h^2} = 1$ are said to be *conjugate hyperbolas*. Investigating the first equation, we see that $h = a$, $k = b$, and the x-axis is the transverse axis. Denoting the quantities a and b for the second hyperbola by a' and b', we see that $h = b'$ and $k = a'$. Hence, $a = b'$ and $b = a'$. The asymptotes of both hyperbolas are $\dfrac{x}{h} - \dfrac{y}{k} = 0$ and $\dfrac{x}{h} + \dfrac{y}{k} = 0$.

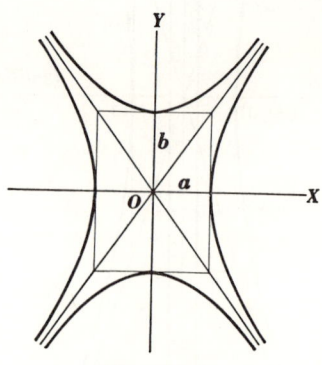

Fig. 5-36

So we see that conjugate hyperbolas have the same asymptotes, as shown in Fig. 5–36. To obtain an equation of the conjugate hyperbola to a given hyperbola in simple form, merely change the sign of the constant term. For example, if the equation of a hyperbola is $9x^2 - 4y^2 + 36 = 0$, the equation of the conjugate hyperbola is $9x^2 - 4y^2 - 36 = 0$.

5–25. *Equilateral hyperbola*

When $a = b$, the hyperbola is said to be *equilateral* or *rectangular*. Since $c^2 = a^2 + b^2$, it follows that for the equilateral hyperbola $c^2 = 2a^2$, or $e = \sqrt{2}$. An equation of an equilateral hyperbola with its foci on the x-axis is

$$x^2 - y^2 = k^2 \tag{5-18}$$

Equations of the asymptotes are $x - y = 0$ and $x + y = 0$. In this case the asymptotes are perpendicular to each other. Other forms of the equation of an equilateral hyperbola, such as $xy = a$, $(x - h)(y - h) = a$, $y = \dfrac{a}{x - h}$, $xy - yk - hy = a$, will be discussed in Chapter 6 under Transformation of the Axes.

EXERCISES

1. Discuss and sketch each of the following hyperbolas and give an equation of the conjugate hyperbola in each case:

   (a) $4x^2 - y^2 + 64 = 0$
   (b) $x^2 - y^2 = 16$
   (c) $4y^2 - x^2 - 25 = 0$
   (d) $3x^2 - y^2 = 48$
   (e) $9x^2 - 16y^2 + 144 = 0$
   (f) $x^2 - 4y^2 + 36 = 0$
   (g) $4y^2 - 9x^2 = 144$
   (h) $3y^2 - 2x^2 = 18$
   (i) $x^2 - y^2 + 25 = 0$
   (j) $y^2 - x^2 + 49 = 0$

2. Find an equation of each of the following hyperbolas having its center at $(0, 0)$:

   (a) $a = 8$, $b = 3$, and foci are on y-axis.
   (b) $a = 7$, $b = 4$, and foci are on x-axis.
   (c) A vertex is at $(4, 0)$, and a focus is at $(5, 0)$.
   (d) A vertex is at $(0, 3)$, and a focus is at $(0, 5)$.
   (e) $b = 3$, $c = 5$, and foci are on x-axis.
   (f) $b = 4$, $c = 5$, and foci are on y-axis.
   (g) $a = 3$, $e = 2$, and foci are on y-axis.
   (h) $a = 6$, $e = \frac{3}{2}$, and foci are on x-axis.
   (i) $c = 12$, $e = \frac{4}{3}$, and foci are on y-axis.
   (j) $c = 4$, $e = 3$, and foci are on x-axis.
   (k) $a = 6$, latus rectum $= 16$, and foci are on y-axis.

(l) $a = 8$, latus rectum $= 4$, and foci are on x-axis.
(m) $b = 3$, latus rectum $= 3$, and foci are on y-axis.
(n) $b = 1$, latus rectum $= 2$, and foci are on x-axis.
(o) $c = 6$, latus rectum $= 10$, and foci are on y-axis.
(p) $c = 4$, latus rectum $= 12$, and foci are on x-axis.
(q) A directrix is $x = \frac{3}{2}$, and $a = 6$.
(r) $c = 3$, hyperbola is equilateral, and foci are on y-axis.
(s) $a = 2$, hyperbola is equilateral, and foci are on x-axis.
(t) A directrix is $y = -4$, and $e = 2$.
(u) An asymptote is $3y - 4x = 0$, and a vertex is at $(6, 0)$.
(v) An asymptote is $x + 2y = 0$, and a vertex is at $(0, -4)$.
(w) The curve contains the points $(-4, 2)$ and $(3, 1)$. Hint: Substitute the coordinates of the given points for x and y in equation (5–16).
(x) The curve contains the points $(7, 6)$ and $(-4, 2)$.
(y) The length of the transverse axis is twice the length of the conjugate axis, foci are on x-axis, and curve contains the point $(3, 1)$.
(z) The length of the conjugate axis equals the length of the transverse axis, and the curve passes through the point $(5, 3)$.

3. Find an equation of the locus of a point moving so that the difference between its distances from the points $(0, 3)$ and $(0, -3)$ is 4.

4. Find an equation of the locus of a point moving so that the difference between its distances from the points $(5, 0)$ and $(-5, 0)$ is 6.

5. Each point of a given set has the property that its distance from the point $(0, -4)$ is twice its distance from the line $y = -1$. Find an equation of the set.

6. A hyperbola having the lines $9x^2 - 5y^2 = 0$ as asymptotes goes through the point $(1, -3)$. Find its equation.

7. What is an equation that represents the set of hyperbolas having the lines $y = \pm \frac{3x}{5}$ for its asymptotes?

8. What is an equation of all the equilateral hyperbolas having the y-axis as the transverse axis?

9. A hyperbola with its center at the origin has the eccentricity 3. What are the equations of the asymptotes if the foci lie on the y-axis; if the vertices lie on the x-axis?

5–26. Construction of the hyperbola

Two methods of constructing a hyperbola will be described here.

Method 1. Let a marker P be tied to a piece of string; let the two loose

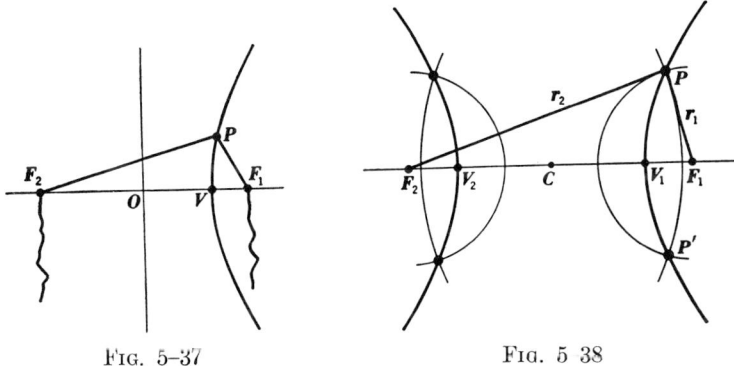

FIG. 5-37   FIG. 5-38

ends of the string be inserted through distinct holes in a piece of cardboard; and let the string be pulled taut so that the marker is between one hole and the mid-point between the two holes. If now the string is fed equally through both holes while P moves, as indicated in Fig. 5-37, it follows that $|F_2P| - |F_1P|$ is a constant and one branch of a hyperbola is generated. Starting again with P in such a position that $|F_1P| - |F_2P|$ is equal to the same constant, we obtain the second branch of the hyperbola. See Section 5-21.

In case it is desired to construct a hyperbola of a certain size, the holes are located at the foci and the difference between the focal radii is made equal to the required length of the transverse axis.

Method 2. With centers at two fixed points or foci F_1 and F_2, Fig. 5-38, draw circles of radii r_1 and r_2 such that $r_1 - r_2 = \pm 2a$. The points of intersection of these circles will be on a hyperbola. The method here is to draw a circle of any radius r_1 with one focus as a center. Now draw a circle of radius $r_2 = r_1 + 2a$, using the other focus as a center. Then at the points of intersection of the two circles the difference between the radii is $2a$.

To construct a hyperbola with certain dimensions, the fixed points are located at the foci and the difference between the radii of the circles is made equal to the required length of the transverse axis.

5-27. *Applications of the hyperbola*

Two practical applications of the hyperbola are as follows:

(1) If two persons located at different positions give the exact times at which they hear the report of a gun, then the difference between the times multiplied by the velocity of sound will give the difference between the distances to the gun from these two positions. Hence, the gun lies on a hyperbola having the positions of the two observers as foci. By means of a third listening post a second hyperbola can be obtained in the same way. Hence, the gun is at one of the intersections of the two curves. This procedure is known as range finding.

(2) The orbits of some comets with respect to the sun are approximately branches of hyperbolas.

5–28. Parametric equations of a hyperbola

In trigonometry, one of the fundamental identities is $\sec^2 \theta - \tan^2 \theta \equiv 1$. Consider a curve whose parametric equations are $x = a \sec \theta$ and $y = b \tan \theta$. In order to eliminate the parameter θ, we rewrite the equations as follows:

$$\frac{x}{a} = \sec \theta \text{ and } \frac{y}{b} = \tan \theta$$

Now squaring both equations and subtracting the second result from the first, we obtain

$$\frac{x^2}{a^2} - \frac{y^2}{b^2} = \sec^2 \theta - \tan^2 \theta = 1$$

But this is an equation of a hyperbola having the x-axis for its transverse axis and the y-axis for its conjugate axis. Hence, points of a hyperbola are represented by the equations

$$\begin{aligned} x &= a \sec \theta \\ y &= b \tan \theta \end{aligned} \tag{5-19}$$

For every point on the hyperbola there is a corresponding value of the angle θ. Hence, equations (5–19) are parametric equations of the hyperbola.

EXERCISES

1. Sketch each of the following hyperbolas from its parametric equations:
   (a) $x = 3 \sec \theta$ and $y = \tan \theta$
   (b) $x = 5 \sec \theta$ and $y = 2 \tan \theta$
   (c) $x = \sec \theta$ and $y = 2 \tan \theta$
   (d) $x = \tan \theta$ and $y = 2 \sec \theta$
   (e) $x = 5 \tan \theta$ and $y = \sec \theta$
   (f) $x = 4 \tan \theta$ and $y = 4 \sec \theta$

2. Eliminate the parameter θ in each of the pairs of equations in Exercise 1, and obtain the corresponding simple equation.

3. Prove that the equations

$$x = \frac{abt}{\pm\sqrt{b^2 t^2 - a^2}} \text{ and } y = \frac{ab}{\pm\sqrt{b^2 t^2 - a^2}}$$

are the parametric equations of a hyperbola. What two points on the curve are *not* defined by these equations?

COMPARISON OF THE CONICS

5–29. Summary

In this chapter we have discussed the parabola, the ellipse, and the hyperbola. For the convenience of the student, equations of the conics are listed again as follows:

1. Parabola:
   (a) If the curve opens right or left, the vertex is at the origin, and the focus is on the x-axis, then the equation is:
   $$Cy^2 + Dx = 0, \text{ where } C \text{ and } D \text{ are not equal to } 0$$
   (b) If the curve opens up or down, the vertex is at the origin, and the focus is on the y-axis, the equation is:
   $$Ax^2 + Ey = 0, \text{ where } A \text{ and } E \text{ are not equal to } 0$$

2. Ellipse:
   $$Ax^2 + Cy^2 + F = 0, \text{ where } A \text{ and } C \text{ have like signs, } A \neq C,$$
   $$\text{and } A \text{ and } C \text{ are not equal to } 0$$
   (a) If A, C, and F all have the same sign, the ellipse is imaginary.
   (b) If $F = 0$, the locus is a point-ellipse or degenerate ellipse.
   (c) If F is different in sign from A and C, the locus is a real ellipse with its center at the origin. In this case, if $A < C$, the x-axis is the major axis; and, if $A > C$, the y-axis is the major axis. If $A = C$, the locus is a circle.

3. Hyperbola:
   $$Ax^2 + Cy^2 + F = 0, \text{ where } A \text{ and } C \text{ are not equal to } 0 \text{ and } A$$
   $$\text{and } C \text{ have opposite signs}$$
   (a) If $-\dfrac{F}{A}$ is positive, the x-axis is the transverse axis; and, if $-\dfrac{F}{C}$ is positive, the y-axis is the transverse axis. If $|A| = |C|$, the locus is a rectangular, or equilateral, hyperbola.
   (b) If $F = 0$, the locus is a pair of lines. This locus is a degenerate hyperbola.

REVIEW EXERCISES

1. Discuss completely each of the following equations and sketch the curve:
   (a) $3x^2 + 14y = 0$
   (b) $x^2 - y^2 + 16 = 0$
   (c) $4x^2 + 9y^2 = 36$
   (d) $5y^2 + 16x = 0$
   (e) $9x^2 - 4y^2 = 36$
   (f) $4x^2 + y^2 - 144 = 0$

2. Find a simple equation of the conic satisfying each of the following conditions:
   (a) A focus is at $(0, 10)$, and a vertex is at $(0, 6)$.
   (b) $e = \frac{1}{5}$, latus rectum $= 2$, and foci are on x-axis.
   (c) $e = 1$, focus is on y-axis, and curve contains the point $(-4, -1)$.
   (d) A directrix is $y = 3$, $e = \frac{2}{3}$, and foci are on y-axis.
   (e) Ends of a latus rectum are at $(\pm 6, 3)$, and $e = 1$.
   (f) Latus rectum $= 18$, one asymptote is $y = -\dfrac{2x}{3}$, and foci are on $y =$ axis.
   (g) $e > 1$, and the curve passes through the points $(-1, 2)$ and $(3, 4)$.
   (h) $e > 1$, and the curve passes through the points $(4, 2)$ and $(-7, 4)$.
   (i) The curve is an equilateral hyperbola and contains the point $(-3, 2)$.
   (j) $e < 1$, and the conic contains the points $(-1, 4)$ and $(3, 1)$.

3. Show that the distance from a point on an equilateral hyperbola to the center is a mean proportional between the lengths of the focal radii to the point.

4. If e_1 and e_2 are the eccentricities of two conjugate hyperbolas, prove that
$$\frac{1}{e_1^2} + \frac{1}{e_2^2} = 1.$$

5. Prove that $e^2 = 1 + s^2$, where s is the slope of an asymptote of the curve $\frac{x^2}{a^2} - \frac{y^2}{b^2} = 1$ and e is the eccentricity.

6. How wide is a parabolic arch of height 20 feet and span 40 feet at a distance of 15 feet from the top?

7. How high is a parabolic arch of height 30 feet and span 50 feet at a distance of 10 feet from one end?

8. Show that the length of a latus rectum of a hyperbola equals $2e$ times the distance between the directrix and the center of the latus rectum.

9. Find the eccentricity of an ellipse if the length of the latus rectum is three-fourths the length of the major axis.

10. What is the eccentricity of an ellipse if the length of the latus rectum is two-thirds the length of the minor axis?

11. Two vertices of a triangle are $P_1(2k, k)$ and $P_2(-2k, k)$. Find the locus of the third vertex $P(x, y)$, if the slope of P_2P is 1 unit less than the slope of P_1P.

12. What is the locus of the mid-points of the ordinates of the conic $y^2 = 4ax$?

13. Show that the distance to the mid-point of any focal chord of a parabola from the directrix is one-half the length of the chord.

14. What is the simple equation of the hyperbola having the lines $y = \pm \frac{3x}{5}$ for its asymptotes and the point $(2, 0)$ as one focus?

15. What is the simple equation of the hyperbola having $y = -\frac{3x}{4}$ as one asymptote and passing through the point $(2, -1)$?

16. Show that the eccentricity of an equilateral hyperbola is equal to the ratio of the length of a diagonal of any square to the length of its side.

17. Express the tangent of the angle between the asymptotes of the hyperbola $\frac{x^2}{a^2} - \frac{y^2}{b^2} = 1$ in terms of the eccentricity.

18. Express the cosine of the angle between the asymptotes of the hyperbola $\frac{x^2}{a^2} - \frac{y^2}{b^2} = 1$ in terms of the eccentricity.

19. Show that the length of the semi-minor axis of an ellipse is a mean proportional of the length of the semi-major axis and half of the length of the latus rectum.

20. Show that the perpendicular distance from an asymptote of a hyperbola to either focus is equal to the length of the semi-conjugate axis.

21. Prove that the product of the distances from a point of an equilateral hyperbola to its foci is equal to the square of the distance to the point from the center of the hyperbola.

22. Show that the common chord of a parabola, and the circle whose center is the vertex of the parabola and whose radius is equal to three halves the distance from the vertex to the focus, bisects the line-segment joining the vertex with the focus.

23. An arbitrary point P of a parabola, not the vertex, is joined with the vertex O, and a second line is drawn through P, perpendicular to OP meeting the axis in Q. Prove that the projection of PQ on the axis is equal to the length of the latus rectum.

24. The end-points of a chord of a parabola, which subtends a right angle at the vertex, are on opposite sides of the axis and at distances from the axis whose product is the square of the length of the latus rectum. Verify the above.

25. Express the latus rectum of an ellipse in terms of b and e; in terms of e and c.

26. The lines joining the ends of the minor axis with a point P of an ellipse meet the major axis in the points R and Q. Show that the semi-major axis is a mean proportional between the distances from the center to R and Q.

27. Prove that the segment of a directrix between the points of intersection of the lines joining the vertices with a point on an ellipse subtends a right angle at the corresponding focus.

28. Prove that the angle subtended by the segments drawn from the vertices of the hyperbola $\dfrac{x^2}{a^2} - \dfrac{y^2}{b^2} = 1$ to each of the points $(0, \pm b)$ is a right angle when, and only when, the hyperbola is rectangular.

29. What is the condition that the segment between the foci of an ellipse subtends a right angle at the points $(0, \pm b)$?

30. Show that an asymptote, a directrix, and the line through the corresponding focus perpendicular to the asymptote are concurrent.

31. Let F be a focus and ℓ the corresponding directrix of a hyperbola. A line through a point P of the hyperbola parallel to an asymptote meets ℓ in the point K. Prove that the triangle FPK is isosceles.

32. Let F be a focus and ℓ the corresponding directrix of a hyperbola. Prove that the segment cut from ℓ by the lines joining the vertices with an arbitrary point on the hyperbola subtends a right angle at F.

33. Prove that the angles subtended at the vertices of an equilateral hyperbola by a chord parallel to the conjugate axis are supplementary.

34. A line through an arbitrary point P on a hyperbola parallel to the conjugate axis meets the asymptotes in A and B. Show that the product of the segments in which P divides AB is constant.

SUPPLEMENTARY MATERIAL

In Section 5–1, we told how the early mathematicians studied the plane sections of a right circular cone and called them conics. Then in the sections that followed we defined a parabola, an ellipse, and a hyperbola; and we claimed that these curves were actually conics. In this supplementary discussion we shall prove that this claim is true.

(a) *The parabola.* Consider a right circular cone V–$ABCD$, Fig. 5–39. The plane VAB contains the vertex V and a diameter AB of the base. Draw plane ODC perpendicular to plane VAB and intersecting it in the line OL which is parallel to VA. The plane ODC intersects the surface of the cone in the *curve DOC*. Through O and any other point P of the curve DOC draw planes each of which is parallel to the base. The segments OS and EF are the diameters of the circular sections in which these planes intersect the cone. The segment TP is perpendicular to the plane VAB since it lies in planes EPF and DOC and, hence, it is also perpendicular to the lines LO and EF.

Let DOC be the xy-plane, O the origin, and OL the positive x-axis. Then P is at (x, y), where $x = OT$ and $y = TP$. In circle EPF, we have

(1) $$(TP)^2 = (ET)(TF)$$

By our construction it follows that $ET = SO$ and $\dfrac{TF}{OT} = \dfrac{SO}{SV}$. Hence, equation (1) becomes

$$y^2 = (ET)(TF) = \frac{(SO)(SO)}{SV}(OT)$$

or
$$y^2 = \frac{(SO)^2}{SV} x$$

Since SO and SV are constant, regardless of the position of P, we shall represent $\dfrac{(SO)^2}{SV}$ by $4a$. Now we have $y^2 = 4ax$, which is a simple equation of a parabola.

(b) *The ellipse.* Again consider the cone V–$ABCD$; and, as before, let the plane VAB contain the vertex and a diameter AB of the base. In Fig. 5–40, let the plane MPN be any plane perpendicular to the plane VAB and meeting VA at the point M and VB at the point N. The point P is any point, other than M and N, lying on the curve of intersection of the cone and the plane MPN. Through O, the mid-point of MN, draw a plane parallel to the base and intersecting plane VAB in EF. Through P draw a plane parallel to the base and intersecting plane VAB in GH. Both EF and GH are diameters of circles. Now project P on GH at K. Then KP is perpendicular to plane VAB and to the lines GH and MN.

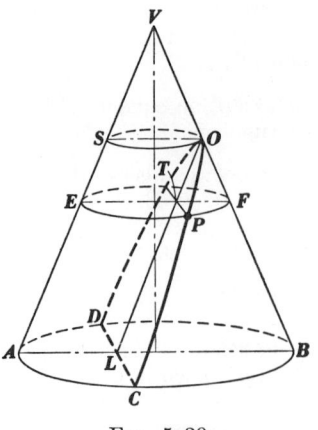

FIG. 5–39

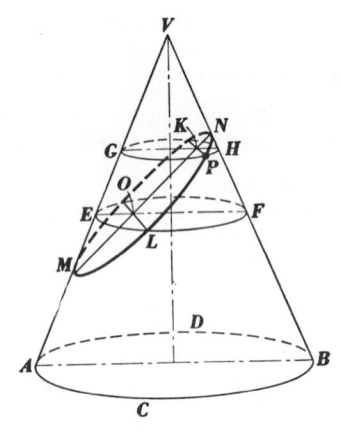

FIG. 5–40

In the circles GPH and ELF, we have

(2) $\qquad (KP)^2 = (GK)(KH) \text{ and } (OL)^2 = (EO)(OF)$

Since triangle GKM is similar to triangle EOM, we have

(3) $$\frac{GK}{EO} = \frac{MK}{MO}$$

Using triangles KHN and OFN, we have

(4) $$\frac{KH}{OF} = \frac{KN}{ON}$$

From equation (2) it follows that

(5) $$\frac{(KP)^2}{(OL)^2} = \frac{(GK)(KH)}{(EO)(OF)}$$

Now using equations (3) and (4), we obtain

(6) $$\frac{(KP)^2}{(OL)^2} = \left(\frac{MK}{MO}\right)\left(\frac{KN}{ON}\right)$$

If we let $OL = b$ and $MO = a$, equation (6) becomes

(7) $$\frac{(KP)^2}{b^2} = \frac{(MK)(KN)}{a^2}$$

Now let ON be the positive x-axis and let OL be the positive y-axis. Then the coordinates of P are $x = OK$ and $y = KP$. From Fig. 5–40 it should be clear that $MK = a + x$ and $KN = a - x$. Substituting these expressions in equation (7), we obtain

$$\frac{y^2}{b^2} = \frac{a^2 - x^2}{a^2}$$

or $\qquad\qquad\qquad\qquad \dfrac{x^2}{a^2} + \dfrac{y^2}{b^2} = 1$

This is a simple equation of an ellipse.

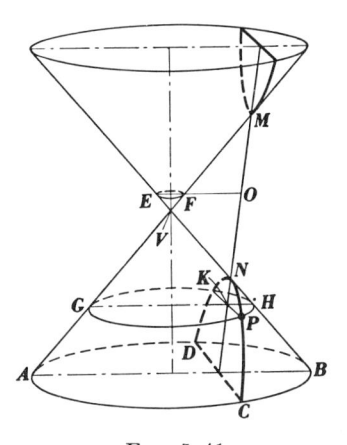

Fig. 5–41

(c) The hyperbola. Again consider the cone V–$ABCD$; and, as shown in Fig. 5-41, let the plane MPN be perpendicular to the plane VAB, meeting the prolongation of VA at the point M and VB at the point N. The point P is any point, other than M and N, lying on the curve of intersection of the cone and the plane MPN. Through O, the mid-point of MN, draw a plane parallel to the base and intersecting the plane VAB in EF. Through P draw a plane parallel to the base and intersecting plane VAB in GH. Both EF and GH are diameters of the circular sections in which the planes intersect the cone. Project P to the plane VAB at K. Then KP is perpendicular to the plane VAB and to the segments GH and MN.

From the circle GPH, we have

(8) $$(KP)^2 = (GK)(KH)$$

Since the triangles GKM and FOM are similar, we have

(9) $$\frac{GK}{FO} = \frac{MK}{MO}$$

Using the triangles KHN and EON, we have

(10) $$\frac{KH}{EO} = \frac{NK}{ON}$$

From equations (8), (9), and (10), we obtain

(11) $$(KP)^2 = \left(\frac{MK}{MO}\right)\left(\frac{NK}{ON}\right)(FO)(EO)$$

Now, if we let O be the origin, ON be the positive x-axis, and a line through O parallel to KP be the positive y-axis, the coordinates of P are $x = OK$ and $y = KP$. Then equation (11) becomes

(12) $$y^2 = \left(\frac{MK}{MO}\right)\left(\frac{NK}{ON}\right)(FO)(EO)$$

If we let $|MO| = |ON| = a$, and $(FO)(EO) = b^2$, equation (12) becomes

(13) $$y^2 = \frac{b^2}{a^2}(MK)(NK)$$

But, $MK = x + a$ and $NK = x - a$. Hence, we have

$$y^2 = \frac{b^2}{a^2}(x^2 - a^2)$$

from which

$$a^2y^2 = b^2x^2 - a^2b^2$$

or

$$\frac{x^2}{a^2} - \frac{y^2}{b^2} = 1$$

This is a simple equation of the hyperbola.

Hence, the section of a cone cut by a plane is an ellipse, a parabola, a hyperbola, a circle, a line or a pair of lines, or possibly just a point.

Transformation of the Axes

6-1. Introduction

So far in our study of the conics we have considered only cases in which the equations appear in simple form. When this is not the case, it is possible by a change of the axes to simplify the equation. This changing of the axes is known as "transforming" the axes. It is an extremely useful technique in the study of more complicated equations.

6-2. Translation of the axes

When the axes have been shifted in such a way that they remain parallel to their original positions, we say that they have been *translated*. Consider a point P, Fig. 6-1, in the plane whose coordinates are x and y with reference to one pair of axes and are x' and y' with reference to a second pair of axes, parallel to the first pair. Let O' be the intersection of the second pair of axes and let h and k be the coordinates of O' with reference to the first pair of axes. We shall denote the original pair of axes by OX and OY and the second pair of axes by $O'X'$ and $O'Y'$. When the same point P is referred to the second pair of axes, it is designated as P'.

Using vectors, we have

$$\overrightarrow{OP} = \overrightarrow{OO'} + \overrightarrow{O'P'}$$

Since $\overrightarrow{OP} = [x, y]$, $\overrightarrow{OO'} = [h, k]$, and $\overrightarrow{O'P'} = [x', y']$, we have

$$[x, y] = [h, k] + [x', y'] = [h + x', k + y']$$

Hence, formulas for translating the axes are

$$x = x' + h$$
$$y = y' + k \quad (6-1)$$

If we solve equations (6-1) for x' and y', we obtain the inverse form of equations (6-1). Thus,

$$x' = x - h$$
$$y' = y - k \quad (6-2)$$

218 ANALYTIC GEOMETRY IN THE PLANE

EXAMPLE 6-1. What do the coordinates of the point $P = (5, -3)$ become after the axes are translated so that the point $O' = (-2, 3)$ is a new origin?

Solution. As shown in Fig. 6-2, $x = 5$, $y = -3$, $h = -2$, and $k = 3$. Hence, by equations (6-2),

$$x' = 5 + 2 \text{ and } y' = -3 - 3$$

Thus, the coordinates of the point P after the axes have been translated are $x' = 7$ and $y' = -6$.

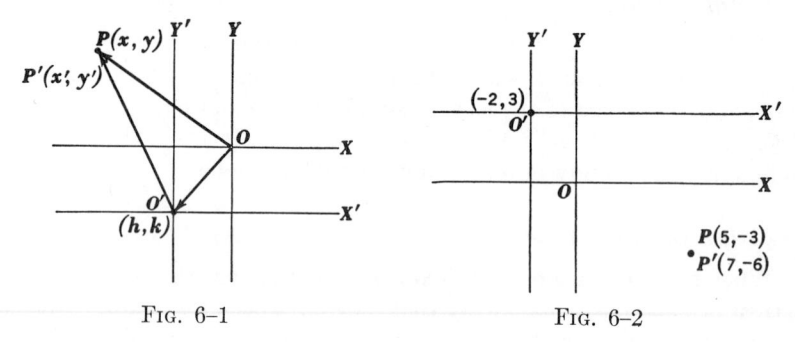

FIG. 6-1   FIG. 6-2

Let us represent an equation of a curve by $f(x, y) = 0$. By equations (6-1) the coordinates x and y are transformed into $x' + h$ and $y' + k$; and so the equation $f(x, y) = 0$ becomes $f(x' + h, y' + k) = 0$. Since the point $P = (x, y)$ whose coordinates satisfy $f(x, y) = 0$ is the same as the point $P' = (x', y')$ whose coordinates satisfy $f(x' + h, y' + k) = 0$, the two equations represent the same locus.

Conversely, if the equation is given in terms of x' and y' as $f(x', y') = 0$, then by equations (6-2) we have $x' = x - h$ and $y' = y - k$; and, hence, the equation becomes $f(x - h, y - k) = 0$. Any point whose coordinates x' and y' satisfy $f(x', y') = 0$ is the same as the point whose coordinates x and y satisfy $f(x - h, y - k) = 0$; and so the two equations again represent the same locus.

EXAMPLE 6-2. What does the equation $2x^2 + y^2 - 12x + 2y + 3 = 0$ become when the axes are translated so that the point $(3, -1)$ is a new origin?

Solution. By equations (6-1), $x = x' + 3$ and $y = y' - 1$. Substituting these values for x and y in the given equation, we obtain

$$2(x' + 3)^2 + (y' - 1)^2 - 12(x' + 3) + 2(y' - 1) + 3 = 0$$

Expanding and collecting like terms, we have

$$2x'^2 + y'^2 + x'(12 - 12) + y'(-2 + 2) + 18 + 1 - 36 - 2 + 3 = 0$$

or

$$2x'^2 + y'^2 - 16 = 0$$

Dividing through by 16, we obtain

$$\frac{x'^2}{8} + \frac{y'^2}{16} = 1$$

This is a simple equation of an ellipse. We see that $a = 4$, $b = 2\sqrt{2}$, $c = 2\sqrt{2}$, $e = \frac{\sqrt{2}}{2}$, and $\frac{a}{e} = \frac{8}{\sqrt{2}} = 4\sqrt{2}$. It is now a simple matter to discuss the curve with reference to the original axes. Knowing that the major axis is the y'-axis and that the center is $(3, -1)$, we readily find the coordinates of the various points with reference to the original axes. As shown in Fig. 6-3, the vertices are at $(3, 3)$ and $(3, -5)$; the foci are at $(3, -1+2\sqrt{2})$ and $(3, -1-2\sqrt{2})$; the points where the curve intersects the minor axis are at $(3+2\sqrt{2}, -1)$ and $(3-2\sqrt{2}, -1)$; and equations of the directrices are $y = -1+4\sqrt{2}$ and $y = -1-4\sqrt{2}$.

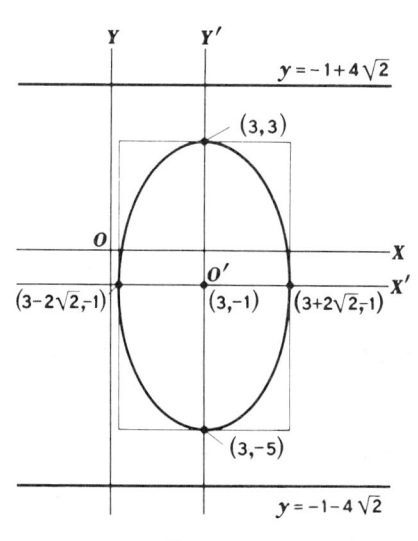

FIG. 6-3

EXERCISES

1. Find the coordinates of each of the following points P when the axes are translated so that O' is a new origin:
    (a) $P = (3, -1)$ and $O' = (-1, 4)$
    (b) $P = (-2, -4)$ and $O' = (0, 3)$
    (c) $P = (1, 3)$ and $O' = (-2, -1)$
    (d) $P = (-5, 2)$ and $O' = (3, 4)$
    (e) $P = (0, 0)$ and $O' = (2, -4)$
    (f) $P = (-4, 3)$ and $O' = (1, 1)$
    (g) $P = (-1, -2)$ and $O' = (-6, 3)$
    (h) $P = (4, 1)$ and $O' = (-1, -1)$

220   ANALYTIC GEOMETRY IN THE PLANE

2. In each of the following cases, the coordinates of a point with respect to a new set of axes and the coordinates of a new origin are given, and it is required to find the original coordinates:

(a) $P' = (3, -2)$ and $O' = (2, 1)$ (e) $P' = (x', y')$ and $O' = (h, k)$
(b) $P' = (-2, 4)$ and $O' = (-1, 0)$ (f) $P' = (2, -2)$ and $O' = (-1, 1)$
(c) $P' = (1, 3)$ and $O' = (3, -2)$ (g) $P' = (-4, -3)$ and $O' = (2, -1)$
(d) $P' = (0, 0)$ and $O' = (4, -3)$ (h) $P' = (-7, 6)$ and $O' = (-5, -4)$

3. Transform each of the following equations by using the given point as a new origin; and discuss the equation and draw the graph of each curve:

(a) $x^2 + y^2 - 2x + 4y + 1 = 0$ and $O' = (1, -2)$
(b) $y^2 + 4x + 4y - 8 = 0$ and $O' = (3, -2)$
(c) $x^2 + 2x - 3y - 11 = 0$ and $O' = (-1, -4)$
(d) $x^2 + 16y^2 + 4x - 8y + 4 = 0$ and $O' = (-2, \frac{1}{4})$
(e) $x^2 - 4y^2 - 6x - 16y - 23 = 0$ and $O' = (3, -2)$
(f) $4x^2 + y^2 - 8x + 6y - 3 = 0$ and $O' = (1, -3)$
(g) $x^2 - 9y^2 - 72y - 135 = 0$ and $O' = (0, -4)$
(h) $2x^2 - y^2 - 20x - 4y + 100 = 0$ and $O' = (5, -2)$

6–3. Simplification of the equation $Ax^2 + Cy^2 + Dx + Ey + F = 0$, where both A and C are not zero

First we assume that neither A nor C is 0. Let us now determine the best choice for a new origin in order that the equation $Ax^2 + Cy^2 + Dx + Ey + F = 0$ may be reduced by a translation to as simple a form as possible. Upon replacement of x by $x' + h$ and y by $y' + k$, the equation becomes

$$A(x'+h)^2 + C(y'+k)^2 + D(x'+h) + E(y'+k) + F = 0$$

Expanding and collecting like terms, we have

$$Ax'^2 + Cy'^2 + (2Ah+D)x' + (2Ck+E)y' + Ah^2 + Ck^2 + Dh + Ek + F = 0$$

Since h and k are at our disposal, let us choose them such that $2Ah + D = 0$ and $2Ck + E = 0$. Then $h = \dfrac{-D}{2A}$ and $k = \dfrac{-E}{2C}$, and the transformed equation becomes

$$Ax'^2 + Cy'^2 + F' = 0$$

where $F' = Ah^2 + Ck^2 + Dh + Ek + F$.

If $F' \neq 0$, the equation $Ax'^2 + Cy'^2 + F' = 0$ can be written as follows:

$$\frac{x'^2}{-\dfrac{F'}{A}} + \frac{y'^2}{-\dfrac{F'}{C}} = 1 \tag{6-3}$$

Thus, we see that the equation $Ax^2 + Cy^2 + Dx + Ey + F = 0$ represents an ellipse when A and C are of like sign; and it represents a hyperbola when A and C have unlike signs.

If $F' = 0$, the equation represents either a point-ellipse or a pair of lines. Hence, when $F' = 0$, the conic is degenerate.

Now we will consider the case when either A or C is 0. If we assume that $A = 0$, then $C \neq 0$; and the equation becomes $Cy^2 + Dx + Ey + F = 0$. Applying the transformation equations $x = x' + h$ and $y = y' + k$, we obtain

$$C(y'+k)^2 + D(x'+h) + E(y'+k) + F = 0$$

Expanding and collecting like terms, we have

$$Cy'^2 + (2Ck + E)y' + Dx' + Ck^2 + Dh + Ek + F = 0$$

On inspecting this equation, we see that, if k is chosen so that $2Ck + E = 0$, or $k = -\dfrac{E}{2C}$, the y' term is eliminated. Now, with this value of k, we choose h so that $Ck^2 + Dh + Ek + F = 0$. We have $C\left(-\dfrac{E}{2C}\right)^2 + Dh + E\left(-\dfrac{E}{2C}\right) + F = 0$. If $D \neq 0$,

$$Dh = \frac{E^2}{2C} - \frac{E^2}{4C} - F = \frac{E^2}{4C} - F$$

and

$$h = \frac{\dfrac{E^2}{4C} - F}{D}$$

Hence, if we choose the new origin to be $O' = \left(\dfrac{\dfrac{E^2}{4C} - F}{D}, -\dfrac{E}{2C}\right)$, the transformed equation becomes

$$Cy'^2 + Dx' = 0 \qquad (6\text{-}4)$$

Since this is an equation of a parabola, it follows that $Cy^2 + Dx + Ey + F = 0$ represents a parabola. The parabola opens to the right if C and D have unlike signs and opens to the left if C and D have like signs.

If $D = 0$, the equation $Cy^2 + Dx + Ey + F = 0$ becomes $Cy^2 + Ey + F = 0$. If this quadratic equation has real factors, the locus is a pair of lines which are either distinct and parallel or coincident. If the quadratic equation has imaginary factors, the locus does not exist. Although this case does not yield a conic section, we shall nevertheless say that the equation represents a degenerate conic, which is real or imaginary.

By a similar argument the equation $Ax^2 + Dx + Ey + F = 0$ can be reduced to the form

$$Ax'^2 + Ey' = 0 \qquad (6\text{-}5)$$

This is an equation of a parabola which opens upward if A and E have

222 ANALYTIC GEOMETRY IN THE PLANE

unlike signs and opens downward if A and E have like signs. If $E = 0$, we again have either a pair of lines or an imaginary locus. See Exercise 1 at the end of this section.

The method presented here for simplifying the equation $Ax^2 + Cy^2 + Dx + Ey + F = 0$ may be applied at times to an equation of any degree. This will be illustrated by an example at the end of this section.

For the second-degree equation the method of *completing the square* is more practical and achieves the same result. We shall give the argument first for the case when neither A nor C is zero. We rewrite the equation as $Ax^2 + Dx + Cy^2 + Ey + F = 0$. Now we complete the squares on the x and y terms, respectively, and we obtain

$$A\left(x^2 + \frac{D}{A}x + \frac{D^2}{4A^2}\right) + C\left(y^2 + \frac{E}{C}y + \frac{E^2}{4C^2}\right) = \frac{D^2}{4A} + \frac{E^2}{4C} - F$$

or

$$A\left(x + \frac{D}{2A}\right)^2 + C\left(y + \frac{E}{2C}\right)^2 = \frac{D^2C + E^2A - 4ACF}{4AC}$$

Using the transformation equations $x' = x + \frac{D}{2A}$ and $y' = y + \frac{E}{2C}$, and letting $F' = -\frac{D^2C + E^2A - 4ACF}{4AC}$, we have $Ax'^2 + Cy'^2 = -F'$. Since $x = x' - \frac{D}{2A}$ and $y = y' - \frac{E}{2C}$, it follows from equations (6-1) that $h = -\frac{D}{2A}$ and $k = -\frac{E}{2C}$.

If $C = 0$ and $A \neq 0$, we write the equation as $Ax^2 + Dx = -Ey - F$. Completing the square on the x-terms, we obtain

$$A\left(x^2 + \frac{D}{A}x + \frac{D^2}{4A^2}\right) = -Ey - F + \frac{D^2}{4A}$$

or

$$A\left(x + \frac{D}{2A}\right)^2 = -E\left(y + \frac{F}{E} - \frac{D^2}{4AE}\right)$$

If we let $x' = x + \frac{D}{2A}$ and $y' = y + \frac{F}{E} - \frac{D^2}{4AE}$, we obtain $Ax'^2 = -Ey'$. Since $x = x' - \frac{D}{2A}$ and $y = y' - \frac{F}{E} + \frac{D^2}{4AE}$, it follows that $h = -\frac{D}{2A}$ and $k = -\frac{F}{E} + \frac{D^2}{4AE}$.

EXAMPLE 6-3. Simplify the equation $y = x^3 - 3x^2 + 2x + 2$.

Solution. If we replace x by $x' + h$ and y by $y' + k$, we have

$$y' + k = (x' + h)^3 - 3(x' + h)^2 + 2(x' + h) + 2$$

or

$$y' = x'^3 + (3h - 3)x'^2 + (3h^2 - 6h + 2)x' + h^3 - 3h^2 + 2h + 2 - k$$

In simplifying an equation, it is usually best to eliminate first the terms of highest possible degree. So we let $3h-3=0$ and obtain $h=1$. This eliminates the x'^2 term. Now we choose k so that, when $h=1$, we have $h^3 - 3h^2 + 2h + 2 - k = 0$; and this gives us $1 - 3 + 2 + 2 - k = 0$, or $k = 2$. Hence, if the point $(1, 2)$ is selected as the new origin, the equation becomes

$$y' = x'^3 - x'$$

EXAMPLE 6-4. Simplify and discuss the equation $x^2 + 4y^2 + 2x + 16y + 13 = 0$.

Solution. Since the equation is of the second degree, we use the method of completing the square. We have

$$x^2 + 2x + 4y^2 + 16y = -13$$

from which
$$x^2 + 2x + 1 + 4(y^2 + 4y + 4) = -13 + 17$$

or
$$(x+1)^2 + 4(y+2)^2 = 4$$

Letting $x' = x + 1$ and $y' = y + 2$, and dividing by 4, we obtain

$$\frac{x'^2}{4} + \frac{y'^2}{1} = 1$$

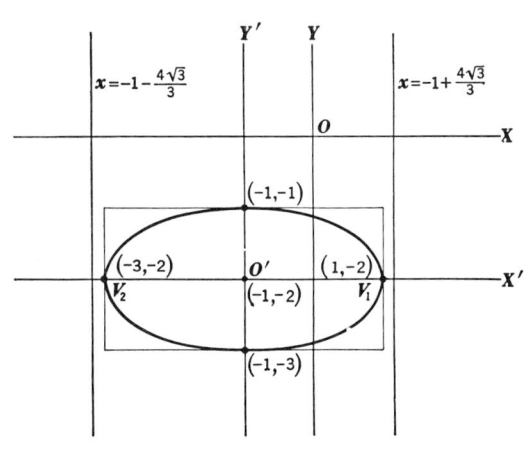

FIG. 6-4

This is an equation of an ellipse having the x'-axis for its major axis and the y'-axis for its minor axis. The new origin, and hence the center, is at $(-1, -2)$. From the properties of an ellipse, we find that $a = 2$, $b = 1$, $c = \sqrt{3}$, $e = \frac{\sqrt{3}}{2}$ and $\frac{a}{e} = \frac{4}{\sqrt{3}}$. This curve is shown in Fig. 6-4. The solution for y in terms of x can easily be obtained from $y' = \frac{1}{2}\sqrt{4-(x')^2}$ and $y' = -\frac{1}{2}\sqrt{4-(x')^2}$ which are equivalent to $y = -2 + \frac{1}{2}\sqrt{4-(x+1)^2}$ and $y = -2 - \frac{1}{2}\sqrt{4-(x+1)^2}$. The domain and range of the relation defined by $x^2 + 2x + 4y^2 + 16y = -13$ are the intervals $[-3, 1]$ and $[-3, -1]$ respectively. These are obtained from Fig. 6-4.

EXAMPLE 6-5. Simplify and discuss the equation $x^2 + 4y - 6x + 17 = 0$.

Solution. We have
$$x^2 - 6x = -4y - 17$$
or
$$x^2 - 6x + 9 = -4y - 17 + 9$$
Hence,
$$(x-3)^2 = -4(y+2)$$

If we let $x' = x - 3$ and $y' = y + 2$, our equation becomes $x'^2 = -4y'$ with the new origin at $(3, -2)$. This represents a parabola opening downward with its vertex at $(3, -2)$ and $a = 1$. The focus is at $(3, -3)$, the ends of the latus rectum are at $(5, -3)$ and $(1, -3)$, and the directrix has for its equation $y = -1$. The curve is shown in Fig. 6-5.

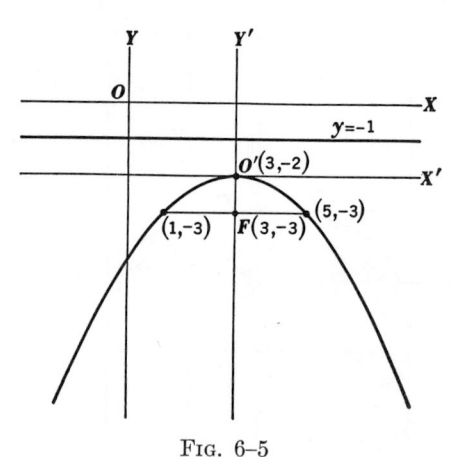

FIG. 6-5

EXERCISES

Simplify and discuss each of the following equations and draw its graph when it exists:

1. $Ax^2 + Dx + Ey + F = 0$
2. $y = x^3 + 6x^2 - 3x - 12$
3. $9x^2 + 4y^2 + 72x = 0$
4. $9x^2 + 25y^2 + 36x - 150y + 36 = 0$
5. $4x = 8y - y^2$
6. $y = 3 + (x-4)^3$

7. $x^2+4y^2-2x-24y+21=0$
8. $x^2+8y-4x-20=0$
9. $9x^2-16y^2+18x-64y-66=0$
10. $9x^2-4y^2+24y-18x+45=0$
11. $2x^2-3y^2+12x-12y-42=0$
12. $y^2-12y+12x+48=0$
13. $25x^2+9y^2-72y+100x+19=0$
14. $x^2-2x+7y+29=0$
15. $4x^2-y^2+24x+2y+36=0$
16. $16x^2-y^2-32x+8y=100$

17. In Exercises 4, 8, 10 and 12, solve for y in terms of x, sketch the graph of each equation obtained, and state the domain and range of the relation defined by each equation obtained.

6–4. The standard equations of the conics

Parabola. Let (h, k) be the vertex of a parabola, as indicated in Fig. 6–6; let $y = k$ be an equation of its axis; and let $(h-a, k)$ be the focus. Then an equation of the directrix is $x = h + a$ and the ends of the latus rectum are at $(h-a, k+2a)$ and $(h-a, k-2a)$.

Let the line $y = k$ be the x'-axis and let the line $x = h$ be the y'-axis. Referred to the new axes, an equation of the parabola is $y'^2 = -4ax'$. Since $y' = y - k$ and $x' = x - h$, an equation of the parabola referred to the original axes is

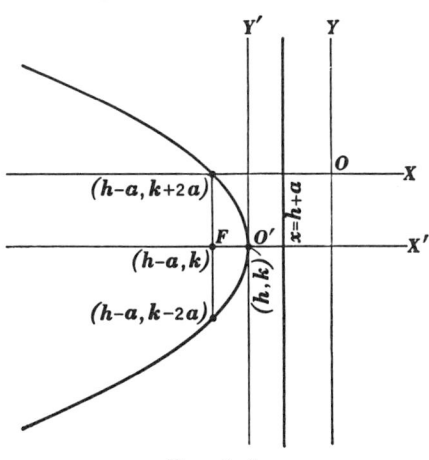

$$(y-k)^2 = -4a(x-h) \qquad (6\text{-}5)$$

FIG. 6–6

This equation is known as the *standard equation* of the parabola having its vertex at (h, k), having the line $y = k$ for its axis, and opening to the left.

In a similar manner we have three other standard equations for the parabola. If the vertex is at (h, k), the axis is $y = k$, and the curve opens to the right, the equation is

$$(y-k)^2 = 4a(x-h) \qquad (6\text{-}6)$$

If the vertex is at (h, k), the axis is $x = h$, and the curve opens upward, the equation is
$$(x-h)^2 = 4a(y-k) \tag{6-7}$$

If the vertex is at (h, k), the axis is $x = h$, and the curve opens downward, the equation is
$$(x-h)^2 = -4a(y-k) \tag{6-8}$$

Ellipse. Let the center of an ellipse be at the point (h, k), as in Fig. 6-7; let the major axis be the line $y = k$; and let the minor axis be the line $x = h$. Then its foci are at $(h+c, k)$ and $(h-c, k)$; its vertices are at $(h+a, k)$ and $(h-a, k)$; the points where the curve intersects its minor axis are at $(h, k+b)$ and $(h, k-b)$; and equations of the directrices are $x = h + \dfrac{a}{e}$ and $x = h - \dfrac{a}{e}$.

If we choose $x = h$ to be the y'-axis and $y = k$ to be the x'-axis, an equation of the ellipse referred to the new axes is
$$\frac{x'^2}{a^2} + \frac{y'^2}{b^2} = 1$$

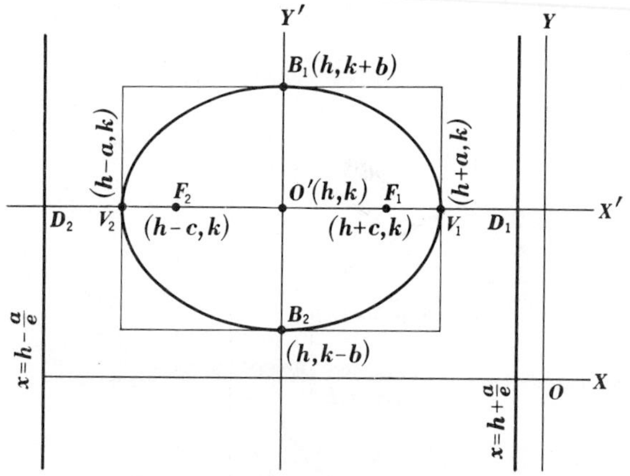

FIG. 6-7

Since $x' = x - h$ and $y' = y - k$, this equation becomes
$$\frac{(x-h)^2}{a^2} + \frac{(y-k)^2}{b^2} = 1 \tag{6-9}$$

If the major axis is chosen as $x = h$ and the minor axis as $y = k$, the equation is
$$\frac{(y-k)^2}{a^2} + \frac{(x-h)^2}{b^2} = 1 \tag{6-10}$$

Equation (6-9) and equation (6-10) are known as *standard equations* of the ellipse.

Hyperbola. Let the center of a hyperbola be at (h, k), as in Fig. 6-8; let the transverse axis be the line $y = k$; and let the conjugate axis be the line $x = h$. Then the foci are at $(h+c, k)$ and $(h-c, k)$; the vertices are at $(h+a, k)$ and $(h-a, k)$; equations of the directrices are $x = h + \frac{a}{e}$ and $x = h - \frac{a}{e}$; and equations of the asymptotes are $y - k = \frac{b}{a}(x-h)$ and $y - k = -\frac{b}{a}(x-h)$.

If we choose $y = k$ to be the x'-axis and $x = h$ to be the y'-axis, an equation of the hyperbola referred to the new axes is

$$\frac{x'^2}{a^2} - \frac{y'^2}{b^2} = 1$$

Since $x' = x - h$ and $y' = y - k$, the equation referred to the original axes is

$$\frac{(x-h)^2}{a^2} - \frac{(y-k)^2}{b^2} = 1 \tag{6-11}$$

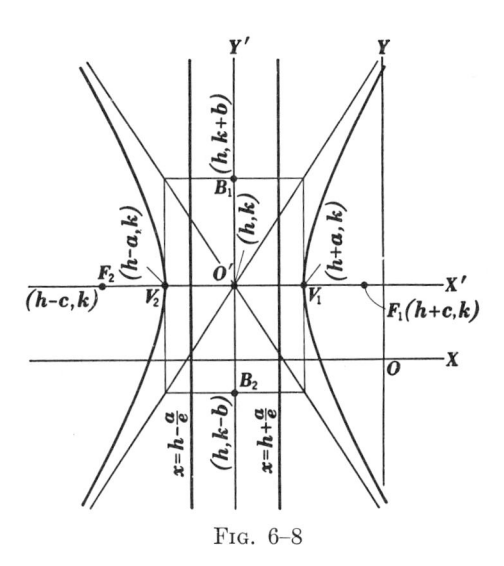

FIG. 6-8

If $x = h$ is the transverse axis and $y = k$ is the conjugate axis, we have

$$\frac{(y-k)^2}{a^2} - \frac{(x-h)^2}{b^2} = 1 \tag{6-12}$$

Equation (6-11) and equation (6-12) are known as *standard equations* of the hyperbola.

EXAMPLE 6-6. Given $V = (-3, -4)$, $F = (-3, -1)$, and $e = 1$; find the standard equation of the conic.

Solution. Since $e=1$, the conic is a parabola. An equation of the axis is $x=-3$, and $a=3$. If we choose the x'-axis to be $y=-4$ and the y'-axis to be $x=-3$, the simple equation is

$$x'^2 = 12y'$$

The curve is shown in Fig. 6-9. Since $x'=x+3$ and $y'=y+4$, the standard equation is

$$(x+3)^2 = 12(y+4)$$

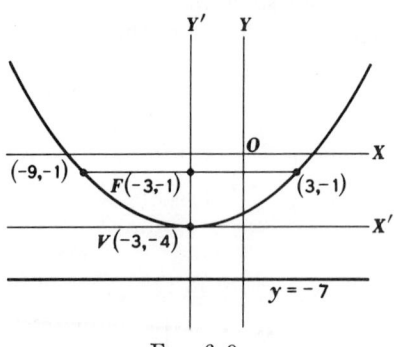

Fig. 6-9

EXAMPLE 6-7. Find the standard equation of the conic that has its vertices at $(2, -1)$ and $(2, 5)$ and $e = \frac{1}{2}$.

Solution. Since $e<1$, the conic is an ellipse. The center of the ellipse is at $(2, 2)$ and $a=3$. In this case, $e = \frac{c}{a}$ or $\frac{1}{2} = \frac{c}{3}$ and it follows that $c = \frac{3}{2}$. Hence $b = \sqrt{a^2 - c^2} = \sqrt{9 - \frac{9}{4}} = \sqrt{\frac{27}{4}} = \frac{3\sqrt{3}}{2}$. Choosing the lines $x=2$ and $y=2$ to be the y'-axis and x'-axis, respectively, we obtain the simple equation of the ellipse to be

$$\frac{y'^2}{9} + \frac{x'^2}{\frac{27}{4}} = 1$$

The curve is shown in Fig. 6-10. Since $y' = y-2$ and $x' = x-2$, the standard equation is

$$\frac{(y-2)^2}{9} + \frac{(x-2)^2}{\frac{27}{4}} = 1$$

EXAMPLE 6-8. If $a=6$, the length of the latus rectum is 24, $e>1$, the center is at $(-1, 3)$, and the vertices lie on the line $y=3$, what is the standard equation of the conic?

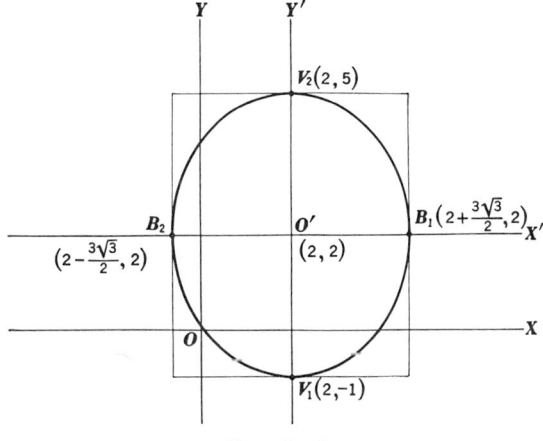

FIG. 6–10

Solution. Since $e > 1$, the conic is a hyperbola. We also know that $\dfrac{2b^2}{a} = 24$ or $\dfrac{2b^2}{6} = 24$. From this we obtain $b^2 = 72$ and $b = 6\sqrt{2}$. Since $c^2 = a^2 + b^2$, we have $c^2 = 36 + 72 = 108$ and $c = 6\sqrt{3}$.

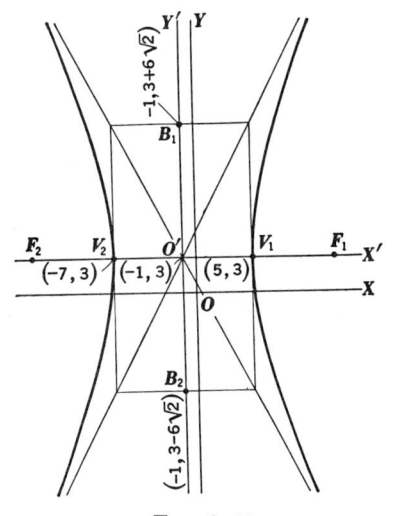

FIG. 6–11

Referred to the x'-axis and y'-axis, a simple equation of the hyperbola is

$$\frac{x'^2}{36} - \frac{y'^2}{72} = 1$$

230   ANALYTIC GEOMETRY IN THE PLANE

The curve is shown in Fig. 6–11. Since $x' = x+1$ and $y' = y-3$, the standard equation that we are seeking is

$$\frac{(x+1)^2}{36} - \frac{(y-3)^2}{72} = 1$$

Equations of the asymptotes are:

$$\frac{x+1}{1} - \frac{y-3}{\sqrt{2}} = 0 \text{ and } \frac{x+1}{1} + \frac{y-3}{\sqrt{2}} = 0$$

EXERCISES

1. Derive equations (6–6), (6–7), (6–8), (6–10), and (6–12).
2. Find a standard equation of each of the following conics and sketch each:
   (a) $e = \frac{1}{2}$, and vertices are at $(-3, 4)$ and $(5, 4)$
   (b) Center is at $(-3, -1)$, $a = 4$, $b = 5$, and foci are on $y = -1$
   (c) Focus is at $(3, 2)$, directrix is $x = 0$, and $e = 1$
   (d) Foci are at $(2, 1)$ and $(2, 7)$, and one vertex is at $(2, 3)$
   (e) Center is at $(5, -2)$, a directrix is $y = -6$, and $a = 2$
   (f) Vertices are at $(-1, -3)$ and $(-1, 9)$ and a focus is at $(-1, -1)$
   (g) Latus rectum $= 16$, vertex is at $(4, 3)$, $e = 1$, and axis is $y = 3$
   (h) Directrix is $x = 6$, vertex is at $(8, -3)$, and $e = 1$
   (i) Directrix is $y = 6$, vertex is at $(8, -3)$, and $e = 1$
   (j) Center is at $(-2, 3)$, $e = 3$, $c = 6$, and vertices are on $x = -2$
   (k) Vertices are at $(5, -3)$ and $(5, 9)$, $b = 4$, and $e > 1$
   (l) Vertices are at $(1, 8)$ and $(1, -2)$, and a focus is at $(1, 5)$
   (m) Foci are at $(2, -2)$ and $(2, 10)$, and $e = \frac{3}{4}$
   (n) $a = 4$, $b = 5$, center is at $(-2, -3)$, and transverse axis is $x = -2$
   (o) Focus is at $(3, 4)$, vertex is at $(3, -1)$, and $e = 1$
   (p) $a = 6$, $b = 3$, major axis is $x = -1$, and minor axis is $y = 4$
   (q) Vertex is at $(5, -1)$, $e = 1$, and ends of latus rectum are at $(1, 1)$ and $(9, 1)$
   (r) Vertices are at $(\pm 3, -4)$, and a focus is at $(-5, -4)$
   (s) Foci are at $(-2, 8)$ and $(-2, -4)$, and vertex is at $(-2, 6)$
   (t) Center is at $(-2, 2)$, a vertex is at $(-2, 8)$, length of latus rectum $= 3$, and $e < 1$
   (u) Center is at $(2, 4)$, a focus is at $(2, 14)$, and $e = 2$
3. Prove that $(y-k)^2 = -4a(x-h)$, $(x-h)^2 = 4a(y-k)$, and $(x-h)^2 = -4a(y-k)$ are equations of parabolas. (Hint: Show that they can be reduced to simple forms by an appropriate transformation of the axes.)
4. Prove that $\dfrac{(y-k)^2}{a^2} + \dfrac{(x-h)^2}{b^2} = 1$ is an equation of an ellipse.

5. Prove that $\dfrac{(y-k)^2}{a^2} - \dfrac{(x-h)^2}{b^2} = 1$ is an equation of a hyperbola.

6–5. Rotation of the axes

Let us see how the coordinates x and y of a point P are transformed if the origin is kept fixed and the axes are rotated through an angle θ, as indicated in Fig. 6–12. The point P has two pairs of coordinates. Its coordinates are x and y with respect to the old axes and are x' and y' with respect to the new axes.

Let α denote the angle between OX' and OP; β, the angle between OY' and OP; θ, the angle between OX and OX'; and $[l, m]$, the direction cosines of OP with respect to the original axes. The direction cosines of OX' with respect to the original axes are $[\cos \theta, \cos(-90°+\theta)]$ or $[\cos \theta, \sin \theta]$, and the direction cosines of OY' with respect to the original axes are $[\cos(90°+\theta), \cos \theta]$ or $[-\sin \theta, \cos \theta]$.

By equation (2–6),
$$\cos \alpha = l \cos \theta + m \sin \theta$$

Multiplying through by $|OP|$, we obtain $|OP|\cos \alpha = |OP|\, l \cos \theta + |OP|\, m \sin \theta$. But $|OP|\cos \alpha = x'$, $|OP|\, l = x$, and $|OP|\, m = y$. Hence, our equation becomes

$$x' = x \cos \theta + y \sin \theta \tag{6–13}$$

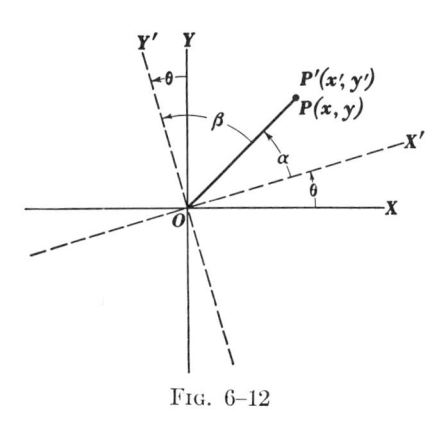

Fig. 6–12

In the same way, we obtain
$$\cos \beta = -l \sin \theta + m \cos \theta$$
or
$$|OP|\cos \beta = -|OP|\, l \sin \theta + |OP|\, m \cos \theta$$
Since $|OP|\cos \beta = y'$,
$$y' = -x \sin \theta + y \cos \theta \tag{6–14}$$

232   ANALYTIC GEOMETRY IN THE PLANE

Solving equations (6–13) and (6–14) for x and y, we obtain:

$$x = \frac{\begin{vmatrix} x' & \sin\theta \\ y' & \cos\theta \end{vmatrix}}{\begin{vmatrix} \cos\theta & \sin\theta \\ -\sin\theta & \cos\theta \end{vmatrix}} = \frac{x'\cos\theta - y'\sin\theta}{1}$$

$$y = \frac{\begin{vmatrix} \cos\theta & x' \\ -\sin\theta & y' \end{vmatrix}}{\begin{vmatrix} \cos\theta & \sin\theta \\ -\sin\theta & \cos\theta \end{vmatrix}} = \frac{x'\sin\theta + y'\cos\theta}{1}$$

or
$$x = x'\cos\theta - y'\sin\theta \qquad (6\text{–}15)$$
and
$$y = x'\sin\theta + y'\cos\theta \qquad (6\text{–}16)$$

As in the case of translation of the axes, transformations of equations involving rotation of the axes may be carried out by substitution.

The following chart is useful for remembering equations (6–13) to (6–16):

	x'	y'
x	$\cos\theta$	$-\sin\theta$
y	$\sin\theta$	$\cos\theta$

We read this as follows:
$$x = [x', y'] \cdot [\cos\theta, -\sin\theta] \qquad (6\text{–}17)$$
$$y = [x', y'] \cdot [\sin\theta, \cos\theta] \qquad (6\text{–}18)$$
$$x' = [x, y] \cdot [\cos\theta, \sin\theta] \qquad (6\text{–}19)$$
$$y' = [x, y] \cdot [-\sin\theta, \cos\theta] \qquad (6\text{–}20)$$

In each case, the second member is the dot product of two vectors.

If we let l_1 and m_1 be the direction cosines of OX' with respect to the original axes and we let l_2 and m_2 be the direction cosines of OY' with respect to the original axes, the formulas for rotation are

$$x = l_1 x' + l_2 y' \qquad (6\text{–}21)$$
$$y = m_1 x' + m_2 y' \qquad (6\text{–}22)$$
$$x' = l_1 x + m_1 y \qquad (6\text{–}23)$$
$$y' = l_2 x + m_2 y \qquad (6\text{–}24)$$

The student should verify these formulas.

EXAMPLE 6–9. What do the coordinates of a point $P = (3, -2)$ become when the axes are rotated through an angle of 30°?

TRANSFORMATION OF THE AXES

Solution. We have

	x'	y'
x	$\cos 30°$	$-\sin 30°$
y	$\sin 30°$	$\cos 30°$

or

	x'	y'
x	$\dfrac{\sqrt{3}}{2}$	$-\dfrac{1}{2}$
y	$\dfrac{1}{2}$	$\dfrac{\sqrt{3}}{2}$

Thus,

$$x' = x \cos 30° + y \sin 30° = 3 \cos 30° - 2 \sin 30° = \frac{3\sqrt{3} - 2}{2}$$

$$y' = -x \sin 30° + y \cos 30° = -3 \sin 30° - 2 \cos 30° = \frac{-3 - 2\sqrt{3}}{2}$$

EXAMPLE 6–10. What equation represents the graph of $3x^2 - 2xy + 3y^2 - 32 = 0$ after the axes are rotated through the positive acute angle $\theta = \arcsin \dfrac{1}{\sqrt{2}}$?

Solution. Since $\sin \theta = \dfrac{1}{\sqrt{2}}$, we have $\theta = 45°$ and $\cos \theta = \dfrac{1}{\sqrt{2}}$. Hence, using equations (6–15) and (6–16), we obtain

$$x = \frac{x' - y'}{\sqrt{2}} \quad \text{and} \quad y = \frac{x' + y'}{\sqrt{2}}$$

Hence, the given equation becomes

$$3\left(\frac{x'-y'}{\sqrt{2}}\right)^2 - 2\left(\frac{x'-y'}{\sqrt{2}}\right)\left(\frac{x'+y'}{\sqrt{2}}\right) + 3\left(\frac{x'+y'}{\sqrt{2}}\right)^2 - 32 = 0$$

Expanding, we obtain

$$\frac{3}{2}(x'^2 - 2x'y' + y'^2) - \frac{2}{2}(x'^2 - y'^2) + \frac{3}{2}(x'^2 + 2x'y' + y'^2) - 32 = 0$$

Multiplying through by 2 and collecting like terms, we have

$$(3 - 2 + 3)x'^2 + (-6 + 6)x'y' + y'^2(3 + 2 + 3) - 64 = 0$$

or

$$4x'^2 + 8y'^2 - 64 = 0$$

When simplified this becomes

$$x'^2 + 2y'^2 = 16$$

234 ANALYTIC GEOMETRY IN THE PLANE

We recognize this to be an equation of an ellipse. The major axis is the x'-axis, the minor axis is the y'-axis, $a = 4$, and $b = 2\sqrt{2}$. The curve is shown in Fig. 6–13.

Referred to the x'-axis and y'-axis, the vertices are at $(\pm 4, 0)$ and the points where the curve intersects the minor axis are at $(0, \pm 2\sqrt{2})$.

Since $a = 4$ and $b = 2\sqrt{2}$, it follows that $c = \sqrt{16 - 8} = 2\sqrt{2}$, $e = \dfrac{\sqrt{2}}{2}$, and $\dfrac{a}{e} = \dfrac{8}{\sqrt{2}} = 4\sqrt{2}$.

Let us discuss the ellipse with reference to the x-axis and y-axis. We have

	x'	y'
x	$\dfrac{1}{\sqrt{2}}$	$-\dfrac{1}{\sqrt{2}}$
y	$\dfrac{1}{\sqrt{2}}$	$\dfrac{1}{\sqrt{2}}$

Substituting $x' = \pm 4$ and $y' = 0$, we obtain $x = \pm 2\sqrt{2}$ and $y = \pm 2\sqrt{2}$ as the coordinates of the vertices. Substituting $x' = 0$ and $y' = \pm 2\sqrt{2}$, we find the coordinates of the points where the curve intersects the minor axis

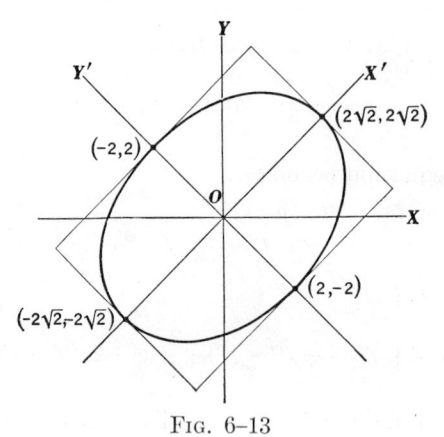

FIG. 6–13

to be $x = \mp 2$ and $y = \pm 2$. With respect to the new axes, the equations of the directrices are $x' = 4\sqrt{2}$ and $x' = -4\sqrt{2}$. Replacing x' by $\dfrac{x + y}{\sqrt{2}}$, which we obtain from equation (6–13), we find the equations of the directrices to be

$$\frac{x+y}{\sqrt{2}} = 4\sqrt{2} \quad \text{and} \quad \frac{x+y}{\sqrt{2}} = -4\sqrt{2}$$

or

$$x+y = 8 \quad \text{and} \quad x+y = -8$$

EXERCISES

1. Find the coordinates of each of the following points after the axes have been rotated through the given angles:
   (a) $(3, -1)$ and $\theta = 30°$
   (b) $(3, 4)$ and $\theta = 30°$
   (c) $(-1, 2)$ and $\theta = 45°$
   (d) $(-2, 3)$ and $\theta = 60°$
   (e) $(-2, -5)$ and $\theta = 60°$
   (f) $(5, 12)$ and $\theta = 45°$

2. Rotate the axes through the given angle and determine what each of the following equations becomes:
   (a) $x^2 - 2xy + y^2 - 3x = 0$ and $\theta = 45°$
   (b) $10x^2 + 24xy + 17y^2 - 144 = 0$ and $\tan \theta = \frac{4}{3}$
   (c) $5x^2 + 4xy + 8y^2 = 36$ and $\theta = \cos^{-1}\frac{1}{\sqrt{5}}$
   (d) $3x^2 + 6xy - 5y^2 = 24$ and $\theta = \cos^{-1}\frac{3}{\sqrt{10}}$
   (e) $16x^2 - 24xy + 9y^2 = 25$ and $\tan 2\theta = \frac{-24}{7}$
   (f) $x^2 + 4xy - 2y^2 = 4$ and $\tan 2\theta = \frac{4}{3}$
   (g) $5x^2 + 12xy + 10y^2 = 5$ and $\tan \theta = \frac{3}{2}$
   (h) $16x^2 - 24xy + 9y^2 = 64$ and $\tan 2\theta = \frac{-24}{7}$

6–6. Simplification of the equation $Ax^2 + Bxy + Cy^2 + Dx + Ey + F = 0$, where $B \neq 0$

By means of a rotation of axes it is possible to choose a positive acute angle θ such that the xy term is eliminated from the equation $Ax^2 + Bxy + Cy^2 + Dx + Ey + F = 0$. To obtain such an angle θ, we proceed as follows: From equations (6–15) and (6–16), we have $x = x' \cos \theta - y' \sin \theta$ and $y = x' \sin \theta + y' \cos \theta$. Substitution of these expressions for x and y in the general equation yields

$$A(x' \cos \theta - y' \sin \theta)^2 + B(x' \cos \theta - y' \sin \theta)(x' \sin \theta + y' \cos \theta)$$
$$+ C(x' \sin \theta + y' \cos \theta)^2 + D(x' \cos \theta - y' \sin \theta) +$$
$$E(x' \sin \theta + y' \cos \theta) + F = 0$$

236 ANALYTIC GEOMETRY IN THE PLANE

Expanding this and collecting like terms, we obtain

(1) $(A \cos^2 \theta + B \cos \theta \sin \theta + C \sin^2 \theta)x'^2 + \{-2A \sin \theta \cos \theta + B(\cos^2 \theta - \sin^2 \theta) + 2C \sin \theta \cos \theta\}x'y' + (A \sin^2 \theta - B \sin \theta \cos \theta + C \cos^2 \theta)y'^2 + (D \cos \theta + E \sin \theta)x' + (-D \sin \theta + E \cos \theta)y' + F = 0$

Our problem is to determine θ in order that the coefficient of the $x'y'$ term will be zero. Hence, we must have $-2A \sin \theta \cos \theta + B(\cos^2 \theta - \sin^2 \theta) + 2C \sin \theta \cos \theta = 0$. Using the identities $\sin 2\theta \equiv 2 \sin \theta \cos \theta$ and $\cos 2\theta \equiv \cos^2 \theta - \sin^2 \theta$, we have

$$-A \sin 2\theta + B \cos 2\theta + C \sin 2\theta = 0$$

or

(2) $\qquad (C - A) \sin 2\theta + B \cos 2\theta = 0$

In order to solve this equation for θ, we note first that if $\sin 2\theta = 0$ we must have $B = 0$. Since $B \neq 0$, it follows that $\sin 2\theta \neq 0$. Thus, we can divide each term of equation (2) by $\sin 2\theta$ and obtain

$$(C - A) + B \cot 2\theta = 0$$

or
$$\cot 2\theta = \frac{A - C}{B} \qquad (6\text{--}25)$$

Knowing $\cot 2\theta$, we find that

$$\cos 2\theta = \frac{A - C}{\pm \sqrt{B^2 + (A - C)^2}} \qquad (6\text{--}26)$$

We choose the sign before the radical so that $\cot 2\theta$ and $\cos 2\theta$ will have the same sign. This insures us that θ is a positive acute angle. Now, by means of the identities $\sin \theta \equiv \sqrt{\frac{1 - \cos 2\theta}{2}}$ and $\cos \theta \equiv \sqrt{\frac{1 + \cos 2\theta}{2}}$, we can apply equations (6–15) and (6–16).

If $A = C$, then $\cot 2\theta = 0$, $2\theta = 90°$, and $\theta = 45°$. Therefore, when $A = C$, the positive acute angle through which the axes should be rotated in order to eliminate the xy term is 45°.

EXAMPLE 6–11. Rotate the axes in such a way as to eliminate the xy term from the equation $5x^2 - 6\sqrt{3}\, xy - y^2 + 16 = 0$, and draw its graph.

Solution. We have $\cot 2\theta = \frac{6}{-6\sqrt{3}} = -\frac{1}{\sqrt{3}}$. Hence, $\cos 2\theta = -\frac{1}{2}$, $\sin \theta = \sqrt{\frac{1 + \frac{1}{2}}{2}} = \frac{\sqrt{3}}{2}$, and $\cos \theta = \sqrt{\frac{1 - \frac{1}{2}}{2}} = \frac{1}{2}$. So we have

TRANSFORMATION OF THE AXES 237

	x'	y'
x	$\frac{1}{2}$	$-\frac{\sqrt{3}}{2}$
y	$\frac{\sqrt{3}}{2}$	$\frac{1}{2}$

or $$x = \frac{x' - \sqrt{3}\, y'}{2} \quad \text{and} \quad y = \frac{\sqrt{3}\, x' + y'}{2}$$

Now the equation becomes

$$5\left(\frac{x' - \sqrt{3}\, y'}{2}\right)^2 - 6\sqrt{3}\left(\frac{x' - \sqrt{3}\, y'}{2}\right)\left(\frac{\sqrt{3}\, x' + y'}{2}\right) - \left(\frac{\sqrt{3}\, x' + y'}{2}\right)^2 + 16 = 0$$

or
$$\frac{5}{4}(x'^2 - 2\sqrt{3}\, x'y' + 3y'^2) - \frac{6\sqrt{3}}{4}(\sqrt{3}\, x'^2 - 2x'y' - \sqrt{3}\, y'^2)$$
$$- \frac{1}{4}(3x'^2 + 2\sqrt{3}\, x'y' + y'^2) + 16 = 0$$

Collecting like terms, we obtain

$$x'^2\left(\frac{5}{4} - \frac{18}{4} - \frac{3}{4}\right) + x'y'\left(\frac{-10\sqrt{3}}{4} + \frac{12\sqrt{3}}{4} - \frac{2\sqrt{3}}{4}\right) + y'^2\left(\frac{15}{4} + \frac{18}{4} - \frac{1}{4}\right) + 16 = 0$$

or $$-4x'^2 + 8y'^2 + 16 = 0$$

This reduces to $x'^2 - 2y'^2 - 4 = 0$. The graph is drawn in Fig. 6–14.

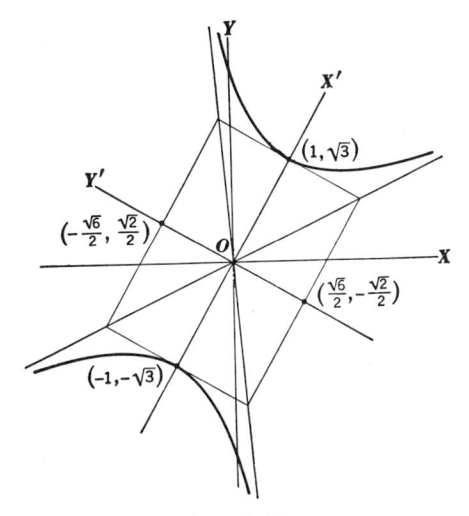

FIG. 6–14

Equations of the asymptotes are $x' - \sqrt{2}\, y' = 0$ and $x' + \sqrt{2}\, y' = 0$. From equations (6–13) and (6–14), we obtain $x' = \dfrac{x + \sqrt{3}\, y}{2}$ and $y' = \dfrac{-\sqrt{3}\, x + y}{2}$; and so, referred to the original axes, the equations of the asymptotes are

$$x + \sqrt{3}\, y + \sqrt{6}\, x - \sqrt{2}\, y = 0 \text{ and } x + \sqrt{3}\, y - \sqrt{6}\, x + \sqrt{2}\, y = 0$$

or $\quad (1 + \sqrt{6})x + (\sqrt{3} - \sqrt{2})y = 0$ and $(1 - \sqrt{6})x + (\sqrt{3} + \sqrt{2})y = 0$

When both the linear terms and the xy term appear in the equation, it is necessary both to translate and to rotate the axes in order to obtain the equation in its *simple* form. Applying equations (6–1) to the given equation $Ax^2 + Bxy + Cy^2 + Dx + Ey + F = 0$, we obtain

$$A(x'+h)^2 + B(x'+h)(y'+k) + C(y'+k)^2 + D(x'+h) + E(y'+k) + F = 0$$

Expanding and collecting like terms, we have

$$Ax'^2 + Bx'y' + Cy'^2 + x'(2Ah + Bk + D) + y'(Bh + 2Ck + E) + Ah^2 + Bhk + Ck^2 + Dh + Ek + F = 0$$

To eliminate the x' and y' terms, we choose the coordinates h and k of the new origin so that $2Ah + Bk + D = 0$ and $Bh + 2Ck + E = 0$. Solving these equations for h and k, we obtain

$$h = \frac{\begin{vmatrix} -D & B \\ -E & 2C \end{vmatrix}}{\begin{vmatrix} 2A & B \\ B & 2C \end{vmatrix}} \quad \text{and} \quad k = \frac{\begin{vmatrix} 2A & -D \\ B & -E \end{vmatrix}}{\begin{vmatrix} 2A & B \\ B & 2C \end{vmatrix}} \qquad (6\text{–}27)$$

From these results we see that it is possible to eliminate the linear terms from a second-degree equation in which the xy term appears by means of a translation of the axes, provided $\begin{vmatrix} 2A & B \\ B & 2C \end{vmatrix} = 4AC - B^2 \neq 0$. After the linear terms are removed the equation is of the form

$$Ax'^2 + Bx'y' + Cy'^2 + F' = 0 \qquad (6\text{–}28)$$

where $F' = Ah^2 + Bhk + Ck^2 + Dh + Ek + F$. Hence, to get F' we merely substitute h for x and k for y in the equation $Ax^2 + Bxy + Cy^2 + Dx + Ey + F = 0$.

If $4AC - B^2 = 0$, we cannot solve for h and k by use of equations (6–27); and, in this case, we rotate the axes and eliminate the xy term before translating. Experience has shown that, when $4AC - B^2 \neq 0$, it is best to translate the axes and eliminate the x-term and y-term first, and then to eliminate the xy-term by a rotation of axes.

EXAMPLE 6–12. Simplify the equation $x^2 - 4xy + y^2 + 10x - 8y + 7 = 0$; and discuss the curve.

TRANSFORMATION OF THE AXES 239

Solution. Since $4AC - B^2 = -12$, we shall translate the axes first. Using equations (6–27), we have

$$h = \frac{\begin{vmatrix} -10 & -4 \\ 8 & 2 \\ 2 & -4 \\ -4 & 2 \end{vmatrix}}{\begin{vmatrix} 2 & -4 \\ -4 & 2 \end{vmatrix}} = \frac{12}{-12} = -1$$

and

$$k = \frac{\begin{vmatrix} 2 & -10 \\ -4 & 8 \\ 2 & -4 \\ -4 & 2 \end{vmatrix}}{\begin{vmatrix} \end{vmatrix}} = \frac{-24}{-12} = 2$$

After translation of the axes to the point $(-1, 2)$ as a new origin, the given equation becomes, from equation (6–28),

$$x'^2 - 4x'y' + y'^2 + (-1)^2 - 4(-1)(2) + (2)^2 + 10(-1) - 8(2) + 7 = 0$$

This reduces to

$$x'^2 - 4x'y' + y'^2 + 1 + 8 + 4 - 10 - 16 + 7 = 0$$

or

(1) $$x'^2 - 4x'y' + y'^2 - 6 = 0$$

Next we proceed to rotate the axes and eliminate the $x'y'$ term. Since $A = C$, we know that $2\theta = 90°$ and $\theta = 45°$. Hence, we have

	x''	y''
x'	$\cos 45°$	$-\sin 45°$
y'	$\sin 45°$	$\cos 45°$

or

$$x' = \frac{x'' - y''}{\sqrt{2}} \quad \text{and} \quad y' = \frac{x'' + y''}{\sqrt{2}}$$

Now, equation (1) becomes

$$\left(\frac{x'' - y''}{\sqrt{2}}\right)^2 - 4\left(\frac{x'' - y''}{\sqrt{2}}\right)\left(\frac{x'' + y''}{\sqrt{2}}\right) + \left(\frac{x'' + y''}{\sqrt{2}}\right)^2 - 6 = 0$$

Expanding, we obtain

$$\frac{x''^2 - 2x''y'' + y''^2}{2} - 4\frac{x''^2 - y''^2}{2} + \frac{x''^2 + 2x''y'' + y''^2}{2} - 6 = 0$$

This reduces to

$$-2x''^2 + 6y''^2 - 12 = 0$$

or

$$x''^2 - 3y''^2 + 6 = 0$$

240   ANALYTIC GEOMETRY IN THE PLANE

Rewriting the equation in simple form, we have

$$\frac{y''^2}{2} - \frac{x''^2}{6} = 1$$

The curve is a hyperbola. Also, $a = \sqrt{2}$, $b = \sqrt{6}$, $c = \sqrt{8}$, $e = \frac{c}{a} = 2$, $\frac{a}{e} = \frac{\sqrt{2}}{2}$, and the transverse axis is the y''-axis, as shown in Fig. 6–15.

The equations of the asymptotes referred to the rotated axes are

$$x'' - \sqrt{3}\, y'' = 0 \quad \text{and} \quad x'' + \sqrt{3}\, y'' = 0$$

From equations (6–13) and (6–14), $x'' = \frac{x'+y'}{\sqrt{2}}$ and $y'' = \frac{-x'+y'}{\sqrt{2}}$. Hence, the asymptotes become

$$\frac{x'+y'}{\sqrt{2}} - \sqrt{3}\,\frac{-x'+y'}{\sqrt{2}} = 0 \quad \text{and} \quad \frac{x'+y'}{\sqrt{2}} + \sqrt{3}\,\frac{-x'+y'}{\sqrt{2}} = 0$$

or $\quad (1+\sqrt{3})x' + (1-\sqrt{3})y' = 0 \quad \text{and} \quad (1-\sqrt{3})x' + (1+\sqrt{3})y' = 0$

Finally, $x' = x+1$ and $y' = y-2$. Hence, the equations of the asymptotes with respect to the original axes are

$$(1+\sqrt{3})(x+1) + (1-\sqrt{3})(y-2) = 0$$
and
$$(1-\sqrt{3})(x+1) + (1+\sqrt{3})(y-2) = 0$$
or
$$(1+\sqrt{3})x + (1-\sqrt{3})y - 1 + 3\sqrt{3} = 0$$
and
$$(1-\sqrt{3})x + (1+\sqrt{3})y - 1 - 3\sqrt{3} = 0$$

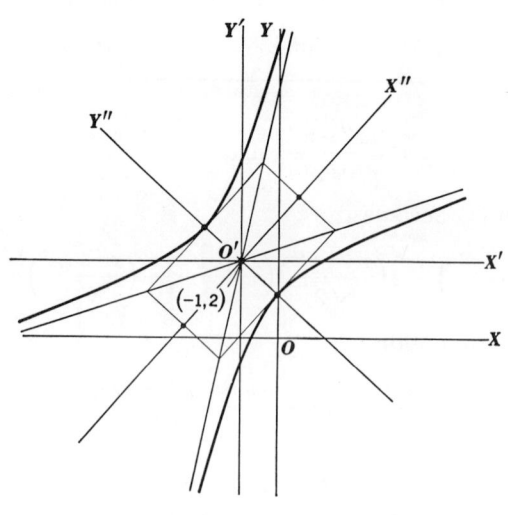

FIG. 6–15

6–7. Equilateral hyperbola

The equilateral hyperbola in standard form was discussed in Section 5–25. Consider the equation $x^2 - y^2 = k^2$. By rotating the axis through an angle whose measure is 45°, we obtain the equation $x'y' + \frac{k^2}{2} = 0$. We now let $\frac{k^2}{2} = a$, and have the equation

$$x'y' + a = 0 \qquad (6\text{--}29)$$

This is a common form of an equation of an equilateral hyperbola.

Now consider any equation of the form

$$Bxy + Dx + Ey + F = 0 \qquad (6\text{--}30)$$

A translation of axes yields

$$Bx'y' + F' = 0 \qquad (6\text{--}31)$$

Let $\frac{F'}{B} = a$, and (6–31) becomes $x'y' + a = 0$. Since, due to the translation, $x' = x - h$ and $y' = y - k$, it follows that

$$(x - h)(y - k) + a = 0 \qquad (6\text{--}32)$$

The asymptotes of the hyperbola $x'y' + a = 0$ are $x' = 0$ and $y' = 0$, which are the rotated axes OX' and OY'. The asymptotes of the hyperbola defined by (6–31) or (6–32) are the translated axes $O'X'$ and $O'Y'$. (See Fig. 6–16 to 6–20.)

In equation (6–30) if $E = 0$, we have $Bxy + Dx + F = 0$. This can be written in the form $x = \frac{a}{y - k}$. Similarly, when $D = 0$, (6–30) can be reduced to $y = \frac{a}{x - h}$.

$x^2 - y^2 = k^2$

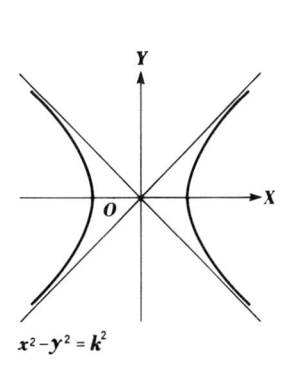

Fig. 6–16

$xy = a$

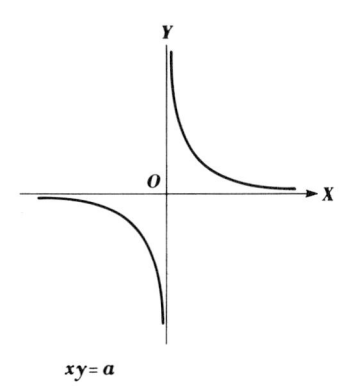

Fig. 6–17

242   ANALYTIC GEOMETRY IN THE PLANE

Two common forms of the equilateral hyperbola are
$$x^2 - y^2 = k^2 \quad \text{and} \quad Bxy + Dx + Ey + F = 0$$
In particular, the second form can always be reduced to one of the following:
$$xy = a, \ (x-h)(y-k) = a, \ y = \frac{a}{x-h}, \quad \text{or} \quad x = \frac{a}{y-k}$$
The graphs of the different forms of an equilateral hyperbola are given in Fig. 6–16 to 6–20.

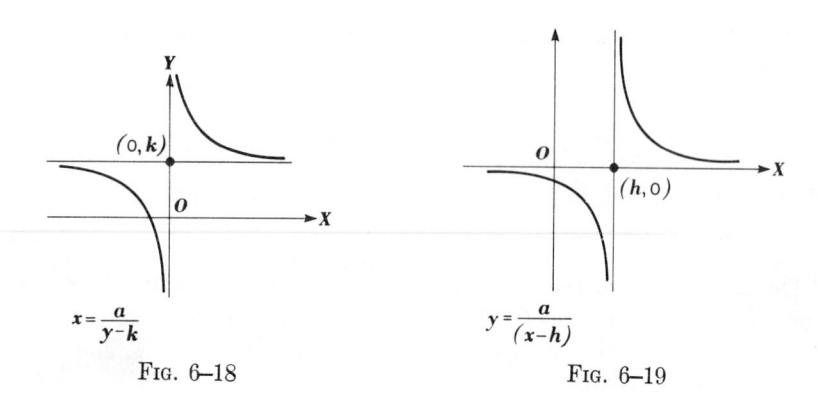

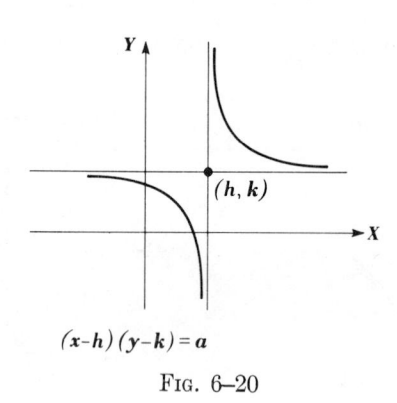

$x = \dfrac{a}{y-k}$

FIG. 6–18

$y = \dfrac{a}{(x-h)}$

FIG. 6–19

$(x-h)(y-k) = a$

FIG. 6–20

EXERCISES

Simplify and discuss each of the following equations and draw the graph:

1. $5x^2 - 2\sqrt{3}\,xy - y^2 + 16 = 0$

2. $337x^2 + 168xy + 288y^2 = 3600$
3. $41x^2 + 24xy + 34y^2 - 188x - 16y + 221 = 0$
4. $16x^2 - 24xy + 9y^2 - 360y - 20x = 0$
5. $24xy - 7x^2 - 120x - 144 = 0$
6. $7y^2 - 6xy - x^2 + 58y - 2x - 9 = 0$
7. $x^2 - 6xy - 7y^2 + 30x + 38y + 97 = 0$
8. $3x^2 + 24xy - 4y^2 - 5y = 0$
9. $x^2 + 2xy + y^2 + 8x - 8y + 4 = 0$
10. $17x^2 + 12xy + 8y^2 - 46x - 28y + 17 = 0$
11. Remove the xy-term, identify the conic, and draw the graph of each of the following equations:

(a) $xy = 2$
(b) $8xy + 1 = 0$
(c) $5x^2 - 6xy + 5y^2 = 8$
(d) $2x^2 + 3xy + 2y^2 = 7$
(e) $7x^2 + 12xy - 2y^2 = 10$
(f) $3x^2 - 10xy + 3y^2 + 32 = 0$
(g) $4x^2 + 6xy - 4y^2 = 5a^2$
(h) $5x^2 + 2\sqrt{3}xy + 7y^2 - 16 = 0$
(i) $4xy - 3y^2 = 8$
(j) $19x^2 + 6xy + 11y^2 = 20$
(k) $x^2 + 3xy + 5y^2 = 22$
(l) $10x^2 + 24xy + 17y^2 = 144$
(m) $x^2 - 2xy + y^2 - 3x = 0$
(n) $16x^2 - 24xy + 9y^2 = 25$
(o) $x^2 + 4xy - 2y^2 = 4$
(p) $5x^2 + 12xy + 10y^2 = 5$
(q) $13x^2 - 10xy + 13y^2 - 72 = 0$
(r) $5x^2 - 8xy + 5y^2 = 9$
(s) $5x^2 + 12xy = 4$
(t) $x^2 - 4xy + 4y^2 = 6y$

12. Verify that each of the equations

$$x^{\frac{1}{2}} + y^{\frac{1}{2}} = 2^{\frac{1}{2}}, \quad x^{\frac{1}{2}} - y^{\frac{1}{2}} = 2^{\frac{1}{2}}, \quad \text{and} \quad x^{\frac{1}{2}} - y^{\frac{1}{2}} = -2^{\frac{1}{2}}$$

represents a different arc of the same parabola and identify the arc which each equation represents. (Hint: rationalize each equation.)

6–8. Invariants of the second-degree equation

Consider any quadratic equation

(1) $$Ax^2 + Bxy + Cy^2 + Dx + Ey + F = 0$$

After a translation or a rotation of the axes, the given equation is converted into a new quadratic equation

(2) $$A'x'^2 + B'x'y' + C'y'^2 + D'x' + E'y' + F' = 0$$

Equation (2) is known as the *transform* of equation (1), provided the coefficients have not been altered by clearing of fractions.

244   ANALYTIC GEOMETRY IN THE PLANE

There are certain functions of the coefficients A, B, C, D, E, and F of equation (1), say $\phi(A, B, C, D, E, F)$, which have the property that

$$\phi(A, B, C, D, E, F) = \phi(A', B', C', D', E', F')$$

These functions are called *invariants* of the second-degree equation. Three of them will be discussed in this section; they are $(A+C)$, $(4AC - B^2)$, and a quantity designated as Δ.

After a rotation of axes, we have, from equation (1) in Section 6-6:

$$A' = A \cos^2 \theta + B \cos \theta \sin \theta + C \sin^2 \theta$$
$$B' = -(A-C) \sin 2\theta + B \cos 2\theta$$
$$C' = A \sin^2 \theta - B \cos \theta \sin \theta + C \cos^2 \theta$$
$$D' = D \cos \theta + E \sin \theta$$
$$E' = -D \sin \theta + E \cos \theta$$
$$F' = F$$

Since a translation of the axes does not affect the terms of highest degree in an equation, it follows that any function of the coefficients of the terms of highest degree is invariant under translation. Hence, the function $(A+C)$ obviously is invariant under translation. Also, if we add A' and C', we obtain $A'+C' = A(\cos^2 \theta + \sin^2 \theta) + C(\cos^2 \theta + \sin^2 \theta) = A+C$. Hence, $(A+C)$ is invariant under both translation and rotation of the axes.

The function $(4AC - B^2)$ is known as the *characteristic* of the second-degree equation. It is obvious that this function also is invariant under translation, since only the coefficients of the highest-degree terms appear in the characteristic. When the axes are rotated, we have already seen that

(3) $$A'+C' = A+C$$

We also obtain

$$A' - C' = A(\cos^2 \theta - \sin^2 \theta) + 2B \cos \theta \sin \theta + C(\sin^2 \theta - \cos^2 \theta)$$

or

(4) $$A' - C' = (A-C) \cos 2\theta + B \sin 2\theta$$

From equation (3) it follows that

$$A'^2 + 2A'C' + C'^2 = A^2 + 2AC + C^2$$

and, from equation (4), we have

$$A'^2 - 2A'C' + C'^2 = (A-C)^2 \cos^2 2\theta + 2B(A-C) \sin 2\theta \cos 2\theta + B^2 \sin^2 2\theta$$

Subtracting, we obtain

(5)
$$4A'C' = -(A-C)^2 \cos^2 2\theta + (A+C)^2 - 2B(A-C) \sin 2\theta \cos 2\theta - B^2 \sin^2 2\theta$$

From the expression for B', we have

(6) $\quad -B'^2 = -(A-C)^2\sin^2 2\theta + 2B(A-C)\sin 2\theta \cos 2\theta - B^2\cos^2 2\theta$

Adding equations (5) and (6), we obtain

$$4A'C' - B'^2 = -(A-C)^2 + (A+C)^2 - B^2 = 4AC - B^2$$

Hence, $4AC - B^2$ is invariant under both translation and rotation of the axes.

The invariant of the second-degree equation designated by Δ is called the *discriminant* of the equation and is defined by the relation

$$\Delta = \begin{vmatrix} A & \dfrac{B}{2} & \dfrac{D}{2} \\ \dfrac{B}{2} & C & \dfrac{E}{2} \\ \dfrac{D}{2} & \dfrac{E}{2} & F \end{vmatrix}$$

We shall show that Δ is invariant under a translation. Applying equations (6–1) to the general equation $Ax^2 + Bxy + Cy^2 + Dx + Ey + F = 0$, we have: $A' = A$; $B' = B$; $C' = C$; $D' = 2Ah + Bk + D$; $E' = Bh + 2Ck + E$; and $F' = Ah^2 + Bhk + Ck^2 + Dh + Ek + F$. Hence,

(7) $\quad \Delta' = \begin{vmatrix} A' & \dfrac{B'}{2} & \dfrac{D'}{2} \\ \dfrac{B'}{2} & C' & \dfrac{E'}{2} \\ \dfrac{D'}{2} & \dfrac{E'}{2} & F' \end{vmatrix}$

$$= \begin{vmatrix} A & \dfrac{B}{2} & \dfrac{2Ah+Bk+D}{2} \\ \dfrac{B}{2} & C & \dfrac{Bh+2Ck+E}{2} \\ \dfrac{2Ah+Bk+D}{2} & \dfrac{Bh+2Ck+E}{2} & Ah^2+Bhk+Ck^2+Dh+Ek+F \end{vmatrix}$$

Now, by multiplying the elements of the first column in equation (7) by $-h$, multiplying the elements of the second column by $-k$, and adding the sum of these to the elements of the third column, we obtain

$$(8) \qquad \Delta' = \begin{vmatrix} A & \dfrac{B}{2} & \dfrac{D}{2} \\ \dfrac{B}{2} & C & \dfrac{E}{2} \\ \dfrac{2Ah+Bk+D}{2} & \dfrac{Bh+2Ck+E}{2} & \dfrac{Dh}{2}+\dfrac{Ek}{2}+F \end{vmatrix}$$

Also, by multiplying the elements of the first row in equation (8) by $-h$, multiplying the elements of the second row by $-k$, and adding the sum of these to the elements of the third row, we obtain

$$\Delta' = \begin{vmatrix} A & \dfrac{B}{2} & \dfrac{D}{2} \\ \dfrac{B}{2} & C & \dfrac{E}{2} \\ \dfrac{D}{2} & \dfrac{E}{2} & F \end{vmatrix} = \Delta$$

Hence, the discriminant Δ is invariant under translation.

6–9. The characteristic $4AC - B^2$ and the discriminant Δ

We saw in Section 6–6 that it is always possible to find a positive acute angle such that the xy term can be eliminated from the general second-degree equation by a rotation of the axes through this angle. Hence, the equation $Ax^2 + Bxy + Cy^2 + Dx + Ey + F = 0$ can be written in the form $A'x'^2 + C'y'^2 + D'x' + E'y' + F' = 0$, where $F = F'$. Now we know from the new equation that, when A' and C' have like signs, we have an ellipse; when A' and C' have unlike signs, we have a hyperbola; and, when either $A' = 0$ or $C' = 0$, we have a parabola. But, in Section 6–8, we saw that $4AC - B^2$ is invariant under rotation. Hence, for a certain angle of rotation, $4AC - B^2 = 4A'C'$. From this it follows that the type of conic represented by a second-degree equation may be determined by the value of the characteristic of the equation.

The conic is a parabola (or two real or imaginary parallel lines), if $4AC - B^2 = 4A'C' = 0$.

The conic is an ellipse (or a point ellipse or an imaginary ellipse), if $4AC - B^2 = 4A'C' > 0$.

The conic is a hyperbola (or two intersecting lines), if $4AC - B^2 = 4A'C' < 0$.

It is important for the student to note that, if $4AC - B^2 = 0$, the terms $Ax^2 + Bxy + Cy^2$ form a perfect square or the negative of a perfect square. Since $4AC - B^2 = 0$ is the condition that the second-degree equation represent a parabola, it follows that $Ax^2 + Bxy + Cy^2 + Dx + Ey + F = 0$ represents a parabola whenever the *second-degree terms form a perfect square*.

If $4AC-B^2\neq 0$, we can eliminate the x and y terms from the general equation by a translation and the equation becomes $Ax'^2+Bx'y'+Cy'^2+F'=0$. Since Δ is invariant under translation, we have for the new equation:

$$\Delta = \begin{vmatrix} A & \frac{B}{2} & 0 \\ \frac{B}{2} & C & 0 \\ 0 & 0 & F' \end{vmatrix} = F'\left(AC-\frac{B^2}{4}\right)$$

If $\Delta=0$ but $4AC-B^2\neq 0$, we must have $F'=0$. The general equation then becomes $Ax'^2+Bx'y'+Cy'^2=0$. If $4AC-B^2<0$, the equation has two real linear factors and therefore it represents a pair of distinct lines. This is the case of a degenerate hyperbola. If $4AC-B^2>0$, the equation is satisfied by one and only one point, namely, that for which $x'=y'=0$. Hence, the locus is a single point; that is, it is a point ellipse (or a point circle). This is the case of a degenerate ellipse.

If $4AC-B^2=0$, we can always eliminate one linear term by a translation. To see this we recall that, after applying equations (6–1) to the general quadratic equation, it becomes

$$Ax'^2+Bx'y'+Cy'^2+D'x'+E'y'+F'=0$$

where $D'=2Ah+Bk+D$ and $E'=Bh+2Ck+E$ (see Section 6–8). In order to eliminate one linear term it is sufficient to choose h and k such that either $D'=0$ or $E'=0$. This is possible since both A and C cannot be zero (if both were zero, B would also be zero).

Let us suppose that $A\neq 0$ and that the x-term has been eliminated. Then the general equation becomes

$$Ax'^2+Bx'y'+Cy'^2+E'y'+F'=0$$

and

$$\Delta = \begin{vmatrix} A & \frac{B}{2} & 0 \\ \frac{B}{2} & C & \frac{E'}{2} \\ 0 & \frac{E'}{2} & F' \end{vmatrix} = \left(-\frac{E'}{2}\right)\left(\frac{AE'}{2}\right)+F'\left(AC-\frac{B^2}{4}\right) = -\frac{AE'^2}{4}$$

If $\Delta=0$, it follows that $\dfrac{AE'^2}{4}=0$. Since $A\neq 0$, we must have $E'=0$; and the equation becomes $Ax'^2+Bx'y'+Cy'^2+F'=0$. If $A>0$ but $4AC-B^2=0$, then $C>0$ and $Ax'^2+Bx'y'+Cy'^2$ is a perfect square. Hence, if

248   ANALYTIC GEOMETRY IN THE PLANE

F' is positive, we have no graph; while, if F' is negative, we have a pair of parallel lines. If $A<0$, we have the same argument for the equivalent equation $-Ax'^2 - Bx'y' - Cy'^2 - F' = 0$. If $F' = 0$, the equation represents two coincident lines.

In case $A = 0$, then $C \neq 0$; and a similar argument can be applied. Hence, the degenerate parabola is a pair of parallel lines, a pair of coincident lines, or no graph at all.

When $\Delta = 0$, the second-degree equation represents a degenerate conic. We see that the characteristic $4AC - B^2$ identifies the type of the conic; and the discriminant Δ determines whether or not it is degenerate.

The following table should be found useful:

$4AC - B^2$	Δ	Locus
0	$\neq 0$	Non-degenerate parabola
0	0	Degenerate parabola: two real or imaginary parallel lines
>0	$\neq 0$	Non-degenerate ellipse (or circle), real or imaginary
>0	0	Degenerate ellipse: point ellipse (or circle)
<0	$\neq 0$	Non-degenerate hyperbola
<0	0	Degenerate hyperbola: two distinct intersecting lines

The invariant properties developed in this chapter may be used to shorten the work required in simplying equations of the second degree. The following chart outlines the steps required to obtain the coefficients of the simplified equation. Matrix multiplication is employed here and should be reviewed. (See Appendix A.)

PROBLEM: Simplify by rotation and translation

$$Ax^2 + Bxy + Cy^2 + Dx + Ey + F = 0$$

NOTE: As indicated by the chart the linear equations resulting from the matrix equation for translation are

(1) $Ah + \dfrac{B}{2} k + \dfrac{D}{2} = 0$

(2) $\dfrac{B}{2} h + Ck + \dfrac{E}{2} = 0$

(3) $\dfrac{D}{2} h + \dfrac{E}{2} k + F = F'$

Equations (1) and (2) are obviously the same as the equations used to find h and k (p. 238, 6–27); however, it is not obvious that (3) is the same as that given for F' on p. 238, namely

Transformation of the Axes

```
                    ┌─────────────┐
                    │ Discriminant│
                    │      Δ      │
                    └──────┬──────┘
              =0           │          ≠ 0
     ┌─────────────────┐   │   ┌──────────────────┐
     │ Factor and sketch│   │   │  Characteristic  │
     │ without rotation │       │    4AC - B²      │
     │  or translation  │       └──────────┬───────┘
     └──────────────────┘         =0       │      ≠ 0
                              ┌────────┐   │   ┌───────────────┐
                              │PARABOLA│       │ CENTRAL CONIC │
                              └────────┘       └───────────────┘
```

Rotation by Invariants

$\cot 2\theta = \dfrac{A-C}{B}$

$\cos 2\theta = \dfrac{A-C}{\pm\sqrt{B^2 + (A-C)^2}}$

$A' + C' = A + C$

$A' - C' = (A-C)\cos 2\theta + B \sin 2\theta$

$B' = 0$

$[D'\ E'] = [D\ E] \begin{bmatrix} \cos\theta & -\sin\theta \\ \sin\theta & \cos\theta \end{bmatrix}$

$F' = F$

Translate by completing the square

Translation by the Invariant Method
(See following note)

$\begin{bmatrix} A & B/2 & D/2 \\ B/2 & C & E/2 \\ D/2 & E/2 & F \end{bmatrix} \begin{bmatrix} h \\ k \\ l \end{bmatrix} = \begin{bmatrix} 0 \\ 0 \\ F' \end{bmatrix}$

$A' = A$
$B' = B$
$C' = C$
$D' = 0$
$E' = 0$

Rotate by the Invariant Method

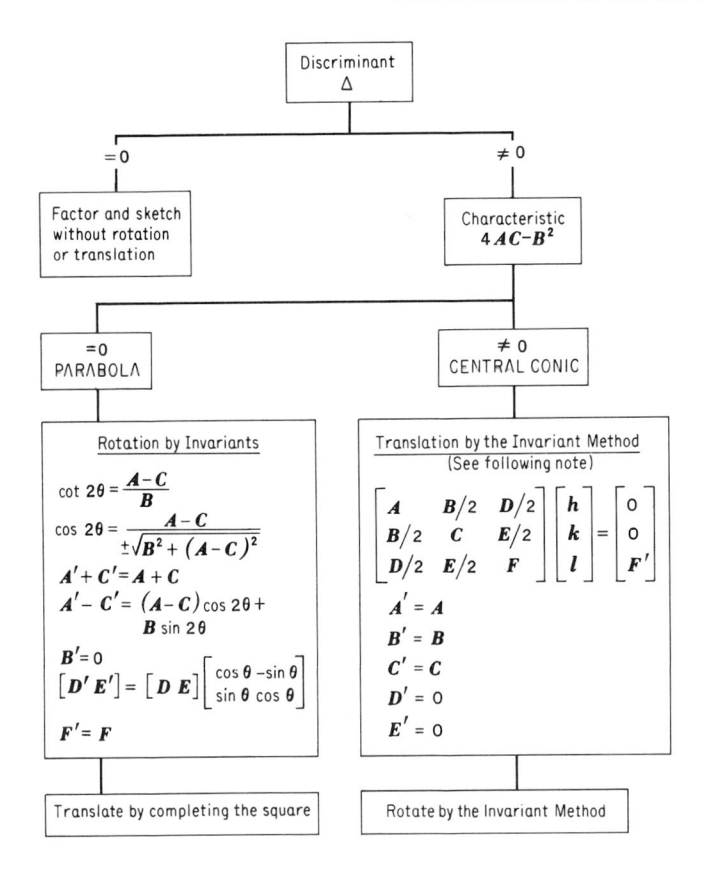

(4) $\qquad F' = Ah^2 + Bhk + Ck^2 + Dh + Eh + F$

To show that (3) and (4) are the same, we express the right side of (4) as follows:

(5) $F' = h\left(Ah + \dfrac{B}{2}k + \dfrac{D}{2}\right) + k\left(\dfrac{B}{2}h + Ck + \dfrac{E}{2}\right) + \dfrac{D}{2}h + \dfrac{E}{2}k + F$

Using equations (1) and (2), we obtain

$$F' = h \cdot (0) + k \cdot (0) + \dfrac{D}{2}h + \dfrac{E}{2}k + F$$

or $\qquad\qquad F' = \dfrac{D}{2}h + \dfrac{E}{2}k + F$

Therefore, equation (3) is consistent with the equation for F' given on p. 238.

EXERCISES

Using the characteristic and the discriminant, determine the type of conic represented by each of the following equations and tell whether each is degenerate or not:

1. $xy - 2y - x + 2 = 0$
2. $xy - 2y - x - 2 = 0$
3. $x^2 + y^2 + 2x - 4y + 5 = 0$
4. $x^2 - 8xy + y^2 = 0$
5. $x^2 - 3xy + y^2 - 10x + 10y + 10 = 0$
6. $x^2 - 8xy + y^2 - 10x + 10y - 2 = 0$
7. $43x^2 + 48xy + 57y^2 + 124x - 18y + 133 = 0$
8. $2x^2 + 72xy + 23y^2 + 60x - 170y - 175 = 0$
9. $5x^2 + 120xy - 30y^2 - 100x + 300y - 100 = 0$
10. $x^2 - 6xy - 7y^2 - 38y + 30x + 97 = 0$

6–10. General transformation of coordinates

In Fig. 6–21 are represented the conditions when coordinates are transformed by both translation and rotation of the axes. One system of axes consisting of OX and OY has the origin O, and another system consisting of $O'X'$ and $O'Y'$ has the origin O'. Also, the axes of one system are not parallel to those of the other system. However, both systems of axes must have the same character (see Section 2–2); that is, both must be left-handed or both must be right-handed. The coordinates of the origin O' with respect to the axes OX and OY are h and k; the axes $O'X''$ and $O'Y''$ have the origin O' and are parallel, respectively, to OX and OY; and the angle between the axes $O'X''$ and $O'X'$ or between the axes $O'Y''$ and $O'Y'$ is θ.

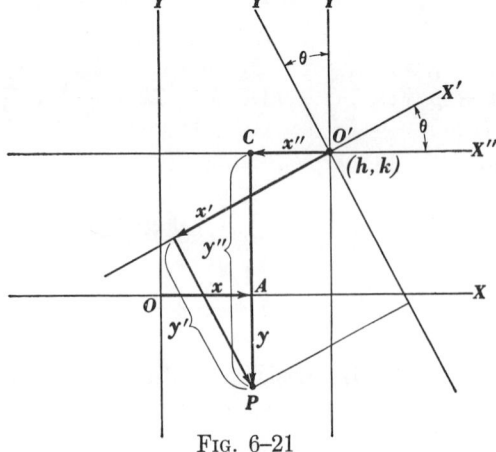

Fig. 6–21

through the four points. If we substitute the coordinates of P_5 in equation (1), we obtain a linear relationship between k_1 and k_2 from which we can determine k_1 as a multiple of k_2 (or vice versa). Substituting this value of k_1 in equation (1) and factoring out k_2, we obtain an equation of the conic through the five points.

The argument just given may apply even if three of the points are on a line. As long as the substitution of the coordinates of P_5 in equation (1) does not yield $k_1(0) + k_2(0) = 0$, it is possible to find a unique conic through the five points. The exact condition is that *no more* than three of the points lie on a line, since the points may then be numbered so that P_5 does not lie on $l_1 = 0$ or $l_2 = 0$.

EXAMPLE 6–13. Find an equation of the conic that contains the points $P_1(1, -1)$, $P_2(1, 1)$, $P_3(-1, 0)$, $P_4(2, 1)$, and $P_5(-1, 2)$.

Solution. An equation of the line P_1P_2 is $x = 1$, and an equation of the line P_3P_5 is $x = -1$; so an equation of a degenerate conic containing P_1, P_2, P_3, and P_5 is $(x-1)(x+1) = 0$. Also, an equation of the line P_1P_3 is $x + 2y + 1 = 0$, and an equation of the line P_2P_5 is $x + 2y - 3 = 0$; so an equation of another degenerate conic through P_1, P_2, P_3, and P_5 is $(x + 2y + 1)(x + 2y - 3) = 0$. Hence, an equation of the pencil of conics through the four points is

$$k_1(x^2 - 1) + k_2(x + 2y + 1)(x + 2y - 3) = 0$$

Since the point $P_4 = (2, 1)$ lies on the conic, the coordinates of this point must satisfy the equation; and so we have $k_1(4-1) + k_2(5)(1) = 0$ or $\dfrac{k_1}{k_2} = -\dfrac{5}{3}$. Letting $k_1 = -5$ and $k_2 = 3$, we have

$$-5x^2 + 5 + 3x^2 + 12xy + 12y^2 - 6x - 12y - 9 = 0$$

or
$$x^2 - 6xy - 6y^2 + 3x + 6y + 2 = 0$$

Check the result by substituting the coordinates of the given points in the equation.

If the conic is a parabola, four points are sufficient to determine its equation; since we already have one condition involving the coefficients which must be satisfied, namely, $4AC - B^2 = 0$.

EXAMPLE 6–14. What is an equation of the parabola that contains the points $P_1 = (1, 0)$, $P_2 = (0, 0)$, $P_3 = (0, 1)$, and $P_4 = (6, 3)$?

Solution. Consider the four lines that contain P_1P_2, P_3P_4, P_1P_3, and P_2P_4. Their equations are: for P_1P_2, $y = 0$; for P_3P_4, $x - 3y + 3 = 0$; for P_1P_3, $x + y - 1 = 0$; for P_2P_4, $x - 2y = 0$. Hence, equations of two degenerate conics through the four points are $y(x - 3y + 3) = 0$ and $(x + y - 1)$

$(x-2y) = 0$. So an equation of the system of conics through the four points is

$$k_1(xy - 3y^2 + 3y) + k_2(x^2 - xy - 2y^2 - x + 2y) = 0$$

Expanding and collecting like terms, we have

$$k_2 x^2 + (k_1 - k_2)xy + (-3k_1 - 2k_2)y^2 - k_2 x + (3k_1 + 2k_2)y = 0$$

whence $A = k_2$, $B = k_1 - k_2$, and $C = -3k_1 - 2k_2$. Since $4AC - B^2 = 0$, we have

$$4k_2(-3k_1 - 2k_2) - (k_1^2 - 2k_1 k_2 + k_2^2) = 0$$

Expanding and collecting terms, we obtain $k_1^2 + 10 k_1 k_2 + 9 k_2^2 = 0$, or $(k_1 + 9k_2)(k_1 + k_2) = 0$. From this it follows that $k_1 = -9k_2$ or $k_1 = -k_2$; and so we see that there are *two* parabolas which contain the four points. If we let $k_1 = -9$ and $k_2 = 1$, we obtain

$$x^2 - 10xy + 25y^2 - x - 25y = 0$$

If we let $k_1 = -1$ and $k_2 = 1$, we obtain

$$x^2 - 2xy + y^2 - x - y = 0$$

Check the results.

When the conic is symmetric with respect to the origin, three points are sufficient to find its equation, since now the equation must be of the form $Ax^2 + Bxy + Cy^2 + F = 0$ and there are only three essential arbitrary constants.

EXAMPLE 6–15. Find an equation of the conic that is symmetric with respect to the origin and contains the points $P_1 = (1, 1)$, $P_2 = (3, 0)$, and $P_3 = (0, -1)$.

Solution. Since the points lie on the conic whose equation is $Ax^2 + Bxy + Cy^2 + F = 0$, we must have

(1) $\qquad A + B + C + F = 0 \quad$ or $\quad A + B + C = -F$
(2) $\qquad 9A \qquad\qquad\quad + F = 0 \quad$ or $\quad 9A \qquad\quad = -F$
(3) $\qquad\qquad\qquad C + F = 0 \quad$ or $\qquad\qquad C = -F$

Since the determinant of the coefficients of A, B, and C is not equal to zero, we can solve for A, B, and C in terms of F. We find that $A = -\dfrac{F}{9}$, $B = \dfrac{F}{9}$, and $C = -F$. Hence, our equation becomes

$$-\frac{F}{9}x^2 + \frac{F}{9}xy - Fy^2 + F = 0$$

which simplifies to

$$x^2 - xy + 9y^2 - 9 = 0$$

The student should determine the type of conic represented by this equation and should check the result.

254   ANALYTIC GEOMETRY IN THE PLANE

Finally, if an axis of the conic is parallel to the x-axis or the y-axis, there is no xy term in the equation. In this case four points are sufficient to determine the conic unless other conditions are imposed (such as that the conic is a parabola), in which case three or fewer points suffice.

EXAMPLE 6-16.   Find an equation of the conic that contains the points $P_1 = (1, 2)$, $P_2 = (-1, 0)$, $P_3 = (2, 1)$, and $P_4 = (-2, 3)$ and whose axes are parallel to the x- and y-axes.

Solution. Using the method of pencils of conics, we have:

The equation of the line containing P_1P_2 is $l_1 = x - y + 1 = 0$.
The equation of the line containing P_3P_4 is $l_2 = x + 2y - 4 = 0$.
The equation of the line containing P_1P_3 is $l_3 = x + y - 3 = 0$.
The equation of the line containing P_2P_4 is $l_4 = 3x + y + 3 = 0$.

Now forming the pencil $k_1 l_1 l_2 + k_2 l_3 l_4 = 0$, we have

(1) $k_1(x - y + 1)(x + 2y - 4) + k_2(x + y - 3)(3x + y + 3) = 0$

Since there can be no xy term in the equation of the conic, the coefficient of this term must be zero. Expanding equation (1) and setting the coefficient of the xy term equal to zero, we obtain $k_1 + 4k_2 = 0$, or $\dfrac{k_1}{k_2} = -\dfrac{4}{1}$.

So in equation (1) we let $k_1 = -4$ and $k_2 = 1$, and we obtain

$$-4(x - y + 1)(x + 2y - 4) + (x + y - 3)(3x + y + 3) = 0$$

When expanded this becomes

$$9y^2 - x^2 + 6x - 24y + 7 = 0$$

Hence, the conic through the given four points is a hyperbola. The student should check the result.

EXAMPLE 6-17.   What is an equation of the parabola whose axis is parallel to the x-axis and which contains the points $P_1 = (-1, 2)$, $P_2 = (1, 1)$, and $P_3 = (0, -2)$?

Solution. Since the axis of the parabola is parallel to the x-axis, the equation is of the form $Cy^2 + Dx + Ey + F = 0$. Since the curve is to contain the three given points, we have

$$4C - D + 2E + F = 0$$
$$C + D + E + F = 0$$
$$4C + 0D - 2E + F = 0$$

If we solve for C, D, and E in terms of F, we obtain

$$C = \frac{\begin{vmatrix} -F & -1 & 2 \\ -F & 1 & 1 \\ -F & 0 & -2 \\ 4 & -1 & 2 \\ 1 & 1 & 1 \\ 4 & 0 & -2 \end{vmatrix}}{} = \frac{\begin{vmatrix} -F & -1 & 2 \\ -2F & 0 & 3 \\ -F & 0 & -2 \\ 4 & -1 & 2 \\ 5 & 0 & 3 \\ 4 & 0 & -2 \end{vmatrix}}{} = -\frac{7}{22}F$$

$$D = \frac{\begin{vmatrix} 4 & -F & 2 \\ 1 & -F & 1 \\ 4 & -F & -2 \\ 4 & -1 & 2 \\ 1 & 1 & 1 \\ 4 & 0 & -2 \end{vmatrix}}{} = \frac{\begin{vmatrix} 3 & -F & 1 \\ 0 & -F & 0 \\ 3 & -F & -3 \end{vmatrix}}{-22} = -\frac{12}{22}F$$

$$E = \frac{\begin{vmatrix} 4 & -1 & -F \\ 1 & 1 & -F \\ 4 & 0 & -F \\ 4 & -1 & 2 \\ 1 & 1 & 1 \\ 4 & 0 & -2 \end{vmatrix}}{} = \frac{\begin{vmatrix} 4 & -1 & -F \\ 5 & 0 & -2F \\ 4 & 0 & -F \end{vmatrix}}{-22} = -\frac{3}{22}F$$

Replacing these values in the equation $Cy^2 + Dx + Ey + F = 0$, we have

$$-\frac{7F}{22}y^2 - \frac{12F}{22}x - \frac{3F}{22}y + F = 0$$

or
$$7y^2 + 12x + 3y - 22 = 0$$

The student should check the result.

EXERCISES

1. Find an equation of the conic that contains the following points:
    (a) $P_1 = (2, 1)$, $P_2 = (-1, 0)$, $P_3 = (1, -1)$, $P_4 = (3, 2)$, and $P_5 = (0, -2)$
    (b) $P_1 = (1, 2)$, $P_2 = (0, 2)$, $P_3 = (-1, 1)$, $P_4 = (-3, -1)$, and $P_5 = (0, -1)$
    (c) $P_1 = (0, 0)$, $P_2 = (3, 0)$, $P_3 = (0, 1)$, $P_4 = (2, -1)$, and $P_5 = (-1, 3)$
    (d) $A = (-2, -1)$, $B = (1, 0)$, $C = (1, 3)$, $D = (5, 1)$, and $E = (4, -2)$

2. Find an equation of the parabola that contains the following points:
    (a) $P_1 = (2, 1)$, $P_2 = (-1, 1)$, $P_3 = (1, 2)$, and $P_4 = (0, 3)$
    (b) $P_1 = (-1, 0)$, $P_2 = (0, 1)$, $P_3 = (2, -1)$, and $P_4 = (4, 0)$
    (c) $A = (5, -2)$, $B = (0, 1)$, $C = (2, 0)$, and $D = (-1, 1)$
    (d) $P_1 = (0, 0)$, $P_2 = (3, 1)$, $P_3 = (1, 3)$, and $P_4 = (6, 3)$

3. Find an equation of the conic that is symmetric with respect to the

origin and contains the following points:

(a) $P_1 = (3, -1)$, $P_2 = (2, 4)$, and $P_3 = (0, 2)$
(b) $P_1 = (1, 2)$, $P_2 = (-1, -3)$, and $P_3 = (3, 0)$
(c) $A = (1, 1)$, $B = (2, -3)$, and $C = (0, 2)$
(d) $P = (2, -1)$, $Q = (0, 1)$, and $R = (-1, 3)$

4. Find an equation of the conic that has its axes parallel to the x- and y-axes and contains the following points:

(a) $P_1 = (2, 1)$, $P_2 = (1, -2)$, $P_3 = (-3, 0)$, and $P_4 = (0, 1)$
(b) $P_1 = (-1, 0)$, $P_2 = (-3, 2)$, $P_3 = (2, 4)$, and $P_4 = (3, -2)$
(c) $L = (2, 1)$, $M = (-1, 1)$, $N = (0, -1)$, and $D = (1, 5)$
(d) $P = (-1, 1)$, $Q = (3, 3)$, $R = (3, -1)$, and $O = (0, 0)$

5. Find an equation of the parabola that has its axis parallel to the x-axis and contains the following points:

(a) $P_1 = (2, 1)$, $P_2 = (-1, -1)$, and $P_3 = (1, 0)$
(b) $P_1 = (1, -1)$, $P_2 = (-2, 0)$, and $P_3 = (-1, 3)$
(c) $A = (-1, 0)$, $B = (-2, 3)$, and $C = (1, 1)$
(d) $L = (-2, 2)$, $M = (-4, 1)$, and $N = (1, -1)$

6. Find an equation of the parabola that has its axis parallel to the y-axis and contains the following points:

(a) $P_1 = (0, 0)$, $P_2 = (3, -4)$, and $P_3 = (-1, 2)$
(b) $P_1 = (3, -2)$, $P_2 = (-1, 0)$, and $P_3 = (0, 2)$
(c) $P = (-4, 1)$, $Q = (1, -1)$, and $R = (-2, 2)$
(d) $A = (1, 2)$, $B = (4, 3)$, and $C = (2, 0)$

6–12. Use of set symbolism for conics (Supplementary)

The concepts and notation presented in Chapter 1, Section 1–13, can conveniently be used when studying the conics. The purpose of this section is to illustrate how these concepts and the symbolism are used. The material presented is supplementary.

The statement

$$f = \{(x,y) \mid y = f(x) \text{ and } x^2 + y^2 + 4x - 6y - 12 = 0\} \quad (6\text{–}34)$$

reads in words: "f is the set of all real ordered pairs (x,y) such that f is a function and x and y satisfy the equation $x^2 + y^2 + 4x - 6y - 12 = 0$."

If the statement had been given as

$$f = \{(x,y) \mid x^2 + y^2 + 4x - 6y - 12 = 0\}$$

then the set of real ordered pairs (x,y) would be a relation that is not a function. This will become more clear in the discussion that follows.

Since it is given that $y = f(x)$ in (6–34), the defining equation can be written as $x^2 + [f(x)]^2 + 4x - 6[f(x)] - 12 = 0$. We can express $f(x)$ explicitly in terms of x by treating $x^2 + y^2 + 4x - 6y - 12 = 0$ as a quadratic in y and solving for y. This gives

$$y = f(x) = -3 + \sqrt{21-4x-x^2} \quad \text{or} \quad y = f(x) = -3 - \sqrt{21-4x-x^2}$$

The original f may be represented by either f_1 or f_2, where

$$f_1 = \{(x,y) \mid y = f_1(x) = -3 + \sqrt{21-4x-x^2}\}$$

and

$$f_2 = \{(x,y) \mid y = f_2(x) = -3 - \sqrt{21-4x-x^2}\}$$

At this point it is suggested that the sections in Chapter 1 dealing with relations, functions, domain, range and real number intervals be re-read.

It should be noted that if f is defined as $\{(x,y) \mid x^2 + y^2 + 4x - 6y - 12 = 0\}$, the relation may be expressed as $f = f_1 \cup f_2$.

EXAMPLE 6–18. Discuss completely the conic defined by

$$g = \{(x,y) \mid y = g(x) \quad \text{and} \quad 9x^2 + 4y^2 - 18x + 16y - 11 = 0\}$$

Solution. By completing the squares, the equation $9x^2 + 4y^2 - 18x + 16y - 11 = 0$ becomes

$$\frac{(x-1)^2}{4} + \frac{(y+2)^2}{9} = 1$$

The curve is an ellipse with its center at $(1, -2)$ and with a vertical major axis. Important constants are: $a = 3$, $b = 2$, $c = \sqrt{a^2 - b^2} = \sqrt{5}$, $e = \frac{c}{a} = \frac{\sqrt{5}}{3}$, $\frac{a}{e} = \frac{9\sqrt{5}}{5}$, and the length of the latus rectum is $\frac{2b^2}{a} = \frac{8}{3}$.

Solving for y we obtain $y = g(x) = -2 + \frac{3}{2}\sqrt{3+2x-x^2}$ or $y = g(x) = -2 - \frac{3}{2}\sqrt{3+2x-x^2}$. Hence, the original function g is either

$$g_1 = \left\{(x,y) \mid y = g(x) = -2 + \frac{3}{2}\sqrt{3+2x-x^2}\right\}$$

or

$$g_2 = \left\{(x,y) \mid y = g(x) = -2 - \frac{3}{2}\sqrt{3+2x-x^2}\right\}$$

Since the example specifies that $g = \{(x,y) \mid y = g(x)$ and $9x^2 + 4y^2 - 18x + 16y - 11 = 0\}$, then g is either g_1 or g_2. However, if g had been defined as $g = \{(x,y) \mid 9x^2 + 4y^2 - 18x + 16y - 11 = 0\}$, then g is $g_1 \cup g_2$, the union of g_1 and g_2. It should be observed that both g_1

258   ANALYTIC GEOMETRY IN THE PLANE

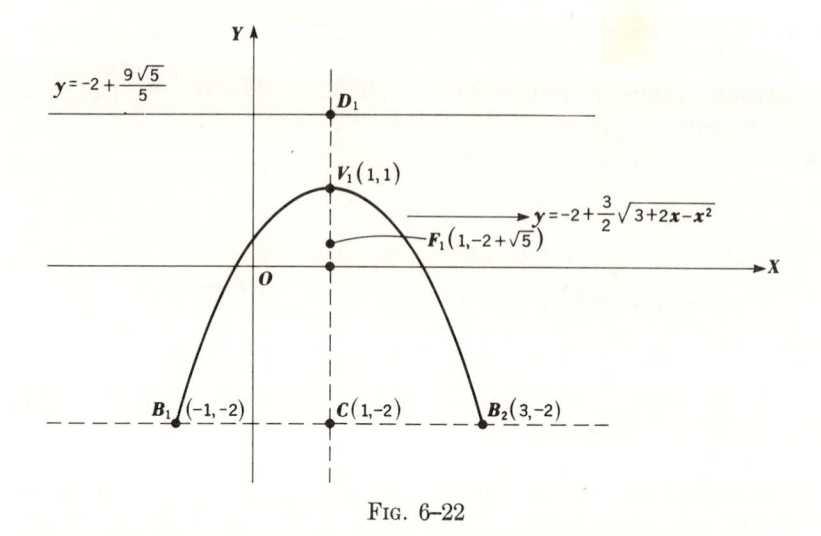

FIG. 6-22

and g_2 are functions, while their union is a relation that is not a function.

The graph of $g = g_1 = \{(x,y) \mid y = g(x) = -2 + \frac{3}{2}\sqrt{3+2x-x^2}\}$, is given in Fig. 6-22.

From Fig. 6-22, it is evident that the domain and range of g are respectively the real number intervals $[-1,3]$ and $[-2,1]$. The equation of the directrix passing through D_1 is

$$y = -2 + \frac{a}{e} = -2 + \frac{9\sqrt{5}}{5}$$

The graph of $g = g_2 = \{(x,y) \mid y = g(x) = -2 - \frac{3}{2}\sqrt{3+2x-x^2}\}$ is given in Fig. 6-23.

From Fig. 6-23, it is evident that the domain and range of g are respectively the real number intervals $[-1,3]$ and $[-5,-2]$.

The equation of the directrix through D_2 is

$$y = -2 - \frac{a}{e} = -2 - \frac{9\sqrt{5}}{5}$$

It is noted that the equations of the directrices might have been stated in set notation as

$$\{(x,y) \mid y = f(x) = -2 + \frac{9\sqrt{5}}{5}\}$$

and

$$\{(x,y) \mid y = f(x) = -2 - \frac{9\sqrt{5}}{5}\}$$

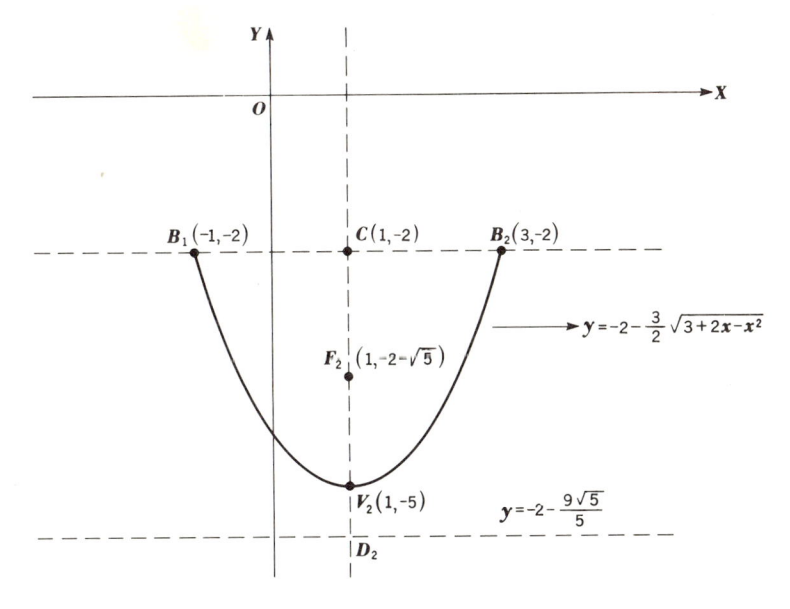

Fig. 6–23

If g had been defined as $g = \{ (x,y) \mid 9x^2 + 4y^2 - 18x + 16y - 11 = 0 \}$, then the resulting graph would be the entire ellipse, namely the union of the sets of ordered pairs (x,y) in Figs. 6–22 and 6–23. The domain and range of this relation g would be the real number intervals $[-1,3]$ and $[-5,1]$, respectively.

The type of discussion used here can be used effectively in a discussion of any of the various relations or functions resulting from the general second degree equation in x and y.

EXERCISES

In each of the following exercises identify the conic and find:
(a) the center (or vertex)
(b) the eccentricity
(c) the equation of the directrices
(d) the equation of the asymptotes, if they exist
(e) equations for y in terms of x
(f) a defining condition in terms of a statement which involves a focus, eccentricity and directrix
(g) a defining condition involving the focal radii if the conic is an ellipse or an hyperbola.

Also, sketch the graph of each.

1. $f = \{ (x,y) \mid x^2 - 6x - 4y + 9 = 0 \}$

2. $f = \{(x,y) \mid y = f(x) \text{ and } y^2 + 3x + 8y + 22 = 0\}$

3. $f = \{(x,y) \mid y = f(x) \text{ and } \dfrac{(x+3)^2}{9} + \dfrac{(y-1)^2}{4} = 1\}$

4. $f = \{(x,y) \mid \dfrac{(x+2)^2}{4} + \dfrac{(y-1)^2}{9} = 1\}$

5. $f = \{(x,y) \mid 4x^2 - 9y^2 - 24x + 18y - 9 = 0\}$

6. $f = \{(x,y) \mid 4x^2 - 9y^2 - 24x + 18y + 63 = 0\}$

7. $f = \{(x,y) \mid y = f(x) \text{ and } \dfrac{x^2}{25} - \dfrac{y^2}{9} = 1\}$

8. $f = \{(x,y) \mid \dfrac{x^2}{25} + \dfrac{y^2}{9} = 1\}$

9. $f = \{(x,y) \mid y = f(x) \text{ and } 4x^2 + 9y = 0\}$

10. $f = \{(x,y) \mid y = f(x) \text{ and } 4y^2 + 9x = 0\}$

7

Polar Coordinates

7-1. Introduction

Before beginning the study of this chapter, the student should review the following topics from trigonometry:

1. Relations between radian and degree measures of angles.
2. Definitions of the trigonometric functions.
3. Trigonometric functions of $0°$, $30°$, $45°$, $60°$, $90°$, $180°$, $270°$, and $360°$.
4. Trigonometric functions of $90° \pm \theta$, $180° \pm \theta$, $270° \pm \theta$, and $-\theta$ in terms of the trigonometric functions of θ.
5. The behavior of the trigonometric functions of θ.

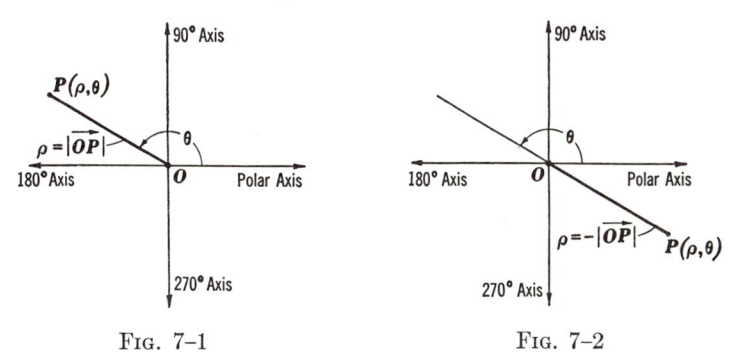

Fig. 7-1

Fig. 7-2

7-2. Polar coordinates of a point

Let us fix in the plane a point O called the *pole* and from it fix a half-line (or ray) extending horizontally to the right, as indicated in Fig. 7-1. This fixed ray is called the *polar axis*. It is convenient to consider three other axes: the $90°$-axis, which is a half-line extending upward from the pole and perpendicular to the polar axis; the $180°$-axis, which is a half-line extending horizontally to the left from the pole; and the $270°$-axis, which is a half-line extending downward from the pole and perpendicular to the polar axis.

262 ANALYTIC GEOMETRY IN THE PLANE

Let P be a point distinct from O. Draw the line through O and P. Then the vector $\overrightarrow{OP}$ lies on the terminal ray of certain angles (all differing by multiples of 360°); the opposite ray extended is the terminal ray for certain other angles (differing from the first angles by odd multiples of 180°). Now associate with P any one of the angles just described. We shall call this angle a *vectorial angle* and shall denote it by θ. If θ is generated in a *counter-clockwise* sense, it is said to be *positive*; if it is generated in a *clockwise* sense, it is said to be *negative*.

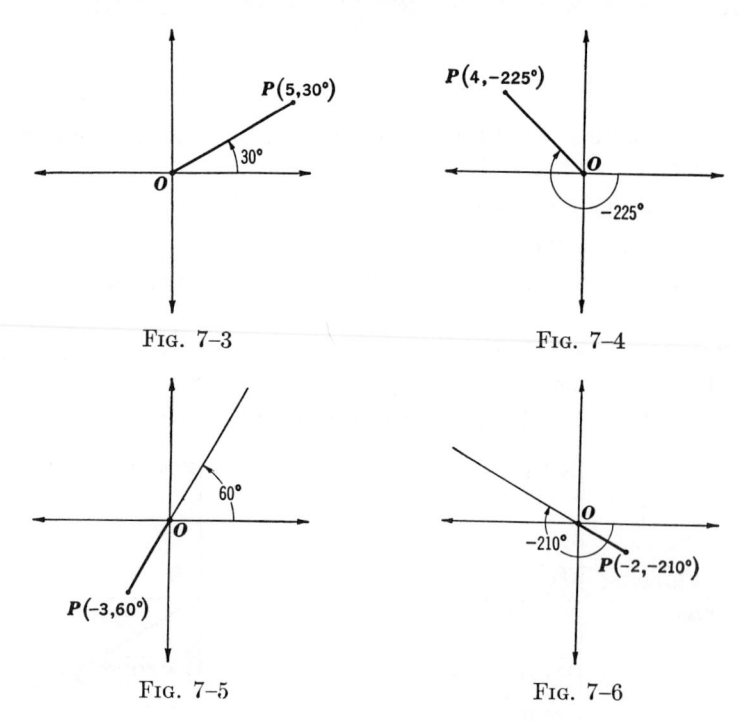

FIG. 7–3

FIG. 7–4

FIG. 7–5

FIG. 7–6

If the terminal ray of angle θ contains $\overrightarrow{OP}$, we associate with P a second number, which is $\rho = |OP|$, as in Fig. 7–1. If $\overrightarrow{OP}$ has a direction opposite to that of the terminal ray of angle θ, we associate with P a second number $\rho = -|OP|$, as in Fig. 7–2. The number ρ is called a *radius vector*.* Then (ρ, θ) are the *polar coordinates* of P. If P is the pole, then $(0, \theta)$ are its polar coordinates for all values of θ.

In Figs. 7–3, 7–4, 7–5, and 7–6 are illustrated various positions of points in the plane and their corresponding coordinates. In cartesian

---
* Other systems of polar coordinates are possible in which the radius vector is always taken greater than or equal to 0. It is important for the student to note that, although ρ is called a *radius vector*, it is *not* a vector.

POLAR COORDINATES   263

coordinates each pair of numbers determines a unique point, and with each point there is associated a unique pair of numbers. In polar coordinates each pair of numbers determines a unique point, but the converse is not true. With each point there are associated an infinite number of pairs of numbers. For example, as shown in Fig. 7–7, the point $P = (6, 30°)$ has also the following polar coordinates: $(-6, 210°)$, $(-6, -150°)$, or $(6, -330°)$. In fact, all pairs of polar coordinates of P are included in the following: $(6, 30° + k\,360°)$, $(-6, 210° + k\,360°)$, where k is any positive or negative integer or zero.

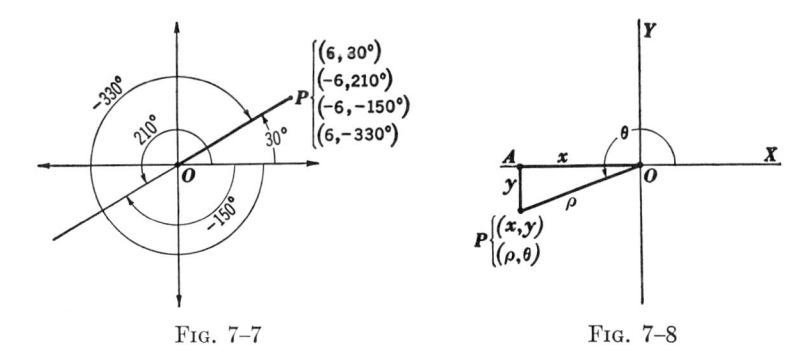

FIG. 7–7   FIG. 7–8

EXERCISES

1. Explain carefully how to name a point in the plane when polar coordinates are used.

2. Give the coordinates of each of the following points in three (in general) other ways such that the magnitude of the angle is less than or equal to 360°:

   (a) $(5, 60°)$
   (b) $(-3, 150°)$
   (c) $(6, 180°)$
   (d) $(-1, -30°)$
   (e) $\left(2, \dfrac{\pi}{4}\right)$
   (f) $\left(-1, \dfrac{5\pi}{6}\right)$
   (g) $\left(3, \dfrac{3\pi}{2}\right)$
   (h) $(-2, 240°)$
   (i) $(-6, -120°)$
   (j) $(3, -300°)$
   (k) $(1, 90°)$
   (l) $(2, -\pi)$
   (m) $\left(4, -\dfrac{2\pi}{3}\right)$
   (n) $(2, 0)$
   (o) $(0, -315°)$

7–3. The relation between polar coordinates and cartesian coordinates

Let us introduce the x-axis and the y-axis so that the origin coincides with the pole, the positive x-axis coincides with the polar axis, and the positive y-axis coincides with the 90°-axis, as shown in Fig. 7–8. Then it is possible to express x and y in terms of ρ and θ.

264   ANALYTIC GEOMETRY IN THE PLANE

Since the vectorial angle θ is in standard position, we have

$$x = \rho \cos \theta \quad \text{and} \quad y = \rho \sin \theta \tag{7-1}$$

By squaring and adding, and using the definition of $\tan \theta$ for $x \neq 0$, we have

$$\rho = \pm \sqrt{x^2 + y^2} \tag{7-2}$$

$$\tan \theta = \frac{y}{x} \tag{7-3}$$

If $x = 0$ and $y \neq 0$, then $\tan \theta = \frac{y}{x}$ is not defined; but we know that $\theta = (2k+1)90°$ where k is a positive or negative integer or zero. If $x = 0$ and $y = 0$, then $\tan \theta$ is undefined; but now we know that the point is the pole, the polar coordinates of which are 0 and θ for *any* angle θ. Because of these peculiarities which arise, it is better in general to use equations (7-1). If $\rho \neq 0$, it is always possible to choose θ such that ρ is positive by selecting θ to be an angle having the point P on its terminal side.

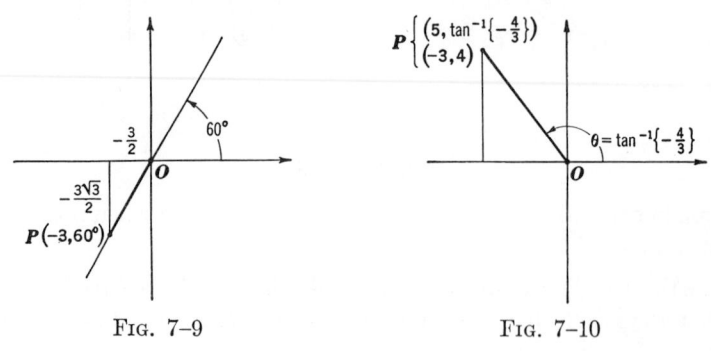

FIG. 7-9   FIG. 7-10

EXAMPLE 7-1. What are the cartesian coordinates of the point $P = (-3, 60°)$?

Solution. We know that $\rho = -3$ and $\theta = 60°$; so, as shown in Fig. 7-9, $x = -3 \cos 60° = -\frac{3}{2}$ and $y = -3 \sin 60° = -\frac{3\sqrt{3}}{2}$. Hence, the cartesian coordinates of the point are $\left(-\frac{3}{2}, -\frac{3\sqrt{3}}{2}\right)$.

EXAMPLE 7-2. What are polar coordinates of the point P, Fig. 7-10, whose cartesian coordinates are $(-3, 4)$?

Solution. We have $\rho = \sqrt{9 + 16} = 5$ and $\tan \theta = -\frac{4}{3}$. Since we have chosen ρ positive, we must choose θ so that $90° < \theta < 180°$. Hence, particular polar coordinates of the point are $\left(5, \tan^{-1}\left\{-\frac{4}{3}\right\}\right)$. In general, the point P has the polar coordinates $\left(5, \tan^{-1}\left\{-\frac{4}{3}\right\} + 2k\pi\right)$, for $k = 0, \pm 1, \pm 2$, and so on.

As when cartesian coordinates are used, equations of the form $f(\rho, \theta) = 0$ represent certain loci, usually curves. Specifically, the locus is the set of those points P having some pair of polar coordinates satisfying $f(\rho, \theta) = 0$.

Consider the locus $F(x, y) = 0$. Let P be on this locus and let it have the cartesian coordinates x and y and the polar coordinates ρ and θ. Hence, equations (7–1) hold and we have $F(\rho \cos \theta, \rho \sin \theta) = 0$, so that P lies on the locus of this last equation. Conversely, any point P on the locus of this equation lies on the original locus. The last equation therefore represents the same set of points $F(x, y) = 0$.

In a later section we shall discuss some specific curves. Now the problem will be to transform equations from polar coordinates to cartesian coordinates and from cartesian coordinates to polar coordinates.

EXAMPLE 7–3. What does the equation $x^2 - y^2 = a^2$ become when transformed to polar coordinates?

Solution. Replacing x by $\rho \cos \theta$ and y by $\rho \sin \theta$, the equation becomes

$$\rho^2 \cos^2 \theta - \rho^2 \sin^2 \theta = a^2$$

or

$$\rho^2 (\cos^2 \theta - \sin^2 \theta) = a^2$$

Since $\cos 2\theta \equiv \cos^2 \theta - \sin^2 \theta$, we have

$$\rho^2 \cos 2\theta = a^2$$

This is a polar equation for an equilateral hyperbola.

EXAMPLE 7–4. Transform the equation $\rho = \dfrac{4}{2 - \sin \theta}$ to cartesian coordinates.

Solution. Clearing of fractions, we have

$$2\rho - \rho \sin \theta = 4$$

Since $y = \rho \sin \theta$, it follows that $2\rho - y = 4$. To eliminate ρ, we write the equation as $2\rho = y + 4$. Squaring both sides of the equation, we obtain $4\rho^2 = (y+4)^2$. Since $\rho^2 = x^2 + y^2$,

$$4x^2 + 4y^2 = y^2 + 8y + 16$$

or

$$4x^2 + 3y^2 - 8y - 16 = 0$$

This is an equation of an ellipse. Hence, any point in the set of points defined by $\rho = \dfrac{4}{2 - \sin \theta}$ is on the ellipse. Let us see if the converse is true;

266   ANALYTIC GEOMETRY IN THE PLANE

that is, if all points on the ellipse are also on the given locus. By equations (7–1), the equation $4x^2+3y^2-8y-16=0$ becomes

$$4\rho^2 \cos^2 \theta + 3\rho^2 \sin^2 \theta - 8\rho \sin \theta - 16 = 0$$

or

$$4\rho^2(1-\sin^2 \theta) + 3\rho^2 \sin^2 \theta - 8\rho \sin \theta - 16 = 0$$

Hence,

$$4\rho^2 - \rho^2 \sin^2 \theta - 8\rho \sin \theta - 16 = 0$$

This can be written

$$4\rho^2 - (\rho \sin \theta + 4)^2 = 0$$

Factoring, we obtain

$$(2\rho + \rho \sin \theta + 4)(2\rho - \rho \sin \theta - 4) = 0$$

or

$$\rho = \frac{4}{2-\sin \theta} \quad \text{and} \quad \rho = -\frac{4}{2+\sin \theta}$$

If $P = (\rho, \theta)$ is a point on one locus, then $Q = (-\rho, \theta + 180°)$ is a point on the other locus. To see this, replace ρ by $-\rho$ and θ by $\theta + 180°$ in the equation $\rho = \frac{4}{2-\sin \theta}$. We thus obtain

$$-\rho = \frac{4}{2-\sin(180°+\theta)} \quad \text{or} \quad \rho = -\frac{4}{2+\sin \theta}$$

Since $P = Q$, the two equations represent the same set of points. Therefore, $\rho = \frac{4}{2-\sin \theta}$ is a polar equation of an ellipse and the operation of squaring did not introduce any extraneous points.

EXAMPLE 7–5. Transform the equation $\rho = \sin 2\theta$ to cartesian coordinates.

Solution. Since $\sin 2\theta \equiv 2 \sin \theta \cos \theta$, we have

$$\rho = 2 \sin \theta \cos \theta$$

Since $\rho \sin \theta = y$ and $\rho \cos \theta = x$, we multiply both sides of the equation by ρ^2 (this does not affect the locus since the pole satisfies the equation $\rho = \sin 2\theta$) and we obtain

$$\rho^3 = 2(\rho \sin \theta)(\rho \cos \theta)$$

or

$$\rho^3 = 2yx$$

Squaring both sides, we obtain

$$\rho^6 = 4y^2x^2$$

or

$$(x^2+y^2)^3 = 4y^2x^2$$

To verify the result, we transform back to polar coordinates, obtaining $\rho^6 = 4\rho^4 \sin^2 \theta \cos^2 \theta$ or $\rho = \pm \sin 2\theta$. But, if $P = (\rho, \theta)$ satisfies the equation

$\rho = -\sin 2\theta$, then $Q = (-\rho, \theta + 180°)$ satisfies the equation $\rho = \sin 2\theta$. Since $P = Q$, the two equations represent the same set of points.

The student should note that, in transforming from polar coordinates to cartesian coordinates, it is advisable first to eliminate the terms involving the trigonometric functions and then to eliminate the terms that involve only ρ. It is also important to transform the result back to polar coordinates and to check for extraneous loci.

EXERCISES

1. Transform each of the following equations to polar coordinates:

    (a) $x^2 + y^2 = 16$
    (b) $y = 6$
    (c) $y^2 - x^2 = 4$
    (d) $x^2 + y^2 - 2x + 3y - 6 = 0$
    (e) $2x^2 + y^2 = 16$
    (f) $y^2 = 4ax$
    (g) $x^2 - 32y - 256 = 0$
    (h) $3y^2 + 2x = 0$

2. Transform each of the following equations to cartesian coordinates:

    (a) $\rho \sin \theta = 4$
    (b) $\rho = 3 \cos \theta$
    (c) $\rho = 4$
    (d) $\theta = \tan^{-1} \frac{1}{2}$
    (e) $\rho = \cos \theta$
    (f) $\rho = 3 \cos 2\theta$
    (g) $\rho = \dfrac{3}{1 - 2 \cos \theta}$
    (h) $\rho = 2(1 - \sin \theta)$

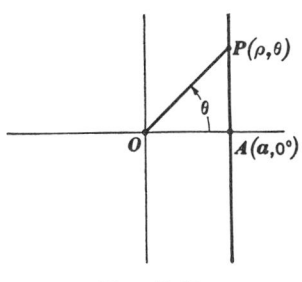

FIG. 7-11

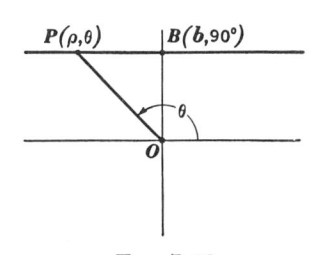

FIG. 7-12

7–4. The line

Just as we derived equations in cartesian coordinates, so also we can derive equations in polar coordinates. We shall start with the line.

(a) *Vertical lines.* Consider a line parallel to the 90°-axis and intersecting the polar axis at the point $(a, 0°)$, as shown in Fig. 7–11. Let $P = (\rho, \theta)$ be an arbitrary point on the line. Since θ is in standard position, we have

$$\rho \cos \theta = a \qquad (7\text{--}4)$$

268   ANALYTIC GEOMETRY IN THE PLANE

If (ρ_0, θ_0) is a point whose coordinates satisfy the equation $\rho \cos \theta = a$, it follows that $\rho_0 \cos \theta_0 = a$, or $x = a$. Hence, the point lies on the vertical line containing $(a, 0°)$.

(b) *Horizontal lines.* Consider a line parallel to the polar axis and intersecting the 90°-axis at the point $(b, 90°)$, as shown in Fig. 7–12. Again θ is in standard position and, hence,

$$\rho \sin \theta = b \qquad (7\text{–}5)$$

If (ρ_0, θ_0) is a point whose coordinates satisfy the equation $\rho \sin \theta = b$, we have $\rho_0 \sin \theta_0 = b$, or $y = b$. Hence, the point lies on the horizontal line containing $(b, 90°)$.

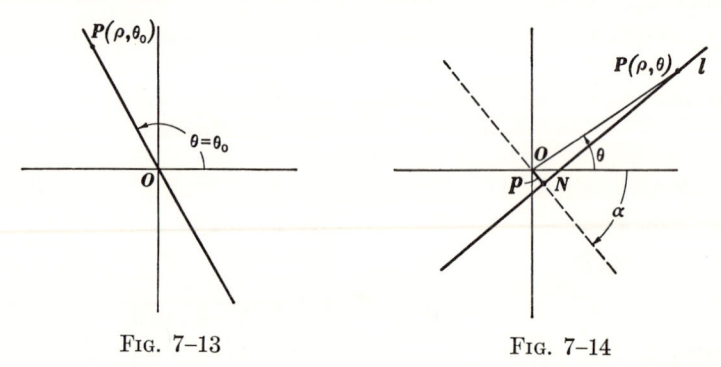

FIG. 7–13   FIG. 7–14

(c) *Lines through the pole.* Consider a line passing through the pole and making with the polar axis an angle θ_0, as shown in Fig. 7–13. For any point $P = (\rho, \theta)$ on the line, we have

$$\theta = \theta_0 \qquad (7\text{–}6)$$

If (ρ_0, θ_0) is a point whose coordinates satisfy the equation $\theta = \theta_0$, then, since θ is a constant, the point lies on the line passing through the pole and having the equation $\theta = \theta_0$.

(d) *General lines.* Consider any line in the plane, as the line l in Fig. 7–14. Let $P = (\rho, \theta)$ be an arbitrary point on the line. Let $ON = p$ be the normal intercept of the line, where $p \neq 0$, and let α be an angle which the normal vector makes with the positive x-axis. If $P = (\rho, \theta)$ is an arbitrary point on the line, we have

$$\rho \cos (\theta - \alpha) = p \qquad (7\text{–}7)$$

The general equation of the line in polar coordinates is not too important, but the student should learn thoroughly the forms for the horizontal line, the vertical line, and the line through the pole. However, it is interesting to notice that the polar equation $\rho \cos (\theta - \alpha) = p$ is another form of the equation $x \cos \alpha + y \sin \alpha = p$, which is the normal equation of a line in cartesian coordinates. Expanding equation (7–7) with the aid of trigonometry, we obtain

$$\rho \cos \theta \cos \alpha + \rho \sin \theta \sin \alpha = p$$

or
$$x \cos \alpha + y \sin \alpha = p$$

EXERCISES

Identify the locus of each of the following equations, draw its graph, and transform it to cartesian coordinates:

1. $\rho \sin \theta = -4$
2. $\rho = 2 \sec \theta$
3. $\rho \cos (\theta - 30°) = 4$
4. $\rho \cos (\theta + 45°) = -2$
5. $\rho = -\csc \theta$
6. $\rho \cos \theta = -3$
7. $\rho = -3 \csc \theta$
8. $\rho \sin (45° + \theta) = 3$
9. $\rho \sin (30° + \theta) = -2$
10. $\rho = -4 \sec \theta$
11. $3\rho \sin \theta = 5$
12. $\rho \cos (\theta + 60°) = -5$
13. $\rho = -3 \csc (60 - \theta)$
14. $\rho = \sec (45° + \theta)$
15. $\theta = 120°$
16. $\tan \theta = 2$
17. $3 \sin \theta = 0$
18. $2 \cos \theta = 0$
19. $5\rho \sin \theta + 3\rho \cos \theta = 15$
20. $4\rho \cos \theta - \rho \sin \theta = 4$
21. $\tan \theta = -1$
22. $\theta = 30°$
23. $\rho \cos \theta - \rho \sin \theta = 2$
24. $\rho \sin \theta + 2 \rho \cos \theta = -2$
25. $\rho \cos \theta = \frac{1}{2}$
26. $\rho \sin \theta = \dfrac{-1}{\sqrt{2}}$
27. $2\rho \cos \theta - 3\rho \sin \theta = 6$
28. $\rho \sin \theta = 4 + 6 \rho \cos \theta$

7–5. The circle

(a) *Circles with center at the pole.* A circle with its center at the pole is represented by what is probably the simplest of all polar equations. If the radius of the circle is denoted by a, then an arbitrary point on the circumference is at (a, θ), as shown in Fig. 7–15. Since the radius vector is constant and always equal to a, an equation of the circle is

$$\rho = a \tag{7-8}$$

In discussing the polar coordinates of a point, we noticed that it is possible to name the same point in a variety of different ways. This situation leads to variations in the equations of some curves. For example, since any point $(-a, \theta)$ lies on the circumference of a circle whose radius is a and whose center is the pole, another equation of the circle just discussed is

$$\rho = -a \tag{7-9}$$

(b) *Circles containing the pole with center on the polar axis.* Let the

270   ANALYTIC GEOMETRY IN THE PLANE

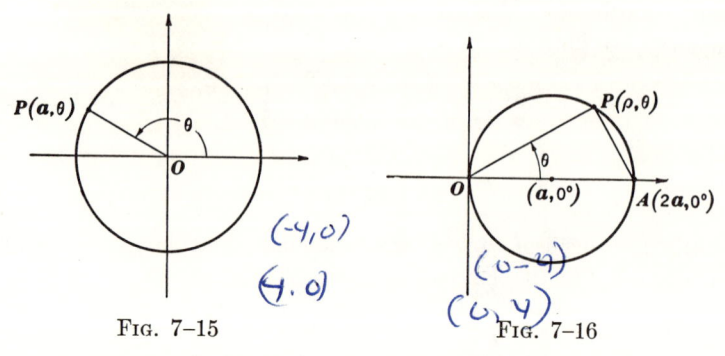

FIG. 7-15

FIG. 7-16

center of a circle be the point $(a, 0°)$ and let the radius be a, as indicated in Fig. 7-16. Angle OPA is a right angle, since it is inscribed in a semicircle. Hence,

$$\frac{\rho}{2a} = \cos \theta$$

or
$$\rho = 2a \cos \theta \tag{7-10}$$

(c) *Circles containing the pole with center on the 90°-axis.* Let the center of a circle be the point $(b, 90°)$ and let the radius be b, as represented in Fig. 7-17. Since angle OPB is a right angle,

$$\frac{\rho}{2b} = \sin \theta$$

or
$$\rho = 2b \sin \theta \tag{7-11}$$

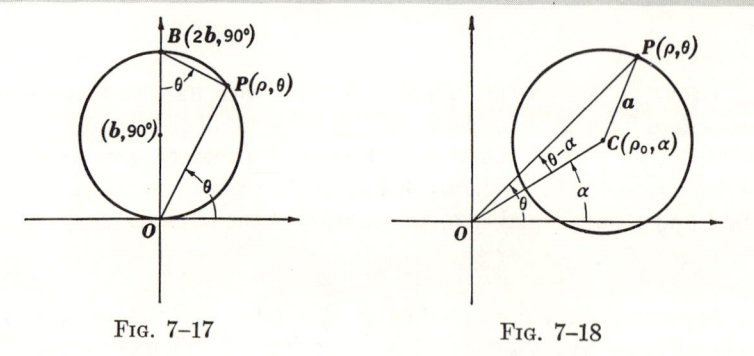

FIG. 7-17

FIG. 7-18

(d) *General circles.* Consider the center of a circle to be $C = (\rho_0, \alpha)$ and the radius to be a, as shown in Fig. 7-18. Since α is the angle between OC and the polar axis, we obtain from the law of cosines the equation

$$a^2 = \rho^2 + \rho_0^2 - 2\rho\,\rho_0 \cos(\theta - \alpha) \tag{7-12}$$

Of the various polar equations of circles, the important ones are:

POLAR COORDINATES   271

$\rho = a$ for a circle of radius a with its center at the pole
$\rho = 2a \cos \theta$ for a circle of radius a with its center at $(a, 0°)$
$\rho = 2b \sin \theta$ for a circle of radius b with its center at $(b, 90°)$

EXERCISES

1. Derive a polar equation of each of the following circles:
   (a) Center is at $(-a, 0°)$ and radius is a
   (b) Center is at $(-b, 90°)$ and radius is b
   (c) Center is at $(0, 0°)$ and radius is r
   (d) Center is at (a, α) and radius is r

2. Derive a polar equation of each of the following circles:
   (a) Center is at $(-6, 0°)$ and radius is 6
   (b) Center is at the pole and radius is 3
   (c) Center is at $(4, 90°)$ and radius is 4
   (d) Center is at $(-2, 120°)$ and radius is 3

3. Show that $\rho = a \cos \theta + b \sin \theta$ is an equation of a circle passing through the pole and the points $(a, 0°)$ and $(b, 90°)$.

4. Transform each of the following polar equations to cartesian coordinates and sketch the curve:
   (a) $\rho = -3 \sin \theta$
   (b) $\rho = 2 \cos \theta$
   (c) $\rho = 2 \cos \theta - 3 \sin \theta$
   (d) $\rho = -4$
   (e) $\rho \csc \theta = 2$
   (f) $\rho \sec \theta = -3$

5. Transform each of the following cartesian equations to polar coordinates and sketch the curve:
   (a) $x^2 + y^2 = 6x$
   (b) $x^2 + y^2 = 4$
   (c) $x^2 + y^2 - 3x + y = 0$
   (d) $x^2 + y^2 - 2y = 0$
   (e) $x^2 + y^2 - 2x + 4y = 0$
   (f) $x^2 + y^2 - 4x = 0$
   (g) $x^2 + y^2 - 8x - 2y + 6 = 0$
   (h) $3x^2 + 3y^2 + 5y = 0$

6. Find the center and radius of each of the following circles and draw its graph:
   (a) $\rho = 3 \cos \theta - 2 \sin \theta$
   (b) $\rho + 2 \cos \theta = 3 \sin \theta$
   (c) $\rho^2 - 4\rho \sin \theta + 2\rho \cos \theta - 6 = 0$
   (d) $\rho^2 + 10 = 6\rho \sin \theta + 4\rho \cos \theta$
   (e) $\rho^2 + 2\rho \sin \theta = 6\rho \cos \theta + 4$
   (f) $\rho = 4 \sin \theta - 2 \cos \theta$
   (g) $\rho^2 = 2 - 4\rho \cos \theta + 6\rho \sin \theta$

7. A circle whose radius is 6 has its center at the point $(10, 0°)$. Show that its equation is $\rho^2 - 20\rho \cos \theta + 64 = 0$.

8. A circle has its center at the point (6, 90°). If its radius is 4, show that its equation is $\rho^2 - 12\rho \sin\theta + 20 = 0$.

9. Show that an equation of a circle whose radius is r and having a center at the point $(h, 0°)$ is $\rho^2 - 2h\rho \cos\theta + h^2 - r^2 = 0$.

10. What is the polar equation of a circle that has its center at $\left(k, \dfrac{\pi}{2}\right)$ and whose radius is r given $k > r$?

7–6. Intercept points

It is possible to obtain a fairly accurate graph of any curve represented by the equation $\rho = f(\theta)$ by locating the points where the curve intersects the axes, by ascertaining how the curve is related to the axes with respect to symmetry, by determining the values of the angles for which $f(\theta)$ is not defined, by observing where $f(\theta)$ increases or decreases, and by plotting a few points.

The points where a curve cuts the polar axis, the 90°-axis, the 180°-axis, and the 270°-axis are called the *intercept points* of the curve. To find them, we let $\theta = 0° + k\ 360°$, $90° + k\ 360°$, $180° + k\ 360°$, and $270° + k\ 360°$, where k is an integer.

EXAMPLE 7–6. What are the intercept points of the curve whose equation is $\rho = 3 \sin\theta$?

Solution. From equation (7–11) we know this is an equation of a circle with radius $\tfrac{3}{2}$ and having its center at $(\tfrac{3}{2}, 90°)$. To find the intercept points, construct the following table:

θ	0°	90°	180°	270°
$\rho = 3 \sin\theta$	0	3	0	-3

The points indicated in the table are the only intercept points of the curve, since $\sin\theta$ has a period of 360°. In this example there are only *two* intercept points, since (0, 0°) and (0, 180°) are the same point, and (3, 90°) and (-3, 270°) also are the same point.

7–7. Symmetry

If the image of every point of a curve in a given line is also on the curve, we say that the curve is symmetric with respect to the line (see Section 4–4).

(a) *Polar axis and 180°-axis.* The locus of an equation $f(\rho, \theta) = 0$ is symmetric with respect to the polar axis or the 180°-axis if the following condition exists: For any point (ρ, θ) on the curve, there is another point $(\rho, 2k\pi - \theta)$ or a point $(-\rho, \{2k+1\}\pi - \theta)$ which is also on the curve, where k is any positive or negative integer or zero. In Fig. 7–19 are shown two points P and P' which are located in the specified relative positions.

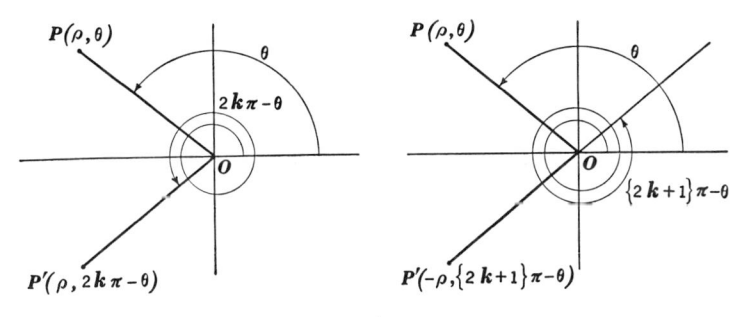

FIG. 7–19

To test a curve for symmetry with respect to the polar axis, one must show that, if a point P is on the curve, its image point P' is also on the curve. Simple sufficiency tests are as follows:

Step I. Replace θ by $2k\pi - \theta$ in the given equation. If the resulting equation becomes equivalent to the original equation for some integer k, the curve is symmetric with respect to the polar axis or the $180°$-axis.

Step II. Replace ρ by $-\rho$ and replace θ by $\{2k+1\}\pi - \theta$. If the resulting equation is equivalent to the original equation for some integer k, the curve is symmetric with respect to the polar axis or the $180°$-axis.

EXAMPLE 7–7. Test each of the following curves for symmetry with respect to the polar axis: $\rho = \sin \frac{\theta}{2}$; $\rho = 4 \sin 2\theta$.

Solution. For $\rho = \sin \frac{\theta}{2}$ we have, using step I, $\rho = \sin \left(\frac{2k\pi - \theta}{2}\right) = \sin \left(k\pi - \frac{\theta}{2}\right)$. When k is an odd integer, $\rho = \sin \frac{\theta}{2}$. Therefore, the curve is symmetric with respect to the polar axis.

For $\rho = 4 \sin 2\theta$ we have, using step I, $\rho = 4 \sin 2(2k\pi - \theta) = 4 \sin 2(-\theta) = -4 \sin 2\theta$. Hence, this equation is not equivalent to the original. Using step II, we find that $-\rho = 4 \sin 2(\{2k+1\}\pi - \theta) = 4 \sin 2(\pi - \theta) = 4 \sin (-2\theta) = -4 \sin 2\theta$, or $-\rho = -4 \sin 2\theta$. Since this equation is equivalent to the original equation, the curve is symmetric with respect to the polar axis.

274   ANALYTIC GEOMETRY IN THE PLANE

In general, $\cos(-k\theta) \equiv \cos k\theta$. It is therefore obvious that, when ρ is a function of $\cos k\theta$, where k is any constant, the curve is symmetric with respect to the polar axis.

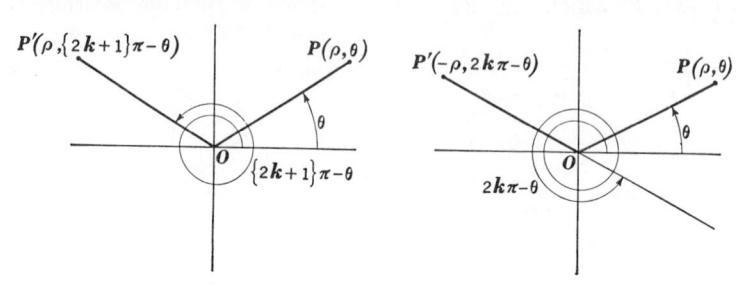

FIG. 7–20

(b) *90°-axis and 270°-axis.* The locus of an equation $f(\rho, \theta) = 0$ is symmetric with respect to the 90°-axis or the 270°-axis if the following condition holds true: For any point (ρ, θ) on the curve, there is another point $(\rho, \{2k+1\}\pi - \theta)$ or a point $(-\rho, 2k\pi - \theta)$ which is also on the curve, where k is any positive or negative integer or zero. In Fig. 7–20 are shown two points P and P' which are located in the specified relative positions.

Simple sufficiency conditions for symmetry with respect to the 90°-axis or the 270°-axis are as follows:

Step I. Replace θ by $(2k+1)\pi - \theta$, where k is an integer.

Step II. Replace ρ by $-\rho$ and replace θ by $2k\pi - \theta$, where k is an integer.

If, for some integer k, the equation obtained in either step I or step II is equivalent to the original equation, the curve is symmetric with respect to the 90°-axis and the 270°-axis.

EXAMPLE 7-8. Test each of the following curves for symmetry with respect to the 90°-axis: $\rho = \cos \frac{\theta}{2}$; $\rho = 4 \sin 2\theta$.

Solution. For $\rho = \cos \frac{\theta}{2}$ we have, using step I, $\rho = \cos \frac{(2k+1)\pi - \theta}{2} = \cos\left(k\pi + \frac{\pi}{2} - \frac{\theta}{2}\right) = \pm \sin \frac{\theta}{2}$. Hence, the equation is not equivalent to the original.

Using step II, we have $-\rho = \cos \frac{2k\pi - \theta}{2} = \cos\left(k\pi - \frac{\theta}{2}\right)$. When $k = 1$, $\cos\left(k\pi - \frac{\theta}{2}\right) = -\cos \frac{\theta}{2}$. Therefore, $-\rho = -\cos \frac{\theta}{2}$. Since this equation is equivalent to the original equation, the curve is symmetric with respect to the 90°-axis.

For $\rho = 4 \sin 2\theta$ we find that, using step I, $\rho = 4 \sin 2(\{2k+1\}\pi - \theta) =$

4 sin $(4k\pi + 2\pi - 2\theta)$. For $k = 1$, $\rho = 4 \sin (2\pi - 2\theta) = -4 \sin 2\theta$; and the equation is not equivalent to the original.

Using step II, we have $-\rho = 4 \sin 2(2k\pi - \theta)$. For $k = 1$ in this case, $-\rho = 4 \sin 2(-\theta) = -4 \sin 2\theta$. Since this equation is equivalent to the original equation, the curve is symmetric with respect to the 90°-axis.

In general, $\sin (\pi - \theta) \equiv \sin \theta$. Hence, it is obvious that, if ρ is a function of $\sin \theta$, the curve is symmetric with respect to the 90°-axis and the 270°-axis.

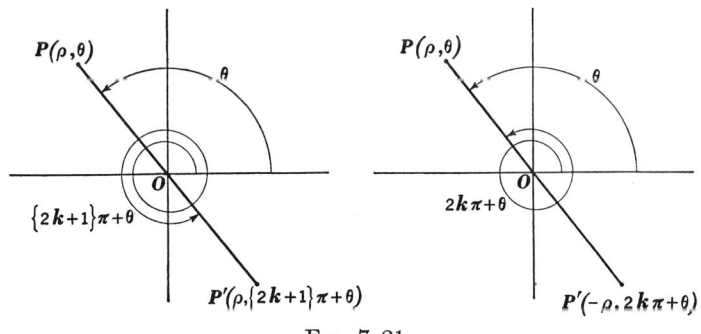

FIG. 7-21

(c) *Pole*. Two points P and P' are images with respect to a point C, provided C is the mid-point of the segment PP'. The locus of an equation $f(\rho, \theta) = 0$ is symmetric with respect to the pole if the following condition is satisfied: For any point (ρ, θ) on the curve there is another point $(\rho, \{2k+1\}\pi+\theta)$ or a point $(-\rho, 2k\pi+\theta)$ which is also on the curve, where k is any positive or negative integer or zero. In Fig. 7-21 are shown two points P and P' which are located in the specified relative positions.

The following sufficiency tests may be used for symmetry with respect to the pole:

Step I. Replace θ by $(2k+1)\pi + \theta$.

Step II. Replace ρ by $-\rho$ and replace θ by $2k\pi + \theta$.

If, for some integer k, the equation obtained in either step I or step II becomes equivalent to the original equation, the curve is symmetric with respect to the pole.

EXAMPLE 7-9. Test each of the following curves for symmetry with respect to the pole: (a) $\rho = 3 \sin \theta$; (b) $\rho = \cos 2\theta$; (c) $\rho^2 = 4 \sin \theta$.

Solution. (a) Neither step I nor step II gives an equivalent equation. This result suggests that the curve may not be symmetric with respect to the pole. If we can find one point on the curve whose image is not on the curve, we shall know that the curve is not symmetric with respect to the pole. Consider the point $\left(3, \dfrac{\pi}{2}\right)$ which is on the curve. The image of this

point with respect to the pole is at $\left(-3, 2k\pi+\frac{\pi}{2}\right)$ or at $\left(3, \{2k+1\}\pi+\frac{\pi}{2}\right)$. But
$$3 \sin\left(2k\pi+\frac{\pi}{2}\right) = 3 \neq -3$$
and
$$3 \sin\left(\{2k+1\}\pi+\frac{\pi}{2}\right) = -3 \neq 3$$

Hence, the image is not on the curve. This fact shows that the curve is not symmetric with respect to the pole.

(b) Using step I, we have $\rho = \cos 2(\{2k+1\}\pi+\theta) = \cos(2\pi+2\theta) = \cos 2\theta$. Since the original equation is unaltered, the curve is symmetric with respect to the pole.

(c) Using step II, we find that $(-\rho)^2 = 4 \sin(2k\pi+\theta) = 4 \sin \theta$, or $\rho^2 = 4 \sin \theta$; and the original equation is unchanged. Hence, the curve is symmetric with respect to the pole.

It is obvious that if ρ appears throughout an equation with an even exponent the curve is symmetric with respect to the pole.

In concluding this section on symmetry we note the following facts:

1. A curve is not necessarily symmetric with respect to the polar axis, the 90°-axis, or the pole.
2. A curve may be symmetric with respect to just one of the three.
3. A curve that is symmetric with respect to any two of the three is automatically symmetric with respect to the third. Hence, after tests have been made for two types of symmetry and have been found to work, there is no need to test for the third type, except to use this test as a possible check on the previous work.

7–8. Extent

Here we are interested in obtaining those values of θ for which ρ is not defined. We must be careful to distinguish whether it is not defined because it becomes imaginary or because it becomes infinite. In the first case we shall merely say that ρ is *undefined*; while in the second case we shall say that ρ is *infinite* and shall write $\rho = \infty$. The student must be sure to understand that $\rho = \infty$ implies that $|\overrightarrow{OP}|$ increases indefinitely as θ approaches the value considered.

We shall also be interested in determining the values of θ for which ρ has a maximum value if they can be obtained by inspection.

EXAMPLE 7–10. Find the extent of each of the following curves:
(a) $\rho = 4 \sin \theta$; (b) $\rho = \dfrac{2}{1-2 \cos \theta}$; (c) $\rho^2 = 4 \sin \theta$.

Solution. (a) Since the maximum value of $\sin \theta$ is 1, the maximum value of ρ is 4. But $\sin \theta = 1$ when $\theta = 90°$. Hence, $\theta = 90°$ gives maximum ρ. There are no values of θ for which ρ is undefined.

(b) As $1 - 2 \cos \theta$ approaches 0, $|\rho|$ increases indefinitely. Hence, when $1 - 2 \cos \theta = 0$, it follows that $\rho = \infty$. But the equation $1 - 2 \cos \theta = 0$ gives $\cos \theta = \tfrac{1}{2}$; from which $\theta = 60°$ or $300°$.

(c) Since $\sin \theta = 1$ gives the greatest value of ρ^2, it follows that $\theta = 90°$ gives the maximum value of ρ. If $\sin \theta$ is negative, ρ is undefined. From trigonometry we know that $\sin \theta$ is negative when $180° < \theta < 360°$. Hence, for these values of θ, ρ is undefined.

7–9. Sketching curves representing $\rho = f(\theta)$

By putting together all the facts discussed previously and by plotting some points, we are able to sketch a graph of a curve. This procedure is illustrated in the examples that follow:

EXAMPLE 7–11. Sketch the curve whose equation is $\rho = 1 - 2 \cos \theta$.

Solution. Intercepts: It suffices to consider $\theta = 0°$, $90°$, $180°$, $270°$, as follows:

θ	0°	90°	180°	270°
$\rho = 1 - 2 \cos \theta$	-1	1	3	1

Symmetry with respect to polar axis: Since ρ is a function of $\cos \theta$, the curve is symmetric with respect to the polar axis.

Symmetry with respect to 90°-axis: The point $(-1, 0)$ is on the curve. The image of this point on the 90°-axis is the point at $(-1, \{2k+1\}\pi)$ or at $(1, 2k\pi)$. Substituting these values in the given equation, we obtain:

$$1 - 2 \cos (2k+1)\pi = 3 \neq -1 \text{ and } 1 - 2 \cos 2k\pi = -1 \neq 1$$

Hence, the image of $(-1, 0°)$ is not on the curve and it is not symmetric with respect to the 90°-axis.

Symmetry with respect to pole: The curve is not symmetric with respect to the pole.

Extent: The greatest value ρ can have is obtained when $\cos \theta = -1$. Hence, ρ assumes its maximum value 3 when $\theta = 180°$, and the corresponding point is at $(3, \pi)$.

Plotting: Since $\cos \theta$ has a period 2π and the curve is symmetric with respect to the polar axis, we shall construct the table only for θ between 0 and π.

278 ANALYTIC GEOMETRY IN THE PLANE

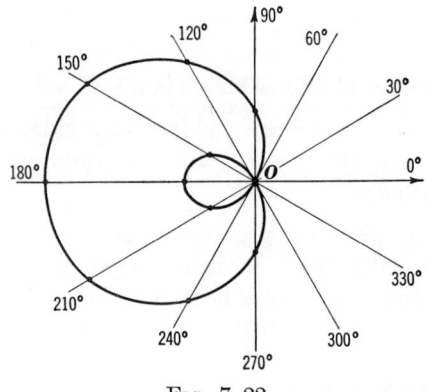

FIG. 7-22

θ	0°	30°	60°	90°	120°	150°	180°
$\rho = 1 - 2\cos\theta$	-1	-0.7	0	1	2	2.7	3

The curve is shown in Fig. 7-22. This curve is called a *limacon*.

EXAMPLE 7-12. Sketch the curve $\rho^2 = 4 \sin 2\theta$.

Solution. Intercepts:

θ	0°	90°	180°	270°
$\rho = \pm 2\sqrt{\sin 2\theta}$	0	0	0	0

Symmetry: Since the exponent of ρ is even, the curve is **symmetric** with respect to the pole.

The point $\left(2, \dfrac{\pi}{4}\right)$ satisfies the equation $\rho^2 = 4 \sin 2\theta$. The image of this point in the polar axis is at $\left(2, 2k\pi - \dfrac{\pi}{4}\right)$ or at $\left(-2, \{2k+1\}\pi - \dfrac{\pi}{4}\right)$. Since these coordinates do not satisfy the given equation, the curve is not symmetric with respect to the polar axis. The student should check this.

The curve cannot be symmetric with respect to the 90°-axis.

Extent: When $\sin 2\theta = 1$, ρ^2 has its greatest value. But $\sin 2\theta = 1$ when $2\theta = 90°$ or $450°$ or when $\theta = 45°$ or $225°$. Hence, the end points of the maximum radius vectors are at $(2, 45°)$ and $(2, 225°)$. If $\sin 2\theta$ is negative, $\rho^2 = 4 \sin 2\theta$ is negative and ρ is undefined. Now we know that $\sin 2\theta$ is negative when $180° < 2\theta < 360°$ or when $540° < 2\theta < 720°$. Hence, for the values of θ for which $90° < \theta < 180°$ and $270° < \theta < 360°$, ρ is undefined.

POLAR COORDINATES 279

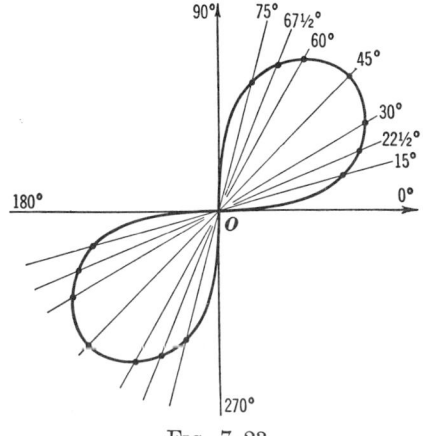

FIG. 7-23

Plotting: Since sin $2\theta \equiv \sin 2(\theta + \pi)$, it follows that sin 2θ assumes all its values as θ varies from 0 to π. Since ρ is undefined for values of θ such that $90° < \theta < 180°$ and since the curve is symmetric with respect to the pole, we shall construct the table for $0° \leq \theta \leq 90°$. We let 2θ take on the values $0°, 30°, 45°, 60°$, etc., since by using these angles ρ can be found without the use of tables. We then plot the points from their coordinates θ and ρ.

2θ	0°	30°	45°	60°	90°	120°	135°	150°	180°
$\rho = \pm 2\sqrt{\sin 2\theta}$	0	± 1.4	± 1.6	± 1.9	± 2	± 1.9	± 1.6	± 1.4	0
θ	0	15°	$22\frac{1}{2}°$	30°	45°	60°	$67\frac{1}{2}°$	75°	90°

The curve is shown in Fig. 7-23. This curve is called a *lemniscate*.

EXAMPLE 7-13. Sketch the curve $\rho = \dfrac{2}{1-\cos\theta}$.

Solution. Intercepts:

θ	0	90°	180°	270°
$\rho = \dfrac{2}{1-\cos\theta}$	undefined	2	1	2

Symmetry: Polar axis, yes; 90°-axis, no; pole, no.

Extent: When $1 - \cos\theta = 0$, ρ is undefined; but, as θ *approaches* 0° and 360°, $|\rho|$ increases indefinitely. Hence, $\rho = \infty$ when $\theta = 0°$ or $360°$.

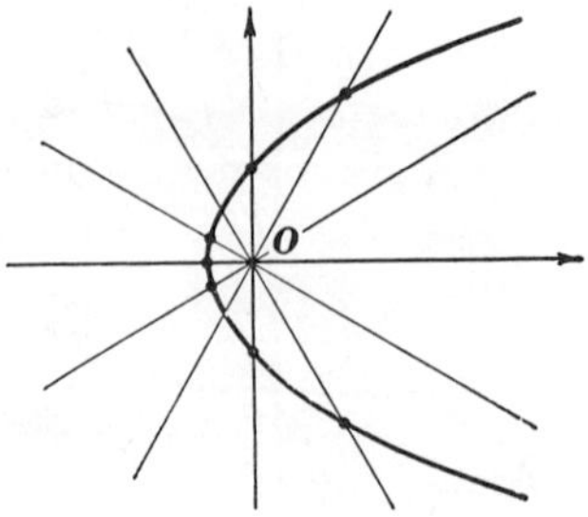

FIG. 7-24

Plotting: Since the period of $\cos \theta$ is 2π and since the curve is symmetric with respect to the polar axis, we construct the table through the range for which $0° < \theta \leq 180°$.

θ	0	30°	60°	90°	120°	150°	180°
$\rho = \dfrac{2}{1-\cos\theta}$	undefined	14.9	4	2	1.3	1.1	1

The curve is a parabola, as shown in Fig. 7-24.

EXERCISES

1. Sketch each of the following curves and transform its equation to cartesian coordinates:

   (a) $\rho = 1 + 2\cos\theta$
   (b) $\rho = 3 - 3\sin\theta$
   (c) $\rho = 3 - \cos\theta$
   (d) $\rho = 3\sin 2\theta$
   (e) $\rho = \cos 3\theta$
   (f) $\rho = \frac{1}{2}\cos 4\theta$
   (g) $\rho^2 = 4\cos 2\theta$
   (h) $\rho = -\sin 2\theta$
   (i) $\rho = 2\cos^2\theta$
   (j) $\rho = \sec\theta$
   (k) $\rho = -4\sin\theta$
   (l) $\rho = 3$
   (m) $\rho = -3\theta$
   (n) $\rho = \dfrac{3}{\theta}$
   (o) $\rho = -\dfrac{2}{1-\cos\theta}$
   (p) $\rho = \dfrac{3}{1+\cos\theta}$
   (q) $\rho = \dfrac{1}{2\cos\theta - 1}$
   (r) $\rho = \dfrac{3}{3-\cos\theta}$
   (s) $\rho = \dfrac{4}{2-\sin\theta}$
   (t) $\rho = 1 - 2\sin\theta$
   (u) $\rho = 2\cos\theta + 2$
   (v) $\rho = 2 + \sin\theta$
   (w) $\rho = 2\cos 2\theta$

POLAR COORDINATES   281

(x) $\rho = 2 \sin 3\theta$
(y) $\rho = -3 \sin 4\theta$
(z) $\rho^2 = 16 \sin 2\theta$
(aa) $\rho^2 = -16 \cos 2\theta$
(ab) $\rho = -\sin^2 \frac{\theta}{2}$
(ac) $\rho = 2 \csc \theta$
(ae) $\theta = \tan^{-1}(-\frac{1}{2})$
(ad) $\rho = 2 \cos \theta$
(af) $\rho = 2\theta$

(ag) $\rho\theta = -4$
(ah) $\rho = \dfrac{1}{1+\sin \theta}$
(ai) $\rho = -\dfrac{4}{3-3 \sin \theta}$
(aj) $\rho = -\dfrac{3}{3 \cos \theta + 1}$
(ak) $\rho = -\dfrac{2}{3+\sin \theta}$
(al) $\rho = \dfrac{3}{3+\cos \theta}$

2. Sketch each of the following curves by assigning specific numerical values to the arbitrary constants. Transform each equation to cartesian coordinates. The names of the curves are written beside the equations.

(a) $\rho = 2r \sin \theta$, *circle*
(b) $\rho = 2r \cos \theta$, *circle*
(c) $\rho \cos \theta = -6$, *line*
(d) $\rho = 3 \csc \theta$, *line*
(e) $\rho = a \pm a \cos \theta$, *cardioid*
(f) $\rho = a \pm a \sin \theta$, *cardioid*
(g) $\rho = a \pm b \cos \theta$, *limacon* (treat the cases for which $a > b$ and $a < b$)
(h) $\rho = a \pm b \sin \theta$, *limacon* (treat the cases for which $a > b$ and $a < b$)
(i) $\rho^2 = a^2 \cos 2\theta$, *lemniscate*
(j) $\rho^2 = a^2 \sin 2\theta$, *lemniscate*
(k) $\rho = a \cos k\theta$, where k is a positive or negative integer, *rose*
(l) $\rho = a \sin k\theta$, where k is a positive or negative integer, *rose*
m) $\rho = \dfrac{k}{a \pm a \cos \theta}$, *parabola*
(n) $\rho = \dfrac{k}{a \pm a \sin \theta}$, *parabola*
(o) $\rho = \dfrac{k}{a \pm b \cos \theta}$ $\begin{cases} ellipse \text{ for } a > b \\ hyperbola \text{ for } a < b \end{cases}$
(p) $\rho = \dfrac{k}{a \pm b \sin \theta}$ $\begin{cases} ellipse \text{ for } a > b \\ hyperbola \text{ for } a < b \end{cases}$
(q) $\rho = k\theta$, *spiral of Archimedes*
(r) $\rho = \dfrac{k}{\theta}$, *hyperbolic spiral*
(s) $\rho^2 \theta = k$, *lituus*

7–10. The conic

The conic has already been defined in Chapter 5. If a focus is placed at the pole and the corresponding directrix is taken to be the line $\rho \cos \theta = -p$, as indicated in Fig. 7–25, it follows from the definition of a conic that

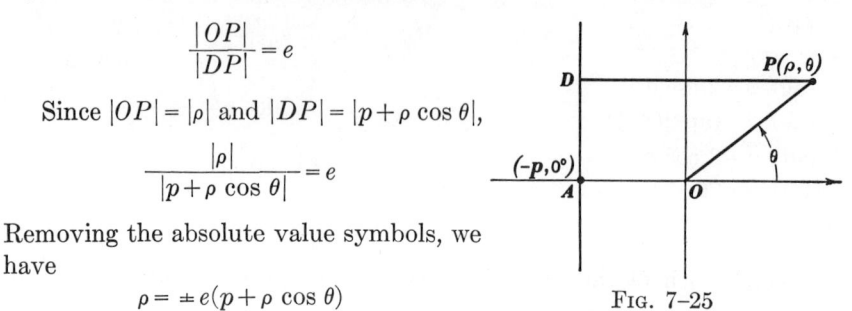

$$\frac{|OP|}{|DP|} = e$$

Since $|OP| = |\rho|$ and $|DP| = |p + \rho \cos \theta|$,

$$\frac{|\rho|}{|p + \rho \cos \theta|} = e$$

Removing the absolute value symbols, we have

$$\rho = \pm e(p + \rho \cos \theta)$$

Fig. 7–25

Using the positive sign, we obtain

$$\rho = e(p + \rho \cos \theta)$$

which becomes

$$\rho(1 - e \cos \theta) = ep$$

or

$$\rho = \frac{ep}{1 - e \cos \theta} \qquad (7\text{–}13\text{a})$$

Using the negative sign, we have

$$-\rho = ep + e\rho \cos \theta$$

from which

$$\rho = \frac{-ep}{1 + e \cos \theta} \qquad (7\text{–}13\text{b})$$

If $P = (\rho, \theta)$ is any point on the locus of equation (7–13a), or the coordinates ρ and θ of P satisfy equation (7–13a), then the coordinates $-\rho$ and $(\theta + \pi)$, which are also coordinates of P, satisfy equation (7–13b); and point P therefore lies on the locus of equation (7–13b). Conversely, if P is on the locus of equation (7–13b), then P also is on the locus of equation (7–13a). Hence, these two equations represent the same set. If $e > 1$, the conic is a hyperbola; if $e < 1$, the conic is an ellipse; and if $e = 1$, the conic is a parabola.

If a focus of a conic is placed at the pole and the corresponding directrix is the horizontal line whose equation is $\rho \sin \theta = -p$, an equation of the conic is

$$\rho = \frac{ep}{1 - e \sin \theta} \qquad (7\text{–}14\text{a})$$

or

$$\rho = \frac{-ep}{1 + e \sin \theta} \qquad (7\text{–}14\text{b})$$

The curve corresponding to equation (7–13a) or equation (7–13b) is symmetric with respect to the polar axis. Since the pole is the focus, we can draw the following conclusions: If $e<1$, the polar axis lies along the major axis; if $e>1$, the polar axis lies along the transverse axis; and, if $e=1$, the polar axis lies along the axis of the parabola. The curve corresponding to equation (7–14a) or equation (7–14b) is symmetric with respect to the 90°-axis. Hence, for different values of e, the 90°-axis lies along the major axis, the transverse axis, or the axis of the conic. This information is extremely useful when the conic is to be sketched.

EXAMPLE 7–14. Sketch the conic $\rho = \dfrac{5}{2+3\cos\theta}$.

Solution. Since $e = \dfrac{3}{2} > 1$, the conic is a hyperbola.

Intercepts:

θ	0°	90°	180°	270°
$\rho = \dfrac{5}{2+3\cos\theta}$	1	2.5	-5	2.5

Symmetry: Polar axis, yes; 90°-axis, no; pole, no.

Extent: As $\cos\theta$ approaches $-\tfrac{2}{3}$, $|\rho|$ increases indefinitely. Hence, $\rho = \infty$ when $\theta = \cos^{-1}(-\tfrac{2}{3})$.

Sketching: The transverse axis is the polar axis, and the vertices are at $(1, 0°)$ and at $(-5, 180°)$. Hence, the center of the hyperbola is at $(3, 0°)$. Also we know that the conjugate axis is perpendicular to the polar axis and contains the point $(3, 0°)$; and $a = 2$, $c = 3$, and $b = \sqrt{9-4} = \sqrt{5}$. See Fig. 7–26 for the graph.

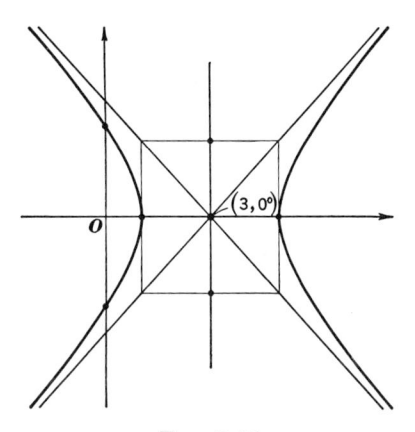

FIG. 7–26

284   ANALYTIC GEOMETRY IN THE PLANE

In general, either of the following equations represents a conic having a focus at the pole and a vertical line or a horizontal line as the directrix:

$$\rho = \frac{k_1}{k_2 + k_3 \cos \theta} \qquad (7\text{-}15\text{a})$$

or

$$\rho = \frac{k_1}{k_2 + k_3 \sin \theta} \qquad (7\text{-}15\text{b})$$

where k_1, k_2, and k_3 are positive or negative constants and $e = \left|\frac{k_3}{k_2}\right|$.

EXERCISES

1. Derive an equation for a conic having a focus at the pole and each of the following lines for a directrix:

   (a) $\rho \sin \theta = -p$
   (b) $\rho \cos \theta = p$
   (c) $\rho \sin \theta = p$

2. Identify and sketch each of the following conics:

   (a) $\rho = \dfrac{4}{1 - \sin \theta}$
   (b) $\rho = \dfrac{2}{2 + \cos \theta}$
   (c) $\rho = -\dfrac{3}{1 + \cos \theta}$
   (d) $\rho = \dfrac{3}{1 - 2 \sin \theta}$
   (e) $\rho = \dfrac{15}{3 - 4 \cos \theta}$
   (f) $\rho = -\dfrac{2}{3 + 3 \sin \theta}$
   (g) $\rho = -\dfrac{1}{3 - \sin \theta}$
   (h) $\rho = \dfrac{8}{4 - 3 \cos \theta}$

3. Transform the equation of each of the conics of Exercise 2 to cartesian coordinates.

4. Find the foci, vertices, and directrices of each of the following conics, draw its graph, and transform the equation to cartesian coordinates:

   (a) $3\rho - \rho \cos \theta - 6 = 0$
   (b) $2\rho + \rho \cos \theta + 5 = 0$
   (c) $\rho = 3\rho \sin \theta - 9$
   (d) $\rho - 8 = 4\rho \sin \theta$
   (e) $2\rho = 3 - 4\rho \cos \theta$
   (f) $4\rho + 2\rho \sin \theta = 9$

7–11. Locus problems

Sometimes it is easier to derive an equation of a curve in polar coordinates than it is in cartesian coordinates. The work of obtaining the equation is often simplified by a judicious choice for the pole.

EXAMPLE 7–15. Find an equation for the locus of the vertex P of a triangle whose base OA is of length $a>0$ and which has the angle at P equal to one-half of the angle at O.

Solution. The conditions are represented for one triangle in Fig. 7-27. Let $\angle AOP = \theta$. Then $\angle OPA = \frac{\theta}{2}$, and the exterior angle to $\angle OAP$ is equal to $\frac{3\theta}{2}$. Using the law of sines, we have

$$\frac{\rho}{\sin \angle OAP} = \frac{a}{\sin \frac{\theta}{2}}$$

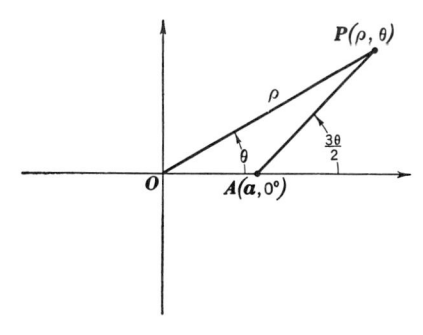

Fig. 7-27

Since $\angle OAP = 180° - \frac{3\theta}{2}$, we obtain the following result for ρ:

$$\rho = \frac{a \sin \frac{3\theta}{2}}{\sin \frac{\theta}{2}} = \frac{a \left(3 \sin \frac{\theta}{2} - 4 \sin^3 \frac{\theta}{2}\right)}{\sin \frac{\theta}{2}}*$$

or
$$\rho = a\left(3 - 4 \sin^2 \frac{\theta}{2}\right) = a\left(3 - 4 \cdot \frac{1 - \cos \theta}{2}\right)$$

Hence, $\rho = a(1 + 2 \cos \theta)$

Note that the locus does not include points for which $\theta = k\pi$, where k is an integer.

EXERCISES

1. Find equations of the set of points P and P' constructed as follows and as shown in Fig. 7-28. Through a fixed point O a line is drawn cutting a fixed line CD at a point Q. On this line points P and P' are chosen such that $|QP| = |QP'| = b > 0$. (Hint: Let O be the pole, and let the fixed line be $\rho \cos \theta = a$, where $a > b$.)

---
* $\sin 3x \equiv 3 \sin x - 4 \sin^3 x$.

2. Find an equation of the set of the mid-points of chords of the circle $\rho = 2r \cos \theta$ drawn from the pole.

3. The chord OQ of the circle $\rho = a \cos \theta$ drawn from the pole O is extended to P so that $|QP|$ equals the diameter of the circle. Find the locus of P.

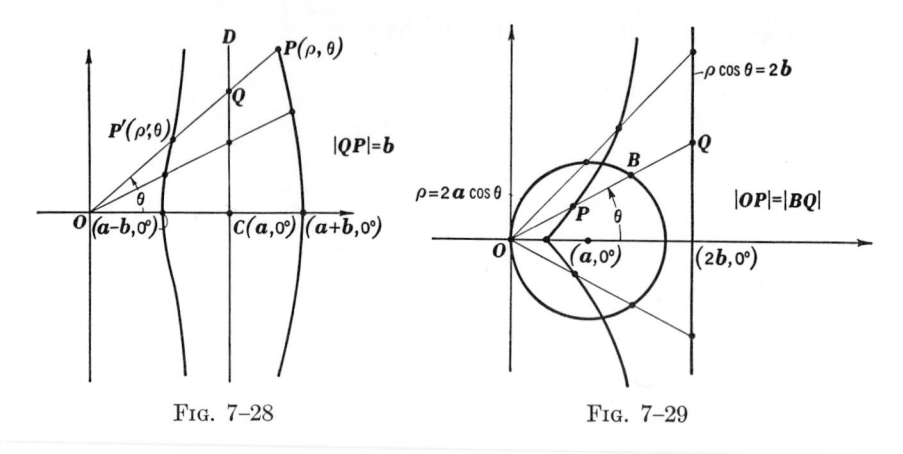

FIG. 7–28

FIG. 7–29

4. The chord OQ of the circle $\rho = a \cos \theta$ drawn from the pole is extended to P so that $|QP|$ equals the length of the perpendicular distance from Q to the polar axis. Find the locus of P.

5. The chord OQ of the circle $\rho = a \sin \theta$ drawn from the pole is extended to P so that $|QP|$ is equal to the radius of the circle. Find the locus of P.

6. As indicated in Fig. 7–29, lines are drawn from the pole O on the circle $\rho = 2a \cos \theta$ to meet the line $\rho \cos \theta = 2b$ at Q, where $a < b$. On any such line OQ, lay off $|OP| = |BQ|$. Find the locus of P.

7–12. Intersection of curves in polar coordinates

The problem of finding where two curves intersect when their equations are given in polar coordinates often involves the solving of a trigonometric equation. In any event, we solve the equations simultaneously. However, not all the points of intersection may be found in this way. It may happen that a point is on two curves, and yet no single pair of its coordinates satisfies both equations. In this case the simultaneous solution fails to give all the points of intersection, since a point may lie on both curves and yet have different pairs of coordinates which satisfy the equations of the curves. This is especially true of the pole, since the coordinates of the pole can be 0 and any angle θ. For example, consider the circles whose equations are $\rho = a \cos \theta$ and $\rho = a \sin \theta$. Both contain the pole; yet the coordinates 0 and $\frac{\pi}{2}$ satisfy $\rho = a \cos \theta$, while the coordinates 0 and 0° satisfy $\rho = a \sin \theta$.

What then is a method for finding all the points of intersection of two curves? Let us represent the equations of two curves by $f(\rho, \theta) = 0$ and $g(\rho, \theta) = 0$.* First, if there exist two angles θ_1 and θ_2 such that $f(0, \theta_1) = 0$ and $g(0, \theta_2) = 0$, then the pole lies on both curves. Here $f(0, \theta_1) = 0$ means that ρ has been replaced by 0 and θ has been replaced by θ_1 in the equation $f(\rho, \theta) = 0$. Hence, to test whether or not the pole is a common point on both curves, we replace ρ by 0 in the equations $f(\rho, \theta) = 0$ and $g(\rho, \theta) = 0$ and solve the equations separately for θ. If one or the other of the equations has no (real) solution θ, the pole does not lie on that curve. If each equation has a solution θ, the pole is an intersection point.

In the same way, if there exists a point $P = (\rho_0, \theta_0)$, *other than the pole*, such that $f(\rho_0, \theta_0) = 0$ and also such that either $g(\rho_0, \theta_0 + 2k\pi) = 0$ or $g(-\rho_0, \theta_0 + \{2k+1\}\pi) = 0$ for some *integer* k, then the point P is a point of intersection of the two curves. This follows since the points (ρ_0, θ_0), $(\rho_0, \theta_0 + 2k\pi)$, and $(-\rho_0, \theta_0 + \{2k+1\}\pi)$, where k is an integer, have the same position.

Hence, to find all the points of intersection (other than the pole) of two curves whose equations are $f(\rho, \theta) = 0$ and $g(\rho, \theta) = 0$, we solve each of the following pairs of equations simultaneously:

(Pair 1) $\qquad\qquad f(\rho, \theta) = 0$ and $g(\rho, \theta + 2k\pi) = 0$

(Pair 2) $\qquad\qquad f(\rho, \theta) = 0$ and $g(-\rho, \theta + \{2k+1\}\pi) = 0$

We then determine every integer k for which solutions exist.

EXAMPLE 7–16. Find all the points of intersection of the circles $\rho = a \cos \theta$ and $\rho = a \sin \theta$.

Solution. The circles are shown in Fig. 7–30. We have already seen that the pole lies on both curves. Now, considering the equations $f(\rho, \theta) = \rho - a \cos \theta = 0$ and $g(\rho, \theta) = \rho - a \sin \theta = 0$ and using the first pair of simultaneous equations, we have

$$a \cos \theta = a \sin (\theta + 2k\pi) = a \sin \theta$$

Since $\cos \theta = 0$ does not yield a solution of this equation, we may divide through by $\cos \theta$ and obtain

$$\tan \theta = 1$$

Hence, $\theta = \frac{\pi}{4}$ and $\frac{5\pi}{4}$. Substituting these values of θ in the equation $\rho =$

---

* $f(\rho, \theta) = 0$ and $g(\rho, \theta) = 0$ are equations in which only the variables ρ and θ appear. So $f(\rho, \theta) = 0$ and $g(\rho, \theta) = 0$ might be such equations as $\rho^2 - 2\rho\theta + \sin \theta = 0$, $\rho - a \cos \theta = 0$, $\rho \sin \theta + 2 = 0$, and $\rho - \theta = 0$.

288   ANALYTIC GEOMETRY IN THE PLANE

$a \cos \theta$, we obtain $\rho = \dfrac{a\sqrt{2}}{2}$ and $\rho = -\dfrac{a\sqrt{2}}{2}$. Since the point $\left(\dfrac{a\sqrt{2}}{2}, \dfrac{\pi}{4}\right)$ is the same as the point $\left(-\dfrac{a\sqrt{2}}{2}, \dfrac{5\pi}{4}\right)$, we have found another point of intersection

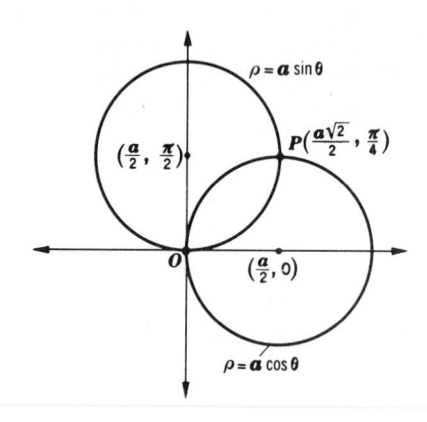

FIG. 7-30

Finally, using the second pair of simultaneous equations, we obtain

$$a \cos \theta = -a \sin (\theta + \{2k+1\}\pi) = a \sin \theta$$

Again we have $\tan \theta = 1$. Hence, the second pair of equations does not give any new points of intersection, and the circles have two points in common, namely, $(0, 0)$ and $\left(\dfrac{a\sqrt{2}}{2}, \dfrac{\pi}{4}\right)$.

EXAMPLE 7-17. Find coordinates of the points of intersection of the cardioid $\rho = 2a(1 - \cos \theta)$ and the circle $\rho = 2a \cos \theta$.

Solution. The curves are shown in Fig. 7-31. Testing for the pole, we have $0 = 2a(1 - \cos \theta)$, or $\cos \theta = 1$, which is satisfied by $\theta = 0$. Also, $0 = 2a \cos \theta$, or $\cos \theta = 0$, which is satisfied by $\theta = \dfrac{\pi}{2}$. Hence, the pole is common to both curves.

Now let $f(\rho, \theta) = \rho - 2a(1 - \cos \theta) = 0$ and $g(\rho, \theta) = \rho - 2a \cos \theta = 0$. Considering the first pair of simultaneous equations, we have

$$2a(1 - \cos \theta) = 2a \cos (\theta + 2k\pi) = 2a \cos \theta$$

Hence,

$$2 \cos \theta = 1, \text{ or } \cos \theta = \tfrac{1}{2}$$

and

$$\theta = \dfrac{\pi}{3} \text{ or } \dfrac{5\pi}{3}$$

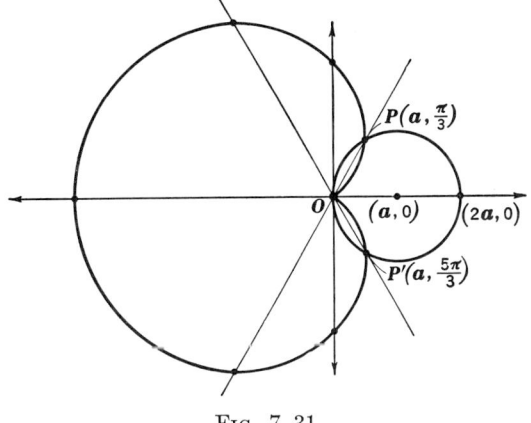

FIG. 7-31

When $\theta = \frac{\pi}{3}$, then $\rho = 2a \cos \frac{\pi}{3} = a$; when $\theta = \frac{5\pi}{3}$, then $\rho = 2a \cos \frac{5\pi}{3} = a$. So two other points on the curves are $\left(a, \frac{\pi}{3}\right)$ and $\left(a, \frac{5\pi}{3}\right)$.

By considering the second pair of simultaneous equations, we obtain
$$2a(1 - \cos \theta) = -2a \cos (\theta + \{2k+1\}\pi) = 2a \cos \theta$$
Hence, we again have $\cos \theta = \frac{1}{2}$ and we find no new points of intersection.

EXAMPLE 7-18. Find the points common to the curves $\rho = \theta$ and $\rho = \pi - \theta$.

Solution. Since $(0, 0)$ satisfies the equation $\rho = \theta$, and since $(0, \pi)$ satisfies the equation $\rho = \pi - \theta$, the pole lies on both curves.

Let $f(\rho, \theta) = \rho - \theta = 0$ and $g(\rho, \theta) = \rho - \pi + \theta = 0$. Considering the first pair of simultaneous equations, we have, after eliminating ρ,
$$\theta = \pi - (\theta + 2k\pi), \text{ or } 2\theta = (1 - 2k)\pi$$

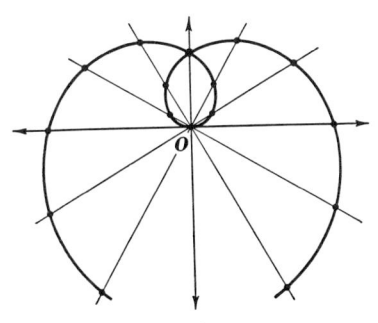

FIG. 7-32

290 ANALYTIC GEOMETRY IN THE PLANE

Hence,
$$\theta = \frac{(1-2k)\pi}{2}$$

Now, as k takes on integral values, we obtain the following results:

k	θ	$\rho = \theta$	$\theta + 2k\pi$	$\rho = \pi - \theta$
0	$\dfrac{\pi}{2}$	$\dfrac{\pi}{2}$	$\dfrac{\pi}{2}$	$\dfrac{\pi}{2}$
1	$\dfrac{-\pi}{2}$	$\dfrac{-\pi}{2}$	$\dfrac{3\pi}{2}$	$\dfrac{-\pi}{2}$
2	$\dfrac{-3\pi}{2}$	$\dfrac{-3\pi}{2}$	$\dfrac{5\pi}{2}$	$\dfrac{-3\pi}{2}$
.	.	.	.	.
.	.	.	.	.
n	$\dfrac{(1-2n)\pi}{2}$	$\dfrac{(1-2n)\pi}{2}$	$\dfrac{(1+2n)\pi}{2}$	$\dfrac{(1-2n)\pi}{2}$

The first two columns of the table give us the values of θ and ρ for the points on the locus of $\rho = \theta$, while the third and fourth columns give us the values of θ and ρ for points on the locus of $\rho = \pi - \theta$. In each row the coordinates give us the same point; and the common curve is shown in Fig. 7–32. Hence, we have here an infinitude of intersections. If we consider the second pair of simultaneous equations, we have, after eliminating ρ,
$$\theta = -\pi + (\theta + \{2k+1\}\pi), \text{ or } 2k\pi = 0 \text{ if } k = 0$$

This tells us that, regardless of the value of θ, whenever (ρ, θ) is a point on the locus of $\rho = \theta$, the point $(-\rho, \pi + \theta)$ is on the locus of $\rho = \pi - \theta$. Since the point (ρ, θ) is identical with the point $(-\rho, \pi + \theta)$, the two curves actually coincide. The intersections found by considering the first pair of simultaneous equations are thus points where the curve *crosses itself*.

EXAMPLE 7–19. Find the points of intersection of the hyperbolic spiral $\rho\theta = 8$ and the line $\theta = 2$.

Solution. The loci are shown in Fig. 7–33. The pole is not a point of intersection. Consider $f(\rho, \theta) = \rho\theta - 8 = 0$ and $g(\rho, \theta) = \theta - 2 = 0$.

Using the first pair of simultaneous equations, we have, after eliminating ρ,
$$\theta + 2k\pi - 2 = 0, \text{ or } \theta = 2 - 2k\pi$$

If $k = 0$, we have $\theta = 2$ and $\rho = 4$.

If $k = 1$, we have $\theta = 2 - 2\pi$ and $\rho = \dfrac{8}{2 - 2\pi} = \dfrac{4}{1 - \pi}$.

POLAR COORDINATES 291

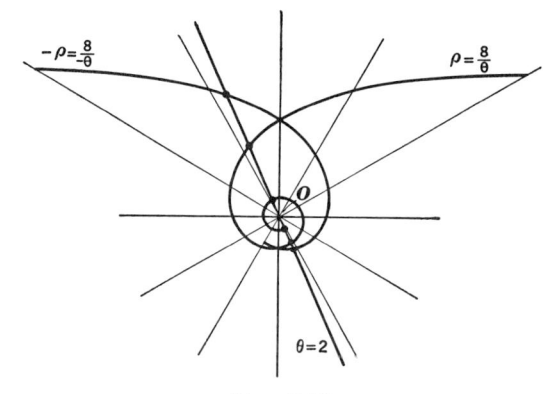

FIG. 7-33

If $k = 2$, we have $\theta = 2 - 4\pi$ and $\rho = \dfrac{8}{2 - 4\pi} = \dfrac{4}{1 - 2\pi}$.

. . . . .
. . . . .

If $k = n$, we have $\theta = 2 - 2n\pi$ and $\rho = \dfrac{4}{1 - n\pi}$.

Using the second pair of simultaneous equations, we obtain

$$\theta + (2k+1)\pi = 2, \text{ or } \theta = 2 - (2k+1)\pi$$

If $k = 0$, we have $\theta = 2 - \pi$ and $\rho = \dfrac{8}{2 - \pi}$.

If $k = 1$, we have $\theta = 2 - 3\pi$ and $\rho = \dfrac{8}{2 - 3\pi}$.

If $k = 2$, we have $\theta = 2 - 5\pi$ and $\rho = \dfrac{8}{2 - 5\pi}$.

. . . . .
. . . . .

If $k = n$, we have $\theta = 2 - (2n+1)\pi$ and $\rho = \dfrac{8}{2 - (2n+1)\pi}$

Hence, there are an infinite number of points of intersection of the line and the spiral.

7-13*. Common curves in polar coordinates

It is important to be able to recognize the polar equation of some of the most common curves. To help facilitate this recognition, and for reference, several polar equations along with the types of curves they represent are presented here.

---

* See Exercise 2 in Section 7-9.

I. Certain Basic Curves

(a) Straight lines:
  Passing through the pole: $\theta = \theta_0$.
  Parallel to the axes: $\rho \sin \theta = a$ $\qquad$ $\rho = a \csc \theta$
  $\qquad\qquad\qquad\qquad\qquad$ or
  $\qquad\qquad\qquad\qquad\rho \cos \theta = a$ $\qquad$ $\rho = a \sec \theta$

(b) Circles:
  Center at the pole: $\rho = a$
  Passing through the pole: $\rho = 2a \cos \theta$
  $\qquad\qquad\qquad\qquad\qquad\rho = 2a \sin \theta$

(c) The x and y conics:

$$\rho = \frac{ep}{1 \pm \rho \sin \theta}$$
$$\rho = \frac{ep}{1 \pm \rho \cos \theta}$$

$e > 1$ ellipse
$e = 1$ parabola
$e < 1$ hyperbola

II. Certain Other Important and Interesting Curves

(a) Straight lines: $\rho \cos(\theta - \alpha) = p$
  $\qquad\qquad\qquad\rho \cos \theta \cos \alpha + \rho \sin \theta \sin \alpha = p$

(b) Circles: $\rho = a \sin \theta \pm b \cos \theta$
  $\qquad\qquad a^2 = \rho^2 + \rho_0^{2'} - 2\rho\rho_0 \cos(\theta - \theta_0)$,

where (ρ_0, θ_0) is center and a is radius.

(c) Parabola: $\rho = \pm a \sec^2 \frac{\theta}{2}, \rho = \pm a \csc^2 \frac{\theta}{2}$

(d) Roses: $\rho = a \cos k\theta$
  $\qquad\qquad\rho = a \sin k\theta$
  Where k is a positive integer, these curves have k leaves when k is odd, and $2k$ leaves when k is even.

(e) Lemniscates: $\rho^2 = a^2 \cos 2\theta$
  $\qquad\qquad\qquad\rho^2 = a^2 \sin 2\theta$

(f) Limacons: $\rho = a \pm b \sin \theta$ $\qquad$ $a = b$, Cardioid
  $\qquad\qquad\quad\rho = a \pm b \cos \theta$ $\qquad$ $a < b$, Nodal Limacon
  $\qquad\qquad\qquad\qquad\qquad\qquad\qquad$ $a > b$, No node, no cusp

(g) Archimedes spiral: $\rho = k\theta$

(h) Hyperbolic spiral: $\rho = \dfrac{k}{\theta}$

EXERCISES

Find coordinates of all points common to each of the following pairs of curves:

1. $\rho = 3 \cos \theta$ and $\rho = 3 \sin \theta$
2. $\rho = \cos 2\theta$ and $\rho = \sin \theta$
3. $\rho = \cos 2\theta$ and $\rho = \cos \theta$
4. $\rho = 2(1 - \sin \theta)$ and $\rho = \cos \theta$
5. $\rho = 2(1 - \sin \theta)$ and $\rho = 2 \cos \theta$
6. $\rho = 1 + \cos \theta$ and $\rho = \cos 2\theta$
7. $\rho = 4 \sin 2\theta$ and $\rho = 2$
8. $\rho = 4 \cos 2\theta$ and $\rho = 1$
9. $\rho\theta = 4$ and $\theta = 2$
10. $\rho\theta = 6$ and $\rho = 2$
11. $\rho = \theta$ and $\rho = \pi + \theta$
12. $\rho = -3\theta$ and $\rho = 2\theta$

REVIEW EXERCISES

1. Locate each of the following points, and rename it in three (in general) other ways such that the absolute value of the angle is less than or equal to $360°$:
   (a) $(-2, 30°)$
   (b) $\left(4, -\dfrac{\pi}{4}\right)$
   (c) $(3, -120°)$
   (d) $(-3, 240°)$
   (e) $(\tfrac{1}{2}, -\pi)$
   (f) $(-5, -315°)$
   (g) $\left(1, \dfrac{3\pi}{2}\right)$
   (h) $(3, 330°)$
   (i) $(-1, 90°)$
   (j) $(2, -60°)$

2. Derive an equation of a circle having its center at $(-6, 90°)$ and a radius equal to 6.

3. Find an equation of the circle that has its center at $(4, \pi)$ and has a radius equal to 4.

4. Derive an equation of the conic for which the eccentricity is e, a focus is at the pole, and the directrix is $\rho \cos \theta = 4$.

5. Derive an equation of the conic for which the eccentricity is e, a focus is at the pole, and the directrix is $\rho \sin \theta = 3$.

6. Sketch each of the following curves and transform its equation to cartesian coordinates:
   (a) $\rho = 3 \cos \theta - 2$
   (b) $\rho = 2 \cos \theta - 3$
   (c) $\rho = \cos 2\theta$
   (d) $\rho = \sin 3\theta$
   (e) $\rho^2 = -16 \cos 2\theta$
   (f) $\rho^2 = -4 \sin 2\theta$
   (g) $\rho = \sec^2 \dfrac{\theta}{2}$
   (h) $\rho = 2 \csc^2 \dfrac{\theta}{2}$
   (i) $\rho = \dfrac{2}{2 - 3 \cos \theta}$
   (j) $\rho = \dfrac{3}{3 - 2 \sin \theta}$
   (k) $\rho = -\dfrac{4}{1 + 2 \sin \theta}$
   (l) $\rho = \dfrac{3}{2 + \cos \theta}$
   (m) $\rho^2 = \cos \theta$
   (n) $\rho^2 = -\sin \theta$
   (o) $\rho = \sin \dfrac{\theta}{2}$
   (p) $\rho = \cos \dfrac{\theta}{2}$

7. Transform each of the following cartesian equations to polar form:
   (a) $x^2 + y^2 = 16$
   (b) $x^2 - y^2 = 4$
   (c) $xy + 8 = 0$
   (d) $y^2 = -16x$
   (e) $x^2 = 4y$
   (f) $2x - 3y + 6 = 0$
   (g) $y = -4$
   (h) $x^2 + y^2 - 3x + 2y - 4 = 0$
   (i) $(x^2 + y^2)^2 = 4xy$
   (j) $y = 3x^2 - 2x + 1$

8. Find all the points of intersection of the following pairs of curves:
   (a) $\rho = 2 + 2 \cos \theta$ and $\rho = 3$
   (b) $\rho = 1 - 2 \cos \theta$ and $\rho = \sin \theta$
   (c) $\rho = \sin 2\theta$ and $\rho = \sin \theta$
   (d) $\rho = \sin^4 \theta$ and $\rho = \sin 2\theta$
   (e) $\rho = \frac{1}{2} \cos \theta$ and $\rho = \sin \theta$
   (f) $\rho = 2$ and $\rho = 1 + \cos \theta$
   (g) $\rho = 2 - 3 \sin \theta$ and $\rho = 2 \sin \theta$
   (h) $\rho = 2 \csc \theta$ and $\rho = 4 \sin \theta$

8
Transcendental and Other Curves

In Chapters 4, 5, and 6, we discussed loci of equations of the second degree, that is, the circles and the conics. In this chapter, we shall study curves represented by equations of degree greater than two and also non-algebraic equations.

A function $y = g(x)$ will be said to be an *algebraic* function of x, provided it satisfies an equation of the form

$$f_0(x)y^n + f_1(x)y^{n-1} + \ldots + f_n(x) = 0 \qquad (8\text{--}1)$$

where the coefficients $f_0(x)$, $f_1(x)$, etc. are *polynomials* in x and n is a positive integer.

A polynomial of degree n in x is of the form

$$f(x) = a_0 + a_1 x + a_2 x^2 + \ldots + a_n x^n \qquad (8\text{--}2)$$

where n is a positive integer and a_0, a_1, etc. are constants, and where $a_n \neq 0$. If a_0 is the only constant which is not zero, then this constant is called a polynomial in x of zero degree. However, if all the coefficients $a_0, a_1, a_2, a_3, \ldots$, are zero, the polynomial reduces to the constant 0, which is called the zero polynomial and has no degree associated with it.

All other functions are called *non-algebraic*, or *transcendental*, functions. Some examples of transcendental functions are $y = \sin x$, $y = a^x$, and $y = \log_a x$. Algebraic and transcendental functions of x and y are similarly defined. Loci of equations involving algebraic (or transcendental) functions are called algebraic (or transcendental) curves.

8–1. The graph of $y = x^n$ and other algebraic curves

First consider the case where n is an even integer. If $x = 0$, then $y = 0$; and the curve passes through the origin. If we replace x by $-x$, the equation is unchanged; since, if n is an even number, $(-x)^n = x^n$. But this means that the curve is symmetric with respect to the y-axis. As x increases or decreases indefinitely, y *increases* indefinitely at a much

296   ANALYTIC GEOMETRY IN THE PLANE

"faster rate" than $|x|$, as shown in Fig. 8-1. When $n = 2$, the curve is a parabola.

Next consider the case where n is an odd integer greater than 1. Again the curve passes through the origin. If we replace x by $-x$ and also replace y by $-y$, we obtain $-y = (-x)^n = -x^n$. Hence, the curve is symmetric with respect to the origin. As x increases indefinitely, y increases indefinitely at a much "faster rate" than x; and, as x decreases indefinitely, y decreases indefinitely at a much "faster rate" than x. A typical graph of the equation $y = x^n$ is shown in Fig. 8-2.

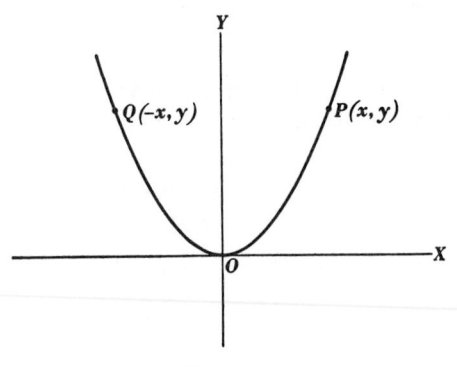

FIG. 8-1

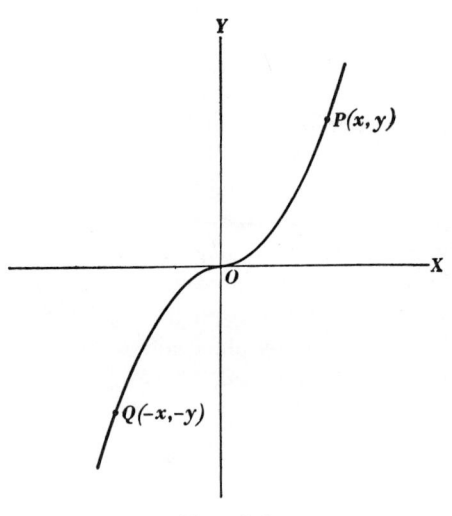

FIG. 8-2

Now consider the case where n is a negative integer, that is, $n = -k$ and k is a positive integer. In this case, $y = x^{-k} = \dfrac{1}{x^k}$. In examining the intercepts we find that, when $x = 0$, y is not defined. However, as $|x|$ approaches zero, $|y|$ increases indefinitely. Hence, the y-axis is a vertical asymptote to the curve. In the same way, when $y = 0$, x is not defined; but, as $|x|$ increases indefinitely, $|y|$ approaches zero. Hence, the x-axis is a horizontal asymptote to the curve.

If k is an even integer and if (x, y) is a point on the locus, then the point $(-x, y)$ lies on the curve, since $y = \dfrac{1}{(-x)^k} = \dfrac{1}{x^k}$. Hence, in this case the curve is symmetric with respect to the y-axis. If k is an odd integer and if (x, y) is a point on the locus, then the coordinates $-x$ and $-y$ satisfy the equation, since $-y = \dfrac{1}{(-x)^k} = -\dfrac{1}{x^k}$, or $y = \dfrac{1}{x^k}$. Therefore, in this case, the curve is symmetric with respect to the origin. We shall illustrate with two examples.

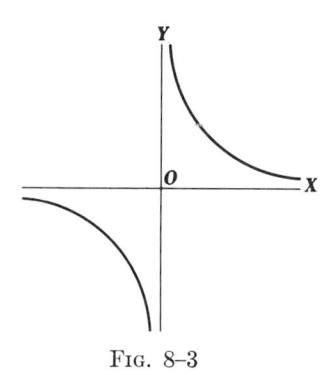

Fig. 8-3

EXAMPLE 8-1. Sketch the graph of $y = x^{-1}$.

Solution. This equation can be written in the form $y = \dfrac{1}{x}$. There are no intercepts. The curve is symmetric with respect to the origin, and the coordinate axes are asymptotes of the curve. Constructing a table of values, we have the following results:

x	-3	-2	-1	$-\tfrac{1}{2}$	0	$\tfrac{1}{2}$	1	2	3
$y = \dfrac{1}{x}$	-0.33	-0.5	-1	-2	undefined	2	1	0.5	0.33

The curve is shown in Fig. 8-3. If we write the equation of this curve in the form $xy = 1$, we recognize it to be an equation of an equilateral hyperbola.

298   ANALYTIC GEOMETRY IN THE PLANE

EXAMPLE 8–2. Sketch the graph of $y = x^{-2}$.

Solution. Rewriting this equation in the form $y = \dfrac{1}{x^2}$, we see that the curve has no intercepts. Also, the coordinate axes are asymptotes of the curve. Since the exponent of x is even, the curve is symmetric with respect to the y-axis. Furthermore, the curve lies entirely above the x-axis.

A table of values follows:

x	-3	-2	-1	$-\frac{1}{2}$	0	$\frac{1}{2}$	1	2	3
$y = x^{-2}$	0.11	0.25	1	4	undefined	4	1	0.25	0.11

The curve is shown in Fig. 8–4.

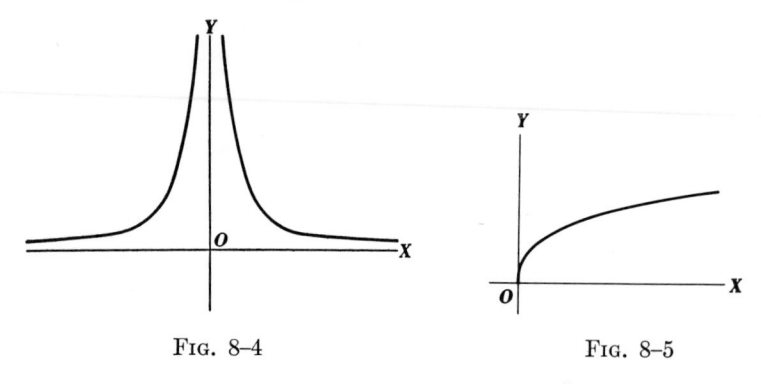

FIG. 8–4   FIG. 8–5

Finally consider the equation $y = x^n$ in which $n = \dfrac{p}{q}$, where p and q are integers prime to each other and $q > 0$. If n is positive, the curve passes through the origin and there are no horizontal or vertical asymptotes. If we rewrite the equation $y = x^{\frac{p}{q}}$ in the form $y = (\sqrt[q]{x})^p$, we can determine the following facts: (a) when q is even, x is never less than 0; (b) when p is even, y is never less than 0.

If n is negative, the equation becomes $y = \dfrac{1}{\sqrt[q]{x^p}}$. Hence, y is not defined when $x = 0$. If p is even, the curve is symmetric with respect to the y-axis. As x increases, y decreases. We again illustrate the conditions with some examples.

EXAMPLE 8–3. Sketch the graph of $y = x^{\frac{1}{2}}$.

Solution. Since $y = \sqrt{x}$, both x and y are positive or zero. The only intercept point on either axis is the origin, and there are no horizontal or vertical asymptotes. The curve is not symmetric with respect to either axis or with respect to the origin.

Constructing a table of values, we have the following results:

x	0	1	2	3	4
$y = \sqrt{x}$	0	1	1.4	1.7	2

The curve is shown in Fig. 8–5.

EXAMPLE 8–4. Sketch the graph of $y = x^{\frac{1}{3}}$.

Solution. When $x = 0$, then $y = 0$; and, when $y = 0$, then $x = 0$. Therefore, the origin is the only intercept point. If (x, y) is a point on the curve, the coordinates $-x$ and $-y$ also satisfy the equation. Hence, the curve is symmetric with respect to the origin. As $|x|$ increases indefinitely, $|y|$ also increases indefinitely; and there are no horizontal or vertical asymptotes.

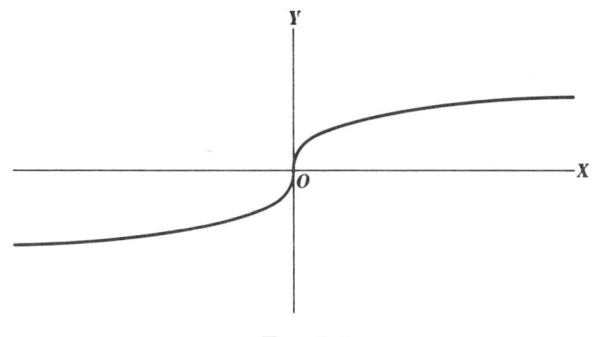

FIG. 8–6

A table of values follows:

x	-8	-6	-2	-1	0	1	2	6	8
$y = x^{\frac{1}{3}}$	-2	-1.8	-1.3	-1	0	1	1.3	1.8	2

The curve is shown in Fig. 8–6.

EXAMPLE 8–5. Sketch the graph of $y = x^{-\frac{2}{3}}$.

Solution. When $x = 0$, y is not defined; and, when $y = 0$, x is not defined. If (x, y) is a point on the graph, then the coordinates $-x$ and y satisfy the equation; and the curve is symmetric with respect to the y-axis. As $|x|$ increases indefinitely, y approaches 0 and is always positive; and so the

300   ANALYTIC GEOMETRY IN THE PLANE

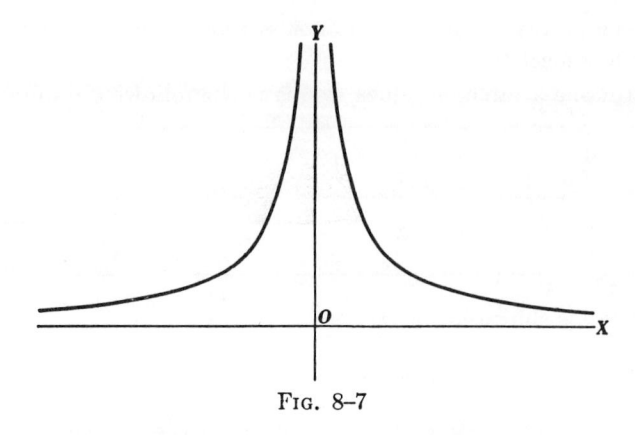

FIG. 8–7

x-axis is a horizontal asymptote. As $|x|$ approaches 0, y increases indefinitely; and so the y-axis is a vertical asymptote.

Constructing a table of values, we have the following results:

x	-3	-2	-1	-0.5	0	0.5	1	2	3
$y = x^{-\frac{2}{3}}$	0.49	0.63	1	1.6	undefined	1.6	1	0.63	0.49

The curve is shown in Fig. 8–7.

In order to indicate the relative positions of the graphs of $y = x^n$ for various values of n, we have shown in Fig. 8–8 the graphs of $y = x$, $y = x^2$, $y = x^{-1}$, $y = x^{\frac{1}{3}}$, and $y = x^{-\frac{1}{2}}$.

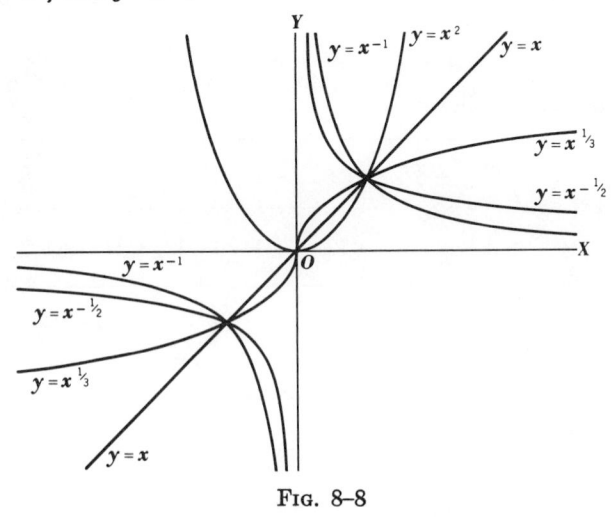

FIG. 8–8

In the previous examples we have seen that by examining an equation (1) for symmetry with respect to the axes and the origin, (2) for horizontal and vertical asymptotes, (3) for x- and y-intercepts, and by plotting a few other points, we are able to obtain a fairly accurate graph of the curve. As to how this works for more complicated equations than $y = x^n$, Section 8–2 and Examples 8–6, 8–7, and 8–8 should be carefully studied.

8–2. Discussion and sketching of curves in rectangular coordinates

A. Purpose. Throughout Chapter 5, the requirement to discuss and sketch a curve was interpreted differently with each conic studied. The items included in each case were sufficient for the purpose, since each curve was recognized from the equation. In more general forms of equations, recognition of the curve in rectangular coordinates is not so immediate. In sketching such curves discussion plays an important role. The following items give a systematic procedure to follow in discussing and sketching a curve whose equation is given in rectangular coordinates.

B. Discussion of a rectangular equation.

I. *Intercepts.* Find the intercepts and list them as points. If there are none, then so indicate.

II. *Symmetry.* Check the equation of the curve for symmetry with respect to the x-axis, y-axis, and the origin. List the conclusion for each one separately. If no symmetry exists, this should be noted.

III. *Extent.* The extent of a curve consists of those real values of the variables for which a real locus exists. This discussion involves the domain and range of the relation or function being studied.

In determining extent, values of any variable which must be excluded fall under the following categories:

(a) Values of one of the variables for which the other variable is *imaginary*. These values must be excluded since for these values no real locus exists. For example, if we first express y in terms of x and determine that there are values of x for which y is imaginary, those values of x must be excluded. Likewise, if x is expressed in terms of y, those values of y for which x is imaginary must be excluded.

(b) Values of one variable which yield the *indeterminate* form $\frac{0}{0}$. When one variable is solved in terms of the other, the values for which both the numerator and denominator are zero must be excluded.

(c) Values of one variable for which the other variable is *undefined*. These will be discussed under Asymptotes, below.

Extent may be expressed either in terms of those permissible values of the variables or in terms of those values of the variables which must be excluded. For example, if no values of the variables are excluded, extent can be expressed as: $-\infty < x < \infty, -\infty < y < \infty$. If certain values of the variables are excluded, extent can be expressed as: x may assume all values except $x = 4$ and $-2 < x < 2$; y may assume all values such that $y \leq 3$.

IV. *Asymptotes.* Find the equations of the vertical and horizontal asymptotes. To do this, determine if one of the variables is undefined for any values of the other variable. Locate the vertical asymptotes by expressing y in terms of x as in Extent, above. Find the values of x for which the denominator is zero, but for which the numerator is not zero. If x_1 is such a value, a vertical line through the point $(x_1, 0)$ will be a vertical asymptote. Locate the horizontal asymptotes by expressing x in terms of y and finding the values of y for which the denominator is zero, but for which the numerator is not zero. If y_1 is such a value, a horizontal line through $(0, y_1)$ will be a horizontal asymptote. If no asymptotes exist, this should be noted.

C. SKETCHING OF CURVES. The following requirements are for sketching any curve:

1. Select a suitable scale if none is given. Draw and label the coordinate axes and origin.
2. Locate and label the coordinates of the intercepts.
3. Draw the horizontal and vertical asymptotes and label each with its equation.
4. Sketch the curve and label it with its equation.

EXAMPLE 8–6. Discuss the equation $x^2 y + 4x^2 - 4 = 0$ and sketch the curve. (Fig. 8–9)

Solution.

I. *Intercepts.* $(x_1, 0)$: $(1, 0)$ and $(-1, 0)$
                           $(0, y_2)$: none

II. *Symmetry* with respect to:

       Axes   Origin
   x: no O: no
   y: yes

III. *Extent.*

$y = \dfrac{4 - 4x^2}{x^2}$ Therefore, x may assume any value except $x = 0$

$x = \pm \left(\dfrac{4}{y+4} \right)^{1/2}$ Therefore, y may assume any value greater than -4.

TRANSCENDENTAL AND OTHER CURVES   303

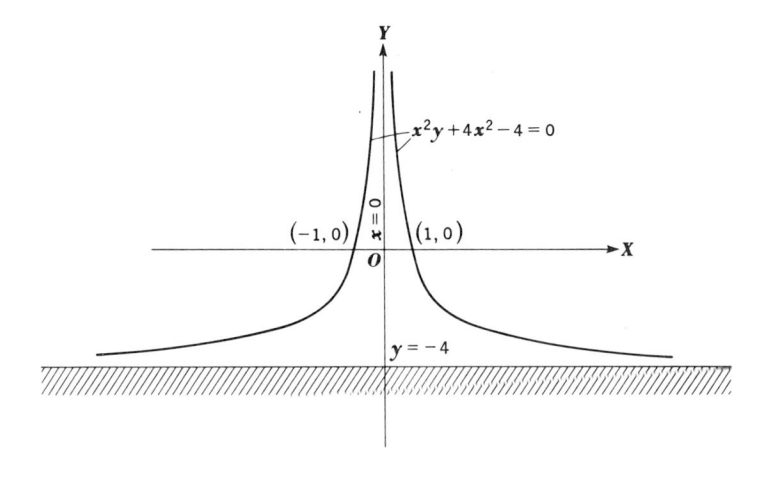

Fig. 8-9

From this study and the graph, Fig. 8-9, it is evident that the domain and range of the function are respectively the real number intervals $(-\infty,0) \cup (0,\infty)$ and $(-4,\infty)$.

IV. *Asymptotes.*
Vertical: $x = 0$
Horizontal: $y = -4$ (from equations in III.)

PROCEDURE FOR ACTUAL SKETCHING OF THE CURVE
(a) Locate intercepts, asymptotes and excluded values (shaded area), and label as appropriate.
(b) Draw the curve for $x > 0$.
(c) Since it is symmetric with respect to the y-axis, reproduce a symmetric branch for $x < 0$.

EXERCISES
Sketch the graph of each of the curves whose equations are as follows:
1. $xy + 10 = 0$
2. $xy - 3x - 4 = 0$
3. $xy + x - 3y = 0$
4. $x^2y - 4y = 6$
5. $xy^2 - 4x = 6y^2$
6. $x^2y - 2x^2 - y - 2 = 0$
7. $x(y^2 - 4) = 8$
8. $x(y + 2) = (x - 3)^2$
9. $y(x^2 + 4x + 4) = x^2 - x + 6$
10. $x(y^2 + 5y - 6) = y^2 - 4y + 4$

304   ANALYTIC GEOMETRY IN THE PLANE

EXAMPLE 8–7. Sketch the graph of $y = \dfrac{x^3 - 9x}{10}$.

Solution. When $y = 0$, we have $x = -3$, 0, and 3; when $x = 0$, $y = 0$. Hence, there are three x-intercepts and one y-intercept.

Symmetry: If we replace x by $-x$ and y by $-y$, we obtain an equation that is equivalent to the original equation. Hence, the curve is symmetric with respect to the origin. It is not symmetric to the x-axis or to the y-axis. Why?

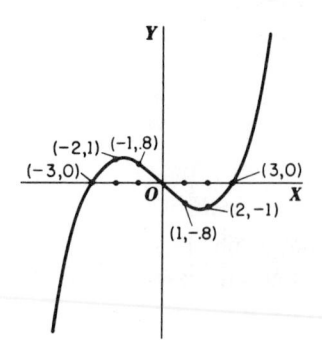

FIG. 8–10

Asymptotes: There are no horizontal or vertical asymptotes. This follows since as x increases, y also increases; and as y increases, x also increases.

If we write the equation as $y = \dfrac{x(x-3)(x+3)}{10}$, we see that when $x < -3$, $y < 0$; when $-3 < x < 0$, $y > 0$; when $0 < x < 3$, $y < 0$; and when $x > 3$, $y > 0$.

Table of values:

x	-4	-3	-2	-1	0	1	2	3	4
y	-2.8	0	1	.8	0	$-.8$	-1	0	2.8

The graph is shown in Fig. 8–10.

EXAMPLE 8–8. Sketch the graph of $y = \dfrac{x^2}{4 - x^2}$.

Solution. The curve intersects the axes only at the origin.

Symmetry: Replacing x by $-x$ leaves the equation unchanged. Therefore, the curve is symmetric with respect to the y-axis. It is not symmetric with respect to the x-axis or to the origin. Why?

Asymptotes: If we write the equation as $y = \dfrac{x^2}{(2-x)(2+x)}$, we see that when $x<-2$, $y<0$; when $-2<x<2$, $y>0$; when $x>2$, $y<0$. As x approaches -2 from either side, $|y|$ increases indefinitely. Since the distance from the line $x=-2$ to the curve approaches zero as y increases,

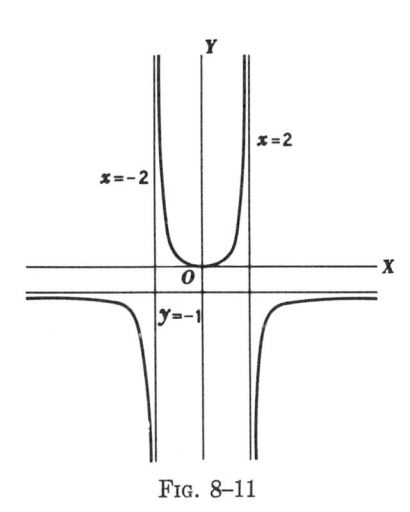

Fig. 8-11

$x=-2$ is a vertical asymptote. As x approaches $+2$ from either side, $|y|$ again increases indefinitely and so $x=2$ is also a vertical asymptote. As $|x|$ increases indefinitely, y approaches -1. Hence, $y=-1$ is a horizontal asymptote.

* Table of values:

x	-4	-3	-2	-1	$-\tfrac{1}{2}$	0	$\tfrac{1}{2}$	1	2	3	4
y	$-\tfrac{4}{3}$	$-\tfrac{9}{5}$	$\pm\infty$	$\tfrac{1}{3}$	$\tfrac{1}{15}$	0	$\tfrac{1}{15}$	$\tfrac{1}{3}$	$\pm\infty$	$-\tfrac{9}{5}$	$-\tfrac{4}{3}$

The curve is shown in Fig. 8-11.

---

* Here the domain of the function is the union of the real number intervals $(-\infty,-2)$, $(-2,2)$, and $(2,\infty)$, and the range is the union of the real number intervals $(-\infty,-1)$ and $[0,\infty)$.

EXERCISES

Sketch the graph of each of the curves whose equations are as follows:

1. $y = x^3$
2. $y = -2x^4$
3. $y = -x^3$
4. $y = \dfrac{3}{x^2}$
5. $y = \dfrac{2}{\sqrt{x}}$
6. $y = -x^{\frac{1}{2}}$
7. $y = \dfrac{1}{\sqrt[3]{x}}$
8. $y = x^{\frac{2}{3}}$
9. $y = \dfrac{2}{x}$
10. $y = x^4$
11. $y = \dfrac{x^5}{15}$
12. $y = \sqrt[4]{x}$
13. $y = -x^{\frac{1}{4}}$
14. $y = -\dfrac{1}{\sqrt{x}}$
15. $y = -x^{\frac{2}{3}}$

Sketch the graphs of the following curves. Examine each equation for intercepts, symmetry, and horizontal and vertical asymptotes:

16. $4y = x^3 - 9x$
17. $10y = 9x - x^3$
18. $y^3 = 8x^2$
19. $8y^3 = -27x^2$
20. $4y = x^3 - 9x^2$
21. $y = x(x^2 - 1)$
22. $5y = x(1 - x^2)$
23. $xy = 6x + 12$
24. $(x - 3)^2 y = 9$
25. $(y - 2)^2 x = 12$
26. $xy^2 = 6(x - 1)$
27. $y(x^2 + 4) = 12$
28. $(x^2 - 9)y = x^2 + 9$
29. $x^2 y = 4$
30. $x^2 y^2 = 16$
31. $xy + 4y = 8$
32. $(x - 2)^2 y = 2x$
33. $y^2 = 4x^2 - x^4$
34. $8y^2 = x^4 - 4x^2$
35. $y^2 (x + 1) = 4x$
36. $y(x - 1)(x^2 - 4) = 4$

37. Discuss and graph representative situations for the equation $y = (x-a)(x-b)(x-c)$ in each of the following cases:

    (a) $a < b < c$ (b) $a < b = c$ (c) $a = b < c$ (d) $a = b = c$.

38. Same as Exercise 37, but replace equation with $y^2 = (x-a)(x-b)(x-c)$.

39. Same as Exercise 37, but replace equation with
$$y = \frac{1}{(x-a)(x-b)(x-c)}.$$

40. Same as Exercise 37, but replace equation with
$$y^2 = \frac{1}{(x-a)(x-b)(x-c)}.$$

8–3. The exponential curves

The equation of an exponential curve is of the general form
$$y = a^x \tag{8-3}$$
where $a > 0$ but is not equal to 1. We shall illustrate exponential curves by means of two examples. In the first, $a > 1$; in the second, $a < 1$.

EXAMPLE 8–9. Sketch the graph of the curve whose equation is $y = 10^x$.

TRANSCENDENTAL AND OTHER CURVES   307

Solution. By determining the outstanding characteristics of the curve and then plotting a few points, we are able to make a fairly good sketch of the curve whose equation is $y = 10^x$. As x increases through positive values, 10^x is positive and increases "much more rapidly" than x. As x is assigned negative values, we see that the values of 10^x remain positive and become smaller as x decreases. Hence, $y = 10^x$ is never negative. Since the distance from the x-axis to the curve approaches zero as x decreases indefinitely, the x-axis is a horizontal asymptote of the curve.

Constructing a table of values, we have the following results:

x	-2	-1	0	1	2
$y = 10^x$	0.01	0.1	1	10	100

Although it is not practical to plot the first and last values, they give information regarding the steepness of the curve, which is shown in Fig. 8–12.

EXAMPLE 8–10. Sketch the curve whose equation is $y = (\frac{1}{2})^x$.

Solution. We find that, as x increases through positive values, the values of $(\frac{1}{2})^x$ remain positive but decrease so as to approach zero. Hence, the x-axis is an asymptote for the curve. As x is assigned negative values, $(\frac{1}{2})^x$ increases much more rapidly than does $|x|$.

A table of values follows:

x	-2	-1	0	1	2
$y = (\frac{1}{2})^x$	4	2	1	0.5	0.25

The curve is shown in Fig. 8–13.

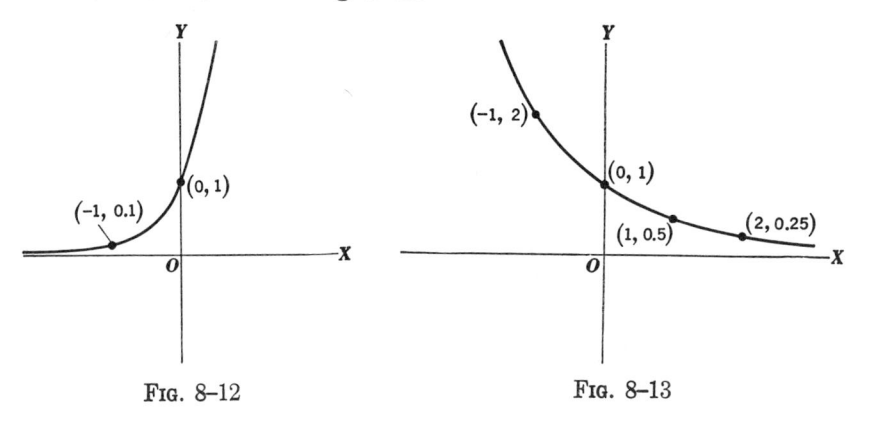

FIG. 8–12

FIG. 8–13

Another important exponential equation is
$$y = e^x \tag{8-4}$$
where e is an irrational* number, the value of which is 2.718281828 ... This number e is important in the calculus as a base for logarithms. The graph of $y = e^x$ is similar to the graph of $y = 10^x$. Another important exponential equation is
$$y = e^{-x^2} \tag{8-5}$$
which represents the *probability* curve (see Exercise 10).

EXERCISES

Draw the graph of each of the curves whose equations are as follows:

1. $y = 2^x$
2. $y = e^{-x}$
3. $y = e^x$
4. $y = 2^{-x}$
5. $y = e^x + e^{-x}$
6. $y = 3^{-x^2}$
7. $y = 10^{-x}$
8. $y = 2^{x^2}$
9. $y = e^x - e^{-x}$
10. $y = e^{-x^2}$
11. $y = -e^{x^2}$
12. $y = (\frac{1}{3})^x$

8–4. The logarithmic curve

The general form of equations for graphs known as *logarithmic* curves is
$$y = \log_a x \tag{8-6}$$
where the base a must be positive and not equal to 1. From the definition of a logarithm we know that the equation $y = \log_a x$ can be rewritten in the form $x = a^y$. But this is the exponential equation with the variables interchanged. The following examples illustrate the conditions.

EXAMPLE 8–11. Sketch the graph of $y = \log_{10} x$.

Solution. In order to sketch this curve, we first write the equation in the form $x = 10^y$. As y increases through positive values, x is positive and increases "much more rapidly" than does y. As y takes on negative

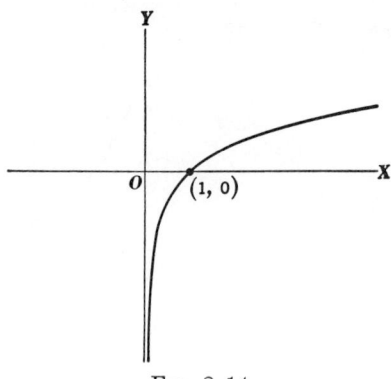

FIG. 8–14

* A real number is *irrational* if it cannot be expressed as the quotient of two integers.

values, the values of $x = 10^y$ remain positive and become smaller as y decreases. Since the distance from the y-axis to the curve approaches zero as y decreases indefinitely, it follows that the y-axis is an asymptote.

A table of values for x and y follows:

y	-2	-1	0	1	2
$x = 10^y$	0.01	0.1	1	10	100

The curve is shown in Fig. 8–14.

EXAMPLE 8–12. Sketch the graph of $y = \log_2 (x^2 - 4)$.

Solution. In order to sketch this curve, we first rewrite the equation in the form
$$x^2 - 4 = 2^y$$
from which we obtain
$$x^2 = 4 + 2^y$$

If $y = 0$, $x = \pm\sqrt{5}$; if $y = 1$, $x = \pm\sqrt{6}$; if $y = -1$, $x = \pm\dfrac{3\sqrt{2}}{2}$. As y increases indefinitely, x increases and decreases indefinitely (since $x = \pm\sqrt{4+2^y}$). As y decreases indefinitely, x^2 approaches 4 and x approaches ± 2. Since 2^y is never negative, x^2 is always greater than 4 and $|x| > 2$. The lines $x = 2$ and $x = -2$ are asymptotes of the curve. Since x appears to an even power in the given equation, the curve is symmetric with respect to the y-axis, as shown in Fig. 8–15.

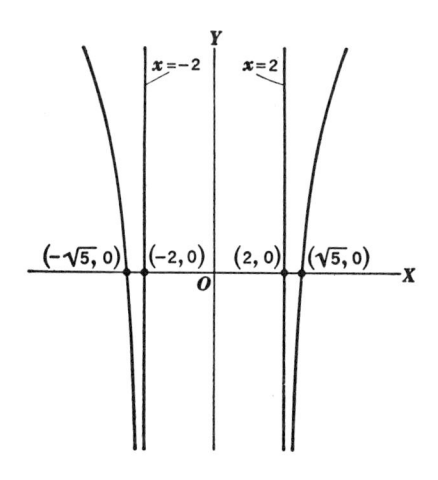

FIG. 8–15

EXERCISES

Sketch each of the curves whose equations are as follows:

1. $y = \log_2 x$
2. $y = \log_e x$
3. $y = \log_{10}(x-4)$
4. $y = \log_2(x^2-4)$
5. $y = \log_3 x^2$
6. $y = \log_2(2-x)$
7. $y = \log_3(x+2)$
8. $y = \log_2 x^2$
9. $y = \log_{10}(4-x)$
10. $y = \log_2(x^2+4)$
11. $y = \log_{10}(4-x^2)$
12. $y = \log_e(x+\sqrt{x^2+1})$

8–5. The trigonometric curves

A *trigonometric* curve is one corresponding to any of the following equations:

$$y = \sin x \quad (8\text{-}7)$$
$$y = \cos x \quad (8\text{-}8)$$
$$y = \tan x \quad (8\text{-}9)$$
$$y = \cot x \quad (8\text{-}10)$$
$$y = \sec x \quad (8\text{-}11)$$
$$y = \csc x \quad (8\text{-}12)$$

EXAMPLE 8–13. Sketch the graph of $y = \sin x$.

Solution. If we let one unit on the x-axis be 1 radian and we recall that $\sin k\pi = 0$, where k is a positive or negative integer or zero, it follows that the curve crosses the x-axis at the points $(k\pi, 0)$. Since $|\sin x|$ is never greater than 1, the curve lies between or touches the lines $y = 1$ and $y = -1$. Also, we know from trigonometry that $\sin x$ repeats itself at intervals of 2π. Hence, the curve is called a *periodic* curve and the period is 2π. The greatest absolute value of a periodic quantity is called the *amplitude* of the quantity. Thus, the amplitude of the sine is 1.

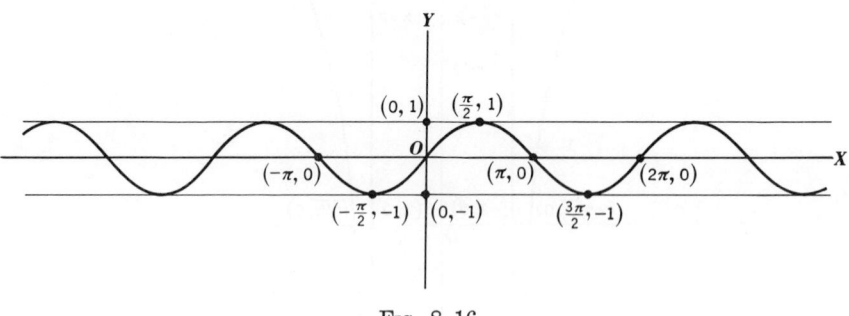

FIG. 8–16

Next, we construct a table of values of x and y over the interval from 0 to 2π for x. The results follow:

x	0	$\frac{\pi}{6}$	$\frac{\pi}{3}$	$\frac{\pi}{2}$	$\frac{2\pi}{3}$	$\frac{5\pi}{6}$	π	$\frac{7\pi}{6}$	$\frac{4\pi}{3}$	$\frac{3\pi}{2}$	$\frac{5\pi}{3}$	$\frac{11\pi}{6}$	2π
$y = \sin x$	0	0.5	0.87	1	0.87	0.5	0	-0.5	-0.87	-1	-0.87	-0.5	0

Now, by plotting the points obtained from the table and joining them by a smooth curve, we obtain a graph of $y = \sin x$, as shown in Fig. 8–16. That portion of the curve contained within an interval of repetition is sometimes called a *cycle* of the curve.

The curves for the other trigonometric functions are also periodic. The functions $y = \cos x$, $y = \sec x$, and $y = \csc x$ each have a period equal to 2π, while the functions $y = \tan x$ and $y = \cot x$ have a period equal to π. The amplitude of $y = \cos x$ is 1, as the maximum absolute value of the cosine is 1. The maximum values of the other four functions here mentioned are undefined, and those functions therefore do not have definite amplitudes.

EXAMPLE 8–14. Sketch the graph of $y = \tan x$.

Solution. Since the period of $\tan x$ is π, we arrange the following table of values over the interval from $x = 0$ to $x = \pi$:

x	0	$\frac{\pi}{6}$	$\frac{\pi}{4}$	$\frac{\pi}{3}$	$\frac{\pi}{2}$	$\frac{2\pi}{3}$	$\frac{3\pi}{4}$	$\frac{5\pi}{6}$	π
$y = \tan x$	0	0.58	1	1.7	undefined	-1.7	-1	-0.58	0

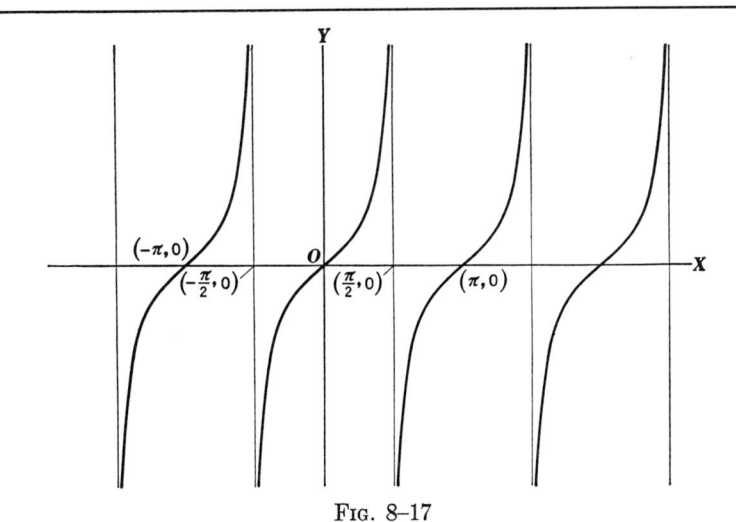

FIG. 8–17

312 ANALYTIC GEOMETRY IN THE PLANE

We know from trigonometry that, as the angle increases from 0 to $\frac{\pi}{2}$, tan x increases indefinitely. As x increases from $\frac{\pi}{2}$ to π, tan x again *increases* through negative values to 0. The curve is shown in Fig. 8–17.

EXAMPLE 8–15. Sketch the graph of $y = \sec x$.

Solution. Since $\sec x \equiv \frac{1}{\cos x}$, the period of sec x is 2π. Constructing a table of values, we obtain the following results:

x	0	$\frac{\pi}{6}$	$\frac{2\pi}{6}$	$\frac{3\pi}{6}$	$\frac{4\pi}{6}$	$\frac{5\pi}{6}$	$\frac{6\pi}{6}$	$\frac{7\pi}{6}$	$\frac{8\pi}{6}$	$\frac{9\pi}{6}$	$\frac{10\pi}{6}$	$\frac{11\pi}{6}$	$\frac{12\pi}{6}$
$y = \sec x$	1	1.2	2	undefined	-2	-1.2	-1	-1.2	-2	undefined	2	1.2	1

We also know that sec $x \leq -1$ or $1 \leq \sec x$. Hence, the curve lies outside of or touches the strip included between the lines $y = 1$ and $y = -1$, as shown in Fig. 8–18.

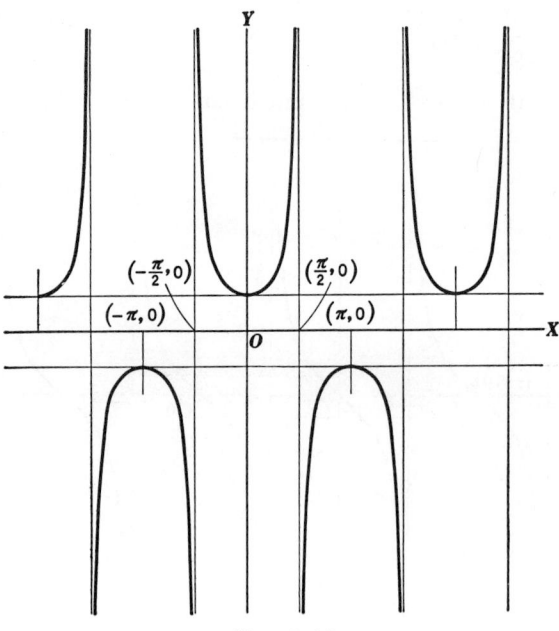

FIG. 8–18

EXERCISES

Sketch each of the curves whose equations are as follows:

1. $y = \cos x$
2. $y = \cot x$
3. $y = \csc x$
4. $y = \sin x$
5. $y = \tan x$
6. $y = \sec x$
7. What are equations for the vertical asymptotes of the following curves:
   (a) $y = \tan x$; (b) $y = \cot x$; (c) $y = \sec x$; (d) $y = \csc x$.

8–6. Curves represented by $y = a \sin (bx + \phi)$, where a, b, and ϕ are constants

If x is replaced in the equation by $x + \dfrac{2k\pi}{b}$, where k is an integer, we obtain $y = a \sin\left(b\left\{x + \dfrac{2k\pi}{b}\right\} + \phi\right) = a \sin(bx + \phi + 2k\pi)$, or $y = a \sin(bx + \phi)$. Since y repeats itself whenever x is replaced by $x + \dfrac{2k\pi}{b}$, it is periodic. The smallest interval of repetition is defined to be *the period*. In this example, we obtain the smallest interval of repetition when $k = 1$, in which case the period is $\dfrac{2\pi}{b}$. The amplitude is $|a|$, and the angle ϕ is called the phase angle. When ϕ is positive, it is called a *lead* angle; when it is negative, it is called a *lag* angle.

Since $bx + \phi = 0$ when $x = -\dfrac{\phi}{b}$, it is convenient to translate the axes to the new origin $O' = \left(-\dfrac{\phi}{b}, 0\right)$ before sketching. Then the formulas for translation become $x = x' - \dfrac{\phi}{b}$ and $y = y'$ and the equation reduces to $y' = a \sin bx'$. This is a curve similar to the sine curve with amplitude a and period $\dfrac{2\pi}{b}$. Hence, to sketch the curve whose equation is $y = a \sin (bx + \phi)$, we first translate the axes to $\left(-\dfrac{\phi}{b}, 0\right)$ as a new origin. Then from this point we mark off an *interval* equal in length to the period; and in this interval we draw a cycle of a curve similar to the sine curve with amplitude a.

EXAMPLE 8–16. Sketch the curve whose equation is $y = 2 \sin (3x - \pi)$.

Solution. The period is $\dfrac{2\pi}{3}$, the amplitude is 2, and the phase angle is $-\pi$. Since $3x - \pi = 0$ when $x = \dfrac{\pi}{3}$, we translate the axes to the new origin $\left(\dfrac{\pi}{3}, 0\right)$. Then the equation reduces to $y' = 2 \sin 3x'$. Now mark off along

314   ANALYTIC GEOMETRY IN THE PLANE

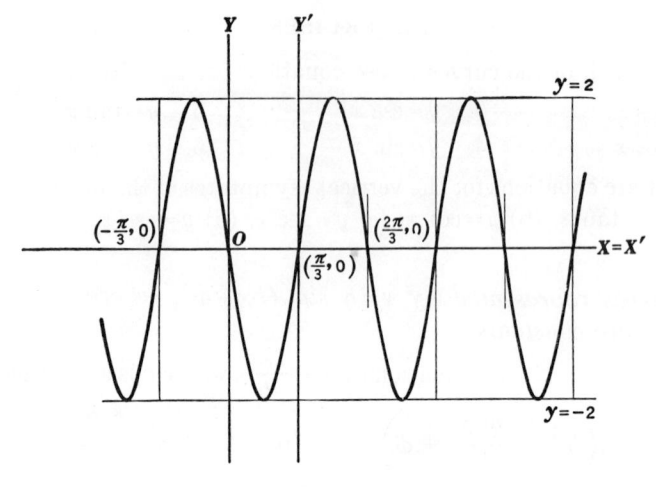

FIG. 8–19

the x-axis and to the right of the new origin an interval with a length equal to $\frac{2\pi}{3}$, as in Fig. 8–19; and sketch the curve. In this example we find that, when $x = 0$, then $y = 2 \sin(-\pi) = 0$. Hence, the curve passes through the origin. However, this is not generally true.

The curve whose equation is of the form $y = a \cos(bx + \phi)$ is sketched by proceeding as just described for a sine curve.

EXAMPLE 8–17. Sketch the curve whose equation is $y = 3 \cos(2x + 2)$.

Solution. The period of the curve is π, the amplitude is 3, and the phase angle is 2. Since $2x + 2 = 0$ when $x = -1$, we translate the axes to

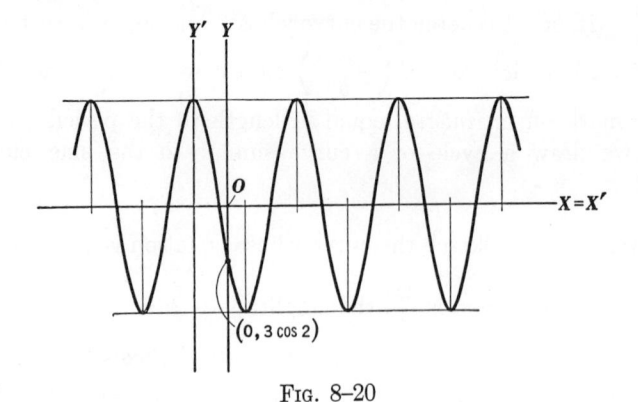

FIG. 8–20

the point $(-1, 0)$ as a new origin. The formulas for translation then become $x = x' - 1$ and $y = y'$. Now the equation reduces to $y' = 3 \cos 2x'$. From the new origin mark off an interval of length π along the x-axis, as shown in Fig. 8–20, and sketch a curve similar to the cosine curve with an amplitude equal to 3. When $x = 0$, it is found that $y = 3 \cos 2$. Hence, this curve cuts the y-axis at the point $(0, 3 \cos 2)$.

8–7. Curves represented by other equations involving trigonometric functions

To illustrate the procedure, we will consider equations of the following two types: $y = x f(x)$ and $y = f_1(x) + f_2(x)$, where $f(x)$ is a trigonometric function of x and either $f_1(x)$ or $f_2(x)$ is also such a function.

EXAMPLE 8–18. Sketch the curve whose equation is $y = x \sin x$.

Solution. Since $|\sin x|$ is never greater than 1, $x \sin x$ is never greater than $|x|$ nor less than $-|x|$. Hence, the graph of $y = x \sin x$ lies between or touches the lines $y = x$ and $y = -x$. Furthermore, when $x = k\pi$, for k equal to a positive or negative integer or 0, the curve crosses the x-axis. When $x = \dfrac{k\pi}{2}$, for k equal to an *odd* integer, the curve touches either the line $y = x$ or the line $y = -x$; if $k = 0$, the curve intersects both lines. The curve is shown in Fig. 8–21.

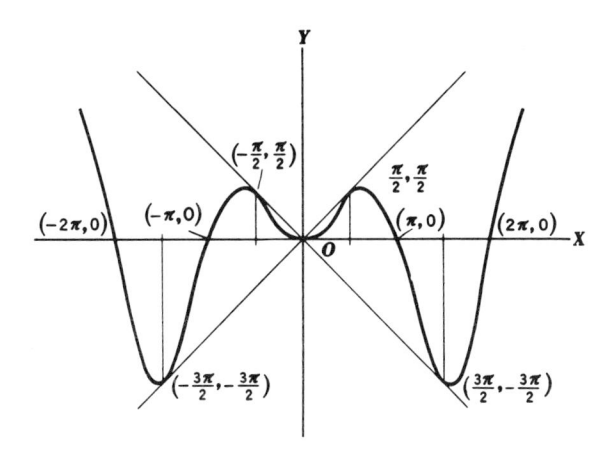

FIG. 8–21

316 ANALYTIC GEOMETRY IN THE PLANE

EXAMPLE 8–19. Sketch the graph of $y = x + \sin x$.

Solution. We let $f_1(x) = x$ and $f_2(x) = \sin x$, and we draw the graphs of $y = x$ and $y = \sin x$. Then, for every value of x, we construct the value $f_1(x) + f_2(x)$; that is, we locate the point whose ordinate is the sum of the ordinates of the points on the two curves having the abscissa x. The resulting points constitute the graph of $y = f_1(x) + f_2(x) = x + \sin x$, which is shown in Fig. 8–22.

From the given equation it is clear that the curve repeatedly crosses the line $y = x$, since $y = x + \sin x$ becomes $y = x$ whenever $x = k\pi$, for k equal to an integer.

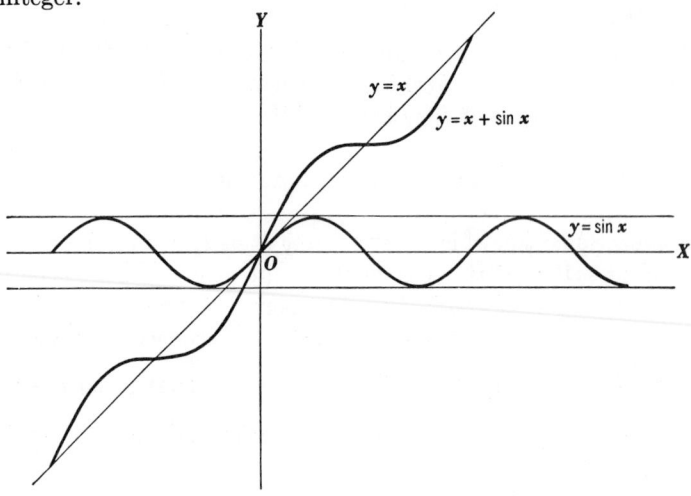

FIG. 8–22

EXERCISES

Sketch each of the following curves and, where possible, give the period, amplitude, and phase angle:

1. $y = \sin 3x$
2. $y = 2 \cos 2x$
3. $y = \tan \dfrac{x}{2}$
4. $y = \tfrac{1}{2} \cot 2x$
5. $y = \sec \dfrac{x}{3}$
6. $y = 2 \csc 3x$
7. $y = 3 \sin (2x + \pi)$
8. $y = 2 \cos (3x - 2)$
9. $y = -\tan (x + \pi)$
10. $y = 2 \cot (2x - 3)$
11. $y = -\cos \left(x - \dfrac{\pi}{2}\right)$
12. $y = \tfrac{1}{2} \sin (3x - \pi)$
13. $y = -2 \sin 3x$
14. $y = -\tan x$
15. $y = \cos \dfrac{3x}{2}$
16. $y = xe^x$
17. $y = 2x \sin x$
18. $y = xe^{-x}$
19. $y = x \cos x$
20. $y = x + \sin x$
21. $y = x - \sin x$
22. $y = x + \cos x$
23. $y = x - \cos x$
24. $y = e^x \sin x$
25. $y = e^x \cos x$
26. $y = x^2 e^x$
27. $y = x^2 e^{-x}$

8–8. The graphs of the inverse trigonometric functions

We symbolize the expression "*y is any angle whose sine is x*" by the notation $y = \text{arc sin } x$, or $y = \sin^{-1} x$. We call $\sin^{-1} x$ or arc sin x the *inverse sine of x*. The equation $y = \sin^{-1} x$ defines a relation but not a function since, corresponding to a given value of x, there is more than one value of y. Indeed, there is an infinite number of possible values of y. In order to sketch the curve of $y = \sin^{-1} x$, we rewrite the equation in the form $x = \sin y$. But this is the sine curve with the variables interchanged.

From the equation $x = \sin y$, it is obvious that x is a periodic function of y with period 2π and amplitude 1. Since $|x|$ is never greater than 1, the curve lies between the lines $x = 1$ and $x = -1$. Now, starting at the origin, we mark off intervals 2π units in length along the y-axis, where each unit is 1 radian. Then in each interval we draw a cycle of the sine curve passing through the points $(0, 0)$, $\left(1, \frac{\pi}{2}\right)$, $(0, \pi)$, $\left(-1, \frac{3\pi}{2}\right)$, and $(0, 2\pi)$. The curve is shown in Fig. 8–23.

In application, it is frequently necessary that to each value of x there correspond at most one value of y. In such a case, that portion of the curve is chosen for which y takes on the values from $-\frac{\pi}{2}$ to $\frac{\pi}{2}$ while x takes on the values from -1 to 1, as indicated in Fig. 8–24. This portion of the curve is called the *principal part* of the inverse-sine curve. Correspondingly, the single value of y such that (x, y) lies on the principal part of the curve $y = \text{arc sin } x$, for which $-1 \leq x \leq 1$, is called the *principal arc-sine function*. When the principal value is intended, the function is often written with a capital initial letter. So $y = \text{Arc sin } x$ or $y = \text{Sin}^{-1} x$ implies that $-\frac{\pi}{2} \leq y \leq \frac{\pi}{2}$.

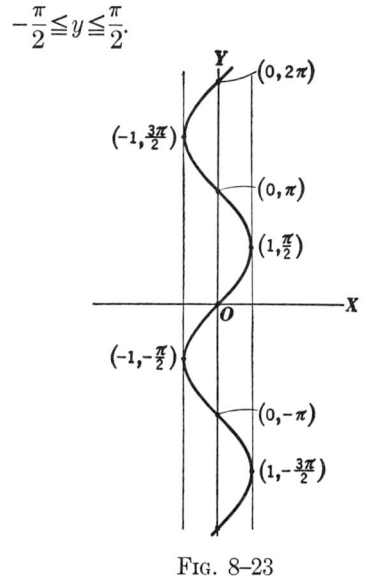

FIG. 8–23

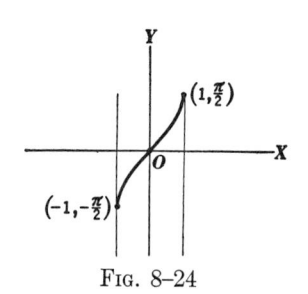

FIG. 8–24

318   ANALYTIC GEOMETRY IN THE PLANE

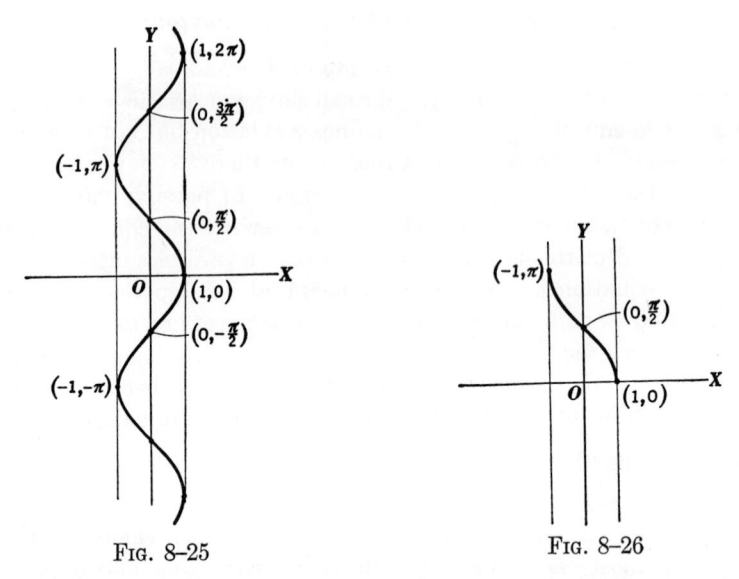

FIG. 8-25

FIG. 8-26

In the same way, $y = \text{arc cos } x$ or $y = \cos^{-1} x$ means "*y is any angle whose cosine is x.*" To draw the graph of $y = \cos^{-1} x$, we rewrite the equation in the form $x = \cos y$ and we draw the cosine curve along the y-axis, as shown in Fig. 8-25.

In order to select a single-valued part of the function $x = \cos y$, we choose that portion of the curve which lies between $y = 0$ and $y = \pi$, as indicated in Fig. 8-26. The function of x so defined is called the *principal arc-cosine function* and will be symbolized with a capital initial letter. So $y = \text{Arc cos } x$ or $y = \text{Cos}^{-1} x$ implies that $0 \leq y \leq \pi$, in order that for each value of x there will be a unique value of y.

The graphs of the other inverse functions are illustrated in the examples that follow.

EXAMPLE 8-20.   Sketch the graph of $y = \tan^{-1} x$.

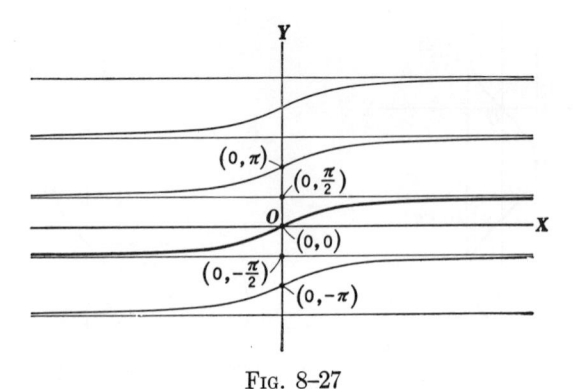

FIG. 8-27

Solution. We rewrite the given equation in the form $x = \tan y$, and then we sketch the tangent curve along the y-axis, as shown in Fig. 8–27.

The principal part of the curve for the inverse tangent is that portion of the curve from $-\frac{\pi}{2}$ to $\frac{\pi}{2}$ on the y-axis. This part is shown in Fig. 8–27 by the heavy line. It is important to notice that the end points $-\frac{\pi}{2}$ and $\frac{\pi}{2}$ are not included. We write the equation of this part as $y = \text{Arc} \tan x$ or $y = \text{Tan}^{-1} x$ for $-\frac{\pi}{2} < y < \frac{\pi}{2}$.

EXAMPLE 8–21. Sketch the graph of $y = \cot^{-1} x$.

Solution. We rewrite this equation in the form $x = \cot y$, and we sketch the curve along the y-axis, as shown in Fig. 8–28. The principal part lies between $y = 0$ and $y = \pi$ and is represented by the heavy line. Its equation is written as $y = \text{Cot}^{-1} x$ or $y = \text{Arc} \cot x$ over the interval $0 < y < \pi$. Again we notice that the end points are not included.

EXAMPLE 8–22. Sketch the graph of $y = \sec^{-1} x$.

Solution. We rewrite this equation in the form $x = \sec y$, and then we sketch the secant curve along the y-axis, as shown in Fig. 8–29. This time the principal part is that portion of the curve in the intervals from 0 to $\pi/2$ and from $-\pi$ to $-\pi/2$ where in the first interval $\pi/2$ is not included and in the second interval $-\pi/2$ is not included. Its equation is

$$y = \text{Arc} \sec x \text{ or } y = \text{Sec}^{-1} x \begin{cases} \text{for } x \geq 1, 0 \leq y < \pi/2 \\ \text{for } x \leq -1, -\pi \leq y < -\pi/2 \end{cases}$$

EXAMPLE 8–23. Sketch the graph of $y = \csc^{-1} x$.

Solution. Rewrite this equation in the form $x = \csc y$, and then sketch the graph, as shown in Fig. 8–30. The principal part is that

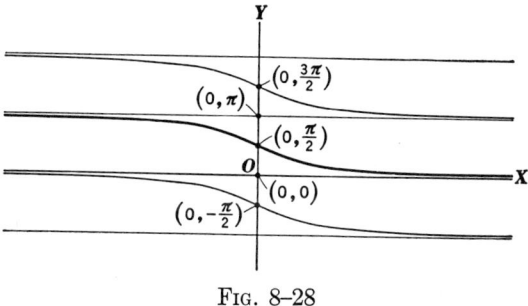

FIG. 8–28

320   ANALYTIC GEOMETRY IN THE PLANE

portion of the curve for which y is contained in the intervals from $-\pi$ to $-\pi/2$ and from 0 to $\pi/2$, where in the first interval $-\pi$ is not included and in the second interval 0 is not included. Its equation is

$$y = \text{Arc csc } x \text{ or } y = \text{Csc}^{-1} x \begin{cases} \text{for } x \geq 1, 0 < y \leq \pi/2 \\ \text{for } x \leq -1, -\pi < y \leq -\pi/2 \end{cases}$$

EXAMPLE 8–24. Sketch the graph of $y = 3 \sin^{-1} 2x$.

Solution. First we rewrite this equation in the form $2x = \sin \frac{y}{3}$ or $x = \frac{1}{2} \sin \frac{y}{3}$. We know that x is periodic, with period 6π; and that the amplitude is $\frac{1}{2}$. To sketch the curve, we draw the lines $x = \frac{1}{2}$ and $x = -\frac{1}{2}$. Then we mark off on the y-axis an interval equal to 6π, starting from the origin; and in this compartment we draw a cycle of a curve similar to the sine curve. The curve is shown in Fig. 8–31.

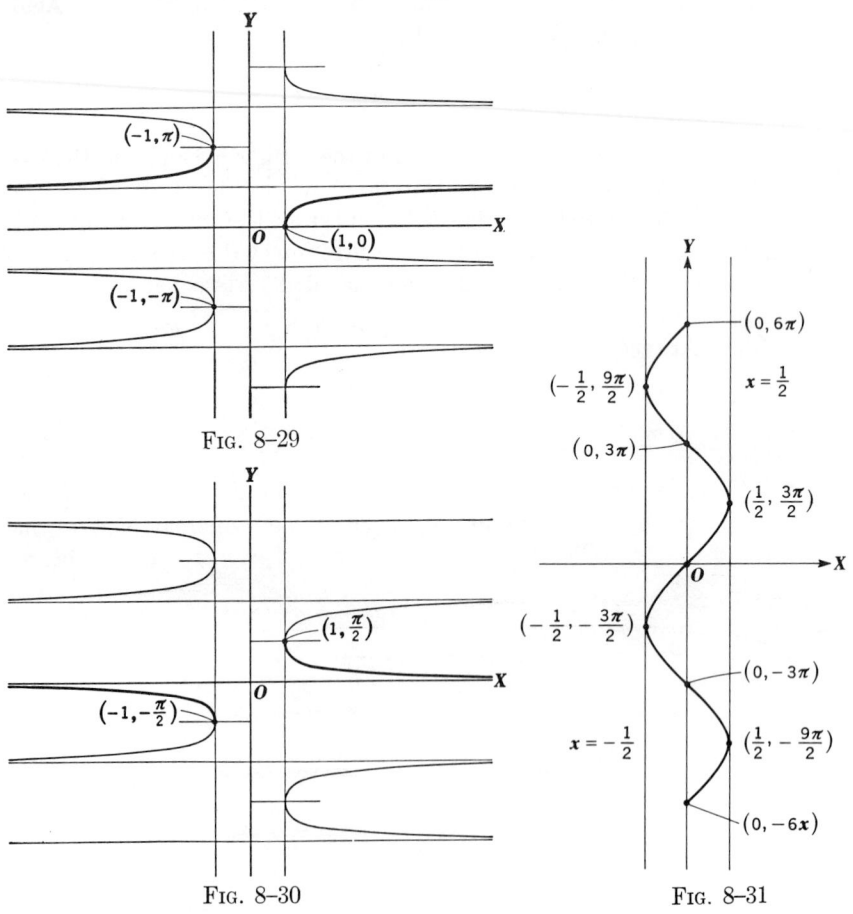

FIG. 8–29

FIG. 8–30

FIG. 8–31

TRANSCENDENTAL AND OTHER CURVES   321

EXERCISES

1. Evaluate each of the following:
   (a) $\text{Sin}^{-1}(-1)$
   (b) $\text{Tan}^{-1}(-1)$
   (c) $\text{Cos}^{-1}(-\frac{1}{2})$
   (d) $\text{Arc sec }\sqrt{2}$
   (e) $\text{Arc cos}\left(\frac{-\sqrt{3}}{2}\right)$
   (f) $\text{Arc tan }\frac{1}{\sqrt{3}}$
   (g) $\text{Cot}^{-1}\left(-\frac{1}{\sqrt{3}}\right)$
   (h) $\text{Sin}^{-1}\frac{\sqrt{2}}{2}$
   (i) $\text{Sec}^{-1}\frac{2\sqrt{3}}{3}$

2. Sketch the graph of each of the following equations:
   (a) $y = \text{Arc cos } x$
   (b) $y = \text{Sin}^{-1} x$
   (c) $y = \csc^{-1} x$
   (d) $y = \text{arc sec } x$
   (e) $y = \text{Arc tan } x$
   (f) $y = \text{Cot}^{-1} x$

3. Sketch each of the following curves:
   (a) $y = 2 \sin^{-1} 3x$
   (b) $y = \frac{1}{2} \cos^{-1} 2x$
   (c) $y = 3 \tan^{-1} \frac{x}{2}$
   (d) $y = \text{arc cot } 3x$
   (e) $y = 2 \text{ arc sec } \frac{x}{3}$
   (f) $y = \frac{1}{3} \text{ arc sin } \frac{2x}{3}$
   (g) $y = 3 \cos^{-1} 2x$
   (h) $y = \frac{1}{3} \sin^{-1} \frac{x}{2}$
   (i) $y = \frac{1}{2} \tan^{-1} x$

4. Sketch each of the following curves:
   (a) $x = 3 \cos^{-1} 2y$
   (b) $x = \text{arc tan } 3y$
   (c) $y = 3 \text{ arc sin } (3x - 1)$
   (d) $y = 2 \cos^{-1} (2x - 3)$
   (e) $x = 2 \sin^{-1} 3y$
   (f) $y = 2 \sin^{-1} (x + 3)$

8–9. Parametric equations

We have already discussed parametric equations of the line, the circle, and the conics. There are many other curves which can be conveniently represented by means of a pair of parametric equations and we shall discuss a few of these here.

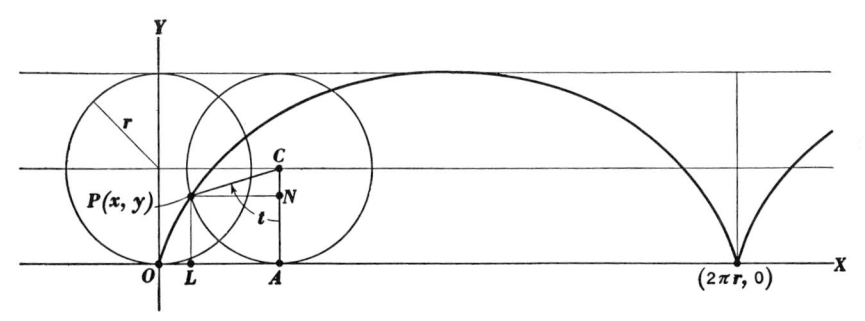

FIG. 8–32

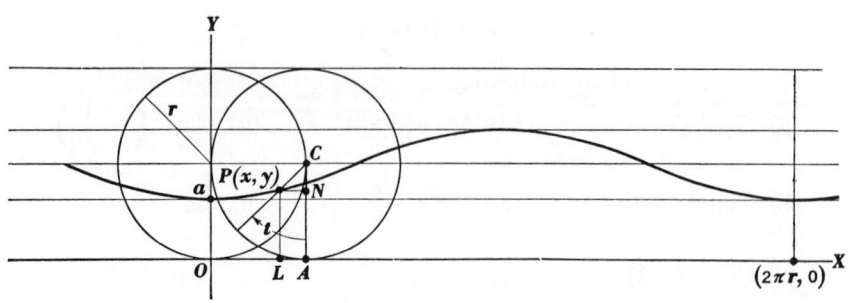

FIG. 8-33

(a) *The cycloids.* Fix a point P on the circumference of a circle and let the circle roll along a straight line without slipping. Suppose that P is at the origin when the center C of the circle is on the y-axis, as shown in Fig. 8-32; and let the x-axis be the line along which the circle rolls. The locus of the point P as the circle rolls along the line is called a *cycloid*.

In Fig. 8-32, $x = \overline{OL} = \overline{OA} - \overline{LA}$. But $\overline{OA}$ = arc $\overline{AP} = rt$, where r is the radius of the circle and t is the radian measure of the angle ACP, measured positively clockwise from AC. Also, $\overline{LA} = \overline{PN} = r \sin t$. Hence, we have

$$x = rt - r \sin t \tag{8-13}$$

Also, $y = \overline{LP} = \overline{AC} - \overline{NC}$. But $\overline{AC} = r$ and $\overline{NC} = r \cos t$. So

$$y = r - r \cos t \tag{8-14}$$

Equations (8-13) and (8-14) are parametric equations of the cycloid.

EXERCISES

1. Fix a point P on the radius of a circle such that $|CP| = a$, where $a < r$. Derive equations of the locus of P in terms of the parameter $t = \angle NCP$ as the circle rolls without slipping along a straight line. This curve, which is shown in Fig. 8-33, is called a *curtate cycloid* or a *trochoid*.

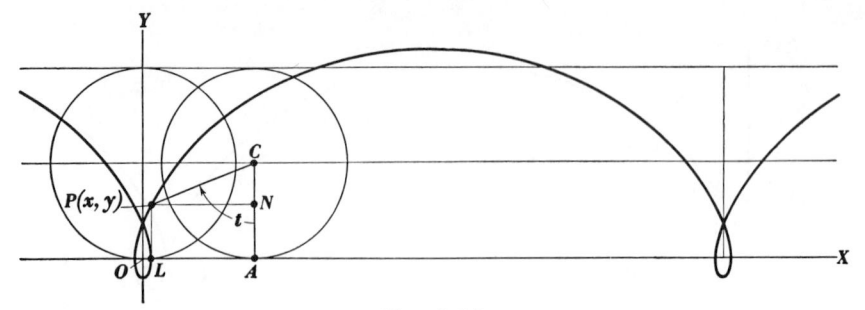

FIG. 8-34

2. Fix a point P on the prolongation of the radius of a circle such that $|CP| = a$, where $a > r$. Derive equations of the locus of P in terms of the parameter $t = \angle NCP$ as the circle rolls without slipping along a straight line. This curve, which is shown in Fig. 8-34, is called a *prolate cycloid* or a *trochoid*.

(b) *The epicycloids.* If a point P is fixed on the circumference of a circle and this circle is rolled without slipping along the circumference of another circle, the locus of P is called an epicycloid. Such a curve is shown in Fig. 8-35.

Let us derive equations of the epicycloid in terms of the angle θ in Fig. 8-35. From the figure, $x = OM = OL + LM$. But $OL = |OC| \cos \theta$ and $|OC| = R + r$, where R is the radius of the fixed circle and r is the radius of the rolling circle. Also $\overline{LM} = \overline{NP} = r \cos \angle CPN$. But $\angle CPN = 180° - (\theta + \phi)$. Hence, $\overline{LM} = r \cos (180° - \{\theta + \phi\}) = -r \cos (\theta + \phi)$. So $x = (R + r) \cos \theta - r \cos (\theta + \phi)$. But we know that arc $\overline{AD} = $ arc $\overline{AP}$. Since arc $\overline{AD} = R\theta$ and arc $\overline{AP} = r\phi$, we have $r\phi = R\theta$ or $\phi = \dfrac{R\theta}{r}$. Now we have

$$x = (R+r) \cos \theta - r \cos \left(\theta + \frac{R\theta}{r}\right)$$

or
$$x = (R+r) \cos \theta - r \cos \left(\frac{R+r}{r}\right)\theta \qquad (8\text{-}15)$$

In the same way, $y = \overline{MP} = \overline{LC} - \overline{NC}$. But $\overline{LC} = (R+r) \sin \theta$ and $\overline{NC} = r \sin (180° - \{\theta + \phi\}) = r \sin (\theta + \phi)$. So $y = (R+r) \sin \theta - r \sin (\theta + \phi)$, or

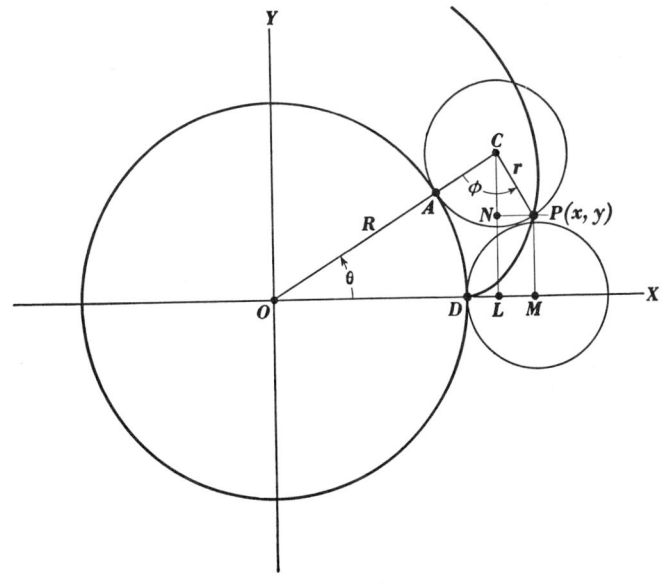

Fig. 8-35

$$y = (R+r) \sin \theta - r \sin \left(\frac{R+r}{r}\right)\theta \qquad (8\text{-}16)$$

Thus, equations (8-15) and (8-16) are parametric equations for the epicycloid.

As an exercise, the student should find an equation of the epicycloid when $r = R$ and also when $r = \frac{R}{2}$.

(c) *The hypocycloids.* If a circle rolls without slipping along the inside of the circumference of a fixed circle, the path of a point P on the circumference of the rolling circle is called a hypocycloid. Such a curve is shown in Fig. 8-36.

Let us obtain parametric equations of the hypocycloid in terms of the angle θ, Fig. 8-36. We have

$$x = \overline{OM} = \overline{OL} + \overline{LM} = (R-r)\cos\theta + r\cos\angle CPN$$

Since $\angle CPN + \theta = \phi$, we have

$$x = (R-r)\cos\theta + r\cos(\phi - \theta)$$

Also, since $r\phi = R\theta$, we have

$$x = (R-r)\cos\theta + r\cos\left(\frac{R-r}{r}\right)\theta \qquad (8\text{-}17)$$

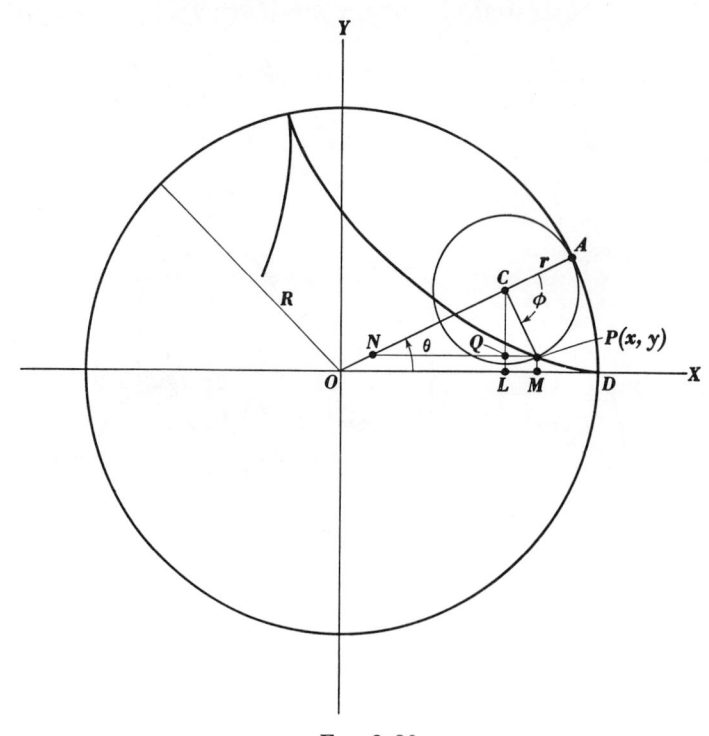

Fig. 8-36

In the same way, $y = \overline{MP} = \overline{LQ} = \overline{LC} - \overline{QC}$. Here, $\overline{LC} = (R-r) \sin \theta$ and $\overline{QC} = r \sin \angle CPN = r \sin (\phi - \theta)$. So we have

$$y = (R-r) \sin \theta - r \sin (\phi - \theta)$$

or
$$y = (R-r) \sin \theta - r \sin \left(\frac{R-r}{r}\right)\theta \qquad (8\text{-}18)$$

Equations (8-17) and (8-18) are parametric equations of the hypocycloid.

EXERCISES

1. Obtain the equations of the hypocycloid by replacing r by $-r$ in equations (8-15) and (8-16).

2. If $r = \dfrac{R}{2}$, prove that the hypocycloid is the diameter of the circle.

3. If $r = \dfrac{R}{4}$, show that the hypocycloid has for its equations $x = R \cos^3 \theta$ and $y = R \sin^3 \theta$.

4. Eliminate the parameter in Exercise 3 and obtain the equation $x^{\frac{2}{3}} + y^{\frac{2}{3}} = R^{\frac{2}{3}}$. This is known as the *hypocycloid of four cusps* or the *astroid*.

(d) *The involutes of the circle.* Let a string be wound about the circumference of a circle. Fasten one end to the circumference and then unwind the string, keeping it taut. The free end traces a curve called an *involute of the circle.* See Fig. 8-37.

From the figure, $x = \overline{OB} = \overline{OE} + \overline{EB} = R \cos \theta + |AP| \sin \theta$. Since $|AP| = \text{arc } AL = R\theta$, we have

$$x = R \cos \theta + R\theta \sin \theta \qquad (8\text{-}19)$$

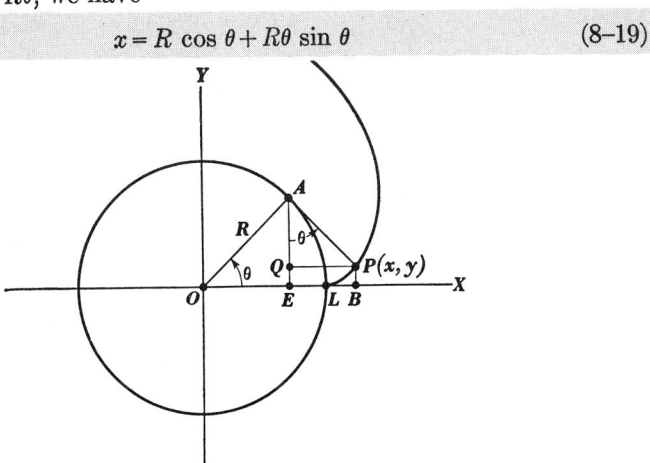

Fig. 8-37

In the same way, we have $y = \overline{BP} = \overline{EQ} = \overline{EA} - \overline{QA} = R \sin \theta - |AP| \cos \theta$, or

$$y = R \sin \theta - R\theta \cos \theta \qquad (8\text{--}20)$$

So equations (8–19) and (8–20) are parametric equations of the involute of the circle.

(e) *The strophoids.* Let a be the distance of a fixed point A from a fixed line YY', as indicated in Fig. 8–38. From A draw any ray meeting YY' at C. Let O be the projection of A on YY'. Mark off two points P and P' on the ray containing AC such that $|P'C| = |PC| = |OC|$. The combined path of the points P and P' is called a *strophoid*.

To obtain parametric equations of the strophoid, let the fixed line be the y-axis, let the coordinates of A be $-a$ and 0, and let the angle OAC be the parameter θ, where $-\frac{\pi}{2} < \theta < \frac{\pi}{2}$. We have $|P'C| = |CP| = |OC|$ and $OC = a \tan \theta$, whence $x^2 = |OM|^2 = |CP|^2 \cos^2 \theta = a^2 \tan^2 \theta \cos^2 \theta = a^2 \sin^2 \theta$.

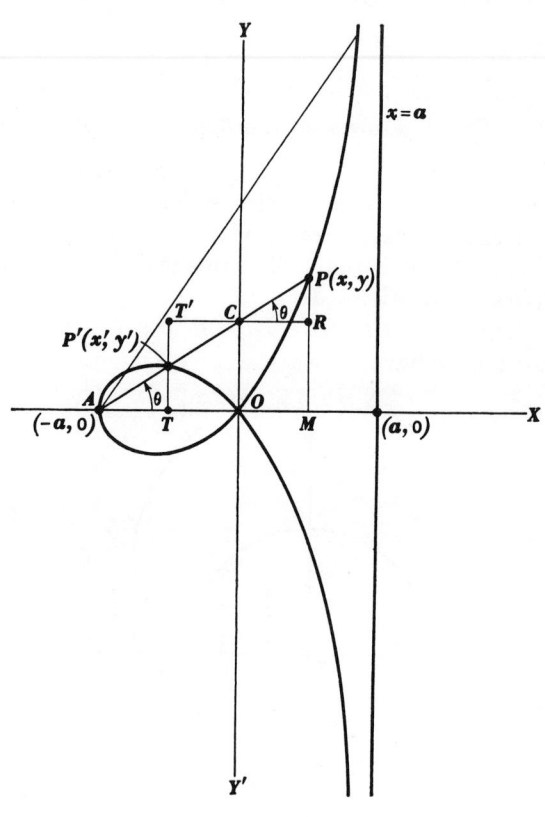

Fig. 8–38

Therefore, $x = \pm a \sin \theta$. Similarly, $x'^2 = a^2 \sin^2 \theta$. But, since x and x' have opposite signs, one of them is $+a \sin \theta$ and the other is $-a \sin \theta$. Let the points P and P' be so designated that, for all values of θ,

$$x = +a \sin \theta \text{ and } x' = -a \sin \theta$$

Also, $y = \overline{MP} = \overline{MR} + \overline{RP} = a \tan \theta + a \tan \theta \sin \theta$. This follows, since $\overline{MR} = \overline{OC} = a \tan \theta$ and $\overline{RP} = |CP| \sin \theta = a \tan \theta \sin \theta$. Likewise, we have $y' = \overline{TP'} = \overline{OC} - \overline{P'T'} = a \tan \theta - a \tan \theta \sin (180° - \theta)$, from which $y' = a \tan \theta (1 - \sin \theta)$. So the point P moves along a curve whose parametric equations are

$$x = a \sin \theta \qquad (8\text{-}21)$$

$$y = a \tan \theta (1 + \sin \theta) \qquad (8\text{-}22)$$

and the point P' moves along a curve whose parametric equations are

$$x' = -a \sin \theta \qquad (8\text{-}23)$$

$$y' = a \tan \theta (1 - \sin \theta) \qquad (8\text{-}24)$$

It should be noted that, if θ is negative, equations (8-21) and (8-22) represent that part of the curve which lies in the third quadrant, while equations (8-23) and (8-24) represent the part of the curve in the fourth quadrant.

By eliminating the parameter from equations (8-21) and (8-22) and from equations (8-23) and (8-24), we obtain a cartesian equation of the curve. The student should show that this becomes

$$y^2 = \frac{x^2(a+x)}{a-x} \qquad (8\text{-}25)$$

Let the coordinates x and y satisfy equation (8-25). It is obvious that the values of x must be such that $-a \leq x \leq a$. Assume first that $x \neq -a$. Also, we will introduce an angle ϕ the value of which is such that $-\frac{\pi}{2} < \phi < \frac{\pi}{2}$ and which is defined by the relation $\phi = \text{Arc} \sin \frac{x}{a}$ or $x = a \sin \phi$. Substituting this value of x in equation (8-25), we obtain

$$y^2 = a^2 \sin^2 \phi \, \frac{a + a \sin \phi}{a - a \sin \phi}$$

Multiplying numerator and denominator by $(1 + \sin \phi)$ and canceling a, we have

$$y^2 = a^2 \sin^2 \phi \, \frac{(1 + \sin \phi)^2}{1 - \sin^2 \phi} = c^2 \tan^2 \phi \, (1 + \sin \phi)^2$$

whence

$$y = \pm a \tan \phi (1 + \sin \phi) \qquad (8\text{-}26)$$

If the positive sign in equation (8-26) holds, then we let $\theta = \phi$ and we assume that equations (8-21) and (8-22) apply. If the negative sign in equation (8–26) is used, we let $\theta = -\phi$, in which case equations (8–23) and (8–24) apply.

It should be noticed that, in order to include the value $-a$ for x, we may agree that equations (8–21) and (8–22) hold when $\theta = -\dfrac{\pi}{2}$ and become $x = -a$ and $y = 0$. If equations (8–21) to (8–24) are considered for angles θ outside the limits $-\dfrac{\pi}{2}$ and $\dfrac{\pi}{2}$, then both pairs of equations represent the same curve. For example, if we replace θ by $(\pi + \theta)$ in equations (8–23) and (8–24), we get $x' = -a \sin(\pi + \theta)$ and $y' = a[\tan(\pi + \theta)][1 - \sin(\pi + \theta)]$ or $x' = a \sin \theta$ and $y' = a \tan \theta (1 + \sin \theta)$, which are the same as equations (8–21) and (8–22). Hence, equation (8–25) is an equation of the strophoid. From this equation, we see that $x = a$ is a vertical asymptote of the curve.

(f) *The witch.* Draw a circle of radius r tangent to the x-axis and the line $y = 2r$, as shown in Fig. 8–39. Let a ray from O meet the circle at R and the line at Q. From Q we drop a perpendicular to the x-axis at D, and we let $P(x, y)$ be the projection of R on DQ. The locus of P is called the *witch of Agnesi* or just the *witch*.

To find the equations of a witch in terms of the parameter $\theta = \angle AOQ$, we have

$$x = \overline{OD} = \overline{AQ} = 2r \tan \theta \tag{8-27}$$

and, since $|OR| = 2r \cos \theta$,

$$y = \overline{DP} = \overline{LR} = |OR| \cos \theta = 2r \cos^2 \theta \tag{8-28}$$

Eliminating the parameter, we obtain

$$y = \frac{2r(4r^2)}{x^2 + 4r^2} = \frac{8r^3}{x^2 + 4r^2} \tag{8-29}$$

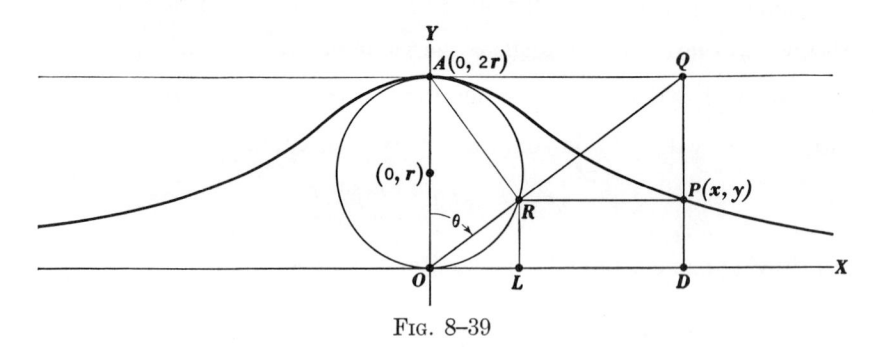

FIG. 8–39

Since the substitution of $-x$ for x does not alter the equation, the curve is symmetric with respect to the y-axis. Also, as x increases indefinitely, y approaches 0. It follows, therefore, that the x-axis is an asymptote for the curve. The curve was named after Maria Agnesi (1718–1799), an Italian mathematician, although it was known earlier.

EXERCISES

In Section 8-9 we discussed some particular curves that are best represented by means of a set of parametric equations. The following exercises represent other curves represented by means of a set of parametric equations. In each exercise draw the curve and find the Cartesian equation. Identify the curve when possible.

1. $x = t^2 + 1, y = t^2 - 1$
2. $x = \cos 2\theta, y = 2\cos^2 \theta$
3. $x = a \sec \phi, y = b \tan \phi$
4. $x = 1 + t^2, y = 4t - t^3$
5. $x = 1 - t^2, y = t + t^3$
6. $x = \sin \alpha, y = \cos 3\alpha$
7. $x = 1 - \cos t, y = \cos 2t$
8. $x = a \sin \theta, y = b \cos \theta$
9. $x = t + 2, y = t^2 - 4$
10. $x = \log_{10} n^3, y = \log_{10} 100n$
11. $x = \sec^2 \phi, y = \tan^2 \phi$
12. $x = a \cos^3 t, y = a \sin^3 t$
13. $x = \sin t + \cos t, y = \sin t$
14. $x = \cos t + \sin t, y = \tfrac{1}{2} \sin 2t$

8-10. The ovals of Cassini

Let $A = (a, 0)$ and $B = (-a, 0)$ be two vertices of a triangle. Consider the locus of the third vertex $P = (x, y)$ if the product of the variable sides of the triangle is a constant c^2. See Fig. 8-40.

By hypothesis, we have $|AP| \, |BP| = c^2$. Since $|AP| = \sqrt{(x-a)^2 + y^2}$ and $|BP| = \sqrt{(x+a)^2 + y^2}$, it follows that $\sqrt{(x-a)^2 + y^2} \sqrt{(x+a)^2 + y^2} = c^2$. Squaring both sides, we obtain

$$[(x-a)^2 + y^2][(x+a)^2 + y^2] = c^4$$

This can be rewritten in the form

$$(x^2 + y^2 + a^2)^2 - 4a^2 x^2 = c^4 \tag{8-30}$$

The graphs of this equation for various values of c are called the *ovals of Cassini*, after the astronomer Cassini (1625–1712).

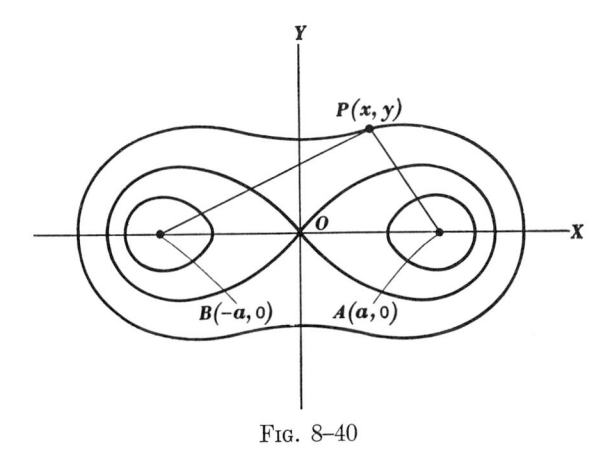

Fig. 8-40

330   ANALYTIC GEOMETRY IN THE PLANE

If $c = a$, equation (8–30) becomes
$$(x^2 + y^2 + a^2)^2 - 4a^2x^2 = a^4$$
This may be converted to
$$(x^2 + y^2)^2 + 2a^2(x^2 + y^2) + a^4 - 4a^2x^2 = a^4$$
or
$$(x^2 + y^2)^2 = 2a^2(x^2 - y^2) \tag{8-31}$$

Transforming equation (8–31) to polar coordinates, we obtain $(\rho^2)^2 = 2a^2(\rho^2 \cos^2\theta - \rho^2 \sin^2\theta)$, or $\rho^2 = 2a^2(\cos^2\theta - \sin^2\theta) = 2a^2 \cos 2\theta$. So we see that, for $a = c$, the ovals of Cassini become the lemniscate.

As an exercise, it may be shown that, when $c < a$, we obtain two non-intersecting ovals; and, when $c > a$, the ovals become a single closed curve.

8–11. The hyperbolic functions

It is not one of the purposes of this book to discuss hyperbolic functions very deeply. However, they are so important in advanced mathematics that they are introduced here.

The hyperbolic functions of x are the hyperbolic sine (sinh x); the hyperbolic cosine (cosh x); the hyperbolic tangent (tanh x); the hyperbolic cotangent (coth x); the hyperbolic secant (sech x); and the hyperbolic cosecant (csch x). Just as there are tables to enable one to evaluate the trigonometric functions, so tables have been compiled that enable us to evaluate the hyperbolic functions. The hyperbolic sine and the hyperbolic cosine are defined as follows:

$$\sinh x = \frac{e^x - e^{-x}}{2} \tag{8-32}$$

$$\cosh x = \frac{e^x + e^{-x}}{2} \tag{8-33}$$

where e is the irrational number 2.718 . . . introduced in Section 8–2. Starting with these two functions, we obtain the other hyperbolic functions, as follows:

$$\tanh x = \frac{\sinh x}{\cosh x} = \frac{e^x - e^{-x}}{e^x + e^{-x}} \tag{8-34}$$

$$\coth x = \frac{1}{\tanh x} = \frac{e^x + e^{-x}}{e^x - e^{-x}} \tag{8-35}$$

$$\operatorname{sech} x = \frac{1}{\cosh x} = \frac{2}{e^x + e^{-x}} \tag{8-36}$$

$$\operatorname{csch} x = \frac{1}{\sinh x} = \frac{2}{e^x - e^{-x}} \tag{8-37}$$

Just as there exist certain identities involving the trigonometric functions, so the hyperbolic functions are related to each other by means of some interesting identities. Some of these follow directly from the definitions and have just been given. Also,

$$\cosh^2 x - \sinh^2 x \equiv 1 \tag{8-38}$$

To prove this, we know that $\cosh x = \dfrac{e^x + e^{-x}}{2}$ and $\sinh x = \dfrac{e^x - e^{-x}}{2}$. Squaring and subtracting, we obtain

$$\cosh^2 x - \sinh^2 x \equiv \frac{e^{2x} + 2 + e^{-2x}}{4} - \frac{e^{2x} - 2 + e^{-2x}}{4} \equiv 1$$

A few other important identities are:

$$\tanh^2 x + \operatorname{sech}^2 x \equiv 1 \tag{8-39}$$
$$\coth^2 x - \operatorname{csch}^2 x \equiv 1 \tag{8-40}$$
$$\sinh 2x \equiv 2 \sinh x \cosh x \tag{8-41}$$
$$\cosh 2x \equiv \cosh^2 x + \sinh^2 x \tag{8-42}$$

The student should prove these identities.

Let us consider the curve whose parametric equations are

$$x = a \cosh t \text{ and } y = a \sinh t$$

where a is a constant and t is the parameter. Squaring and subtracting, we obtain

$$x^2 - y^2 = a^2 \cosh^2 t - a^2 \sinh^2 t$$

or, since $\cosh^2 t - \sinh^2 t \equiv 1$,

$$x^2 - y^2 = a^2$$

But this is an equation of an equilateral hyperbola. Since the equations $x = \cosh t$ and $y = \sinh t$ are parametric equations of an equilateral hyperbola, the name *hyperbolic functions* has been given to the several functions just defined.

EXAMPLE 8–25. Sketch the graph of $y = \sinh x$.

Solution. We have $y = \sinh x = \dfrac{e^x - e^{-x}}{2}$. When $x = 0$, $y = \dfrac{1-1}{2} = 0$. As x increases indefinitely, e^x increases indefinitely and much more rapidly than does x, while e^{-x} approaches 0; hence, y increases indefinitely. As x decreases indefinitely, e^x approaches 0 and e^{-x} increases indefinitely. Since the minus sign precedes e^{-x}, it follows that y decreases indefinitely. If we replace x by $-x$ in the equation $y = \dfrac{e^x - e^{-x}}{2}$, we find that $\sinh(-x) = -\sinh x$. Hence, the curve is symmetric with respect to the origin. It is shown in Fig. 8–41.

EXAMPLE 8–26. Sketch the graph of $y = \cosh x$.

Solution. We have $y = \cosh x = \dfrac{e^x + e^{-x}}{2}$. When $x = 0$, $y = \dfrac{1+1}{2} = 1$.

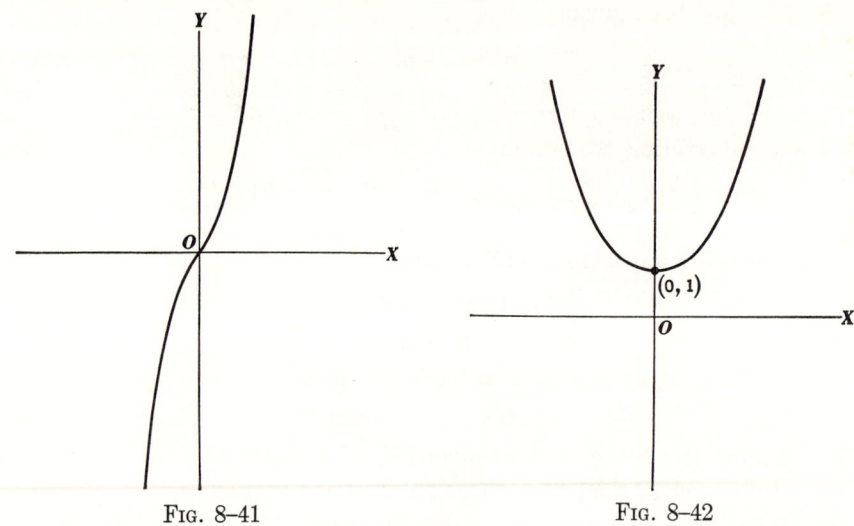

FIG. 8–41   FIG. 8–42

As x increases, y increases more rapidly than x does. If x is replaced by $-x$, the equation is unchanged; and, hence, the curve is symmetric with respect to the y-axis. The value of y is never less than 1. The curve is shown in Fig. 8–42.

This curve assumes the shape in which a uniform chain hangs when held by the ends. It is called a *catenary*.

EXERCISES

1. Sketch each of the following curves:
   (a) $y = \tanh x$
   (b) $y = \coth x$
   (c) $y = \operatorname{sech} x$
   (d) $y = \operatorname{csch} x$

2. Prove each of the following identities:
   (a) $\sinh(-x) \equiv -\sinh x$
   (b) $\cosh(-x) \equiv \cosh x$
   (c) $\tanh(-x) \equiv -\tanh x$
   (d) $\coth(-x) \equiv -\coth x$
   (e) $\cosh 2x \equiv \cosh^2 x + \sinh^2 x$
   (f) $\tanh 2x \equiv \dfrac{2\tanh x}{1 + \tanh^2 x}$

3. Prove each of the following identities:
   (a) $\sinh(x+y) \equiv \sinh x \cosh y + \cosh x \sinh y$
   (b) $\sinh(x-y) \equiv \sinh x \cosh y - \cosh x \sinh y$
   (c) $\cosh(x+y) \equiv \cosh x \cosh y + \sinh x \sinh y$
   (d) $\cosh(x-y) \equiv \cosh x \cosh y - \sinh x \sinh y$

Analytic Geometry of Space

9

The Point and Space Vectors

9–1. The point in space

Consider three fixed planes in space which are mutually perpendicular to each other and intersect in a common point. This common point is called the *origin*, and the fixed planes are called the *reference planes*. Each pair of reference planes intersects in a line, and these three lines are called the *axes*. Thus, the axes are three mutually perpendicular lines intersecting at the origin. In Fig. 9–1 the axes are the lines $X'X$, $Y'Y$, and $Z'Z$.

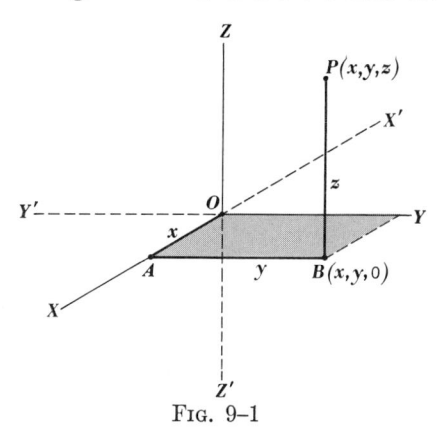

FIG. 9–1

The line $X'X$ is the x-axis; the line $Y'Y$ is the y-axis; and the line $Z'Z$ is the z-axis. We have chosen the positive direction of the x-axis to be "toward" the reader, the positive direction of the y-axis to be to the "right," and the positive direction of the z-axis to be "upward." The type of system of axes indicated here is called a *right-handed* system. If the positive direction of one of the axes is changed, we have a new system of axes, which is also in common use and is known as a *left-handed* system. Two systems of axes will be said to have the same *character* in this respect if one can be moved rigidly so as to coincide with the other, that is, so as to bring the x'-axis into the x-axis, the y'-axis into the y-axis, and the z'-axis into the z-axis.

The three reference planes are also called the xy-plane, the xz-plane, and the yz-plane. We locate a point in space by means of the respective

335

336   ANALYTIC GEOMETRY OF SPACE

directed distances to the point from the yz-plane, the xz-plane, and the xy-plane. We denote the point by P and the directed distances by x, y, and z. The three directed distances x, y, and z are called the *coordinates* of the point P. The three planes divide the space around the origin into eight regions called *octants*. The signs of the coordinates of a point in any octant may be determined by the following rule:

The value of x is positive when P lies "in front" of the yz-plane and is negative when P lies "in back" of that plane. The value of y is positive when P lies "to the right" of the xz-plane and is negative when P lies "to the left" of that plane. The value of z is positive when P lies "above" the xy-plane and is negative when P lies "below" that plane.

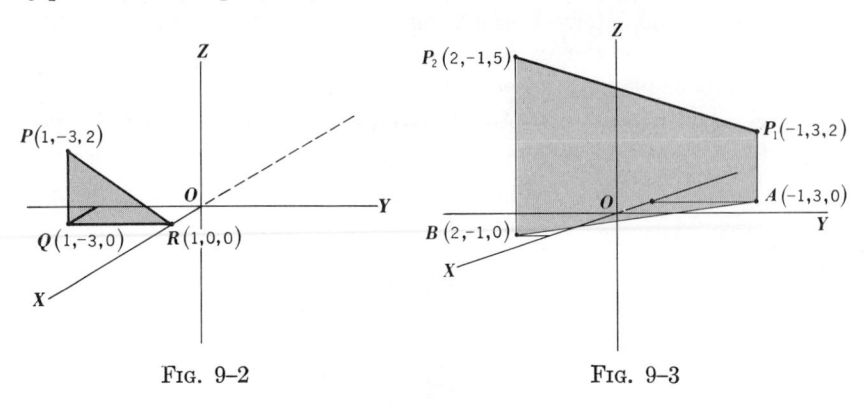

FIG. 9-2   FIG. 9-3

With every point there is associated a unique triple of three numbers (x, y, z), and for every triple of numbers there is a unique point. It is important for the student to remember that x, y, and z represent directed distances and that each may be positive, negative, or zero. If one of the coordinates is 0, the point lies in one of the reference planes; if two of the coordinates are 0, the point lies on one of the axes. The origin is $(0, 0, 0)$.

9-2. Projections

By the *projection of a point on a plane* we mean the foot of the perpendicular dropped from the point to the plane. For example, in Fig. 9-2, the projection of the point $P\ (1, -3, 2)$ on the xy-plane is the point $Q\ (1, -3, 0)$.

By the *projection of a point on a line* is meant the foot of the perpendicular dropped from the point to the line. So the projection of the point $P\ (1, -3, 2)$ on the x-axis is the point $R\ (1, 0, 0)$.

The *projection of a segment* P_1P_2 in Fig. 9-3 on a plane is the segment AB, where A is the projection of P_1 on the plane and B is the projection of P_2 on the plane.

EXAMPLE 9-1. What is the projection of the segment from the point P_1 (-1, 3, 2) to P_2 (2, -1, 5) on the xy-plane?

Solution. The conditions are shown in Fig. 9-3. The projection of P_1 on the xy-plane is A (-1, 3, 0) and the projection of P_2 on the xy-plane is B (2, -1, 0). The projection of P_1P_2 on the xy-plane is the segment from A to B.

If a *positive direction* happens to have been specified on the line joining A and B, as on a coordinate axis, then the *directed* distance $\overline{AB}$ is also referred to as the projection of P_1P_2 on this line.

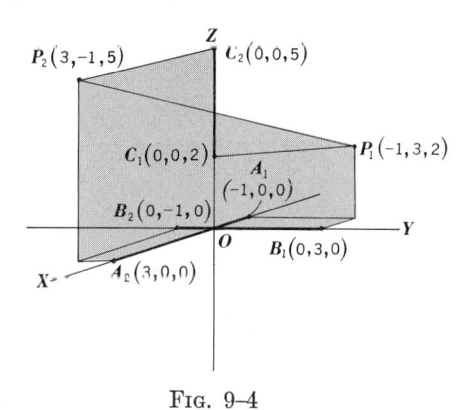

FIG. 9-4

Thus, if $P_1 = (x_1, y_1, z_1)$ and $P_2 = (x_2, y_2, z_2)$ are two given points, then the projections of P_1 and P_2 on the x-axis are $A_1(x_1, 0, 0)$ and $A_2(x_2, 0, 0)$, respectively; and the projection of P_1P_2 on the x-axis is $\overline{A_1A_2}$, where $\overline{A_1A_2} = \overline{A_1O} + \overline{OA_2} = -x_1 + x_2$. The projections of P_1 and P_2 on the y-axis are $B_1(0, y_1, 0)$ and $B_2(0, y_2, 0)$, respectively; and the projection of P_1P_2 on the y-axis is $\overline{B_1B_2} = \overline{B_1O} + \overline{OB_2} = -y_1 + y_2$. The projections of P_1 and P_2 on the z-axis are $C_1(0, 0, z_1)$ and $C_2(0, 0, z_2)$, respectively; and the projection of P_1P_2 on the z-axis is $\overline{C_1C_2} = \overline{C_1O} + \overline{OC_2} = -z_1 + z_2$. We shall introduce the following definitions: $\Delta x = x_2 - x_1$, $\Delta y = y_2 - y_1$, and $\Delta z = z_2 - z_1$. Hence, the projections of the segment P_1P_2 on the x-axis, the y-axis, and the z-axis, respectively, are Δx, Δy, and Δz.

EXAMPLE 9-2. What are the projections of the segment from P_1 (-1, 3, 2) to $P_2(3, -1, 5)$ on the x-axis, y-axis, and z-axis, respectively?

Solution. The conditions are shown in Fig. 9-4. The projections are:

$$\overline{A_1A_2} = \Delta x = 3 - (-1) = 4$$
$$\overline{B_1B_2} = \Delta y = -1 - 3 = -4$$
$$\overline{C_1C_2} = \Delta z = 5 - 2 = 3$$

EXERCISES

1. Locate each of the following points: $(1, 0, -1)$; $(2, 1, -3)$; $(-1, 1, 2)$; $(3, 2, 0)$; $(-4, 2, 3)$; $(0, -1, 2)$; $(0, 0, 5)$; $(0, 1, 0)$; $(6, 0, 0)$; $(-5, -3, -1)$; $(-2, -1, 0)$.

2. Where are the points for which $x = 0$?

3. Where are the points for which $y = 0$?

4. Where are the points for which $z = 0$?

5. Describe the set of points satisfying each of the following conditions:

   (a) $x - y = 0$ (c) $2x + z - 4 = 0$ (e) $z = 4$
   (b) $x + 3y - 6 = 0$ (d) $x - 5 = 0$ (f) $3y - z + 3 = 0$

6. If $P(x, y, z)$ is any point in space, what are the coordinates of a point in space that is symmetric to P with respect to the origin? How many such points are there?

7. The point $(-1, 2, 3)$ is one corner of a rectangular box whose sides are parallel to the reference planes and whose center is at the origin. What are the coordinates of the other vertices of the box?

8. What are the vertices of a cube which has one vertex at the origin and has its sides parallel to the reference axes, if the edges of the cube are of length a?

9. Determine the projections of the following points on the xy-plane, the xz-plane, and the yz-plane, respectively: $(1, -1, 3)$; $(2, 3, -5)$; $(3, -2, -1)$; $(-1, 2, -3)$; $(-2, -5, 1)$; $(3, 0, -1)$; $(-4, 2, 0)$; $(0, 4, 2)$; $(-3, -1, -2)$; $(2, -1, 3)$.

10. Determine the projections of the following segments on the x-axis, y-axis, and z-axis, respectively:

    (a) $P_1(3, -2, 4)$ to $P_2(-1, 1, -1)$ (d) $P_1(4, 2, 1)$ to $P_2(4, -3, 2)$
    (b) $P_1(2, 3, -1)$ to $P_2(-2, 1, 1)$ (e) $P_1(-1, 3, 2)$ to $P_2(4, 3, 0)$
    (c) $P_1(5, -3, 2)$ to $P_2(2, 3, 2)$ (f) $P_1(-3, 2, 5)$ to $P_2(0, 0, 0)$

11. Find the sums of the projections of the sides of the following quadrilaterals on the x-axis, y-axis, and z-axis, respectively, if the sides are the segments P_1P_2, P_2P_3, P_3P_4, and P_4P_1 and if the vertices have the following coordinates:

    (a) $P_1(-1, 2, 3)$, $P_2(3, -1, 4)$, $P_3(5, 2, -1)$, and $P_4(-3, 2, 6)$
    (b) $P_1(0, 0, 0)$, $P_2(1, -2, 3)$, $P_3(-2, 3, 1)$, and $P_4(-5, 1, 2)$
    (c) $P_1(1, -1, 2)$, $P_2(2, 1, 3)$, $P_3(-1, 2, 3)$, and $P_4(-4, 3, -1)$
    (d) $P_1(2, 1, 3)$, $P_2(1, 0, -1)$, $P_3(-2, 1, -2)$, and $P_4(-3, 4, 1)$

9–3. Scalar components of a segment

We define the *scalar components* of the segment from $P_1(x_1, y_1, z_1)$ to $P_2(x_2, y_2, z_2)$ to be the projections of the segment on the x-axis, y-axis, and z-axis, respectively. We shall call P_1 the initial point of the segment and P_2 the terminal point of the segment. The notation P_1P_2 designates the segment from P_1 to P_2. So the scalar components of P_1P_2 are Δx, Δy, and Δz. In order to distinguish scalar components from the coordinates of a point, we shall enclose the scalar components of a segment in *brackets*. If a segment has its initial point at the origin and its terminal point at $P(x, y, z)$, then its scalar components are $[x, y, z]$.

EXAMPLE 9–3. What are the scalar components of the segment from $P_1(1, -3, 2)$ to $P_2(3, 4, -1)$?

Solution. Since $\Delta x = 3 - 1 = 2$, $\Delta y = 4 - (-3) = 7$, and $\Delta z = -1 - 2 = -3$, the scalar components of the segment are $[2, 7, -3]$.

EXAMPLE 9–4. What are the scalar components of the segment from the origin O to $P = (-1, 2, 3)$?

Solution. Since the initial point is the origin, the scalar components of the segment are $[-1, 2, 3]$.

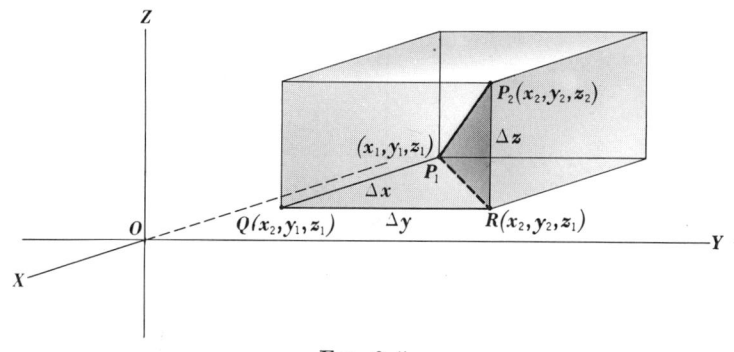

FIG. 9–5

9–4. Length, or magnitude, of a segment

In order to find the length, or magnitude, of the segment from $P_1(x_1, y_1, z_1)$ to $P_2(x_2, y_2, z_2)$, we proceed as follows: Consider P_1P_2 to be the diagonal of a rectangular parallelepiped (a box) with faces parallel to the reference planes, as indicated in Fig. 9–5. Project the segment P_1P_2 on the plane that passes through P_1 and is parallel to the xy-plane, and let R be the projection of P_2 on this plane. Then P_1P_2 is the hypotenuse of the right triangle P_1RP_2, and P_1R is the hypotenuse of the right triangle

340   ANALYTIC GEOMETRY OF SPACE

P_1QR, where Q is the projection of P_1 on the plane that passes through P_2 and is parallel to the yz-plane. Hence, we have

$$|P_1P_2|^2 = |P_1R|^2 + |RP_2|^2 \text{ and } |P_1R|^2 = |P_1Q|^2 + |QR|^2$$

So

$$|P_1P_2|^2 = |P_1Q|^2 + |QR|^2 + |RP_2|^2$$

or, since $\overline{P_1Q} = \Delta x$, $\overline{QR} = \Delta y$, and $\overline{RP_2} = \Delta z$,

$$|P_1P_2|^2 = |\Delta x|^2 + |\Delta y|^2 + |\Delta z|^2$$

Thus,

$$|P_1P_2| = \sqrt{(\Delta x)^2 + (\Delta y)^2 + (\Delta z)^2} \tag{9-1}$$

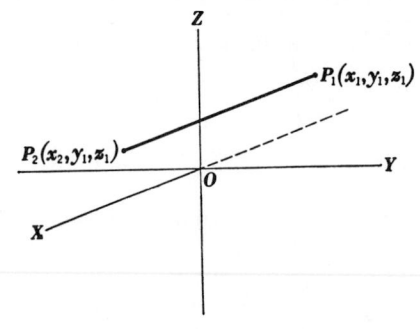

FIG. 9-6

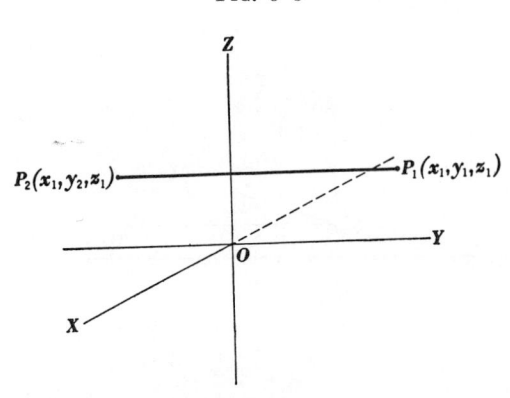

FIG. 9-7

This equation can be expressed by the following rule: The length, or magnitude, of a segment equals the square root of the sum of the squares of the scalar components of the segment.

If the segment is parallel to the x-axis, as in Fig. 9-6, we have $\Delta y = 0$ and $\Delta z = 0$. Hence, $|P_1P_2| = |\Delta x|$.

If the segment is parallel to the y-axis, as in Fig. 9-7, we have $\Delta x = 0$ and $\Delta z = 0$. Hence, $|P_1P_2| = |\Delta y|$.

If the segment is parallel to the z-axis, as in Fig. 9–8, we have $\Delta x = 0$ and $\Delta y = 0$. Hence, $|P_1P_2| = |\Delta z|$.

EXAMPLE 9–5. Find the length of the segment from $P_1(-1, 2, 3)$ to $P_2(2, -3, 1)$.

Solution. Since $\Delta x = 3$, $\Delta y = -5$, and $\Delta z = -2$, the scalar components of P_1P_2 are $[3, -5, -2]$. Hence, $|P_1P_2| = \sqrt{9+25+4} = \sqrt{38}$.

EXAMPLE 9–6. Find the length of the segment from $P_1(-1, -3, 7)$ to $P_2(-1, -3, 4)$.

Solution. Since $\Delta x = 0$ and $\Delta y = 0$, the segment is parallel to the z-axis and $|P_1P_2| = |-3| = 3$.

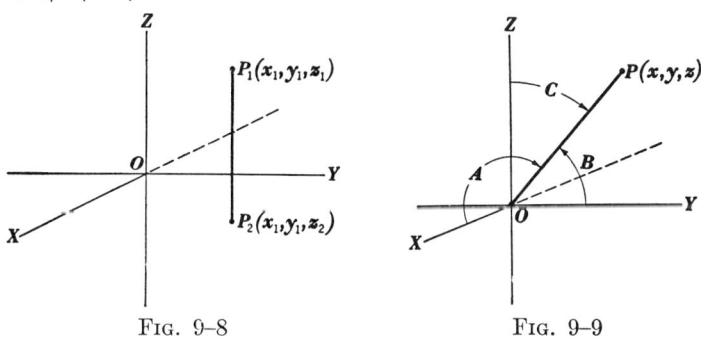

FIG. 9–8   FIG. 9–9

9–5. Direction cosines of a segment

Consider the segment from P_1 to P_2. Through P_1 draw three half-lines (or rays) having the same directions as the *positive* x-axis, y-axis, and z-axis, respectively. *The cosines of the angles A, B, and C that the directed segment makes with these directed rays are defined to be the direction cosines of the segment.* Here A is the angle that the directed segment makes with the x-axis, B is the angle that it makes with the y-axis, and C is the angle that it makes with the z-axis. Note that it does not matter which direction is considered positive in measuring A, B, and C. For illustrative purposes, we have chosen a segment having its initial point at the origin, as shown in Fig. 9–9.

Denoting the direction cosines by l, m, and n, respectively, we have

$$l = \cos A, \ m = \cos B, \text{ and } n = \cos C$$

Since the cosine of an angle is never greater than 1 or less than -1, it follows that $-1 \le l \le 1$, $-1 \le m \le 1$, and $-1 \le n \le 1$.

From Fig. 9–5,

$$l = \frac{\Delta x}{|P_1P_2|}, \ m = \frac{\Delta y}{|P_1P_2|}, \text{ and } n = \frac{\Delta z}{|P_1P_2|}$$

342 ANALYTIC GEOMETRY OF SPACE

If we square each direction cosine and add the squares, we obtain

$$l^2 + m^2 + n^2 = \frac{(\Delta x)^2 + (\Delta y)^2 + (\Delta z)^2}{|P_1P_2|^2} = 1 \tag{9-2}$$

Thus, *the sum of the squares of the direction cosines of a segment is 1.* Also, to find the direction cosines, we divide the scalar components by the magnitude of the segment.

EXAMPLE 9-7. Find the direction cosines of the segment from P_1 $(-2, 3, -1)$ to $P_2(3, 1, -3)$.

Solution. Since $\Delta x = 5$, $\Delta y = -2$, $\Delta z = -2$, and $|P_1P_2| = \sqrt{33}$, we have

$$l = \frac{5}{\sqrt{33}}, \quad m = -\frac{2}{\sqrt{33}}, \quad \text{and } n = -\frac{2}{\sqrt{33}}$$

We check the results by showing that $l^2 + m^2 + n^2 = 1$. Thus,

$$\frac{25}{33} + \frac{4}{33} + \frac{4}{33} = 1$$

EXAMPLE 9-8. Find the direction cosines of the segment from the origin to $P = (3, -1, 2)$.

Solution. Since the initial point is at the origin, it follows that $\Delta x = 3$, $\Delta y = -1$, $\Delta z = 2$, and $|OP| = \sqrt{14}$. Hence,

$$l = \frac{3}{\sqrt{14}}, \quad m = -\frac{1}{\sqrt{14}}, \quad \text{and } n = \frac{2}{\sqrt{14}}$$

Again, we have

$$l^2 + m^2 + n^2 = \frac{9 + 1 + 4}{14} = 1$$

EXERCISES

1. Find the perimeter of each of the following triangles:
   (a) $P_1 = (-1, 2, 3)$, $P_2 = (3, 1, 2)$, and $P_3 = (1, -1, -1)$
   (b) $P_1 = (3, 4, -1)$, $P_2 = (1, 2, -1)$, and $P_3 = (2, -1, 3)$
   (c) $P_1 = (0, 0, 0)$, $P_2 = (2, 1, 3)$, and $P_3 = (-5, 2, -1)$

2. Find the lengths of the six edges of the tetrahedron whose vertices are $P_1 = (2, -1, 3)$, $P_2 = (-1, 3, 2)$, $P_3 = (-3, 2, 1)$, and $P_4 = (1, 0, -1)$.

3. Prove that the following points are vertices of a regular tetrahedron: $(k, 0, 0)$, $(0, k, 0)$, $(0, 0, k)$, (k, k, k). (Hint: A regular tetrahedron is a figure with four faces and six equal edges.)

4. If the length of the segment from the origin O to $P(x, y, z)$ is 1, prove that the direction cosines of OP are $l = x$, $m = y$, and $n = z$.

5. Given the following initial points and scalar components, construct each segment, and find the terminal point, the magnitude, and the direction cosines of the segment:
   (a) $P_1(1, -3, 2)$; scalar components $[2, 1, -1]$
   (b) $P_1(2, 3, -1)$; scalar components $[1, -1, 3]$
   (c) $P_1(0, 0, 3)$; scalar components $[-3, 2, 4]$
   (d) $P_1(4, -2, 5)$; scalar components $[2, 1, -3]$

6. Are $[\frac{1}{2}, \frac{2}{3}, \frac{3}{5}]$ the direction cosines of a segment? Explain your answer.

7. If $l = \frac{1}{2}$ and $m = \frac{3}{4}$, what is n? Is it unique?

8. Prove that the points $(3, 1, 1)$, $(3, -2, 4)$ and $(6, -2, 1)$ are the vertices of an equilateral triangle. Find the fourth vertex of a regular tetrahedron having the given points as three of its vertices. Is it unique?

9. If $l = -\frac{1}{3}$ and $m = -\frac{2}{5}$, find n.

10. If $m = \frac{3}{4}$ and $n = -\frac{1}{2}$, find l.

11. If $l = \frac{2}{5}$ and $n = -\frac{4}{5}$, find m.

12. If a segment OP makes an angle of 30° with the x-axis and an angle of 60° with the y-axis, what angle does it make with the z-axis? Is this angle unique?

13. If a segment OP makes equal angles with the three axes, what are the angles?

14. Find the distance from each of the coordinate axes to the point (x, y, z).

15. What are the images of the point $P(x,y,z)$ in the yz, xz, and xy planes, respectively? (Images with respect to a point and a line are defined on p. 42.)

16. What are the images of the point $P(x,y,z)$ in the x, y, and z axes, respectively?

9–6. Space vectors

We shall define a *space vector* to be the collection, considered as an entity, of all segments in space having the "same direction" and the same magnitude. Any one segment of the collection is sufficient to represent the vector. Algebraically a vector is described by the scalar components of any one of its representative segments. We shall represent a vector **u** by $\mathbf{u} = [u_1, u_2, u_3]$ and we shall call the numbers u_1, u_2, and u_3 the *scalar components* of the vector. The *vector components* of **u** parallel to the x-axis,

344   ANALYTIC GEOMETRY OF SPACE

y-axis, and z-axis are the *vectors* $\mathbf{u}_x = [u_1, 0, 0]$, $\mathbf{u}_y = [0, u_2, 0]$, and $\mathbf{u}_z = [0, 0, u_3]$. The *magnitude* of a space vector is the length of any one of its representative segments. We shall represent the magnitude of a vector $\mathbf{u}$ by $|\mathbf{u}|$. So $|\mathbf{u}| = \sqrt{u_1^2 + u_2^2 + u_3^2}$.

If $\mathbf{u} = [u_1, u_2, u_3]$ and $\mathbf{v} = [v_1, v_2, v_3]$, it follows from the definition of a vector that $\mathbf{u} = \mathbf{v}$ if and only if $u_1 = v_1$, $u_2 = v_2$, and $u_3 = v_3$. That is, *two vectors are equal if and only if their corresponding scalar components are equal.*

The sum of two vectors is defined to be the vector whose scalar components are equal to the sums of the respective scalar components of the original two vectors. So, if $\mathbf{u} = [u_1, u_2, u_3]$ and $\mathbf{v} = [v_1, v_2, v_3]$, and if $\mathbf{w} = \mathbf{u} + \mathbf{v}$, then the sum vector is $\mathbf{w} = [u_1 + v_1, u_2 + v_2, u_3 + v_3]$.

Geometrically, we add two vectors by placing the initial point of a representative segment of the second vector at the terminal point of a representative segment of the first vector. Then the segment obtained by joining the initial point of the first vector to the terminal point of the second vector is a representative segment of the sum vector. In Fig. 9–10, let OP_1 be a representative segment of the first vector $\mathbf{u}$ and let P_1P_2 be a representative segment of the second vector $\mathbf{v}$. When we wish to indicate a vector in terms of a representative segment P_1P_2 of the vector, we shall use the notation $\overrightarrow{P_1P_2}$. Thus, we have

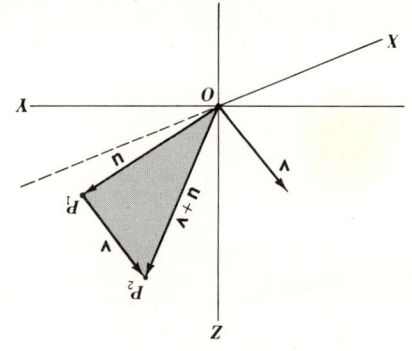

FIG. 9–10

$$\overrightarrow{OP_2} = \overrightarrow{OP_1} + \overrightarrow{P_1P_2} = \mathbf{u} + \mathbf{v} = [u_1, u_2, u_3] + [v_1, v_2, v_3]$$

or

$$\mathbf{u} + \mathbf{v} = [u_1 + v_1, u_2 + v_2, u_3 + v_3] \qquad (9\text{–}3)$$

If $\mathbf{u} = [u_1, u_2, u_3]$, *we shall define* $-\mathbf{u}$ *by the notation* $-\mathbf{u} + [-u_1, -u_2, -u_3]$. *Since* $|\mathbf{u}| = |-\mathbf{u}|$, *it follows that the two vectors* $\mathbf{u}$ *and* $-\mathbf{u}$ *have the same magnitude but opposite directions.*

We shall say that $\mathbf{0} = [0, 0, 0]$ is the *zero vector*, even though it has no direction. The magnitude $|\mathbf{0}|$ of the zero vector is 0. Hence, $\mathbf{u} + \mathbf{0} = [u_1, u_2, u_3] + [0, 0, 0] = [u_1, u_2, u_3] = \mathbf{u}$. By our rule for the addition of vectors, it follows that $\mathbf{u}+(-\mathbf{u}) = [0, 0, 0]$, or the sum of a vector and its negative is the zero vector.

The definition of the vector $-\mathbf{u}$ enables us to define the difference between two vectors in terms of addition. Thus, to subtract one vector from another, change the sign of the one being subtracted and add it to the other. That is,

$$\mathbf{v} - \mathbf{u} = \mathbf{v} + (-\mathbf{u}) = [v_1 - u_1, v_2 - u_2, v_3 - u_3] \tag{9-4}$$

EXAMPLE 9-9. Find the magnitude of the sum vector $\mathbf{u}+\mathbf{v}$, if $\mathbf{u} = [-1, 2, 3]$ and $\mathbf{v} = [2, -1, 1]$.

Solution. We have $\mathbf{u}+\mathbf{v} = [1, 1, 4]$. Hence, $|\mathbf{u}+\mathbf{v}| = \sqrt{18}$.

EXAMPLE 9-10. Find the magnitude of the difference vector $\mathbf{v}-\mathbf{u}$ for the vectors $\mathbf{u}$ and $\mathbf{v}$ of Example 9-9.

Solution. Since $-\mathbf{u} = [1, -2, -3]$, we have

$$\mathbf{v} - \mathbf{u} = \mathbf{v} + (-\mathbf{u}) = [2+1, -1-2, 1-3] = [3, -3, -2]$$

So $|\mathbf{v}-\mathbf{u}| = \sqrt{22}$.

We shall define the vector $k\mathbf{u}$, which is the product of k and the vector $\mathbf{u} = [u_1, u_2, u_3]$, by the notation $k\mathbf{u} = [ku_1, ku_2, ku_3]$. Thus, if $\mathbf{u} = [1, -3, 2]$, we have $2\mathbf{u} = [2, -6, 4]$.

A *unit* vector is any vector of magnitude *one*. By our definition of $k\mathbf{u}$ it is possible to express any non-zero vector $\mathbf{u}$ as a constant times a unit vector which has the "same direction" as $\mathbf{u}$. The constant is the magnitude of $\mathbf{u}$. So, if $\mathbf{u} = [u_1, u_2, u_3]$, we have $\mathbf{u} = |\mathbf{u}| \left[\dfrac{u_1}{|\mathbf{u}|}, \dfrac{u_2}{|\mathbf{u}|}, \dfrac{u_3}{|\mathbf{u}|} \right]$, where $\left[\dfrac{u_1}{|\mathbf{u}|}, \dfrac{u_2}{|\mathbf{u}|}, \dfrac{u_3}{|\mathbf{u}|} \right]$ is a unit vector having the same direction as the vector $\mathbf{u}$. The zero vector is equal to 0 times any unit vector.

EXAMPLE 9-11. Express the vector $\mathbf{u} = [-1, 2, -3]$ as a constant times a unit vector having the same direction as $\mathbf{u}$.

Solution. The magnitude of $\mathbf{u}$ is $\sqrt{14}$. Hence,

$$\mathbf{u} = \sqrt{14} \left[-\dfrac{1}{\sqrt{14}}, \dfrac{2}{\sqrt{14}}, -\dfrac{3}{\sqrt{14}} \right]$$

The *direction cosines of a non-zero vector* are the direction cosines of any representative segment of the vector. So, the direction cosines of the vector $\mathbf{u} = [u_1, u_2, u_3]$ are $l = \dfrac{u_1}{|\mathbf{u}|}$, $m = \dfrac{u_2}{|\mathbf{u}|}$, and $n = \dfrac{u_3}{|\mathbf{u}|}$.

EXAMPLE 9-12. What are the direction cosines of $\mathbf{u} = [2, -1, 3]$?

Solution. Since the magnitude of $\mathbf{u}$ is $\sqrt{14}$, it follows that the direction cosines of **u** are $l = \dfrac{2}{\sqrt{14}}$, $m = -\dfrac{1}{\sqrt{14}}$, and $n = \dfrac{3}{\sqrt{14}}$.

The scalar components of a unit vector are also its direction cosines. The direction cosines of any vector are the scalar components of the unit vector that has the same direction as the original vector.

The point in which a vector intersects the unit sphere with its center at the origin has for coordinates the direction cosines of the original vector.

EXERCISES

1. Define a space vector algebraically and geometrically.

2. Give four geometric representations of each of the following vectors, and find the magnitude of each:
   (a) $[-1, 2, 3]$
   (b) $[3, -1, 2]$
   (c) $[0, 1, 5]$
   (d) $[1, 3, 4]$

3. Determine which pairs of the following vectors are equal:
   (a) $[-1, 2, 5]$
   (b) $[3, -2, 1]$
   (c) $[-3, 2, -1]$
   (d) $[-1, 2, 5]$
   (e) $[-3, 2, -1]$
   (f) $[1, -2, -5]$

4. Give the magnitudes of the sum and difference of each of the following pairs of vectors:
   (a) $\mathbf{u} = [2, -1, 3]$ and $\mathbf{v} = [-1, 1, 1]$
   (b) $\mathbf{u} = [3, 1, -2]$ and $\mathbf{v} = [0, 2, 1]$
   (c) $\mathbf{u} = [-1, 0, 3]$ and $\mathbf{v} = [2, 1, -1]$
   (d) $\mathbf{u} = [5, 3, -2]$ and $\mathbf{v} = [2, -1, 0]$

5. Find the magnitude of each of the following vectors, if $\mathbf{u} = [3, -1, 2]$; and draw a representative segment of each:
   (a) $3\mathbf{u}$
   (b) $-2\mathbf{u}$
   (c) $\tfrac{1}{2}\mathbf{u}$

6. Express each of the following vectors as a constant times a unit vector having the "same direction" as the given vector:
   (a) $[2, -3, 4]$
   (b) $[1, -1, 3]$
   (c) $[3, 2, -1]$

7. When are two vectors said to be equal?

8. What is meant by the zero vector?

9. Are all unit vectors equal? Explain.

10. Find the direction cosines of each of the following vectors:
    (a) $[2, 3, 1]$
    (b) $[-1, 3, -1]$
    (c) $[3, 4, 2]$
    (d) $[3, -4, 0]$

11. Express the non-zero vector $\mathbf{u} = [u_1, u_2, u_3]$ as a linear combination of $\mathbf{i}$, $\mathbf{j}$, and $\mathbf{k}$ where $\mathbf{i} = [1, 0, 0]$, $\mathbf{j} = [0, 1, 0]$, and $\mathbf{k} = [0, 0, 1]$. Hint: Show that $\mathbf{u} = c_1\mathbf{i} + c_2\mathbf{j} + c_3\mathbf{k}$ where c_1, c_2, and c_3 are constants not all zero.

12. Write the following vectors as linear combinations of $\mathbf{i}$, $\mathbf{j}$, and $\mathbf{k}$:
    (a) $[-1, 3, 2]$ (b) $[2, 5, -3]$ (c) $[0, 3, 2]$ (d) $[4, -2, 5]$

13. Given the vectors $\mathbf{u} = [2, -1, 3]$, $\mathbf{v} = [1, 3, -1]$ and $\mathbf{w} = [3, 1, 1]$, find each of the following vectors and their magnitudes:
    (a) $2\mathbf{u} - \mathbf{v} + \mathbf{w}$
    (b) $\mathbf{u} - 3\mathbf{v} + 2\mathbf{w}$
    (c) $\mathbf{w} + 2\mathbf{v} - 3\mathbf{u}$
    (d) $2\mathbf{v} - \mathbf{u} + 2\mathbf{w}$
    (e) $\mathbf{v} - \mathbf{u} - \mathbf{w}$
    (f) $2\mathbf{w} + 3\mathbf{v} + \mathbf{u}$

9–7. Cosine of the angle between two vectors

Let O be the center of a unit sphere, and also let it be the initial point of representative segments of two non-zero vectors $\mathbf{u}$ and $\mathbf{v}$. Let P_1 and P_2 be the points where the representative segments of these vectors intersect this sphere. Then the coordinates of P_1 are l_1, m_1, and n_1 and those of P_2 are l_2, m_2, and n_2, where l_1, m_1, and n_1 are the direction cosines of $\mathbf{u}$ and l_2, m_2, and n_2 are the direction cosines of $\mathbf{v}$. The angle θ between $\mathbf{u}$ and $\mathbf{v}$ is, by definition, the angle between the directed segments OP_1 and OP_2, as shown in Fig. 9–11. Using the law of cosines, we have

$$\cos\theta = \frac{|\overrightarrow{OP_1}|^2 + |\overrightarrow{OP_2}|^2 - |\overrightarrow{P_1P_2}|^2}{2|\overrightarrow{OP_1}|\,|\overrightarrow{OP_2}|}$$

Since $|\overrightarrow{OP_1}| = |\overrightarrow{OP_2}| = 1$, and $|\overrightarrow{P_1P_2}|^2 = (l_2 - l_1)^2 + (m_2 - m_1)^2 + (n_2 - n_1)^2$, we

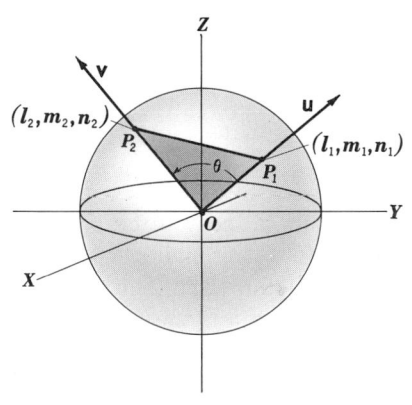

FIG. 9–11

have

$$\cos \theta = \frac{1+1-[(l_2-l_1)^2+(m_2-m_1)^2+(n_2-n_1)^2]}{2}$$

Hence,

$$\cos \theta = \frac{2-[l_2{}^2+m_2{}^2+n_2{}^2+l_1{}^2+m_1{}^2+n_1{}^2-2l_1l_2-2m_1m_2-2n_1n_2]}{2}$$

$$= \frac{2-2+2(l_1l_2+m_1m_2+n_1n_2)}{2}$$

or

$$\cos \theta = l_1l_2+m_1m_2+n_1n_2 \tag{9-5}$$

Rule: *The cosine of the angle θ between two vectors is equal to the sum of the products of the corresponding direction cosines of the two vectors.*

If $\mathbf{u} = [u_1, u_2, u_3]$ and $\mathbf{v} = [v_1, v_2, v_3]$, then

$$\cos \theta = \frac{u_1v_1+u_2v_2+u_3v_3}{|\mathbf{u}|\ |\mathbf{v}|} \tag{9-6}$$

EXAMPLE 9–13. Find the cosine of the angle between the vectors $\mathbf{u} = [2, -1, 3]$ and $\mathbf{v} = [3, 2, 4]$.

Solution. We have

$$\cos \theta = \frac{6-2+12}{\sqrt{14}\ \sqrt{29}} = \frac{16}{\sqrt{14}\ \sqrt{29}}$$

The combination $u_1v_1+u_2v_2+u_3v_3$ is so important and useful that it is given a special name. We call it the *scalar product*, or *dot product*, of the vectors $\mathbf{u}$ and $\mathbf{v}$ and we write it $\mathbf{u} \cdot \mathbf{v}$. *Hence, the scalar, or dot, product of two vectors is equal to the sum of the products of the corresponding scalar components of the two vectors.* Thus,

$$\mathbf{u} \cdot \mathbf{v} = u_1v_1+u_2v_2+u_3v_3 \tag{9-7}$$

Also, if neither $\mathbf{u}$ nor $\mathbf{v}$ is equal to $\mathbf{0}$, the cosine of the angle θ between $\mathbf{u}$ and $\mathbf{v}$ is

$$\cos \theta = \frac{\mathbf{u} \cdot \mathbf{v}}{|\mathbf{u}|\ |\mathbf{v}|} \tag{9-8}$$

From equation (9–8), it follows that

$$\mathbf{u} \cdot \mathbf{v} = |\mathbf{u}||\mathbf{v}| \cos \theta \tag{9-9}$$

We shall *define* two vectors $\mathbf{u}$ and $\mathbf{v}$ to be *perpendicular* to each other when their scalar product is 0. That is, if $\mathbf{u} \cdot \mathbf{v} = 0$, then $\mathbf{u}$ is perpendicular to $\mathbf{v}$. From this definition it follows that the zero vector is perpendicular to every vector. If neither $\mathbf{u}$ nor $\mathbf{v}$ is the zero vector, then the condition that $\mathbf{u} \cdot \mathbf{v} = 0$ implies that $\theta = 90°$ or $270°$.

Two non-zero vectors $\mathbf{u}$ and $\mathbf{v}$ will be defined to be parallel in the same sense or in the opposite sense when the scalar components of $\mathbf{u}$ are propor-

tional to the scalar components of v. That is, if $u_1 = kv_1$, $u_2 = kv_2$, and $u_3 = kv_3$, or if $v_1 = k'u_1$, $v_2 = k'u_2$, and $v_3 = k'u_3$, where k (or k') $\neq 0$, we shall define $\mathbf{u}$ to be *parallel* to $\mathbf{v}$. If k (or k') > 0, the vectors are *parallel in the same sense;* if k (or k') < 0, the vectors are *parallel in the opposite sense.*

The direction cosines of two non-zero vectors are equal if, and only if, the vectors are parallel in the same sense. To see this, let it be assumed that $\dfrac{u_1}{|\mathbf{u}|} = \dfrac{v_1}{|\mathbf{v}|}$, $\dfrac{u_2}{|\mathbf{u}|} = \dfrac{v_2}{|\mathbf{v}|}$, and $\dfrac{u_3}{|\mathbf{u}|} = \dfrac{v_3}{|\mathbf{v}|}$, whence $u_1 = \dfrac{|\mathbf{u}|}{|\mathbf{v}|} v_1$, $u_2 = \dfrac{|\mathbf{u}|}{|\mathbf{v}|} v_2$, and $u_3 = \dfrac{|\mathbf{u}|}{|\mathbf{v}|} v_3$. It then follows that $u_1 = kv_1$, $u_2 = kv_2$, and $u_3 = kv_3$, where $k > 0$; and the vectors are parallel in the same sense.

Conversely, let it be assumed that $u_1 = kv_1$, $u_2 = kv_2$, and $u_3 = kv_3$, where $k > 0$. If we square and add and then solve for k, we obtain:

$$k = \frac{\sqrt{u_1^2 + u_2^2 + u_3^2}}{\sqrt{v_1^2 + v_2^2 + v_3^2}} = \frac{|\mathbf{u}|}{|\mathbf{v}|}$$

Hence, $\dfrac{u_1}{|\mathbf{u}|} = \dfrac{v_1}{|\mathbf{v}|}$, $\dfrac{u_2}{|\mathbf{u}|} = \dfrac{v_2}{|\mathbf{v}|}$, and $\dfrac{u_3}{|\mathbf{u}|} = \dfrac{v_3}{|\mathbf{v}|}$; or the direction cosines are equal.

The student should prove that the direction cosines of two non-zero vectors are *numerically* equal but *opposite* in sign if, and only if, the vectors are parallel in the opposite sense.

EXAMPLE 9-14. Find the value of k if the vector $\mathbf{u} = [-1, 3, k]$ is to be perpendicular to the vector $\mathbf{v} = [2, 1, 3]$.

Solution. Since the vectors are perpendicular, we know that $\mathbf{u} \cdot \mathbf{v} = 0$. Hence,
$$\mathbf{u} \cdot \mathbf{v} = -2 + 3 + 3k = 0$$
or
$$k = -\tfrac{1}{3}$$

EXAMPLE 9-15. If the segment from $P_1 = (3, 1, -1)$ to $P_2 = (-1, 2, 1)$ is perpendicular to the segment from $P_3 = (-3, 2, 4)$ to $P_4 = (x, -2, 3)$, what is x?

Solution. We have $\overrightarrow{P_1P_2} = [-4, 1, 2]$ and $\overrightarrow{P_3P_4} = [x+3, -4, -1]$. In order that the two segments may be perpendicular to each other, it is required that $\overrightarrow{P_1P_2} \cdot \overrightarrow{P_3P_4} = -4(x+3) - 4 - 2 = 0$. This reduces to $-4x - 12 - 4 - 2 = 0$ or $4x = -18$, from which $x = -\dfrac{9}{2}$.

EXAMPLE 9-16. Find x and y if the points $P_1(1, -3, 2)$, $P_2(x, y, 3)$, and $P_3(-1, 2, 1)$ are collinear.

350   ANALYTIC GEOMETRY OF SPACE

Solution. We know that $\overrightarrow{P_1P_2}$ must be parallel to $\overrightarrow{P_1P_3}$. Since $\overrightarrow{P_1P_2} = [x-1, y+3, 1]$ and $\overrightarrow{P_1P_3} = [-2, 5, -1]$, it follows that $x - 1 = -2k$, $y + 3 = 5k$, and $1 = -k$. Since $k = -1$, we obtain $x - 1 = 2$, or $x = 3$; and $y + 3 = -5$, or $y = -8$.

EXERCISES

1. Show that $\mathbf{u} \cdot \mathbf{v} = \mathbf{v} \cdot \mathbf{u}$.

2. Find the scalar product of each of the following pairs of vectors:
   (a) $\mathbf{u} = [-1, 2, -3]$ and $\mathbf{v} = [3, -1, 1]$
   (b) $\mathbf{u} = [2, -2, 5]$ and $\mathbf{v} = [-3, 1, -2]$

3. Find the cosines of the angles of the following triangles whose vertices are:
   (a) $P_1(2, -1, -1)$, $P_2(-3, 4, 2)$, and $P_3(1, -1, 2)$
   (b) $P_1(-1, 2, 1)$, $P_2(2, -1, 3)$, and $P_3(3, 1, 4)$
   (c) $P_1(1, 0, 0)$, $P_2(2, -1, 3)$, and $P_3(0, 1, -1)$
   (d) $P_1(4, 2, 3)$, $P_2(3, 1, -2)$, and $P_3(5, -1, -3)$

4. Prove that the three points in each of the following groups are vertices of a right triangle:
   (a) $P_1(3, 1, -2)$, $P_2(8, 4, 6)$, and $P_3(6, 7, 0)$
   (b) $P_1(0, 4, -2)$, $P_2(-3, -2, -4)$, and $P_3(2, 1, 4)$

5. Show that the points $P_1(-3, -1, 5)$, $P_2(3, -4, 7)$, $P_3(0, 5, -1)$, and $P_4(6, 2, 1)$ are the vertices of a parallelogram.

6. Three vertices of a parallelogram are, in order, $P_1(1, 3, -1)$, $P_2(-1, 2, 3)$, and $P_3(3, -1, 2)$. Find the fourth vertex.

7. Prove that the points $(1, -2, 3)$, $(3, -5, 4)$, and $(7, -11, 6)$ lie on a straight line.

8. Find x and z in order that the points $(-1, 3, -2)$, $(2, -1, 5)$, and $(x, 2, z)$ will lie on a straight line.

9. Find k if the two vectors $[3, -1, 2]$ and $[1, k, 5]$ are perpendicular to each other.

10. If $P_1(3, 1, -1)$, $P_2(2, 4, 5)$, and $P_3(-1, 3, a)$ are vertices of a right triangle, find a.

9–8. The coordinates of a point that divides a segment in a given ratio

In Fig. 9–12, let $P(x, y, z)$ be a point which is on the line containing the fixed points $P_1(x_1, y_1, z_1)$ and $P_2(x_2, y_2, z_2)$ and is so located that $\overrightarrow{P_1P} =$

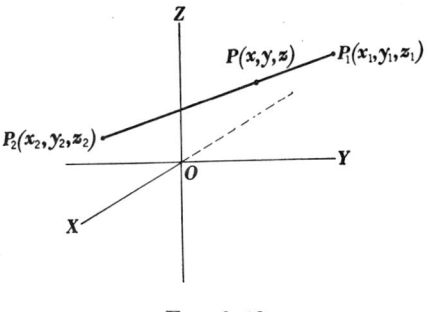

FIG. 9-12

$k \overrightarrow{P_1P_2}$, where k is a given constant. From the properties of vectors we can draw the following conclusions: If $k = 0$, P coincides with P_1; if k is a positive number less than 1, P lies between P_1 and P_2; if $k = 1$, P coincides with P_2; if k is a positive number greater than 1, P lies beyond P_2; if k is a negative number, P lies on the ray through P_1 that does not contain P_2. Since $\overrightarrow{P_1P} = [x - x_1, y - y_1, z - z_1]$ and $\overrightarrow{P_1P_2} = [\Delta x, \Delta y, \Delta z]$, we have

$$[x - x_1, y - y_1, z - z_1] = k[\Delta x, \Delta y, \Delta z]$$

Hence, $x - x_1 = k\, \Delta x$, $y - y_1 = k\, \Delta y$, and $z - z_1 = k\, \Delta z$; or

$$\begin{aligned} x &= x_1 + k\, \Delta x \\ y &= y_1 + k\, \Delta y \\ z &= z_1 + k\, \Delta z \end{aligned} \tag{9-10}$$

Sometimes we are given the ratio of $\overrightarrow{P_1P}$ to $\overrightarrow{PP_2}$. This ratio r is, by definition, that in which P divides the segment P_1P_2. When this is the case, we set up our vector equation as $\overrightarrow{P_1P} = r\overrightarrow{PP_2}$ and proceed as follows: Since $\overrightarrow{P_1P} = [x - x_1, y - y_1, z - z_1]$ and $\overrightarrow{PP_2} = [x_2 - x, y_2 - y, z_2 - z]$, we have $[x - x_1, y - y_1, z - z_1] = r[x_2 - x, y_2 - y, z_2 - z]$. Hence, $x - x_1 = r(x_2 - x)$, $y - y_1 = r(y_2 - y)$, and $z - z_1 = r(z_2 - z)$. Solving for x, y, and z, respectively, we obtain

$$\begin{aligned} x &= \frac{x_1 + rx_2}{1 + r} \\ y &= \frac{y_1 + ry_2}{1 + r} \\ z &= \frac{z_1 + rz_2}{1 + r} \end{aligned} \tag{9-11}$$

Equations (9-10) and (9-11) should not be memorized, but the method of setting up each problem should be understood thoroughly.

EXAMPLE 9-17. We are given two points, $P_1 = (3, -1, 2)$ and $P_2 =$

$(-1, 2, 5)$. Find P if $\overrightarrow{P_1P} = 3\ \overrightarrow{P_1P_2}$.

Solution. Since $\overrightarrow{P_1P} = [x-3, y+1, z-2]$ and $\overrightarrow{P_1P_2} = [-4, 3, 3]$, we obtain $[x-3, y+1, z-2] = 3[-4, 3, 3]$. Hence, $x-3 = -12$, $y+1 = 9$, and $z-2 = 9$; and the coordinates of P are -9, 8, and 11.

EXAMPLE 9-18. Find the coordinates of a point P on the line containing $\overrightarrow{P_1P_2}$ if P is three times as far from $P_2(3, 5, -1)$ as it is from $P_1(-1, 2, 6)$.

Solution. In this problem $3|\overrightarrow{P_1P}| = |\overrightarrow{PP_2}|$. Hence, we could have either $3\ \overrightarrow{P_1P} = \overrightarrow{PP_2}$ or $3\ \overrightarrow{P_1P} = -\overrightarrow{PP_2}$. In other words, there are two answers since P could lie on or outside the segment P_1P_2.

(a) When P lies between P_1 and P_2, we have $3\ \overrightarrow{P_1P} = \overrightarrow{PP_2}$; or

$$3[x+1, y-2, z-6] = [3-x, 5-y, -1-z]$$

Hence, $3(x+1) = (3-x)$, $3(y-2) = (5-y)$, and $3(z-6) = (-1-z)$. Solving for x, y, and z, we obtain

$$4x = 0,\ 4y = 11,\ \text{and}\ 4z = 17$$

The coordinates of P are therefore 0, $\frac{11}{4}$, and $\frac{17}{4}$.

(b) When P lies outside the segment P_1P_2, we have $3\ \overrightarrow{P_1P} = -\overrightarrow{PP_2}$; or

$$3[x+1, y-2, z-6] = -[3-x, 5-y, -1-z]$$

Hence, $3x+3 = -3+x$, $3y-6 = -5+y$, and $3z-18 = 1+z$. Solving for x, y, and z, we obtain

$$2x = -6,\ 2y = 1,\ \text{and}\ 2z = 19$$

It follows that the coordinates of the external point P are -3, $\frac{1}{2}$, and $\frac{19}{2}$.

9–9. The mid-point of a segment

This problem is merely a special application of the relation $\overrightarrow{P_1P} = k\ \overrightarrow{P_1P_2}$, where $k = \frac{1}{2}$; or a special application of the relation $\overrightarrow{P_1P} = r\ \overrightarrow{PP_2}$, where $r = 1$. In either case, we obtain the coordinates of the mid-point to be

$$x = \frac{x_1+x_2}{2},\ y = \frac{y_1+y_2}{2},\ \text{and}\ z = \frac{z_1+z_2}{2}$$

EXAMPLE 9-19. Find the mid-points of the sides of the triangle whose vertices are $P_1 = (-1, 2, 5)$, $P_2 = (3, -4, 1)$, and $P_3 = (5, -2, 1)$.

Solution. The mid-point of side P_1P_2 is
$$M_3 = \left(\frac{3-1}{2}, \frac{-4+2}{2}, \frac{1+5}{2}\right) = (1, -1, 3)$$

The mid-point of side P_1P_3 is
$$M_2 = \left(\frac{5-1}{2}, \frac{-2+2}{2}, \frac{5+1}{2}\right) = (2, 0, 3)$$

The mid-point of side P_2P_3 is
$$M_1 = \left(\frac{3+5}{2}, \frac{-4-2}{2}, \frac{1+1}{2}\right) = (4, -3, 1)$$

EXERCISES

1. Point P is on a segment through $P_1 = (3, -1, 5)$ and $P_2 = (7, 3, -1)$. Find P if $\overrightarrow{P_1P} = \frac{3}{5}\overrightarrow{P_1P_2}$.

2. Find point P on a segment through $P_1 = (3, -1, 5)$ and $P_2 = (7, 3, -1)$, if $\overrightarrow{P_1P} = \frac{3}{5}\overrightarrow{PP_2}$.

3. One end of a segment is at $(6, -1, 2)$, and a point three-fifths of the way along this segment from that end is at $(2, 3, -1)$. Find the coordinates of the other end.

4. What are the mid-points of the sides of a triangle whose vertices are $A(-1, 1, 1)$, $B(2, 3, -1)$, and $C(3, 5, -7)$?

5. If one end of a segment is at $(5, -6, 3)$ and the mid-point of the segment is at $(3, -2, -1)$, what are the coordinates of the other end?

6. Find the coordinates of the points at which the line containing the segment whose end points are $P_1(6, -1, 2)$ and $P_2(1, 3, -4)$ cuts the yz-plane, the xy-plane, and the xz-plane.

7. Prove that the lines joining the mid-points of the opposite sides of any quadrilateral bisect each other.

8. If $P_1(5, 3, -1)$, $P_2(-1, -2, 4)$, and $P_3(7, 1, -6)$ are the mid-points of the sides of a triangle, find the coordinates of the vertices.

9. Find the vertices of a triangle if the mid-points of the sides are $P_1(-1, 2, 3)$, $P_2(3, 6, -5)$, and $P_3(0, -4, -1)$.

REVIEW EXERCISES

1. Where are the points for which $x = 0$?
2. Where are the points for which $y = 0$? Where are those for which $z = 0$?

3. Where are the points for which *both* y and z are zero?
4. Where is the point (0, 0, 0)?
5. Is there any point on the x-axis for which $x = 0$?
6. Is there any point on the z-axis for which $z = 5$?
7. Select a point $P_1(x, y, z)$ and determine how each of the following points is related to P_1: $P_2 = (-x, y, z)$; $P_3 = (-x, -y, z)$; $P_4 = (x, -y, z)$; $P_5 = (x, y, -z)$; $P_6 = (-x, y, -z)$; $P_7 = (-x, -y, -z)$; $P_8 = (x, -y, -z)$.
8. Select a point $P = (x, y, z)$ and determine how each of the following points is related to P: $Q_1 = (0, y, z)$; $Q_2 = (x, 0, z)$; $Q_3 = (x, y, 0)$.
9. Select a point $P = (x, y, z)$ and determine how each of the following points is related to P: $A = (x, 0, 0)$; $B = (0, y, 0)$; $C = (0, 0, z)$.
10. Where are the points for which $|x| > 2$?
11. Where are the points for which $|x| \leq 4$?
12. Where are the points for which $|x| < 10$, $|y| < 10$, and $|z| < 10$?
13. Where are all the points P which have the same x-coordinate?
14. Where are all the points P which have the same x-coordinate and the same y-coordinate?
15. Determine the distance to the point $P = (x, y, z)$ from the z-axis.
16. Determine the scalar components of the segment from $P_1(2, -3, 4)$ to $P_2(3, -3, 8)$.
17. Determine the scalar components of the segment from $P_1(0, 0, 0)$ to $P_2(3, -4, 2)$.
18. Show that the scalar components of the segment from the origin to $P = (x, y, z)$ are x, y, and z.
19. Determine the coordinates of the terminal point of the directed segment whose initial point is $(3, 2, -4)$ and whose scalar components are $-4, 3,$ and -7.
20. Determine the coordinates of the initial point of the directed segment whose terminal point is $(-1, 0, 3)$ and whose scalar components are $-4, 1,$ and 0.
21. Determine a point on the x-axis which is equally distant from the two points $(2, 4, -3)$ and $(-3, 5, 1)$.
22. Show that the points $(3, 0, 0)$, $(-3, 0, 0)$, and $(0, 3\sqrt{3}, 0)$ are the vertices of an equilateral triangle.
23. Show that, if $P_1, P_2, P_3,$ and P_4 are any four points in space, then the mid-points of P_1P_2, P_2P_3, P_3P_4, and P_4P_1 are the vertices of a parallelogram.
24. Determine the point P which lies on the line through $P_1 = (2, 3, 1)$ and $P_2 = (1, 5, -2)$ and the z-coordinate of which is zero. In what ratio does P divide P_1P_2?
25. Determine the mid-points of the six edges of the tetrahedron whose vertices are $(1, 0, 0)$, $(0, 1, 0)$, $(0, 0, 1)$, and $(0, 0, 0)$.

26. Find P on the segment through $P_1 = (3, -1, 4)$ and $P_2 = (-1, 2, -1)$, if $|\overrightarrow{P_1P}| = \frac{2}{3}|\overrightarrow{P_1P_2}|$.

27. Determine the point P which lies on the line through $A = (2, 3, 1)$ and $B = (1, 5, -2)$ and the x-coordinate of which is zero. In what ratio does P divide AB?

28. Determine the point A which lies on the line through $B = (2, 3, 1)$ and $C = (1, 5, -2)$ and the y-coordinate of which is zero. In what ratio does A divide BC?

29. Find the points where the line containing $P_1 = (3, -2, 1)$ and $P_2 = (1, 3, -2)$ intersects the xy-plane; xz-plane; yz-plane.

30. Prove that $\mathbf{u} \cdot (\mathbf{v} + \mathbf{w}) = \mathbf{u} \cdot \mathbf{v} + \mathbf{u} \cdot \mathbf{w}$.

31. Prove that $(\mathbf{u} + \mathbf{v}) \cdot (\mathbf{w} + \mathbf{q}) = \mathbf{u} \cdot \mathbf{w} + \mathbf{u} \cdot \mathbf{q} + \mathbf{v} \cdot \mathbf{w} + \mathbf{v} \cdot \mathbf{q}$.

32. Show that $|\mathbf{u} \cdot \mathbf{v}| \leq |\mathbf{u}| \, |\mathbf{v}|$. When is $|\mathbf{u} \cdot \mathbf{v}| = |\mathbf{u}| \, |\mathbf{v}|$?

10

The Plane

10–1. An equation of a plane

We shall think of a plane as the set of all points P in space such that $\overrightarrow{P_1P}$ is perpendicular to a fixed non-zero vector through a fixed point P_1 on the plane. See Fig. 10–1. We say that the plane is perpendicular to the fixed vector. From the definition of a plane it is clear that, if two points lie on a plane, the line containing the points lies on the plane.

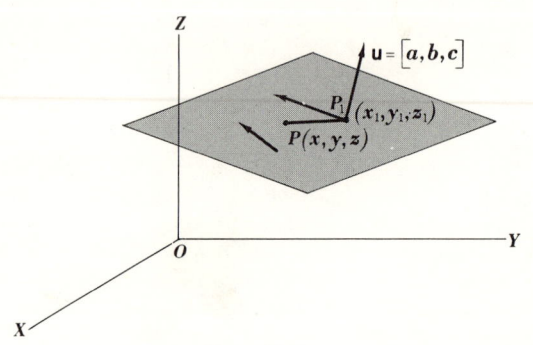

Fig. 10–1

Consider $P_1 = (x_1, y_1, z_1)$ to be a fixed point in space and $\mathbf{u} = [a, b, c]$ to be a non-zero vector represented by a segment having its initial point at P_1; then *the set of all points P in space having the property that $\mathbf{u} \cdot \overrightarrow{P_1P} = 0$ is a plane.* Since $\overrightarrow{P_1P} = [x - x_1, y - y_1, z - z_1]$, we have

$$a(x - x_1) + b(y - y_1) + c(z - z_1) = 0$$

Expanding this, we obtain $ax - ax_1 + by - by_1 + cz - cz_1 = 0$; and, denoting $-ax_1 - by_1 - cz_1$ by d, we find that an equation of a plane in space is of the form

$$ax + by + cz + d = 0 \qquad (10\text{–}1)$$

where a, b, and c are not all 0. Hence, it follows that an equation of a plane is a linear equation.

356

Conversely, any linear equation represents a plane. Let the given equation be $ax+by+cz+d=0$; and let (x_1, y_1, z_1) be any triple of numbers that satisfy the equation. It is clear that such numbers exist. Then, $ax_1+by_1+cz_1+d=0$; and, subtracting this from $ax+by+cz+d=0$, we obtain $a(x-x_1)+b(y-y_1)+c(z-z_1)=0$. But this tells us that the fixed vector $[a, b, c]$ is perpendicular to every vector $[x-x_1, y-y_1, z-z_1]$ for which $x_1, y_1,$ and z_1 satisfy the equation $ax+by+cz+d=0$. Since the point $P_1=(x_1, y_1, z_1)$ is a fixed point and $P=(x, y, z)$ is a variable point in space, it follows that $P=(x, y, z)$ lies on a plane which is perpendicular to the vector $[a, b, c]$ and contains the point $P_1(x_1, y_1, z_1)$.

The vector $\mathbf{u}=[a, b, c]$ is called a *normal* vector to the plane $ax+by+cz+d=0$. In fact, *any vector perpendicular to the plane is called a normal vector to the plane; and any line containing a segment representing a normal vector is a normal line.* The vector $\mathbf{u}=[a, b, c]$ is also called the *coefficient vector* of the equation $ax+by+cz+d=0$.

EXAMPLE 10-1. Find an equation of the plane that passes through the point $(-1, 2, 3)$ and is perpendicular to the vector $\mathbf{u}=[1, -1, -2]$.

Solution. Since $[a, b, c]$ is a vector perpendicular to the plane $ax+by+cz+d=0$, we let $a=1$, $b=-1$, and $c=-2$. Then, we obtain $x-y-2z=k$. Since $P=(-1, 2, 3)$ lies on the plane, we determine k by substituting the coordinates of this point for x, y, and z and obtain $(-1)-(2)-2(3)=k$ or $k=-9$. So, an equation of the required plane is

$$x-y-2z+9=0$$

If one of the coefficients a, b, or c in the linear equation is zero, the equation represents a plane parallel to the axis of the missing variable. To see this, we consider the equation $ax+by+d=0$. A normal vector to this plane is $\mathbf{u}=[a, b, 0]$; and, since the direction cosines of this vector are $\frac{a}{|\mathbf{u}|}$, $\frac{b}{|\mathbf{u}|}$, and 0, it follows that $C=90°$ (see Section 9-5). Hence, any segment representing $\mathbf{u}$ is perpendicular to the z-axis. Since a plane is perpendicular to its normal, it follows that $ax+by+d=0$ is a plane parallel to the z-axis.

If two of the coefficients a, b, and c are zero, the equation represents a plane that is perpendicular to the axis of the variable that appears. Thus, $ax+d=0$ is an equation of a plane perpendicular to the x-axis. To see this we notice that a normal vector to the plane is $\mathbf{u}=[a, 0, 0]$. Hence, $m=n=0$, or $B=C=90°$; and any segment representing $\mathbf{u}$ is perpendicular to both the y-axis and the z-axis. Therefore, any normal is parallel to the x-axis, and the plane is *perpendicular* to the x-axis.

EXERCISES

1. Prove that the set of points satisfying $ax+by+cz+d=0$ has the property that a vector represented by the segment joining any two of its points is perpendicular to the vector $\mathbf{u}=[a, b, c]$.

2. Prove that $ax+cz+d=0$ is an equation of a plane parallel to the y-axis.

3. Prove that $by+cz+d=0$ is an equation of a plane parallel to the x-axis.

4. Prove that $by+d=0$ is an equation of a plane perpendicular to the y-axis.

5. Prove that $cz+d=0$ is an equation of a plane perpendicular to the z-axis.

6. In each of the following cases, find an equation of the plane containing the given point P and having the given normal vector $\mathbf{u}$:
   (a) $P=(-1, 2, -3)$ and $\mathbf{u}=[3, 1, 2]$
   (b) $P=(2, -1, 5)$ and $\mathbf{u}=[-1, 1, 1]$
   (c) $P=(3, 2, -4)$ and $\mathbf{u}=[5, -2, 3]$

7. Sketch each of the following planes by finding the points where the plane cuts the axes:
   (a) $x-y+2z-2=0$
   (b) $3x-y+6z-3=0$
   (c) $x-2y+3z=0$
   (d) $2x+3y-z+6=0$
   (e) $x-3y+2z=0$
   (f) $3x-y+2z-4=0$

8. Find a normal vector to each of the planes in Exercise 7.

9. What is an equation of the plane that contains the mid-point of the segment from $P_1(2, -3, 5)$ to $P_2(4, 5, -3)$ and is perpendicular to the segment?

10. Find an equation of the plane that is perpendicular to the segment from $P_1(1,1, -1)$ to $P_2(3, -5, 5)$ and contains the point P_3 such that $\overrightarrow{P_1P_3}=\frac{5}{2}\overrightarrow{P_1P_2}$.

11. Find an equation of the plane that is parallel to the x-axis and contains the points $P_1(2, -1, 3)$ and $P_2(-1, 0, 5)$.

12. Find an equation of the plane that is parallel to the z-axis and contains the points $P_1(3, -2, 1)$ and $P_2(-1, 2, -1)$.

13. Find an equation of the plane that is parallel to the y-axis and contains the points $P_1(-2, -1, 3)$ and $P_2(1, 1, 1)$.

14. Find equations of the planes that are perpendicular to the segment from $P_1(3, -1, 2)$ to $P_2(-1, 3, 5)$ and contain the points at which the line containing this segment intersects the xy-plane, yz-plane, and xz-plane, respectively.

15. What is an equation of the plane that passes through the origin and is perpendicular to the vector $[-1, 2, 1]$?

16. Find an equation of each of the following planes:

    (a) It is parallel to the xy-plane and contains the point $(3, 2, -5)$.
    (b) It is parallel to the yz-plane and contains the point $(1, -1, 2)$.
    (c) It is perpendicular to the y-axis and also contains the point $(-2, -1, 3)$.

17. What are equations of the reference planes?

18. The vector $\mathbf{u} = [2, -3, 5]$ has its initial point at the origin. Find an equation of the plane having this vector as its normal and containing the terminal point of $\mathbf{u}$.

10–2. Parallel and perpendicular planes

Two planes will be said to be *parallel* when their normal vectors, or coefficients vectors, are parallel. Thus, the planes $a_1x + b_1y + c_1z + d_1 = 0$ and $a_2x + b_2y + c_2z + d_2 = 0$ are parallel if there is a number k such that $a_1 = ka_2$, $b_1 = kb_2$, and $c_1 = kc_2$. Planes which coincide will be considered to be parallel.

Two planes will be said to be *perpendicular* if their normal vectors are perpendicular. Thus, the planes $a_1x + b_1y + c_1z + d_1 = 0$ and $a_2x + b_2y + c_2z + d_2 = 0$ are perpendicular if $a_1a_2 + b_1b_2 + c_1c_2 = 0$.

EXAMPLE 10–2. Find an equation of the plane that is parallel to $3x - 4y + z - 6 = 0$ and contains the point $(3, 2, -6)$.

Solution. Let $ax + by + cz + d = 0$ be an equation of the desired plane. Since the plane is to be parallel to $3x - 4y + z - 6 = 0$, we know that $a = 3k$, $b = -4k$, and $c = k$. A simple choice for a, b, and c, in order that these relations may be true, is: $a = 3$, $b = -4$, and $c = 1$. Then, we obtain $3x - 4y + z = -d$. But $-d$ must have a value such that the point $(3, 2, -6)$ is contained in the plane. Hence, we obtain $3(3) - 4(2) + (-6) = -d = -5$. An equation of the desired plane is $3x - 4y + z + 5 = 0$.

EXAMPLE 10–3. Find b in order that the plane $2x - 3y + 9z - 4 = 0$ will be perpendicular to the plane $3x + by - z + 2 = 0$.

Solution. Since the condition for perpendicularity is $a_1a_2 + b_1b_2 + c_1c_2 = 0$, we know that $2(3) - 3(b) + 9(-1) = 0$ or $b = -1$. Hence, the desired equation is $3x - y - z + 2 = 0$.

360 ANALYTIC GEOMETRY OF SPACE

EXERCISES

1. In each of the following cases, find an equation of the plane which is parallel to the given plane and contains the given point:
   (a) $2x+y-3z+6=0$ and $(2,-2,-5)$
   (b) $3x+y+z-5=0$ and $(0,0,0)$
   (c) $x-2y+2z+6=0$ and $(1,-1,1)$
   (d) $5x+3y-6z-12=0$ and $(2,-3,4)$

2. Find k if the following pairs of planes are perpendicular:
   (a) $2x+y-3z+6=0$ and $x-ky+2z-5=0$
   (b) $x-y+z-3=0$ and $kx+2y-z+5=0$
   (c) $3x-3y+2z-4=0$ and $2x+3y-kz+6=0$
   (d) $kx-y+3kz+6=0$ and $x+6y+z=2$

3. Determine which of the following planes are parallel and which are perpendicular:
   (a) $2x+3y-z+6=0$
   (b) $3x-y+3z-2=0$
   (c) $6x+9y-3z+12=0$
   (d) $x-y-z-2=0$
   (e) $4x+6y-z=0$
   (f) $x+2y-8z=10$

10–3. Intercept equation of a plane

The x-coordinate of the point of intersection of a plane with the x-axis is called the *x-intercept* of the plane; the y-coordinate of the point of intersection of a plane with the y-axis is called the *y-intercept* of the plane; and the z-coordinate of the point of intersection of a plane with the z-axis is called the *z-intercept* of the plane. It is a matter of simple algebra to rewrite the equation $ax+by+cz+d=0$ in the form $\dfrac{x}{-\frac{d}{a}}+\dfrac{y}{-\frac{d}{b}}+\dfrac{z}{-\frac{d}{c}}=1$, provided a, b, c, and d are not equal to 0. If we let $f=-\dfrac{d}{a}$, $g=-\dfrac{d}{b}$, and $h=-\dfrac{d}{c}$, the equation becomes

$$\frac{x}{f}+\frac{y}{g}+\frac{z}{h}=1 \tag{10-2}$$

The intercepts of the plane are f, g, and h, respectively. Equation (10–2) is called the *intercept equation* of a plane since, when the equation is written in this form, the denominator of the x-term is the x-intercept, the denominator of the y-term is the y-intercept, and the denominator of the z-term is the z-intercept.

EXAMPLE 10–4. Write the equation $2x-3y+6z-8=0$ in the intercept form.

Solution. If we transpose the constant 8 to the right-hand side, we have $2x - 3y + 6z = 8$. Now, when we divide through by 8, we obtain

$$\frac{2x}{8} - \frac{3y}{8} + \frac{6z}{8} = 1$$

or

$$\frac{x}{4} + \frac{y}{-\frac{8}{3}} + \frac{z}{\frac{4}{3}} = 1$$

Hence, the x-intercept is 4, the y-intercept is $-\frac{8}{3}$, and the z-intercept is $\frac{4}{3}$.

EXAMPLE 10-5. What is an equation of the plane whose x-intercept is 2, whose y-intercept is -1, and whose z-intercept is -3?

Solution. Using the intercept form, we obtain as an equation of the plane

$$\frac{x}{2} + \frac{y}{-1} + \frac{z}{-3} = 1$$

EXERCISES

1. Write each of the following equations in the intercept form:
   (a) $ax + by + cz + d = 0$
   (b) $2x - 3y + z + 6 = 0$
   (c) $3x - 5y - 2z + 8 = 0$
   (d) $2y - 3x - 4 + z = 0$
   (e) $x + y - 2z + 3 = 0$
   (f) $3x + 2y - 4z - 12 = 0$

2. Find an equation of each of the following planes for which the intercepts are:
   (a) $f = 3$, $g = 4$, and $h = -1$
   (b) $f = -1$, $g = -3$, and $h = 2$
   (c) $f = 2$, $g = \frac{4}{3}$, and $h = -\frac{1}{2}$
   (d) $f = -1$, $g = 2$, and $h = 3$

10-4. Vector product of two vectors

If $\mathbf{u} = [u_1, u_2, u_3]$ and $\mathbf{v} = [v_1, v_2, v_3]$, we define the vector product of $\mathbf{u}$ and $\mathbf{v}$, which is written $\mathbf{u} \times \mathbf{v}$, to be the vector $[u_2v_3 - u_3v_2, u_3v_1 - u_1v_3, u_1v_2 - u_2v_1]$. This is also called the cross product of $\mathbf{u}$ and $\mathbf{v}$. Another way to write this product is

$$\mathbf{u} \times \mathbf{v} = \begin{bmatrix} \begin{vmatrix} u_2 & u_3 \\ v_2 & v_3 \end{vmatrix}, & \begin{vmatrix} u_3 & u_1 \\ v_3 & v_1 \end{vmatrix}, & \begin{vmatrix} u_1 & u_2 \\ v_1 & v_2 \end{vmatrix} \end{bmatrix} \quad (10\text{-}3)$$

An easy method of obtaining the vector product of $\mathbf{u}$ and $\mathbf{v}$ is as follows: First we will form the matrix* $\begin{bmatrix} u_1 & u_2 & u_3 \\ v_1 & v_2 & v_3 \end{bmatrix}$ with the scalar components of $\mathbf{u}$ forming the first row. Next, delete the first column and form

*See Appendix A

362 ANALYTIC GEOMETRY OF SPACE

the determinant $\begin{vmatrix} u_2 & u_3 \\ v_2 & v_3 \end{vmatrix}$; delete the second column and form the determinant $\begin{vmatrix} u_3 & u_1 \\ v_3 & v_1 \end{vmatrix}$; and delete the third column and form the determinant $\begin{vmatrix} u_1 & u_2 \\ v_1 & v_2 \end{vmatrix}$. Write these three determinants, in the order given, as the scalar components of the vector $\mathbf{u} \times \mathbf{v}$.

If we represent $\mathbf{u} \times \mathbf{v}$ by the vector $\mathbf{q}$, this product vector is perpendicular to both $\mathbf{u}$ and $\mathbf{v}$. To prove this, we have

$$\mathbf{u} \cdot \mathbf{q} = u_1 \begin{vmatrix} u_2 & u_3 \\ v_2 & v_3 \end{vmatrix} + u_2 \begin{vmatrix} u_3 & u_1 \\ v_3 & v_1 \end{vmatrix} + u_3 \begin{vmatrix} u_1 & u_2 \\ v_1 & v_2 \end{vmatrix} = \begin{vmatrix} u_1 & u_2 & u_3 \\ u_1 & u_2 & u_3 \\ v_1 & v_2 & v_3 \end{vmatrix} = 0$$

Since $\mathbf{u} \cdot \mathbf{q} = 0$, it follows that $\mathbf{u}$ is perpendicular to $\mathbf{q}$. In exactly the same way, we have

$$\mathbf{v} \cdot \mathbf{q} = \begin{vmatrix} v_1 & v_2 & v_3 \\ u_1 & u_2 & u_3 \\ v_1 & v_2 & v_3 \end{vmatrix} = 0$$

Since $\mathbf{v} \cdot \mathbf{q} = 0$, $\mathbf{v}$ is perpendicular to $\mathbf{q}$. Hence, $\mathbf{q}$ is perpendicular to both $\mathbf{u}$ and $\mathbf{v}$. See Fig. 10–2.

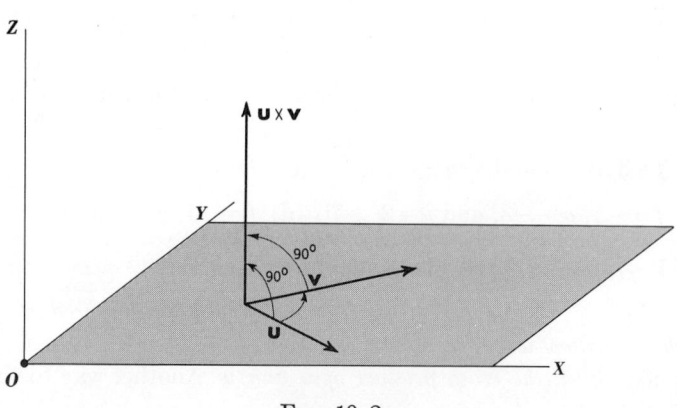

FIG. 10–2

Coplanar vectors: *Three distinct vectors $\mathbf{u}$, $\mathbf{v}$, and $\mathbf{w}$, no one of which is the zero vector, are coplanar (lie in a plane)* if, and only if,* $\begin{vmatrix} u_1 & u_2 & u_3 \\ v_1 & v_2 & v_3 \\ w_1 & w_2 & w_3 \end{vmatrix} = 0.$

---

* Three vectors will be said to lie in a plane if representative segments of the vectors exist which lie in the same plane. Hence, the vectors lie in a plane if there is a vector which is perpendicular to all of them; and the converse is also true.

To prove this, assume that **u**, **v**, and **w** are coplanar. Then **u** is perpendicular to $\mathbf{v} \times \mathbf{w}$, whence $\mathbf{u} \cdot (\mathbf{v} \times \mathbf{w}) = \begin{vmatrix} u_1 & u_2 & u_3 \\ v_1 & v_2 & v_3 \\ w_1 & w_2 & w_3 \end{vmatrix} = 0$. Conversely, if

$\begin{vmatrix} u_1 & u_2 & u_3 \\ v_1 & v_2 & v_3 \\ w_1 & w_2 & w_3 \end{vmatrix} = 0$, we have $\mathbf{u} \cdot (\mathbf{v} \times \mathbf{w}) = 0$. Hence, **u** is perpendicular to $\mathbf{v} \times \mathbf{w}$; so **u**, **v**, and **w** are coplanar.

EXAMPLE 10–6. If $\mathbf{i} = [1, 0, 0]$, $\mathbf{j} = [0, 1, 0]$, and $\mathbf{k} = [0, 0, 1]$, prove that $\mathbf{i} \times \mathbf{j} = \mathbf{k}$.

Solution. The vectors are shown in Fig. 10–3. From equation (10–3),

$$\mathbf{i} \times \mathbf{j} = \left[\begin{vmatrix} 0 & 0 \\ 1 & 0 \end{vmatrix}, \begin{vmatrix} 0 & 1 \\ 0 & 0 \end{vmatrix}, \begin{vmatrix} 1 & 0 \\ 0 & 1 \end{vmatrix} \right] = [0, 0, 1] = \mathbf{k}$$

EXAMPLE 10–7. If **u** and **v** are non-zero vectors, prove that $|\mathbf{u} \times \mathbf{v}|$ is the area of a parallelogram having representative segments of **u** and **v** as adjacent sides.

Solution. We shall choose $\mathbf{u} = [u_1, 0, 0]$ and $\mathbf{v} = [v_1, v_2, 0]$, as indicated in Fig. 10–4. This choice simplifies the work without interfering with the generality of the problem. We know that the area of a parallelogram is equal to the product of the lengths of two adjacent sides and the sine of the angle between these sides. So, area $= |\mathbf{u}|\,|\mathbf{v}| \sin \theta = |\mathbf{u}|\,|\mathbf{v}|\sqrt{1 - \cos^2 \theta}$, where $\cos \theta = \dfrac{\mathbf{u} \cdot \mathbf{v}}{|\mathbf{u}|\,|\mathbf{v}|}$. Hence,

$$\text{Area} = |\mathbf{u}|\,|\mathbf{v}|\sqrt{1 - \left(\frac{\mathbf{u} \cdot \mathbf{v}}{|\mathbf{u}|\,|\mathbf{v}|}\right)^2} = \sqrt{|\mathbf{u}|^2\,|\mathbf{v}|^2 - (\mathbf{u} \cdot \mathbf{v})^2}$$

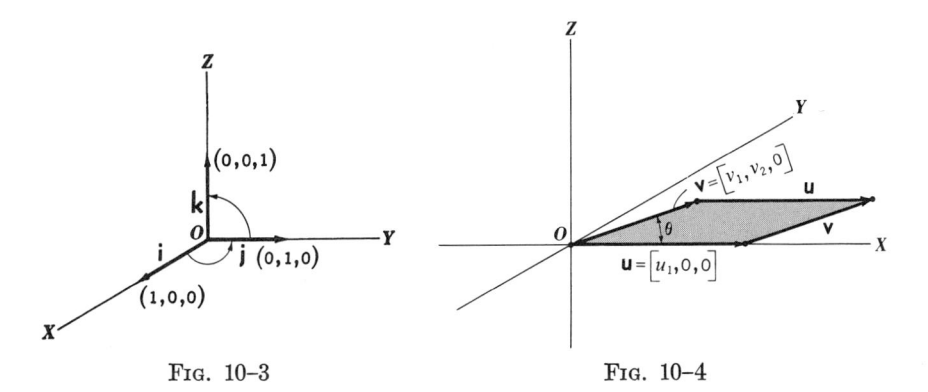

FIG. 10–3

FIG. 10–4

Since $\mathbf{u} \cdot \mathbf{v} = u_1 v_1$, $|\mathbf{u}|^2 = u_1^2$, and $|\mathbf{v}|^2 = v_1^2 + v_2^2$, we obtain

$$\text{Area} = \sqrt{u_1^2 v_1^2 + u_1^2 v_2^2 - u_1^2 v_1^2} = |u_1 v_2|$$

But $\mathbf{u} \times \mathbf{v} = [0, 0, u_1 v_2]$. Therefore, $|\mathbf{u} \times \mathbf{v}| = |u_1 v_2|$ and area = $|\mathbf{u} \times \mathbf{v}|$.

EXAMPLE 10-8. Prove: (a) that $\mathbf{u} \times \mathbf{u} = \mathbf{0}$; (b) that, if $\mathbf{u} \times \mathbf{v} = \mathbf{0}$, then $\mathbf{u} = \mathbf{0}$ or $\mathbf{v} = \mathbf{0}$ or $\mathbf{u}$ is parallel to $\mathbf{v}$.

(a) It follows from equation (10-3) that $\mathbf{u} \times \mathbf{u} = \mathbf{0}$. Thus,

$$\mathbf{u} \times \mathbf{u} = \left[\begin{vmatrix} u_2 & u_3 \\ u_2 & u_3 \end{vmatrix}, \begin{vmatrix} u_3 & u_1 \\ u_3 & u_1 \end{vmatrix}, \begin{vmatrix} u_1 & u_2 \\ u_1 & u_2 \end{vmatrix} \right] = [0, 0, 0]$$

(b) If neither $\mathbf{u}$ nor $\mathbf{v}$ is the zero vector, let us assume that

$$\mathbf{u} \times \mathbf{v} = \left[\begin{vmatrix} u_2 & u_3 \\ v_2 & v_3 \end{vmatrix}, \begin{vmatrix} u_3 & u_1 \\ v_3 & v_1 \end{vmatrix}, \begin{vmatrix} u_1 & u_2 \\ v_1 & v_2 \end{vmatrix} \right] = [0, 0, 0]$$

Hence, $\begin{vmatrix} u_2 & u_3 \\ v_2 & v_3 \end{vmatrix} = 0$, $\begin{vmatrix} u_3 & u_1 \\ v_3 & v_1 \end{vmatrix} = 0$, and $\begin{vmatrix} u_1 & u_2 \\ v_1 & v_2 \end{vmatrix} = 0$. From these relations it follows that there is a number k such that $u_1 = kv_1$, $u_2 = kv_2$, and $u_3 = kv_3$. Therefore, $\mathbf{u}$ is parallel to $\mathbf{v}$.

If $\mathbf{u} = \mathbf{0}$ and $\mathbf{v} \neq \mathbf{0}$, we know that $\mathbf{0} \times \mathbf{v} = [0, 0, 0]$. A similar argument holds if $\mathbf{v} = \mathbf{0}$ and $\mathbf{u} \neq \mathbf{0}$.

10-5. The vector product: A summary

(a) From Section 10-4, we note that $\mathbf{u} \times \mathbf{v}$ is a vector. The magnitude of this vector, as shown in Example 10-7, may be expressed as

$$|\mathbf{u} \times \mathbf{v}| = |\mathbf{u}| \, |\mathbf{v}| \sin \theta, \quad (0 \leq \theta \leq 180°)$$

where θ is the angle between $\mathbf{u}$ and $\mathbf{v}$.

If we let $\mathbf{u} \times \mathbf{v} = \mathbf{q}$, then $\mathbf{q}$ is perpendicular to the plane of $\mathbf{u}$ and $\mathbf{v}$ with its direction determined by the right-handed convention (see Fig. 10-2.)

Therefore, making use of the fact that $\frac{\mathbf{q}}{|\mathbf{q}|}$ is a unit vector in the direction of $\mathbf{q}$ (see Example 9-11, p. 361), we can write

$$\mathbf{u} \times \mathbf{v} = |\mathbf{u}| \, |\mathbf{v}| \sin \theta \frac{\mathbf{q}}{|\mathbf{q}|}, \quad (0 \leq \theta \leq 180°).$$

(b) The unit vectors $\mathbf{i} = [1,0,0]$, $\mathbf{j} = [0,1,0]$, and $\mathbf{k} = [0,0,1]$, introduced in Example 10-6, are of particular importance. These three vectors are unit vectors in the direction of the coordinate axes, as shown in Fig. 10-3. (In the study of vectors and matrices these three vectors are called *basis vectors*.)

Note from Example 10–6 that $\mathbf{i} \times \mathbf{j} = \mathbf{k}$. By using this and examining Fig. 10–3, the student should show that $\mathbf{j} \times \mathbf{k} = \mathbf{i}$; $\mathbf{k} \times \mathbf{i} = \mathbf{j}$; $\mathbf{j} \times \mathbf{i} = -\mathbf{k}$; $\mathbf{i} \times \mathbf{k} = -\mathbf{j}$; $\mathbf{k} \times \mathbf{j} = -\mathbf{i}$.

Note further that $\mathbf{i} \times \mathbf{i} = \mathbf{j} \times \mathbf{j} = \mathbf{k} \times \mathbf{k} = 0$, and, in fact, as is shown in Example 10–8, $\mathbf{u} \times \mathbf{u} = 0$, where $\mathbf{u}$ is any vector.

By using the unit (basis) vectors $\mathbf{i} = [1,0,0]$, $\mathbf{j} = [0,1,0]$, and $\mathbf{k} = [0,0,1]$, we can express any vector $\mathbf{u} = [u_1, u_2, u_3]$ as the sum of vector components:

$$\mathbf{u} = [u_1,0,0] + [0,u_2,0] + [0,0,u_3]$$
$$= u_1[1,0,0] + u_2[0,1,0] + u_3[0,0,1]$$
$$= u_1\mathbf{i} + u_2\mathbf{j} + u_3\mathbf{k}$$

Similarly, $\mathbf{v} = [v_1,v_2,v_3] = v_1\mathbf{i} + v_2\mathbf{j} + v_3\mathbf{k}$. (See Section 9–6, p. 344.)

(c) From Section 10–4, definition (10–3),

$$\mathbf{q} = \mathbf{u} \times \mathbf{v} = \left[\begin{vmatrix} u_2 & u_3 \\ v_2 & v_3 \end{vmatrix}, \begin{vmatrix} u_3 & u_1 \\ v_3 & v_1 \end{vmatrix}, \begin{vmatrix} u_1 & u_2 \\ v_1 & v_2 \end{vmatrix} \right]$$

Writing this vector in terms of the basis vectors $\mathbf{i}$, $\mathbf{j}$ and $\mathbf{k}$, we have:

$$\mathbf{q} = \mathbf{u} \times \mathbf{v} = \begin{vmatrix} u_2 & u_3 \\ v_2 & v_3 \end{vmatrix} \mathbf{i} + \begin{vmatrix} u_3 & u_1 \\ v_3 & v_1 \end{vmatrix} \mathbf{j} + \begin{vmatrix} u_1 & u_2 \\ v_1 & v_2 \end{vmatrix} \mathbf{k}$$

The right-hand member of this last equation suggests that we can write the vector product $\mathbf{u} \times \mathbf{v}$ in the form of a determinant:

$$\begin{vmatrix} \mathbf{i} & \mathbf{j} & \mathbf{k} \\ u_1 & u_2 & u_3 \\ v_1 & v_2 & v_3 \end{vmatrix}$$

This is not a determinant as previously defined since the elements in the first row are vectors, not numbers. However, if we treat this as a determinant and expand it about the first row (Theorem 1, p. 11), we obtain the correct result for the vector product $\mathbf{u} \times \mathbf{v}$. Hence, with this understanding of the use of the above symbol for a determinant, we can write

$$\mathbf{u} \times \mathbf{v} = \begin{vmatrix} \mathbf{i} & \mathbf{j} & \mathbf{k} \\ u_1 & u_2 & u_3 \\ v_1 & v_2 & v_3 \end{vmatrix} = \begin{vmatrix} u_2 & u_3 \\ v_2 & v_3 \end{vmatrix} \mathbf{i} + (-1) \begin{vmatrix} u_1 & u_3 \\ v_1 & v_3 \end{vmatrix} \mathbf{j} + \begin{vmatrix} v_1 & u_2 \\ v_1 & v_2 \end{vmatrix} \mathbf{k}$$

$$= (u_2v_3 - u_3v_2)\mathbf{i} + (u_3v_1 - u_1v_3)\mathbf{j} + (u_1v_2 - u_2v_1)\mathbf{k}$$

This is a useful form to remember.

(d) For example, the vector product, $u \times v$, where $u = [1,3,4]$ and $v = [1,0,2]$ is:

$$\begin{vmatrix} i & j & k \\ 1 & 3 & 4 \\ 1 & 0 & 2 \end{vmatrix} = \begin{vmatrix} 3 & 4 \\ 0 & 2 \end{vmatrix} i + (-1) \begin{vmatrix} 1 & 4 \\ 1 & 2 \end{vmatrix} j + \begin{vmatrix} 1 & 3 \\ 1 & 0 \end{vmatrix} k = 6i + 2j - 3k$$

EXERCISES

1. Given the basis vectors $i = [1,0,0]$, $j = [0,1,0]$, $k = [0,0,1]$, evaluate the following:
   (a) $j \times k$
   (b) $k \times j$
   (c) $i \times i$
   (d) $i \times k$
   (e) $(2j) \times (3k)$
   (f) $(3i) \times (-2k)$

2. If $u = 2i - 3j - k$ and $v = i + 4j - 2k$, find:
   (a) $u \times v$
   (b) $v \times u$
   (c) $(u+v) \times (u-v)$

3. If $u = 3i - j + 2k$ and $v = 2i + j - k$, find:
   (a) $u \times v$
   (b) $v \times u$
   (c) $(u+v) \times (u-v)$

4. Find the area of the triangle having vertices at the points
   $$P(1,3,2), \quad Q(2,-1,1), \quad R(-1,2,3)$$

5. Determine a unit vector perpendicular to the vectors
   $$u = 2i - 6j - 3k \text{ and } v = 4i + 3j - k$$

6. If $u = 3i - j - 2k$ and $v = 2i + 3j + k$, find:
   (a) $|u \times v|$
   (b) $(u + 2v) \times (2u - v)$
   (c) $|(u + v) \times (u - v)|$

7. If $u = 2i + j - 3k$ and $v = i - 2j + k$, find a vector of magnitude 5 perpendicular to both u and v.

8. Show that: $(u \times v) = -(v \times u)$.

9. Find the area of a parallelogram having diagonals
   $$u = 3i + j - 2k \text{ and } v = i - 3j + 4k$$

10. If $u = 4i - j + 3k$ and $v = -2i + j - 2k$, find a unit vector perpendicular to both u and v.

11. Find a unit vector parallel to the xy-plane and perpendicular to the vector $u = 4i - 3j + k$.

12. Show that $u = \frac{2}{3}i - \frac{2}{3}j + \frac{1}{3}k$, $v = \frac{1}{3}i + \frac{2}{3}j + \frac{2}{3}k$, and $w = \frac{2}{3}i + \frac{1}{3}j - \frac{2}{3}k$ are mutually orthogonal unit vectors.

13. Show that $(av) \times bw = ab(v \times w)$, where a and b are scalars.

10–6. Alternating product, or scalar triple product, of three vectors

Let u, v, and w be three vectors. We define the alternating product, or scalar triple product, of these three vectors to be $u \cdot (v \times w)$. A convenient notation for this product is (uvw). It is important to notice that the alternating product is a number. From Section 10–4, definition (10–3),

$$v \times w = \left[\begin{vmatrix} v_2 & v_3 \\ w_2 & w_3 \end{vmatrix}, \begin{vmatrix} v_3 & v_1 \\ w_3 & w_1 \end{vmatrix}, \begin{vmatrix} v_1 & v_2 \\ w_1 & w_2 \end{vmatrix} \right]$$

Then

$$u \cdot (v \times w) = u_1 \begin{vmatrix} v_2 & v_3 \\ w_2 & w_3 \end{vmatrix} + u_2 \begin{vmatrix} v_3 & v_1 \\ w_3 & w_1 \end{vmatrix} + u_3 \begin{vmatrix} v_1 & v_2 \\ w_1 & w_2 \end{vmatrix}$$

In determinant form,

$$u \cdot (v \times w) = \begin{vmatrix} u_1 & u_2 & u_3 \\ v_1 & v_2 & v_3 \\ w_1 & w_2 & w_3 \end{vmatrix}$$

We have already seen in Section 10–4 that, if three vectors are coplaner, their alternating product is zero.

EXERCISES

1. Prove that, in general, there exist vectors u and v for which $u \times v \neq v \times u$.

2. Prove that v is perpendicular to $u \times v$.

3. Prove that, if u is parallel to v, then $(uvw) = 0$, for any vector w.

4. Prove that $u \cdot (v \times w) = (u \times v) \cdot w$.

5. Find k in each of the following cases if the vectors in each group are coplanar:
   (a) $u = [3, -1, k]$, $v = [1, -1, 1]$, and $w = [-2, 3, 2]$
   (b) $u = [-1, 3, 2]$, $v = [2, 1, k]$, and $w = [1, 3, 5]$
   (c) $u = [4, -3, -1]$, $v = [-3, 1, 1]$, and $w = [k, 2, -1]$
   (d) $u = [2, 0, k]$, $v = [1, -1, 1]$, and $w = [0, 1, -3]$

368   ANALYTIC GEOMETRY OF SPACE

6. Find a vector perpendicular to both $\mathbf{u} = [-1, 3, 2]$ and $\mathbf{v} = [4, 3, -2]$.

7. Find two vectors each of which is perpendicular to both $\mathbf{u} = [3, 2, -5]$ and $\mathbf{v} = [-1, 5, 4]$.

8. Prove that $|(\mathbf{uvw})|$ is equal to the volume of a parallelepiped having representative segments of $\mathbf{u}$, $\mathbf{v}$, and $\mathbf{w}$ as adjacent edges. (Hint: Volume equals the area of the base times the altitude.)

9. Let $\mathbf{u}$ and $\mathbf{v}$ be non-zero vectors in the xy-plane. Show that $|\mathbf{u} \times \mathbf{v}|$ is invariant under rotation in this plane.

10. Find the area of each of the parallelograms having representative segments of the following vectors as adjacent sides:
    (a) $\mathbf{u} = [3, -1, 2]$ and $\mathbf{v} = [1, -2, 3]$.
    (b) $\mathbf{u} = [5, -2, 3]$ and $\mathbf{v} = [2, 3, -1]$.
    (c) $\mathbf{w} = [u_1, u_2, u_3]$ and $\mathbf{q} = [q_1, q_2, q_3]$.

11. Three vertices of a parallelogram are $P = (-1, 2, 3)$, $Q = (3, 1, 4)$, and $R = (1, -2, -1)$, in that order. Find the fourth vertex and the area of the parallelogram.

12. Three vertices of a parallelogram are $P_1 = (2, 1, -4)$, $P_2 = (-3, 2, -1)$, and $P_3 = (5, -1, 2)$, in that order. Find the fourth vertex and the area of the parallelogram.

13. Find the volume of each of the parallelepipeds having representative segments of the following vectors as adjacent edges:
    (a) $\mathbf{u} = [2, -1, 3]$, $\mathbf{v} = [1, 2, -1]$, $\mathbf{w} = [-2, 1, 0]$.
    (b) $\mathbf{p} = [3, 2, 1]$, $\mathbf{q} = [1, 0, 2]$, $\mathbf{r} = [-2, 3, -1]$.
    (c) $\mathbf{i} = [1, 0, 0]$, $\mathbf{j} = [0, 1, 0]$, $\mathbf{k} = [0, 0, 1]$.

14. Find $\mathbf{j} \times \mathbf{k}$; $\mathbf{k} \times \mathbf{i}$; and $\mathbf{j} \times \mathbf{i}$.

15. Find $\mathbf{i} \cdot (\mathbf{j} \times \mathbf{k})$; $(\mathbf{j} \times \mathbf{i}) \cdot \mathbf{k}$.

16. Given $P_1 = (2, 3, -3)$, $P_2 = (1, 1, -2)$, and $P_3 = (-1, 1, 4)$:
    (a) Find scalar components of two vectors in the plane determined by these three points.
    (b) Find a vector perpendicular to these two vectors.

17. If $\mathbf{u} = 2\mathbf{i} + 3\mathbf{j} - \mathbf{k}$ and $\mathbf{v} = -\mathbf{i} - 2\mathbf{j} + 3\mathbf{k}$, find $\mathbf{u} + \mathbf{v}$, $\mathbf{u} - \mathbf{v}$, $-2\mathbf{u}$, $|\mathbf{u} + \mathbf{v}|$, $|\mathbf{u} - \mathbf{v}|$, $|-2\mathbf{u}|$.

18. Show that
$$\mathbf{u} \times \mathbf{v} = \begin{vmatrix} \mathbf{i} & \mathbf{j} & \mathbf{k} \\ u_1 & u_2 & u_3 \\ v_1 & v_2 & v_3 \end{vmatrix}$$
where $\mathbf{u} = [u_1, u_2, u_3]$ and $\mathbf{v} = [v_1, v_2, v_3]$.

19. Prove that $(\mathbf{u}+\mathbf{v}) \times \mathbf{w} = (\mathbf{u} \times \mathbf{w}) + (\mathbf{v} \times \mathbf{w})$.
20. Prove that $\mathbf{u} \times (\mathbf{v}+\mathbf{w}) = (\mathbf{u} \times \mathbf{v}) + (\mathbf{u} \times \mathbf{w})$.
21. Prove that $(k\mathbf{u}) \times \mathbf{v} = k(\mathbf{u} \times \mathbf{v})$.
22. Prove that $\mathbf{u} \times (\mathbf{v} \times \mathbf{w}) = (\mathbf{u} \cdot \mathbf{w})\mathbf{v} - (\mathbf{u} \cdot \mathbf{v})\mathbf{w}$.
23. Prove that $(\mathbf{u} \times \mathbf{v}) \times \mathbf{w} = (\mathbf{u} \cdot \mathbf{w})\mathbf{v} - (\mathbf{v} \cdot \mathbf{w})\mathbf{u}$.

10–7. Vector triple products

Let $\mathbf{u}$, $\mathbf{v}$ and $\mathbf{w}$ be three vectors. We define the *vector triple product* of these vectors to be $\mathbf{u} \times (\mathbf{v} \times \mathbf{w})$ or $(\mathbf{u} \times \mathbf{v}) \times \mathbf{w}$. Although either of these is a vector triple product of the given vectors, but the two, in general, are not equal. Symbolically,

$$\mathbf{u} \times (\mathbf{v} \times \mathbf{w}) \neq (\mathbf{u} \times \mathbf{v}) \times \mathbf{w}$$

i.e., vector (cross) multiplication is not associative. To show this, we consider the vectors $\mathbf{u} = 2\mathbf{i} + 3\mathbf{j} - \mathbf{k}$, $\mathbf{v} = \mathbf{i} + \mathbf{j}$, and $\mathbf{w} = -\mathbf{i} + \mathbf{j} + 3\mathbf{k}$. We first compute $\mathbf{u} \times \mathbf{v}$:

$$\mathbf{u} \times \mathbf{v} = \begin{vmatrix} \mathbf{i} & \mathbf{j} & \mathbf{k} \\ 2 & 3 & -1 \\ 1 & 1 & 0 \end{vmatrix} = \mathbf{i} - \mathbf{j} - \mathbf{k}$$

Then

$$(\mathbf{u} \times \mathbf{v}) \times \mathbf{w} = \begin{vmatrix} \mathbf{i} & \mathbf{j} & \mathbf{k} \\ 1 & -1 & -1 \\ -1 & 1 & 3 \end{vmatrix} = -2\mathbf{i} - 2\mathbf{j}$$

Now to find $\mathbf{u} \times (\mathbf{v} \times \mathbf{w})$, we first find $\mathbf{v} \times \mathbf{w}$:

$$\mathbf{v} \times \mathbf{w} = \begin{vmatrix} \mathbf{i} & \mathbf{j} & \mathbf{k} \\ 1 & 1 & 0 \\ -1 & 1 & 3 \end{vmatrix} = 3\mathbf{i} - 3\mathbf{j} + 2\mathbf{k}$$

Then

$$\mathbf{u} \times (\mathbf{v} \times \mathbf{w}) = \begin{vmatrix} \mathbf{i} & \mathbf{j} & \mathbf{k} \\ 2 & 3 & -1 \\ 3 & -3 & 2 \end{vmatrix} = 3\mathbf{i} - 7\mathbf{j} - 15\mathbf{k}$$

This counter example shows that vector multiplication is not associative. It is important to note that the vector triple product is a vector.

EXERCISES

1. Prove: $\mathbf{u} \cdot (\mathbf{v} + \mathbf{w}) = \mathbf{u} \cdot \mathbf{v} + \mathbf{u} \cdot \mathbf{w}$
2. Prove: $\mathbf{u} \times (\mathbf{v} \times \mathbf{w}) = (\mathbf{u} \cdot \mathbf{w})\mathbf{v} - (\mathbf{u} \cdot \mathbf{v})\mathbf{w}$
   $(\mathbf{u} \times \mathbf{v}) \times \mathbf{w} = (\mathbf{u} \cdot \mathbf{w})\mathbf{v} - (\mathbf{v} \cdot \mathbf{w})\mathbf{u}$

370 ANALYTIC GEOMETRY OF SPACE

3. Prove: $(u \times v) \cdot (w \times s) = (u \cdot w)(v \cdot s) - (u \cdot s)(v \cdot w)$
4. Prove: $u \times (v \times w) + v \times (w \times u) + w \times (u \times v) = 0$
5. Prove: $(u \times v) \cdot (w \times s) + (v \times w) \cdot (u \times s) + (w \times u) \cdot (v \times s) = 0$
6. Evaluate:
   (a) $(2i - 3j) \cdot [(i + j - k) \times (3i - k)]$
   (b) $(4i + 2j - k) \cdot [(j - 2k) \times (3i - k)]$
7. Given: $u = i - 2j - 3k$, $v = 2i + j - k$, $w = i + 3j - 2k$. Find:
   (a) $|u \times (v \times w)|$
   (b) $|(u \times v) \times w|$
   (c) $|u \cdot v \times w|$
   (d) $|u \times v \cdot w|$
   (e) $(u \times v) \times (v \times w)$
   (f) $(u \times v)(v \cdot w)$
8. Find the volume of the parallelepiped whose lateral edges are represented by $u = 2i - 3j + 4k$, $v = i + 2j - k$, $w = 3i - j + 2k$.
9. Given points P_1, P_2, P_3 defined by the position vectors $u = x_1 i + y_1 j + z_1 k$, $v = x_2 i + y_2 j + z_2 k$, $w = x_3 i + y_3 j + z_3 k$, find an equation for the plane containing the given points.
10. Find the equation of the plane determined by the points
    (a) $P_1(2,-1,1)$, $P_2(3,2,-1)$, $P_3(-1,3,2)$
    (b) $P_1(0,2,1)$, $P_2(2,2,2)$, $P_3(1,-1,2)$
    Hint: See Example 10–10.

10–8. Determining a plane satisfying three conditions

In the equation $ax + by + cz + d = 0$, there are four arbitrary constants. Since a, b, and c may not all be zero at the same time, we may divide through by whichever one is not zero. Assuming that $a \neq 0$ and dividing through by a, we obtain $x + \frac{b}{a} y + \frac{c}{a} z + \frac{d}{a} = 0$, in which there are only three arbitrary constants, namely, the ratios $\frac{b}{a}, \frac{c}{a}$, and $\frac{d}{a}$. Three conditions are necessary to determine these unknowns; moreover, three conditions will suffice, provided that each condition determines one equation involving the unknowns and provided that the three equations can be solved uniquely. The following examples will illustrate the procedure.

EXAMPLE 10–9. Find an equation of a plane passing through the points $P_1(x_1, y_1, z_1)$, $P_2(x_2, y_2, z_2)$, and $P_3(x_3, y_3, z_3)$, where P_1, P_2, and P_3 are not collinear.

Solution. Let $P(x, y, z)$ be any point in the plane whose equation we are seeking. Then the four points P, P_1, P_2, and P_3 are coplanar. Hence, the vectors $\overrightarrow{P_1P}$, $\overrightarrow{P_1P_2}$, and $\overrightarrow{P_1P_3}$ are coplanar; and we have

$$\begin{vmatrix} x - x_1 & y - y_1 & z - z_1 \\ x_2 - x_1 & y_2 - y_1 & z_2 - z_1 \\ x_3 - x_1 & y_3 - y_1 & z_3 - z_1 \end{vmatrix} = 0$$

If we expand this, we obtain an equation $ax + by + cz + d = 0$, where a, b, and c are the scalar components of the vector $\mathbf{u} = \overrightarrow{P_2P_1} \times \overrightarrow{P_3P_1}$. Since P_1, P_2, and P_3 are not collinear, $\mathbf{u}$ is not the zero vector. Hence, at least one of the coefficients a, b, and c is not zero, and the equation does represent a plane. Moreover, if x_1, y_1, and z_1 replace x, y, and z, the elements of the first row are zero; and, if we replace x, y, and z by x_2, y_2, and z_2 and then by x_3, y_3, and z_3, we find the elements of two rows are alike. Hence, in each of these cases, the determinant vanishes. So, the points P_1, P_2, and P_3 lie on the plane.

EXAMPLE 10–10. Find an equation of the plane containing the points $P_1(-1, 2, 3)$, $P_2(3, 1, 1)$, and $P_3(1, 3, -2)$.

Solution. In this case, $\overrightarrow{P_1P} = [x+1, y-2, z-3]$, $\overrightarrow{P_1P_2} = [4, -1, -2]$, and $\overrightarrow{P_1P_3} = [2, 1, -5]$. So, we have

$$\begin{vmatrix} x+1 & y-2 & z-3 \\ 4 & -1 & -2 \\ 2 & 1 & -5 \end{vmatrix} = 0$$

Expanding this, we get $7(x+1) + 16(y-2) + 6(z-3) = 0$ or

$$7x + 16y + 6z - 43 = 0$$

This is an equation of the plane through the three given points. The student should check by showing that the points P_1, P_2, and P_3 actually lie on the plane.

EXAMPLE 10–11. Find an equation of a plane that contains the point $P_1(x_1, y_1, z_1)$ and is perpendicular to each of the non-parallel planes $a_1x + b_1y + c_1z + d_1 = 0$ and $a_2x + b_2y + c_2z + d_2 = 0$.

Solution. Let $ax + by + cz + d = 0$ be an equation of the plane we seek. Since the desired plane is perpendicular to both $a_1x + b_1y + c_1z + d_1 = 0$ and

$a_2x + b_2y + c_2z + d_2 = 0$, the vectors $\mathbf{u} = [a_1, b_1, c_1]$ and $\mathbf{v} = [a_2, b_2, c_2]$ lie in the plane $ax + by + cz + d = 0$. This is true since, if the planes $ax + by + cz + d = 0$ and $a_1x + b_1y + c_1z + d_1 = 0$ are perpendicular, their respective normals $[a, b, c]$ and $[a_1, b_1, c_1]$ are perpendicular; so, each plane contains the normal vector to the other. The same argument holds for the vector $[a_2, b_2, c_2]$. If $P(x, y, z)$ is an arbitrary point in the plane, a third vector coplanar with $\mathbf{u}$ and $\mathbf{v}$ is $\overrightarrow{P_1P} = [x - x_1, y - y_1, z - z_1]$. Hence,

$$\begin{vmatrix} x - x_1 & y - y_1 & z - z_1 \\ a_1 & b_1 & c_1 \\ a_2 & b_2 & c_2 \end{vmatrix} = 0$$

Expanding, we obtain $ax + by + cz + d = 0$ where $\mathbf{u} \times \mathbf{v} = [a, b, c]$. Since $\mathbf{u}$ is not parallel to $\mathbf{v}$, at least one of the coefficients a, b, and c is not zero.

EXAMPLE 10–12. Find an equation of the plane that contains the point $(-1, 3, 2)$ and is perpendicular to the planes $3x + 2y - z + 6 = 0$ and $x - y + 3z + 9 = 0$.

Solution. Three vectors in the plane are $\overrightarrow{P_1P} = [x+1, y-3, z-2]$, $\mathbf{u} = [3, 2, -1]$, and $\mathbf{v} = [1, -1, 3]$. Therefore,

$$\begin{vmatrix} x+1 & y-3 & z-2 \\ 3 & 2 & -1 \\ 1 & -1 & 3 \end{vmatrix} = 0$$

Expanding this, we obtain $5(x+1) - 10(y-3) - 5(z-2) = 0$ or

$$x - 2y - z + 9 = 0$$

The student should check this result.

EXERCISES

1. Find equations of the planes that contain the following triples of points:
   (a) $P_1(0, 0, 1)$, $P_2(0, 1, 0)$, and $P_3(1, 0, 0)$
   (b) $P_1(3, -5, 4)$, $P_2(-2, 1, 1)$, and $P_3(1, 3, 0)$
   (c) $P_1(2, 3, -1)$, $P_2(-1, 3, 2)$, and $P_3(-1, 1, -1)$
   (d) $P_1(2, -3, 1)$, $P_2(5, 3, 2)$, and $P_3(2, 1, -1)$
   (e) $P_1(0, 0, 0)$, $P_2(0, 1, 0)$, and $P_3(0, 0, 1)$
   (f) $P_1(-3, 1, 1)$, $P_2(1, -3, 2)$, and $P_3(2, -1, -1)$
   (g) $P_1(5, -3, 4)$, $P_2(3, 1, -2)$, and $P_3(0, 1, -3)$
   (h) $P_1(0, 0, 0)$, $P_2(1, 1, 1)$, and $P_3(1, 2, 3)$

2. Find an equation of the plane that is perpendicular to the plane $x - 2y + 3z - 6 = 0$ and contains the points $(3, -1, 4)$ and $(-2, 3, -1)$.

3. Find an equation of the plane that contains the points $(-1, 2, 3)$ and $(2, 3, -1)$ and is perpendicular to the plane $3x - 2y - z - 4 = 0$.

4. Find an equation of the plane that is perpendicular to the planes whose equations are $3x - 4y + z + 6 = 0$ and $2x - y + 3z - 6 = 0$ and that contains the point $(-1, 1, 2)$.

5. Find an equation of the plane that contains the point $(2, 3, -1)$ and is perpendicular to the planes $3x - y - 2z - 4 = 0$ and $x - y + 3z + 6 = 0$.

6. Show that the four points $P_1(1, 1, -11)$, $P_2(5, 0, 9)$, $P_3(5, -5, 25)$, and $P_4(0, 0, -12)$ are coplanar.

7. Find an equation of the plane that passes through $(6, 2, 4)$ and is perpendicular to the line drawn from the origin to that point.

8. Find an equation of a plane that is perpendicular to the segment from $P_1 = (-1, 4, 6)$ to $P_2 = (5, 2, -4)$ and contains the mid-point of the segment.

9. What is an equation of the plane whose intercepts are one-half those of the plane $3x - 2y + 4z - 24 = 0$?

10. A perpendicular to a plane from the origin meets the plane in the point $(3, -2, 1)$. What is an equation of the plane?

11. A line through the point $(3, 5, -2)$ and perpendicular to a plane meets the plane in the point $(2, -1, 3)$. Find an equation of the plane.

12. A line passes through the points $P_1 = (2, -1, 3)$ and $P_2 = (-1, 3, 5)$. Find an equation of the *plane* that is perpendicular to this line and

   (a) Contains the point where the line intersects the xy-plane.
   (b) Contains the point where the line intersects the xz-plane.
   (c) Contains the point where the line intersects the yz-plane.

10–9. Distance from a plane to a point

Let us consider the problem of finding the distance from the plane whose equation is $ax + by + cz + d = 0$ to a point $P_1(x_1, y_1, z_1)$ which is not on the plane. Since P_1 is not on the plane, we know that $ax_1 + by_1 + cz_1 + d \neq 0$. Let $P_2 = (x_2, y_2, z_2)$ be the projection of P_1 on the given plane, as shown in Fig. 10–5. Then we know that $ax_2 + by_2 + cz_2 + d = 0$.

From any point on the plane as the initial point, draw a representative segment of the coefficient vector $\mathbf{u} = [a, b, c]$.

Case 1. We shall suppose, first, that P_1 and the segment of $\mathbf{u}$ lie on the same side of the plane, as in Fig. 10–5. Then the angle between $\mathbf{u}$ and $\overrightarrow{P_2P_1}$ is $0°$, and

374   ANALYTIC GEOMETRY OF SPACE

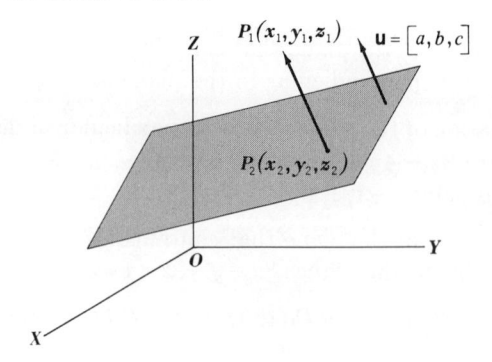

FIG. 10-5

$$\cos 0° = \frac{\mathbf{u} \cdot \overrightarrow{P_2P_1}}{|\mathbf{u}|\ |\overrightarrow{P_2P_1}|}$$

Hence,
$$1 = \frac{a(x_1-x_2)+b(y_1-y_2)+c(z_1-z_2)}{\sqrt{a^2+b^2+c^2}\ |\overrightarrow{P_2P_1}|}$$

Since $d = -ax_2-by_2-cz_2$, we obtain

$$1 = \frac{ax_1+by_1+cz_1+d}{\sqrt{a^2+b^2+c^2}\ |\overrightarrow{P_2P_1}|}$$

from which

$$|\overrightarrow{P_2P_1}| = \frac{ax_1+by_1+cz_1+d}{\sqrt{a^2+b^2+c^2}}$$

Case 2. Now assume that P_1 and $\mathbf{u} = [a, b, c]$ lie on opposite sides of the given plane. Then the angle between the vector $\mathbf{u}$ and the vector $\overrightarrow{P_2P_1}$ is 180°. So, we have

$$\cos 180° = \frac{\mathbf{u} \cdot \overrightarrow{P_2P_1}}{|\mathbf{u}|\ |\overrightarrow{P_2P_1}|}$$

or
$$-1 = \frac{a(x_1-x_2)+b(y_1-y_2)+c(z_1-z_2)}{\sqrt{a^2+b^2+c^2}\ |\overrightarrow{P_2P_1}|}$$

Again replacing $-ax_2-by_2-cz_2$ by d, we obtain

$$-1 = \frac{ax_1+by_1+cz_1+d}{\sqrt{a^2+b^2+c^2}\ |\overrightarrow{P_2P_1}|}$$

from which

$$-|\overrightarrow{P_2P_1}| = \frac{ax_1+by_1+cz_1+d}{\sqrt{a^2+b^2+c^2}}$$

Hence, the *distance* δ from the plane whose equation is $ax+by+cz+d=0$ to the point $P_1 = (x_1, y_1, z_1)$ is

$$\delta = \left| \frac{ax_1+by_1+cz_1+d}{\sqrt{a^2+b^2+c^2}} \right| \qquad (10\text{--}4)$$

When the number $\dfrac{ax_1+by_1+cz_1+d}{\sqrt{a^2+b^2+c^2}}$ is positive, the point P_1 and the vector $\mathbf{u} = [a, b, c]$ *lie on the same side of the plane; when that number is negative, the point and the vector lie on opposite sides of the plane.*

EXAMPLE 10–13. Find the distance from the plane $2x-3y+z-6=0$ to the point $P_1 = (-3, 2, 4)$.

Solution. Using equation (10–4), we obtain

$$\delta = \left| \frac{2(-3)-3(2)+(4)-6}{\sqrt{4+9+1}} \right| = \left| \frac{-14}{\sqrt{14}} \right| = \sqrt{14}$$

EXAMPLE 10–14. Find the distance between the planes $2x-y+3z-4=0$ and $6x-3y+9z+6=0$.

Solution. The given planes are parallel. Therefore, we choose a point on one of them and find the distance from the other plane to this point. A convenient point on the plane $2x-y+3z-4=0$ is $(2, 0, 0)$. Hence,

$$\delta = \left| \frac{6(2)-3(0)+9(0)+6}{\sqrt{36+9+81}} \right| = \frac{18}{\sqrt{126}} = \frac{3\sqrt{14}}{7}$$

EXERCISES

1. In each of the following cases, find the distance from the given plane to the given point:
   (a) $3x-4y+z-4=0$ and $(-1, 2, 3)$
   (b) $x+y-z+2=0$ and $(3, -1, 2)$
   (c) $2x-y+3z+6=0$ and $(1, -1, 1)$
   (d) $5x-3y+4z+12=0$ and $(0, 0, 0)$

2. What is the distance between the planes $2x-3y+z-6=0$ and $2x-3y+z=0$?

3. What is the distance between the planes $2x-y+3z-6=0$ and $6x-3y+9z-9=0$?

4. Find the locus of points equidistant from the points $(2, 3, -1)$ and $(3, -2, 5)$.

5. Find the locus of points equidistant from the planes $2x - 3y + z - 6 = 0$ and $x + y - 2z + 4 = 0$.

6. Find the set of points equidistant from the planes $3x - 2y + z - 6 = 0$ and $x + 3y - 2z + 12 = 0$.

7. Find the set of points equidistant from the planes $x - y + 2z + 3 = 0$ and $2x + y - z - 3 = 0$.

8. If a plane has an x-intercept f, a y-intercept g, a z-intercept h and is a distance p from the origin, prove that

$$\frac{1}{f^2} + \frac{1}{g^2} + \frac{1}{h^2} = \frac{1}{p^2}$$

9. Describe the sets of points having the following properties:
   (a) $2x - 3y + z > 6$
   (b) $x - y - 2z \leq 4$
   (c) $x - y < 0$
   (d) $z \geq 5$
   (e) $2y - 3z < 6$
   (f) $2x \quad 4 \geq 3y$

10. Show that the distance between the parallel planes $ax + by + cz + d_1 = 0$ and $ax + by + cz + d_2 = 0$ is

$$\frac{|d_1 - d_2|}{\sqrt{a^2 + b^2 + c^2}}$$

11. Find two points on the y-axis each of which is two units distant from the plane $2x + y - 2z + 3 = 0$.

12. Find two points on the z-axis each of which is 4 units distant from the plane $x - 2y - 2z + 5 = 0$.

13. Find equations of two planes parallel to the plane $2x - 2y + z - 6 = 0$ and three units from it.

14. Find equations of two planes parallel to the plane $6x - 2y + 3z = 4$ and 4 units from the origin.

15. There are two planes parallel to the plane $3x + 2y - 6z - 12 = 0$ and a distance of 3 units from the origin. Find their equations.

10–10. Angle between two planes

We shall define an angle between two planes whose equations are $a_1x + b_1y + c_1z + d_1 = 0$ and $a_2x + b_2y + c_2z + d_2 = 0$ to be either the angle between the normals $[a_1, b_1, c_1]$ and $[a_2, b_2, c_2]$ or the angle between the normals $[a_1, b_1, c_1]$ and $[-a_2, -b_2, -c_2]$. Such an angle θ may always be chosen so that $0° \leq \theta \leq 180°$; and, except when the normals are perpendicular, θ may be chosen acute or obtuse. From equation (9–6),

$$\cos \theta = \frac{\pm (a_1 a_2 + b_1 b_2 + c_1 c_2)}{\sqrt{a_1^2 + b_1^2 + c_1^2} \ \sqrt{a_2^2 + b_2^2 + c_2^2}} \tag{10-5}$$

It is evident that the angle found by using the plus sign before the parenthesis is the supplement of the angle found by using the minus sign.

EXAMPLE 10–15. Find the acute angle between the planes $2x - 3y + z + 4 = 0$ and $x + y - 3z - 6 = 0$.

Solution. Normal vectors to the two planes are $[2, -3, 1]$ and $[1, 1, -3]$, respectively. If we use the plus sign before the parenthesis in equation (10–5), we obtain

$$\cos \theta = \frac{2 - 3 - 3}{\sqrt{14} \ \sqrt{11}} = \frac{-4}{\sqrt{14} \ \sqrt{11}}$$

Since this result yields an obtuse angle and we want an acute angle, we merely use the minus sign in equation (10–5) by changing the result to $\frac{4}{\sqrt{14} \ \sqrt{11}}$. Thus,

$$\cos \theta = \frac{4}{\sqrt{14} \ \sqrt{11}}, \text{ or } \theta = \text{arc cos} \left(\frac{4}{\sqrt{14} \ \sqrt{11}}\right)$$

10–11. Pencils of planes

By a pencil of planes is meant the set of all the planes that pass through the line of intersection of two given non-parallel planes. Let us assume that $a_1 x + b_1 y + c_1 z + d_1 = 0$ and $a_2 x + b_2 y + c_2 z + d_2 = 0$ are equations of two-non-parallel planes; and let us consider the equation

$$t_1(a_1 x + b_1 y + c_1 z + d_1) + t_2(a_2 x + b_2 y + c_2 z + d_2) = 0 \tag{10-6}$$

where t_1 and t_2 are constants and *both are not zero* at the same time. Since the planes are not parallel, the normal vectors to the planes are not parallel. Therefore, all three coefficients $t_1 a_1 + t_2 a_2$, $t_1 b_1 + t_2 b_2$, and $t_1 c_1 + t_2 c_2$ cannot be zero at the same time; for, if they were, the normal vectors would be parallel. The student should check this by setting the coefficients equal to 0 and eliminating t_1 and t_2. Hence, the linear equation (10–6) is an equation of a plane. Furthermore, any point (x_1, y_1, z_1) that lies on both given planes also lies on the plane represented by equation (10–6). In other words, equation (10–6) is an equation of a plane containing the line of intersection of the given planes. It may also be shown that the equation of any plane through the intersection of two given planes has the form of equation (10–6).

EXAMPLE 10–16. Find an equation of the plane containing the point $(-1, 3, 2)$ and belonging to the pencil defined by $3x - 2y + 6z - 6 = 0$ and $x - y + z + 4 = 0$.

Solution. An equation of the pencil is $t_1(3x - 2y + 6z - 6) + t_2(x - y + z + 4) = 0$. Since the point $(-1, 3, 2)$ must satisfy this equation, we know that
$$t_1[3(-1) - 2(3) + 6(2) - 6] + t_2[-1 - 3 + 2 + 4] = 0$$
From this relation, we obtain $-3t_1 + 2t_2 = 0$, or $\dfrac{t_1}{t_2} = \dfrac{2}{3}$. If we let $t_1 = 2$ and $t_2 = 3$, we obtain
$$2(3x - 2y + 6z - 6) + 3(x - y + z + 4) = 0$$
or
$$9x - 7y + 15z = 0$$
Check the result.

EXAMPLE 10–17. Find an equation of the plane that is perpendicular to the plane $2x - y + z - 4 = 0$ and belongs to the pencil defined by $x - y + 2z - 4 = 0$ and $x + 2y - 3z + 6 = 0$.

Solution. An equation of the pencil is $t_1(x - y + 2z - 4) + t_2(x + 2y - 3z + 6) = 0$. Expanding, we obtain $(t_1 + t_2)x + (2t_2 - t_1)y + (2t_1 - 3t_2)z - 4t_1 + 6t_2 = 0$. Since the plane represented by this equation is to be perpendicular to the plane $2x - y + z - 4 = 0$, it follows that the dot product of their normal vectors must be zero. So
$$2(t_1 + t_2) - (2t_2 - t_1) + 2t_1 - 3t_2 = 0$$
Hence,
$$2t_1 + 2t_2 - 2t_2 + t_1 + 2t_1 - 3t_2 = 0$$
from which
$$5t_1 = 3t_2 \quad \text{or} \quad \dfrac{t_1}{t_2} = \dfrac{3}{5}$$
Choosing $t_1 = 3$ and $t_2 = 5$, we find an equation of the plane to be
$$8x + 7y - 9z + 18 = 0$$

EXERCISES

1. Find the cosine of the acute angle between each of the following pairs of planes:
   (a) $2x - 3y + z - 4 = 0$ and $x + 2y - z + 4 = 0$
   (b) $3x - y - 2z + 6 = 0$ and $x + y - 3z + 3 = 0$
   (c) $x - 2y + z + 2 = 0$ and $2x - 3y + 3z - 6 = 0$

2. (a) Find an equation of the pencil of planes defined by $2x - 3y + z - 6 = 0$ and $x + y - z + 3 = 0$.
   (b) Which plane of the pencil contains the point $(2, -1, 3)$?

(c) Which plane of the pencil contains the origin?
(d) Which one has an x-intercept equal to 3?
(e) Which one has a y-intercept equal to 5?
(f) Which one is perpendicular to the plane $3x - y + 2z - 4 = 0$?
(g) Which one is perpendicular to the plane $x - 2y + 3z + 12 = 0$?
(h) Which one is parallel to the z-axis?
(i) Which one is parallel to the y-axis?

3. Work Exercise 2 if the given planes are $3x + 2y - z - 4 = 0$ and $x - y + z + 5 = 0$.

4. Determine values of t_1 and t_2 which yield the original planes that define the pencil
$$t_1(a_1x + b_1y + c_1z + d_1) + t_2(a_2x + b_2y + c_2z + d_2) = 0$$

5. Find an equation of the plane that passes through the point $(2, 1, -5)$ and contains the line of intersection of the two planes whose equations are $3x + y - 2z + 6 = 0$ and $x - y + z - 3 = 0$.

REVIEW EXERCISES

1. Determine an equation of the plane which passes through $(3, 7, 4)$ and a normal to which passes through $P_1 = (3, 2, 0)$ and $P_2 = (-5, 2, 3)$.

2. Determine an equation of the plane which passes through $(-2, 3, 7)$ and a normal to which passes through $P_1 = (3, -2, 0)$ and $P_2 = (5, 4, -3)$.

3. Determine an equation of the plane which bisects at right angles the segment from $P_1 = (-2, 4, 4)$ to $P_2 = (3, 0, -1)$.

4. Determine an equation of the plane that passes through the origin and is perpendicular to the segment from the origin to $P = (-1, 1, 2)$.

5. Determine an equation of the plane that passes through the origin and is perpendicular to the segment from the origin to $P = (0, 0, 1)$.

6. (a) Show that, if $a = 0$, the plane $ax + by + cz + d = 0$ is parallel to the x-axis.
   (b) Show that, if both $a = 0$ and $d = 0$, the plane contains the x-axis.

7. Say what you can about the plane $ax + by + cz + d = 0$: (a) when $d = 0$; (b) when $b = 0$; (c) when $b = 0$ and $c = 0$; (d) when $c = 0$ and $d = 0$.

8. Find an equation of the plane that contains the points $P_1 = (1, -1, 2)$, $P_2 = (2, 1, 3)$, and $P_3 = (3, 2, 1)$.

9. Find an equation of the plane that contains the points $P_1 = (2, -1, 3)$ and $P_2 = (1, 1, 1)$ and is perpendicular to the plane $3x - y + 2z - 4 = 0$.

10. Find an equation of the plane that contains the point $P_1 = (4, 2, -3)$ and is perpendicular to the planes $2x - y + z - 4 = 0$ and $x - y + z - 2 = 0$.

11. Show that $\begin{vmatrix} x & y & z & 1 \\ f & 0 & 0 & 1 \\ 0 & g & 0 & 1 \\ 0 & 0 & h & 1 \end{vmatrix} = 0$ is an intercept equation of a plane.

12. Find the distance between the planes whose equations are $3x - 2y - 3z + 6 = 0$ and $6x - 4y - 6z + 5 = 0$.

13. (a) Find an equation of the pencil of planes defined by the planes whose equations are $x + 2y + z - 6 = 0$ and $2x - y - 3z + 12 = 0$.
    (b) Which plane of the pencil contains the point $P = (-1, 2, 1)$?
    (c) Which plane of the pencil is perpendicular to the plane $2x + y - z + 4 = 0$?
    (d) Which plane of the pencil has an x-intercept equal to 4?
    (e) Which plane of the pencil has a z-intercept equal to -3?
    (f) Which plane of the pencil is parallel to the z-axis?

14. Work Exercise 13 if the given planes are $3x - y + 2z - 3 = 0$ and $x - y + z - 1 = 0$.

15. Three faces of a box lie in the planes $3x + y - 6 = 0$, $x - 3y = 4$, and $z = 6$. If one vertex is at the point $(8, 6, 1)$, what are equations of the planes containing the other three faces?

16. Find the cosine of the angle which the vector $\mathbf{u} = [2, -1, 3]$ makes with the plane $3x + 2y - z + 6 = 0$. Hint: Find the cosine of the angle between the normal vector and $\mathbf{u}$.

17. What is the cosine of the angle that the plane $x - 2y + z - 4 = 0$ makes with the xy-plane; the yz-plane?

18. For what value of k will the two equations $kx - y + z + 3 = 0$ and $4x - ky + kz + 6 = 0$ represent the same plane?

The Straight Line in Space

11–1. Direction numbers and direction cosines

We shall define *direction numbers* of a line to be the scalar components of any segment, or vector, on the line. A vector will be said to be on a line provided a representative segment of the vector is on the line. Hence, if $P_1 = (x_1, y_1, z_1)$ and $P_2 = (x_2, y_2, z_2)$ are two points on the line, a triple of direction numbers of the line is $[\Delta x, \Delta y, \Delta z]$, where $\Delta x = x_2 - x_1$, $\Delta y = y_2 - y_1$, and $\Delta z = z_2 - z_1$. Another triple would be the negatives of these or any other triple of numbers proportional to them. Hence, if Δx, Δy, and Δz are direction numbers of a line, so are $k\,\Delta x$, $k\,\Delta y$, and $k\,\Delta z$, where $k \neq 0$.

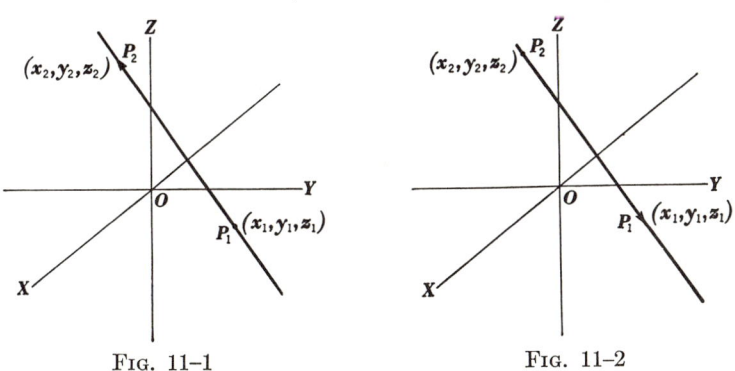

FIG. 11–1   FIG. 11–2

We shall define *direction cosines* of a line to be the direction cosines of any vector on the line. So, if $\overrightarrow{P_1P_2} = [\Delta x, \Delta y, \Delta z]$ is a vector on a given line, as indicated in Fig. 11–1, a triple of direction cosines for the line is

$$l_1 = \frac{\Delta x}{|\overrightarrow{P_1P_2}|}, \quad m_1 = \frac{\Delta y}{|\overrightarrow{P_1P_2}|}, \quad \text{and} \quad n_1 = \frac{\Delta z}{|\overrightarrow{P_1P_2}|}$$

Another triple, for the conditions indicated in Fig. 11–2, is

$$l_2 = -\frac{\Delta x}{|\overrightarrow{P_1P_2}|}, \quad m_2 = -\frac{\Delta y}{|\overrightarrow{P_1P_2}|}, \quad \text{and} \quad n_2 = -\frac{\Delta z}{|\overrightarrow{P_1P_2}|}$$

382   ANALYTIC GEOMETRY OF SPACE

If the direction cosines of the vector $\overrightarrow{P_1P_2}$ in Fig. 11-1 are l_1, m_1, and n_1 and the direction cosines of the opposite vector $\overrightarrow{P_2P_1}$ in Fig. 11-2 are l_2, m_2, and n_2, then $l_2 = -l_1$, $m_2 = -m_1$, and $n_2 = -n_1$.

In general, if $\mathbf{u} = [u_1, u_2, u_3]$ is a vector on the line, a triple of direction cosines of the line is $l = \dfrac{u_1}{|\mathbf{u}|}$, $m = \dfrac{u_2}{|\mathbf{u}|}$, and $n = \dfrac{u_3}{|\mathbf{u}|}$.

EXAMPLE 11-1. What are direction cosines of the line containing the points $P_1 = (-1, 2, 3)$ and $P_2 = (2, 2, -1)$?

Solution. If we consider direction numbers of the line to be the scalar components of $\overrightarrow{P_1P_2}$, we obtain $\overrightarrow{P_1P_2} = [3, 0, -4]$. Then, the direction cosines are $l = \dfrac{3}{5}$, $m = 0$, and $n = -\dfrac{4}{5}$.

If we had used $\overrightarrow{P_2P_1}$ we would have obtained $l = -\dfrac{3}{5}$, $m = 0$, and $n = \dfrac{4}{5}$.

EXAMPLE 11-2. What are direction cosines of the line containing the vector $\mathbf{u} = [3, -1, 2]$?

Solution. A triple of direction cosines is $l = \dfrac{3}{\sqrt{14}}$, $m = -\dfrac{1}{\sqrt{14}}$, and $n = \dfrac{2}{\sqrt{14}}$.

EXERCISES

1. Find direction numbers and direction cosines of the lines through each of the following pairs of points:
   (a) $P_1(-1, 2, 3)$ and $P_2(3, -1, 2)$
   (b) $P_1(5, 3, -2)$ and $P_2(-1, 1, 1)$
   (c) $P_1(3, -1, -1)$ and $P_2(-1, -1, 2)$
   (d) $P_1(0, 0, 1)$ and $P_2(4, -1, 3)$

2. In each of the following cases two of the direction cosines of a line are given and it is required to find a third:
   (a) If $l = \frac{1}{3}$ and $m = -\frac{2}{5}$, what is n?
   (b) If $m = -\frac{1}{2}$ and $n = \frac{2}{3}$, what is l?
   (c) If $l = \frac{3}{5}$ and $m = \frac{1}{2}$, what is n?
   (d) If $l = \frac{4}{5}$ and $n = \frac{2}{3}$, what is m?

3. Could $\frac{1}{2}$, $\frac{1}{3}$, and $-\dfrac{\sqrt{23}}{6}$ be direction cosines of a line? Explain.

4. Could $-\frac{3}{4}$, 0, and 0 be direction cosines of a line? Explain.

5. Could $\frac{3}{5}$, $-\frac{4}{5}$, and 0 be direction cosines of a line? Explain.

6. What are direction cosines of the axes?

11–2. Parametric equations of a line

Consider P_1 and P_2 in Fig. 11–3 to be distinct points of a line, and let P be an arbitrary point of this line. Then, from the ratio formula, $\overrightarrow{P_1P} = t\overrightarrow{P_1P_2}$. When P coincides with P_1, $t=0$; when P coincides with P_2, $t=1$; when P is on the half-line through P_1 containing P_2, t is positive; when P is on the half-line through P_1 not containing P_2, t is negative. As t varies and takes on all real values, P traces out the line containing $\overrightarrow{P_1P_2}$.

Since $\overrightarrow{P_1P} = [x-x_1,\ y-y_1,\ z-z_1]$ and $t\overrightarrow{P_1P_2} = [t(x_2-x_1),\ t(y_2-y_1),\ t(z_2-z_1)]$, we have $x = x_1 + t(x_2-x_1)$, $y = y_1 + t(y_2-y_1)$, and $z = z_1 + t(z_2-z_1)$; or

$$\begin{aligned} x &= x_1 + t\,\Delta x \\ y &= y_1 + t\,\Delta y \\ z &= z_1 + t\,\Delta z \end{aligned} \qquad (11\text{–}1)$$

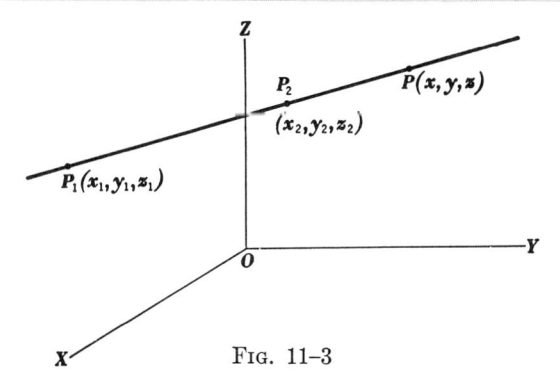

Fig. 11–3

Equations (11–1) are called *parametric equations of a line in space*. The variable t, in terms of which the coordinates x, y, and z are expressed, is called a *parameter*. Equations (11–1) are not *unique* since P_1 may be *any* point on the line and Δx, Δy, and Δz may be *any* triple of direction numbers of the line.

EXAMPLE 11–3. Find parametric equations of the line through the points $P_1(2, -1, 3)$ and $P_2(3, 1, 5)$.

Solution. Since $\Delta x = 1$, $\Delta y = 2$, and $\Delta z = 2$, parametric equations of the line are $x = 2+t$, $y = -1+2t$, and $z = 3+2t$.

EXAMPLE 11–4. What are parametric equations of the x-axis?

Solution. The point $(0, 0, 0)$ is on the x-axis, and a triple of direction numbers is $[1, 0, 0]$. Hence, parametric equations of the x-axis are $x = t$, $y = 0$, and $z = 0$.

11–3. Symmetric equations of a line

If we eliminate t from equations (11–1), we obtain

$$\frac{x-x_1}{\Delta x} = \frac{y-y_1}{\Delta y} = \frac{z-z_1}{\Delta z} \qquad (11\text{-}2)$$

provided Δx, Δy, and Δz are not zero. We shall call equations (11–2) *symmetric equations* of a line. Symmetric equations are not uniquely determined by the line for two reasons: (1) The point (x_1, y_1, z_1) is arbitrary; (2) the denominators may be replaced by numbers proportional to them.

EXAMPLE 11–5. What are symmetric equations of the line that passes through the point $P_1(2, -1, 5)$ and contains the vector $\mathbf{u} = [5, 3, -2]$?

Solution. From equations (11–2),

$$\frac{x-2}{5} = \frac{y+1}{3} = \frac{z-5}{-2}$$

Equations (11–2) are not applicable if any of the direction numbers is zero. If one direction number, say Δz, is zero, symmetric equations are

$$\frac{x-x_1}{\Delta x} = \frac{y-y_1}{\Delta y} \text{ and } z = z_1 \qquad (11\text{-}3)$$

The notation $\frac{x-x_1}{\Delta x} = \frac{y-y_1}{\Delta y}$, $z = z_1$ is in common use. It should be understood that, when a line in space is represented by a pair of symmetric equations, both equations must be taken together.

If two direction numbers are zero, say $\Delta y = 0$ and $\Delta z = 0$, symmetric equations of the line are

$$y = y_1 \text{ and } z = z_1 \qquad (11\text{-}4)$$

11–4. The general equations of a line

We may think of a line in space as the intersection of two planes. Hence, a line is represented by two simultaneous linear equations:

$$\begin{aligned} a_1 x + b_1 y + c_1 z + d_1 &= 0 \\ a_2 x + b_2 y + c_2 z + d_2 &= 0 \end{aligned} \qquad (11\text{-}5)$$

provided the planes represented by the separate equations are not parallel.

To find direction numbers of a line represented by equations (11–5), we proceed as follows: The line is perpendicular to $\mathbf{u} = [a_1, b_1, c_1]$, since it lies in the first plane; and it is also perpendicular to $\mathbf{v} = [a_2, b_2, c_2]$, since it lies in the second plane. Hence, a triple of direction numbers of the line is the same as a triple of scalar components of a vector perpendicular to both $\mathbf{u}$

and **v**. Since such a vector is $\mathbf{u} \times \mathbf{v}$, a triple of direction numbers of the line is

$$\begin{vmatrix} b_1 & c_1 \\ b_2 & c_2 \end{vmatrix}, \begin{vmatrix} c_1 & a_1 \\ c_2 & a_2 \end{vmatrix}, \text{ and } \begin{vmatrix} a_1 & b_1 \\ a_2 & b_2 \end{vmatrix}$$

All three determinants are not zero, since the two planes are not parallel (see Example 10–8).

By assigning an arbitrary value to *one* of the variables in equations (11–5), we can solve for the other two variables and obtain the coordinates of a point (x_1, y_1, z_1) on the line (see Example 11–6). Hence, symmetric equations of the line are

$$\frac{x - x_1}{\begin{vmatrix} b_1 & c_1 \\ b_2 & c_2 \end{vmatrix}} = \frac{y - y_1}{\begin{vmatrix} c_1 & a_1 \\ c_2 & a_2 \end{vmatrix}} = \frac{z - z_1}{\begin{vmatrix} a_1 & b_1 \\ a_2 & b_2 \end{vmatrix}} \qquad (11\text{--}6)$$

In case one of the direction numbers is zero or two of those numbers are zero, we write the equations as explained in Section 11–3. When a point on a line and direction numbers of the line are known, it is a simple matter to write parametric equations of the line. They are

$$x = x_1 + \begin{vmatrix} b_1 & c_1 \\ b_2 & c_2 \end{vmatrix} t$$

$$y = y_1 + \begin{vmatrix} c_1 & a_1 \\ c_2 & a_2 \end{vmatrix} t \qquad (11\text{--}7)$$

$$z = z_1 + \begin{vmatrix} a_1 & b_1 \\ a_2 & b_2 \end{vmatrix} t$$

EXAMPLE 11–6. Find direction numbers of the line represented by the equations $2x - 3y + z - 2 = 0$ and $3x - y + 2z + 4 = 0$; and write equations of the line in both symmetric and parametric forms.

Solution. Let $\mathbf{u} = [2, -3, 1]$ and $\mathbf{v} = [3, -1, 2]$. Then

$$\mathbf{u} \times \mathbf{v} = \left[\begin{vmatrix} -3 & 1 \\ -1 & 2 \end{vmatrix}, \begin{vmatrix} 1 & 2 \\ 2 & 3 \end{vmatrix}, \begin{vmatrix} 2 & -3 \\ 3 & -1 \end{vmatrix} \right] = [-5, -1, 7]$$

Hence, direction numbers of the lines are -5, -1, and 7.

If we let $z = 0$, equations of the line become $2x - 3y - 2 = 0$ and $3x - y + 4 = 0$. Solving for x and y, we obtain

$$x = \frac{\begin{vmatrix} 2 & -3 \\ -4 & -1 \end{vmatrix}}{\begin{vmatrix} 2 & -3 \\ 3 & -1 \end{vmatrix}} = \frac{-14}{7} = -2 \text{ and } y = \frac{\begin{vmatrix} 2 & 2 \\ 3 & -4 \end{vmatrix}}{\begin{vmatrix} 2 & -3 \\ 3 & -1 \end{vmatrix}} = \frac{-14}{7} = -2$$

The point $(-2, -2, 0)$ lies on both planes. (Check this.) So, symmetric equations of the line are
$$\frac{x+2}{-5} = \frac{y+2}{-1} = \frac{z}{7}$$
and parametric equations are
$$x = -2 - 5t,\ y = -2 - t,\ \text{and}\ z = 7t$$

EXERCISES

1. Find parametric equations of the line through each of the following pairs of points:
   (a) $P_1(-3, -2, 1)$ and $P_2(2, -1, 3)$
   (b) $P_1(1, -1, 1)$ and $P_2(1, 0, -1)$
   (c) $P_1(1, 1, -1)$ and $P_2(-2, 0, 3)$
   (d) $P_1(2, -3, 4)$ and $P_2(2, 1, 4)$

2. In each of the following cases find parametric equations of the line that is parallel to the given vector and passes through the given point:
   (a) $\mathbf{u} = [-1, 3, 2]$ and $P_1(5, -1, 2)$
   (b) $\mathbf{u} = [2, 1, -1]$ and $P_1(-3, 1, 1)$

3. Find symmetric equations of the line through each of the following pairs of points:
   (a) $P_1(-3, 4, -1)$ and $P_2(0, -1, 2)$
   (b) $P_1(0, 0, 1)$ and $P_2(3, 0, 1)$
   (c) $P_1(3, -1, -2)$ and $P_2(2, -1, 1)$
   (d) $P_1(7, 4, 5)$ and $P_2(7, 1, 2)$

4. In each of the following cases, find symmetric and parametric equations of the line that passes through the given point and has the given direction numbers.
   (a) $P_1(2, -1, 3)$ and $[-1, 1, 2]$
   (b) $P_1(1, -1, 1)$ and $[0, 3, 1]$
   (c) $P_1(-5, 3, 2)$ and $[2, -1, 4]$
   (d) $P_1(0, 0, 0)$ and $[1, 2, 0]$

5. What are equations of the line that contains the point $(3, 2, 1)$ and is perpendicular to the plane $3x + y - z - 6 = 0$?

6. Find an equation of the plane that passes through the point $(4, 2, -3)$ and is perpendicular to the line $\frac{x-1}{3} = \frac{y+2}{1} = \frac{z}{-4}$.

7. Find direction numbers of the line represented by each of the following pairs of equations and write each line in parametric form:
   (a) $2x - 3y + z - 6 = 0$ and $x - y + 2z + 4 = 0$
   (b) $x + y - 3z + 3 = 0$ and $2x - 3y + 2z - 4 = 0$

(c) $4x - y + 3z - 4 = 0$ and $x - z = 2$

(d) $2x + y - 6 = 0$ and $3x + 2z + 12 = 0$

8. What are equations of the line that passes through $P(2, 3, -1)$ and is parallel to the line represented by the equations $3x + 4y - 2z + 6 = 0$ and $x - 2y - z - 2 = 0$?

9. Find an equation of the plane that passes through the origin and is perpendicular to the line represented by the equations $3x + y - z + 6 = 0$ and $x - y - 3z + 3 = 0$.

10. Find an equation of the plane that contains the line $\dfrac{x-2}{3} = \dfrac{y+1}{-1} = \dfrac{z-1}{2}$ and the point $(5, -4, 3)$.

11. What are equations of the line that contains the point $P(4, -1, 1)$ and is parallel to the line represented by the equations $3x + y - z + 6 = 0$ and $2x - y + z - 4 = 0$?

12. If a line passes through $(3, -2, 6)$ and is perpendicular to and intersects the x-axis, what are its equations?

13. Find equations of the lines that pass through $(-3, 1, 1)$ and are perpendicular to and intersect the y-axis and z-axis, respectively.

14. Find equations of the line that passes through the point $(-1, 2, 3)$, is perpendicular to the line represented by the equations $3x - 2y + z - 4 = 0$ and $x + y - z + 6 = 0$, and also is perpendicular to the line represented by the equations $x - y + 2z - 4 = 0$ and $2x + 3y + 3z - 8 = 0$.

15. By an angle between two lines is meant the angle between any two vectors parallel to the lines. In each of the following cases find the cosine of an angle between the given lines:

(a) $x - y + 3z - 4 = 0$ and $2x + y - z = 0$; and $\dfrac{x-3}{4} = \dfrac{y+1}{1} = \dfrac{z-2}{3}$

(b) $x = 2 - 3t$, $y = 1 + t$, and $z = 2t$; and $x + 3y = 4$ and $y - 2z + 12 = 0$

(c) $\dfrac{x-1}{1} = \dfrac{y+2}{-3}$ and $z - 4 = 0$; and $x = 3 + t$, $y = -1 - t$, and $z = 2 + 3t$

16. If the plane $ax + by + cz + d = 0$ is parallel to the line $\dfrac{x - x_1}{l} = \dfrac{y - y_1}{m} = \dfrac{z - z_1}{n}$, prove that $al + bm + cn = 0$.

17. If the plane $ax + by + cz + d = 0$ is perpendicular to the line $\dfrac{x - x_1}{l} = \dfrac{y - y_1}{m} = \dfrac{z - z_1}{n}$, prove that $\dfrac{a}{l} = \dfrac{b}{m} = \dfrac{c}{n}$.

18. Find an equation of the plane that contains the line $\dfrac{x-x_1}{u_1} = \dfrac{y-y_1}{u_2}$ $= \dfrac{z-z_1}{u_3}$ and is parallel to the line $\dfrac{x-x_2}{v_1} = \dfrac{y-y_2}{v_2} = \dfrac{z-z_2}{v_3}$. (Hint: The normal to the plane must be perpendicular to the vectors $\mathbf{u} = [u_1, u_2, u_3]$ and $\mathbf{v} = [v_1, v_2, v_3]$. Assume that the given lines are not parallel.)

19. Find the shortest distance between the lines $\dfrac{x-x_1}{u_1} = \dfrac{y-y_1}{u_2} = \dfrac{z-z_1}{u_3}$ and the line $\dfrac{x-x_2}{v_1} = \dfrac{y-y_2}{v_2} = \dfrac{z-z_2}{v_3}$. The result will be the absolute value of

$$\frac{\begin{vmatrix} x_2-x_1 & y_2-y_1 & z_2-z_1 \\ u_1 & u_2 & u_3 \\ v_1 & v_2 & v_3 \end{vmatrix}}{|\mathbf{u} \times \mathbf{v}|}$$

(Hint: The shortest distance is measured along the line perpendicular to both of the given lines. Hence, to obtain it we first find an equation of the plane that passes through one of the lines and is parallel to the other line; and we then find the distance from this plane to any point on the other line.)

20. Find an equation of the plane that contains the line $\dfrac{x-3}{-1} = \dfrac{y+2}{2}$ $= \dfrac{z+3}{1}$ and is parallel to the line $\dfrac{x+2}{3} = \dfrac{y-1}{1} = \dfrac{z-4}{-1}$.

21. Find the shortest distance between the two lines of Exercise 20.

11–5. Intersecting planes

Two planes intersect in a line unless they are parallel. Three non-parallel planes which neither pass through a common line nor intersect in parallel lines intersect in a common point. This is the point where the line of intersection of two of the planes pierces the third plane. The problem of finding the point which is common to three given planes is the algebraic problem of solving three linear equations in three unknowns. Here the use of determinants is helpful. If the equations of three planes are $a_1x + b_1y + c_1z + d_1 = 0$, $a_2x + b_2y + c_2z + d_2 = 0$, and $a_3x + b_3y + c_3z + d_3 = 0$, the coordinates of the point of intersection are:

$$x = \frac{\begin{vmatrix} -d_1 & b_1 & c_1 \\ -d_2 & b_2 & c_2 \\ -d_3 & b_3 & c_3 \end{vmatrix}}{\begin{vmatrix} a_1 & b_1 & c_1 \\ a_2 & b_2 & c_2 \\ a_3 & b_3 & c_3 \end{vmatrix}}, \quad y = \frac{\begin{vmatrix} a_1 & -d_1 & c_1 \\ a_2 & -d_2 & c_2 \\ a_3 & -d_3 & c_3 \end{vmatrix}}{\begin{vmatrix} a_1 & b_1 & c_1 \\ a_2 & b_2 & c_2 \\ a_3 & b_3 & c_3 \end{vmatrix}}, \quad \text{and } z = \frac{\begin{vmatrix} a_1 & b_1 & -d_1 \\ a_2 & b_2 & -d_2 \\ a_3 & b_3 & -d_3 \end{vmatrix}}{\begin{vmatrix} a_1 & b_1 & c_1 \\ a_2 & b_2 & c_2 \\ a_3 & b_3 & c_3 \end{vmatrix}} \quad (11\text{–}8)$$

It is understood that
$$\Delta = \begin{vmatrix} a_1 & b_1 & c_1 \\ a_2 & b_2 & c_2 \\ a_3 & b_3 & c_3 \end{vmatrix} \neq 0$$

Furthermore, the condition that $\Delta \neq 0$ is equivalent to the statement that the planes intersect in only one point, since equations (11–8) then have a unique solution.

We shall merely remark here that when $\Delta = 0$ some interesting situations arise. In this case, at least two of the planes may be parallel or even coincident; or the three planes may be collinear (intersect in a line); or the three planes may intersect in parallel lines. These statements are illustrated in the examples that follow.

EXAMPLE 11–7. Find the point of intersection of the planes $2x - y + z - 4 = 0$, $3x + y - 2z = 0$, and $x - 2y + 3z - 6 = 0$.

Solution. Since $\Delta = \begin{vmatrix} 2 & -1 & 1 \\ 3 & 1 & -2 \\ 1 & -2 & 3 \end{vmatrix} = 2$, it follows that the planes intersect in a unique point. The coordinates of this point are

$$x = \frac{\begin{vmatrix} 4 & -1 & 1 \\ 0 & 1 & -2 \\ 6 & -2 & 3 \end{vmatrix}}{\Delta}, \quad y = -\frac{\begin{vmatrix} 2 & 4 & 1 \\ 3 & 0 & -2 \\ 1 & 6 & 3 \end{vmatrix}}{\Delta}, \quad \text{and } z = \frac{\begin{vmatrix} 2 & -1 & 4 \\ 3 & 1 & 0 \\ 1 & -2 & 6 \end{vmatrix}}{\Delta}$$

or $x = 1$, $y = -1$, and $z = 1$. The student should check the result.

EXAMPLE 11–8. Determine whether there is any point common to the three planes $2x - 3y + z - 4 = 0$, $x + y - z + 3 = 0$, and $6x - 9y + 3z + 2 = 0$.

Solution. Since $\Delta = \begin{vmatrix} 2 & -3 & 1 \\ 1 & 1 & -1 \\ 6 & -9 & 3 \end{vmatrix} = 0$, there is no unique point of intersection. In this example, two of the planes are distinct and parallel. Hence, there are no points common to all three planes.

EXAMPLE 11–9. Determine whether or not the planes $2x - 3y + z - 4 = 0$, $x + y - z - 2 = 0$, and $4x - y - z - 8 = 0$ have any points in common.

Solution. We find that $\Delta = \begin{vmatrix} 2 & -3 & 1 \\ 1 & 1 & -1 \\ 4 & -1 & -1 \end{vmatrix} = 0$. Therefore, there is no unique point of intersection. However, since no two of the planes are parallel, each pair intersects in a line. According to Section 11–4, direction numbers of the line of intersection of the planes $2x-3y+z-4=0$ and $x+y-z-2=0$ are 2, 3, and 5; while direction numbers of the line of intersection of the planes $x+y-z-2=0$ and $4x-y-z-8=0$ are -2, -3, and -5. Hence, two of the lines of intersection of the three planes are parallel. A point on the first line is (2, 0, 0); and, since this point is also on the second line, the three planes are collinear.

If the coordinates of the point chosen on the first line had not satisfied the equations of the second line, the planes would have intersected in *distinct* parallel lines.

EXERCISES

In the following exercises find which of the sets of planes have a unique point in common. If they have, find the coordinates of this point. If the three planes of a set do not possess a unique point of intersection, determine how they are related to each other.

1. $x-2y+z+1=0$, $3x+y-2z-9=0$, and $2x-3y-z-2=0$.
2. $2x-3y-2z+5=0$, $x+y-z+3=0$, and $2x+2y-2z-7=0$.
3. $x-y+z-2=0$, $2x+y-z+3=0$, and $4x-y+z-1=0$.
4. $2x+y-z-3=0$, $x-y+z-2=0$, and $4x-y+z+4=0$.
5. $3x-y+z+2=0$, $x+y+z+3=0$, and $x-3y-z-5=0$.
6. $x-3y-z+10=0$, $3x-y+z+2=0$, and $x+y+z-4=0$.

11–6. Three homogeneous linear equations

A linear equation in which the constant term is zero is called *homogeneous*, since every term is of the first degree. We shall now consider the following three homogeneous linear equations:

$$\begin{aligned} a_1x+b_1y+c_1z &= 0 \\ a_2x+b_2y+c_2z &= 0 \\ a_3x+b_3y+c_3z &= 0 \end{aligned} \qquad (11\text{-}9)$$

where none of the vectors $[a_1, b_1, c_1]$, $[a_2, b_2, c_2]$, and $[a_3, b_3, c_3]$ is equal to the zero vector. It is obvious that these three planes have the origin in common. If $\Delta \neq 0$, we may solve for x, y, and z and obtain $x = \dfrac{0}{\Delta}$, $y = \dfrac{0}{\Delta}$, and $z = \dfrac{0}{\Delta}$. Hence, the origin is a unique point of intersection of the three planes. If there is to be *more* than one point of intersection, Δ must be equal to 0.

If two of the planes are parallel, they must coincide; since they have a point (the origin) in common. Hence, if $\Delta = 0$ and two of the planes are

THE STRAIGHT LINE IN SPACE *391*

parallel to each other but are not parallel to the third, the three planes have a line in common.

If $\Delta = 0$ and the three planes are parallel, they coincide.

If $\Delta = 0$ and no two of the planes are parallel, the three planes are collinear. To see this, we proceed as follows: Consider the line represented by the equations

$$a_1 x + b_1 y + c_1 z = 0$$
$$a_2 x + b_2 y + c_2 z = 0 \quad (11\text{-}10)$$

By Section 11-4 we know that $\mathbf{u} = [a_1, b_1, c_1] \times [a_2, b_2, c_2]$ is a vector on the line. Since the line passes through the origin, another vector on the line is $\mathbf{v} = [x, y, z]$, where x, y, and z are the coordinates of any point on the line distinct from the origin. But

$$[x, y, z] = k[a_1, b_1, c_1] \times [a_2, b_2, c_2]$$

where $k \neq 0$. Hence, the coordinates of any point on the line represented by equations (11-10), distinct from the origin, are

$$x = k \begin{vmatrix} b_1 & c_1 \\ b_2 & c_2 \end{vmatrix}, \quad y = k \begin{vmatrix} c_1 & a_1 \\ c_2 & a_2 \end{vmatrix}, \quad \text{and} \quad z = k \begin{vmatrix} a_1 & b_1 \\ a_2 & b_2 \end{vmatrix}$$

If we substitute these coordinates for x, y, and z in the equation of the third plane, we obtain

$$k a_3 \begin{vmatrix} b_1 & c_1 \\ b_2 & c_2 \end{vmatrix} + k b_3 \begin{vmatrix} c_1 & a_1 \\ c_2 & a_2 \end{vmatrix} + k c_3 \begin{vmatrix} a_1 & b_1 \\ a_2 & b_2 \end{vmatrix} = k\Delta = 0$$

This tells us that any point common to two of the planes also lies in the third plane. Hence, the planes are collinear.

It is worth while to note that, when equations (11-9) represent *non-parallel* planes and when $\Delta = 0$, the coordinates of a point on the line of intersection are the same as the scalar components of any one of the vectors $k(\mathbf{u} \times \mathbf{v})$, $k(\mathbf{u} \times \mathbf{w})$, or $k(\mathbf{v} \times \mathbf{w})$, where $\mathbf{u} = [a_1, b_1, c_1]$, $\mathbf{v} = [a_2, b_2, c_2]$, and $\mathbf{w} = [a_3, b_3, c_3]$.

EXAMPLE 11-10. Find l if the planes $2x - 3y + z = 0$, $x + y - lz = 0$, and $3x - y - 2z = 0$ intersect in a line; and find the coordinates of a point (not the origin) on this line.

Solution. The planes are not parallel. In order that they will intersect in a line, the following condition must apply:

$$\begin{vmatrix} 2 & -3 & 1 \\ 1 & 1 & -l \\ 3 & -1 & -2 \end{vmatrix} = 0$$

Expanding by the elements of the second row, we have

$$-1\begin{vmatrix} -3 & 1 \\ -1 & -2 \end{vmatrix} + 1\begin{vmatrix} 2 & 1 \\ 3 & -2 \end{vmatrix} + l\begin{vmatrix} 2 & -3 \\ 3 & -1 \end{vmatrix} = 0, \text{ or } l = 2$$

Using the equations $2x - 3y + z = 0$ and $3x - y - 2z = 0$, we find that a point common to *all* three planes is

$$P = \left(\begin{vmatrix} -3 & 1 \\ -1 & -2 \end{vmatrix}, \begin{vmatrix} 1 & 2 \\ -2 & 3 \end{vmatrix}, \begin{vmatrix} 2 & -3 \\ 3 & -1 \end{vmatrix} \right) = (7, 7, 7)$$

These coordinates should satisfy the equation of the other plane when $l = 2$. Thus, $7 + 7 - 2(7) = 0$. Note that the coordinates of any other point on the line of intersection are $7k$, $7k$, and $7k$, where k has any arbitrary value.

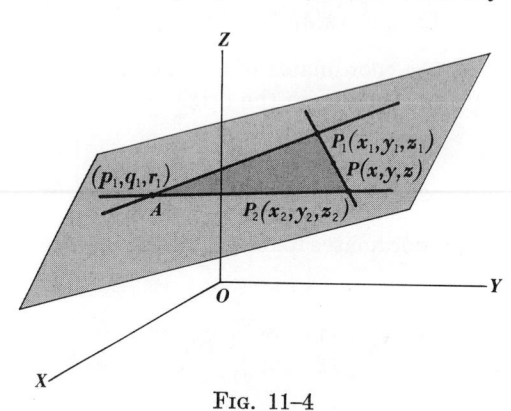

Fig. 11-4

EXERCISES

Solve the following systems of simultaneous equations, obtaining all the solutions in each case:

1. $2x - 3y + 6z = 0$, $x + y - z = 0$, $3x - y + 2z = 0$
2. $x + y - z = 0$, $2x - y + z = 0$, $3x + 2y - 2z = 0$
3. $2x - 3y + z = 0$, $-4x + 6y - 2z = 0$, $x + 2y - z = 0$
4. $3x + 2y - 2z = 0$, $2x + 3y - z = 0$, $8x + 7y - 5z = 0$
5. $2x - y + 3z = 0$, $4x - 2y + 6z = 0$, $-6x + 3y - 9z = 0$
6. If (x_1, y_1, z_1) and (x_2, y_2, z_2) are solutions of three homogeneous linear equations in the three unknowns x, y, z, prove that $(x_1 + x_2, y_1 + y_2, z_1 + z_2)$ is also a solution.

11-7. Parametric equations of a plane

Let us assume that a plane has been determined by a point A on it and by a pair of lines through A, as indicated in Fig. 11-4. Let the coordinates of A be (p_1, q_1, r_1). Also, let $[\Delta_1 x, \Delta_1 y, \Delta_1 z]$ be a non-zero vector on a line through A and an arbitrary point $P_1 = (x_1, y_1, z_1)$; and let $[\Delta_2 x, \Delta_2 y, \Delta_2 z]$ be a non-zero vector on a line through A and an arbitrary point $P_2 = (x_2, y_2, z_2)$. Parametric equations of the two lines are

$$x_1 = p_1 + t_1 \Delta_1 x, \ y_1 = q_1 + t_1 \Delta_1 y, \text{ and } z_1 = r_1 + t_1 \Delta_1 z$$
$$x_2 = p_1 + t_2 \Delta_2 x, \ y_2 = q_1 + t_2 \Delta_2 y, \text{ and } z_2 = r_1 + t_2 \Delta_2 z \quad (11\text{--}11)$$

Now let $P = (x, y, z)$ be any point on the line containing $\overrightarrow{P_1 P_2}$; obviously, P lies in the plane. Then, using the ratio formula $\overrightarrow{P_1 P} = k \ \overrightarrow{P P_2}$, where $k \neq -1$, we obtain:

$$x = \frac{x_1 + k x_2}{1 + k}, \ y = \frac{y_1 + k y_2}{1 + k}, \text{ and } z = \frac{z_1 + k z_2}{1 + k} \quad (11\text{--}12)$$

Substituting in equation (11–12) the values of x_1, x_2, y_1, y_2, z_1, and z_2 from equations (11–11), we obtain:

$$x = \frac{p_1 + t_1 \Delta_1 x + k(p_1 + t_2 \Delta_2 x)}{1 + k}$$

$$y = \frac{q_1 + t_1 \Delta_1 y + k(q_1 + t_2 \Delta_2 y)}{1 + k} \quad (11\text{--}13)$$

$$z = \frac{r_1 + t_1 \Delta_1 z + k(r_1 + t_2 \Delta_2 z)}{1 + k}$$

These equations may be rewritten as follows:

$$x = p_1 + t \Delta_1 x + t' \Delta_2 x$$
$$y = q_1 + t \Delta_1 y + t' \Delta_2 y \quad (11\text{--}14)$$
$$z = r_1 + t \Delta_1 z + t' \Delta_2 z$$

where $t = \dfrac{t_1}{1+k}$ and $t' = \dfrac{k t_2}{1+k}$.

That equations (11–14) represent only points of the plane as t and t' vary is shown in the following manner: Let t and t' be given, and let $t_1 = 2t$ and $t_2 = 2t'$. Substituting these values in equations (11–11), we obtain two points $P_1 = (x_1, y_1, z_1)$ and $P_2 = (x_2, y_2, z_2)$ on the plane. From equation (11–12), with $k = 1$, we obtain a point $P = (x, y, z)$ on the plane, where

$$x = \frac{x_1 + k x_2}{1+k} = \frac{p_1 + 2t \Delta_1 x + p_1 + 2t' \Delta_2 x}{2} = p_1 + t \Delta_1 x + t' \Delta_2 x$$

Similarly,
$$y = q_1 + t \Delta_1 y + t' \Delta_2 y \text{ and } z = r_1 + t \Delta_1 z + t' \Delta_2 z$$

Since these results are the same as equations (11–14), it follows that every point represented by equation (11–14) is on the plane.

Conversely, let x, y, and z be expressed linearly in terms of two parameters t and t', as follows:

$$x = p + a_1 t + a_2 t', \quad y = q + b_1 t + b_2 t', \text{ and } z = r + c_1 t + c_2 t' \qquad (11\text{-}15)$$

where not all determinants of order 2 in $\begin{bmatrix} a_1 & b_1 & c_1 \\ a_2 & b_2 & c_2 \end{bmatrix}$ are zero. Then, x, y, and z are the coordinates of a point on the plane. To prove this we observe that equation (11-15) becomes equation (11-14) if we let $p_1 = p$, $q_1 = q$, $r_1 = r$, $\Delta_1 x = a_1$, $\Delta_1 y = b_1$, $\Delta_1 z = c_1$, $\Delta_2 x = a_2$, $\Delta_2 y = b_2$, and $\Delta_2 z = c_2$.

If $t = t' = 0$, we find that the point (p, q, r) is on the plane. If $t' = 0$, we see that $[a_1, b_1, c_1]$ is a triple of direction numbers of one line on the plane; and, if $t = 0$, then $[a_2, b_2, c_2]$ is a triple of direction numbers of a second line on the plane. Both lines contain the point (p, q, r).

EXAMPLE 11-11. Find parametric equations of a plane containing the points $P_1 = (1, -1, 2)$, $P_2 = (-1, 3, 1)$, and $P_3 = (2, 3, -1)$.

Solution. In this case, $\overrightarrow{P_1 P_2} = [-2, 4, -1]$ and $\overrightarrow{P_1 P_3} = [1, 4, -3]$. Hence, the desired equations are $x = 1 - 2t_1 + t_2$, $y = -1 + 4t_1 + 4t_2$, and $z = 2 - t_1 - 3t_2$.

EXAMPLE 11-12. By eliminating the parameters in Example 11-11, obtain a cartesian equation of the plane.

Solution. From Example 11-11, $-2t_1 + t_2 = x - 1$ and $4t_1 + 4t_2 = y + 1$. Solving for t_1 and t_2, we obtain

$$t_1 = \frac{\begin{vmatrix} x-1 & 1 \\ y+1 & 4 \end{vmatrix}}{\begin{vmatrix} -2 & 1 \\ 4 & 4 \end{vmatrix}} = \frac{4x - 4 - y - 1}{-12} \text{ and } t_2 = \frac{\begin{vmatrix} -2 & x-1 \\ 4 & y+1 \end{vmatrix}}{\begin{vmatrix} -2 & 1 \\ 4 & 4 \end{vmatrix}} = \frac{-2y - 2 - 4x + 4}{-12}$$

Substituting these values for t_1 and t_2 in $z = 2 - t_1 - 3t_2$, we obtain:

$$z = 2 - \frac{4x - 4 - y - 1}{-12} - 3\left(\frac{-2y - 2 - 4x + 4}{-12}\right)$$

Clearing of fractions and collecting terms, we obtain

$$8x + 7y + 12z - 25 = 0$$

Check this result with the points given in Example 11-11.

EXERCISES

1. In each of the following cases, find parametric equations of the plane through the three given points:

    (a) $P_1(-1, 2, 3)$, $P_2(3, 1, 1)$, and $P_3(2, 4, -1)$

(b) $P_1(3, 1, 2)$, $P_2(1, -1, 4)$, and $P_3(1, 3, 1)$
(c) $P_1(5, 2, 3)$, $P_2(1, 1, 2)$, and $P_3(3, 2, -1)$
(d) $P_1(0, 1, -1)$, $P_2(2, 0, 3)$, and $P_3(0, 0, 2)$

2. Find parametric equations of each of the following planes:
   (a) $2x - 3y + 6z - 4 = 0$
   (b) $x + 2y - 3z + 6 = 0$
   (c) $x + y - z - 2 = 0$
   (d) $3x - 2y + z + 4 = 0$

REVIEW EXERCISES

1. Find an equation of the plane that contains the line $\dfrac{x-3}{2} = \dfrac{y+1}{-1} = \dfrac{z-1}{-2}$ and the point $(-1, 1, 2)$.

2. Find equations of the line that passes through the point $(2, -1, 3)$ and is perpendicular to the plane $3x - 4y + z - 6 = 0$.

3. Find an equation of the plane that contains the line $\dfrac{x+1}{2} = \dfrac{y-1}{3} = \dfrac{z+2}{1}$ and is perpendicular to the plane $2x - y + 3z - 6 = 0$.

4. Find the point in which the line represented by the equations $x - y + 2z - 8 = 0$ and $2x + y - z + 2 = 0$ intersects the plane $x + y + z - 1 = 0$.

5. (a) What are equations of the line that contains the point $(1, -1, 2)$ and is parallel to the line represented by the equations $3x - y + 2z - 6 = 0$ and $x + y - z - 4 = 0$.
   (b) Write equations of the line in parametric form.

6. Find k if the following planes have a line in common: $3x - 2y + z = 0$, $kx - y - 2z = 0$, and $x + y - 3z = 0$. Write equations of this line in both symmetric form and parametric form.

7. (a) Find the area of the parallelogram having three of its vertices at $P_1(1, -3, 2)$, $P_2(3, -1, 1)$, and $P_3(-1, 0, 1)$, in that order.
   (b) Where is the fourth vertex?

8. What is the volume of the parallelepiped having for three of its sides representative segments of the vectors $\mathbf{u} = [-1, 2, 3]$, $\mathbf{v} = [1, 1, 1]$, and $\mathbf{w} = [2, -1, -1]$?

9. (a) Find an equation of the plane that passes through the line $\dfrac{x+1}{2} = \dfrac{y+1}{-2} = \dfrac{z-2}{2}$ and is parallel to the line $\dfrac{x-1}{-1} = \dfrac{y-3}{-2} = \dfrac{z-4}{3}$.
   (b) What is the shortest distance between the two lines?

10. Find an equation of the plane that contains the point $(2, -1, 1)$ and the line represented by the equations $2x - y + z - 4 = 0$ and $x + y - 3z + 2 = 0$.

11. Find equations of the planes that bisect the angles formed by the planes $2x - y + 3z + 6 = 0$ and $3x + 2y - z + 12 = 0$.

12

Surfaces and Curves

12–1. Introduction

As a general rule, a single equation *in space* will represent a *surface*. Such an equation may be symbolized by $f(x, y, z) = 0$, where $f(x, y, z)$ is a function of x, y, and z. Examples of $f(x, y, z) = 0$ are $x^2 + y^2 + z^2 - 16 = 0$ and $3x - 2y + 5z + 4 = 0$. Sometimes the equation of the surface may be written in the form $z = f(x, y)$, where $f(x, y)$ is a function of x and y only. Examples of $z = f(x, y)$ are $z = 3x^2 - y^2$ and $z = 2x - 3y + 6$. If the equation is linear, it represents a plane, as has been previously shown. There are exceptions to these general rules, however. Two examples are $x^2 + y^2 + z^2 + 16 = 0$ and $x^2 + y^2 = 0$. Since there is no point $P = (x, y, z)$ whose coordinates satisfy the first of these equations, that equation cannot represent a surface. Furthermore, since the coordinates of every point that satisfies the second equation are 0, 0, and z, this equation represents the z-axis.

A combination of two equations *in space* will be said to represent a *space curve*. This curve is the locus of the points of intersection of the two surfaces represented by the individual equations, just as a line may be represented by the equations of two intersecting planes as we have already seen. If two surfaces do not intersect, there is no meaning to the combination of their equations.

In this chapter, we shall study a few of the more simple surfaces, we shall derive formulas for translating and rotating the axes in space, and we shall introduce other coordinate systems as well as cartesian systems.

12–2. Surfaces of revolution

If a surface is generated by revolving a plane curve about a fixed line in the plane, it is called a *surface of revolution*. Such a surface has the property that the sections cut by planes perpendicular to the fixed line are circles. The fixed line is called the *axis* of the surface. Let us suppose that the plane curve lies in the yz-plane and that the axis of the surface is the y-axis, as indicated in Fig. 12–1. Then, equations of the curve in the plane $x = 0$ are

$$f(y, z) = 0 \text{ and } x = 0 \qquad (12\text{–}1)$$

Assume that the curve does not cross the y-axis, and that no point on it has a negative z-coordinate. Let $P(x, y, z)$ be any point on the surface. Then there is a point L on the curve such that P lies on the circle which passes through L, is perpendicular to the y-axis, and has its center C on the y-axis. It is clear that the y-coordinate of every point on the circle is equal to y. Hence, the coordinates of C are 0, y, and 0 and those of L are 0, y, and z_1, where $z_1 \geq 0$ and $f(y, z_1) = 0$. Evidently, $|\overrightarrow{CL}| = |\overrightarrow{CP}| = z_1$. Since $\overrightarrow{CP} = [x, 0, z]$ and $|\overrightarrow{CP}| = \sqrt{x^2+z^2}$, we have

$$z_1 = \sqrt{x^2+z^2}$$

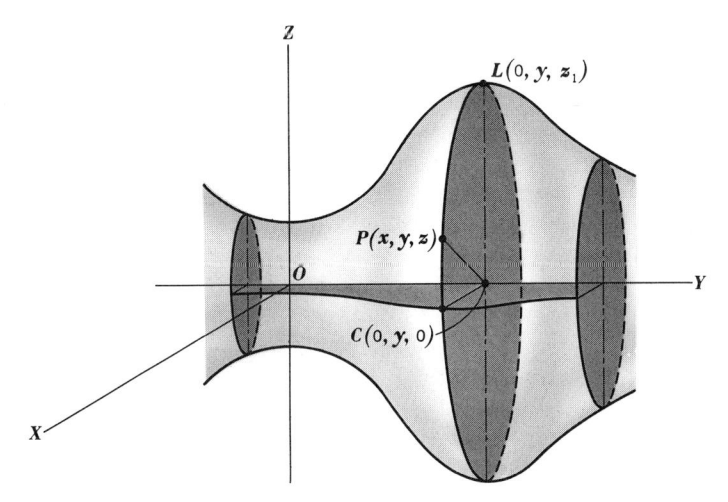

FIG. 12-1

Also, since $f(y, z_1) = 0$, we have

$$f(y, \sqrt{x^2+z^2}) = 0 \qquad (12\text{-}2)$$

This is an equation of the surface of revolution.

If the curve which generates the surface is such that z is never positive, we would have $|\overrightarrow{CL}| = |\overrightarrow{CP}| = -z_1$, and $z_1 = -\sqrt{x^2+z^2}$. Hence, equation (12-2) is replaced by

$$f(y, -\sqrt{x^2+z^2}) = 0 \qquad (12\text{-}3)$$

Rationalization of equations (12-2) and (12-3) leads to an equivalent equation. To see this, we proceed as follows: Let us rewrite equations (12-1) in the form $z = F(y)$ and $x = 0$, where $F(y) \geq 0$. Then equation (12-2) becomes $\sqrt{x^2+z^2} = F(y)$. Since neither $\sqrt{x^2+z^2}$ nor $F(y)$ is less than 0, this is equivalent to

$$x^2+z^2 = F^2(y) \qquad (12\text{-}4)$$

If $F(y) \leq 0$ for all values of y, then equation (12-3) applies, yielding $-\sqrt{x^2+z^2} = F(y)$, or $\sqrt{x^2+z^2} = -F(y)$. Since neither side is less than 0, this is equivalent to $x^2+z^2 = F^2(y)$. Hence, equation (12-4) is again obtained.

EXAMPLE 12-1. Find an equation of the surface of revolution obtained by revolving the curve represented by the equations $z^2 = 16y$ and $x = 0$ about the y-axis.

Solution. The given curve is a parabola in the yz-plane. Consider that part of the curve in Fig. 12-2 for which $z \geq 0$. Let $P = (x, y, z)$ be a point on the surface and associate with it the point $L = (0, y, z_1)$ on the curve, where $z_1 \geq 0$. Then, $|\overrightarrow{CP}| = |\overrightarrow{CL}| = z_1$ and $\overrightarrow{CP} = [x, 0, z]$. Hence,

(1) $$z_1 = \sqrt{x^2+z^2}$$

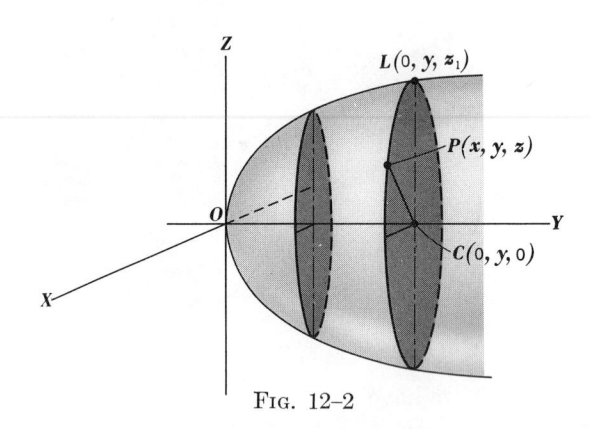

FIG. 12-2

Also, since L is a point on the curve, we know that

(2) $$z_1^2 = 16y$$

Eliminating z_1 from equations (1) and (2), we obtain

$$x^2 + z^2 = 16y$$

which is an equation of the desired surface of revolution.

EXAMPLE 12-2. Find an equation of the surface obtained by revolving the parabola represented by the equations $z^2 = 16y$ and $x = 0$ about the z-axis.

Solution. We let $P = (x, y, z)$ be a point on the surface and we associate with it a point L of the curve, where $L = (0, y_1, z)$, as indicated in Fig.

12-3. Then, $|\overrightarrow{CP}| = |\overrightarrow{CL}| = y_1$, where $y_1 \geq 0$; and $\overrightarrow{CP} = [x, y, 0]$. Hence, $|\overrightarrow{CP}| = \sqrt{x^2 + y^2}$ and

(1) $$y_1 = \sqrt{x^2 + y^2}$$

Also, since L is a point on the curve,

(2) $$z^2 = 16 y_1$$

Eliminating y_1 from equations (1) and (2), we obtain $z^2 = 16\sqrt{x^2 + y^2}$ or

$$z^4 = 256(x^2 + y^2)$$

This is an equation of the desired surface of revolution.

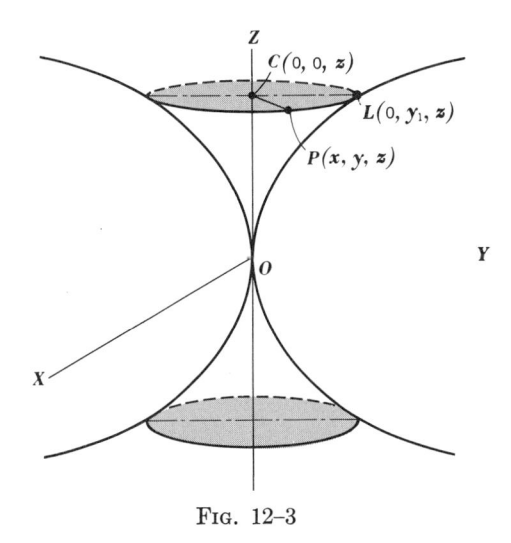

FIG. 12-3

If the plane curve is in the xz-plane, its equations are

$$f(x, z) = 0 \text{ and } y = 0 \qquad (12\text{-}5)$$

By an argument similar to that which has just been given, it can be shown that if $x \geq 0$, an equation of the surface obtained by revolving this curve about the z-axis is

$$f(\sqrt{x^2 + y^2}, z) = 0 \qquad (12\text{-}6)$$

Likewise, if $x \leq 0$, the equation becomes

$$f(-\sqrt{x^2 + y^2}, z) = 0 \qquad (12\text{-}7)$$

If the curve represented by equations (12-5) is revolved about the x-axis and it is assumed that $z \geq 0$, an equation of the surface is

$$f(x, \sqrt{y^2 + z^2}) = 0 \qquad (12\text{-}8)$$

If $z \leq 0$, the equation becomes

$$f(x, -\sqrt{y^2+z^2}) = 0 \qquad (12\text{-}9)$$

Finally, the equations of a curve in the xy-plane are

$$f(x, y) = 0 \text{ and } z = 0 \qquad (12\text{-}10)$$

If such a curve is revolved about the x-axis and $y \geq 0$, an equation of the surface generated is

$$f(x, \sqrt{y^2+z^2}) = 0 \qquad (12\text{-}11)$$

If $y \leq 0$, the equation is

$$f(x, -\sqrt{y^2+z^2}) = 0 \qquad (12\text{-}12)$$

In case the curve represented by equations (12-10) is revolved about the y-axis and $x \geq 0$, an equation of the surface generated is

$$f(\sqrt{x^2+z^2}, y) = 0 \qquad (12\text{-}13)$$

If $x \leq 0$, the equation becomes

$$f(-\sqrt{x^2+z^2}, y) = 0 \qquad (12\text{-}14)$$

EXERCISES

1. A curve in the xz-plane is represented by the equations $f(x, z) = 0$ and $y = 0$. Prove that, if $z \geq 0$, an equation of the surface obtained by revolving this plane curve about the x-axis is $f(x, \sqrt{y^2+z^2}) = 0$.

2. Prove that $f(\sqrt{x^2+y^2}, z) = 0$ is an equation of the surface obtained by revolving the plane curve represented by the equations $f(x, z) = 0$ and $y = 0$ about the z-axis if we assume that $x \geq 0$.

3. In each of the following exercises, find an equation of the surface obtained by revolving the plane curve about the axis indicated, and sketch the surface:

   (a) $x^2 = 16z$ and $y = 0$; x-axis
   (b) $x^2 = 16z$ and $y = 0$; z-axis
   (c) $x^2 + y^2 = 16$ and $z = 0$; x-axis
   (d) $9y^2 - 4z^2 = 36$ and $x = 0$; z-axis
   (e) $9y^2 - 4z^2 = 36$ and $x = 0$; y-axis
   (f) $3x - 4y = 2$ and $z = 0$; x-axis
   (g) $y = \sin x$ and $z = 0$; y-axis
   (h) $y = \cos x$ and $z = 0$; x-axis
   (i) $y = e^x$ and $z = 0$; y-axis
   (j) $x^2 + 4y^2 = 16$ and $z = 0$; x-axis
   (k) $x^2 + 4y^2 = 16$ and $z = 0$; y-axis
   (l) $2y - z + 4 = 0$ and $x = 0$; z-axis

12–3. Cylindrical surfaces

If a line is permitted to move in space in such a way that its direction numbers remain constant and it always intersects a fixed plane curve, the surface it generates is called a *cylinder*. The plane curve is called a *directrix*, and any one of the lines is called a *generator* of the cylinder, as indicated in Fig. 12–4.

When an equation of a surface contains only two variables, it represents a cylinder.* If the equation is considered together with that obtained by setting the other variable equal to zero, the two equations represent a directrix of the surface. Generators are lines which are parallel to the axis of the missing variable and which intersect the directrix. To see this,

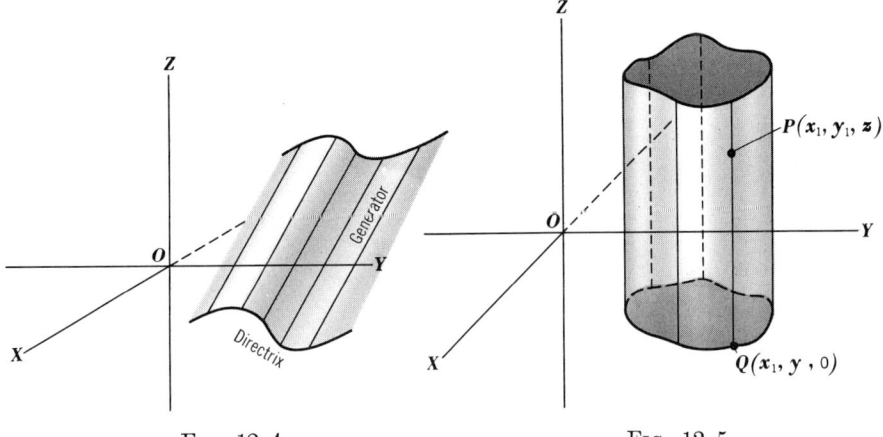

FIG. 12–4.   FIG. 12–5

consider the surface whose equation is $f(x, y) = 0$. The curve represented by the equations $f(x, y) = 0$ and $z = 0$ lies in the xy-plane and may be taken as the directrix of a cylinder whose generators are parallel to the z-axis. It is to be shown that this cylinder is the same surface as is represented by the equation $f(x, y) = 0$.

In Fig. 12–5, let $P(x_1, y_1, z)$ be on the surface $f(x, y) = 0$. Then, $Q(x_1, y_1, 0)$ is on the curve represented by the equations $f(x, y) = 0$ and $z = 0$. Moreover, P lies on the generator represented by the equations $x = x_1$ and $y = y_1$, which is parallel to the z-axis; so, P lies on the cylinder. Conversely, if $P(x_1, y_1, z)$ is on the cylinder, then $Q(x_1, y_1, 0)$ lies on the directrix; hence, $f(x_1, y_1) = 0$, and P is on the surface $f(x, y) = 0$. This proves that the surface $f(x, y) = 0$ is a cylinder as described.

---

* The student should note that nothing has been said to prevent an equation of a cylinder from containing all three variables. For example, the plane $ax + by + cz + d = 0$ is a cylinder in which the directrix is a line. However, if *only two* variables appear, the surface is known to be a cylinder.

EXAMPLE 12-3. Sketch the surface whose equation is $x^2 = 16y$.

Solution. By the argument just given, the equation represents a cylinder with the directrix in the xy-plane and represented by the equations $x^2 = 16y$ and $z = 0$ and with generators parallel to the z-axis. The directrix is a parabola. If in Fig. 12-6 $Q(x_1, y_1, 0)$ is any point of the parabola, then $x_1^2 = 16y_1$; and the coordinates x_1, y_1, and z satisfy the same equation for all values of z. Hence, $P(x_1, y_1, z)$ lies on a line that passes through Q and is parallel to the z-axis; and the surface is generated by allowing this line to move as Q moves along the parabola.

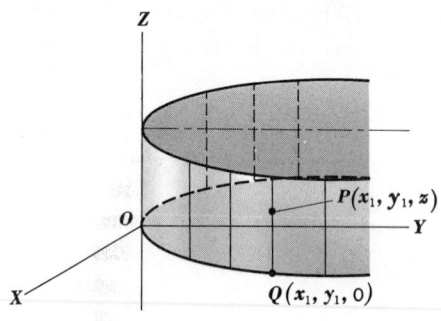

FIG. 12-6

In general, $f(x, z) = 0$ is an equation of a cylinder whose directrix is the curve in the xz-plane represented by the equations $f(x, z) = 0$ and $y = 0$ and whose generators are lines that are parallel to the y-axis and intersect the directrix. Also, $f(y, z) = 0$ is an equation of a cylinder whose directrix is the curve in the yz-plane represented by the equations $f(y, z) = 0$ and $x = 0$ and whose generators are lines that are parallel to the x-axis and intersect the directrix.

EXERCISES

Sketch each of the following surfaces and give the equations of the directrix:

1. $x^2 + y^2 = 16$
2. $y - 3 = 0$
3. $x^2 - 4z = 0$
4. $4x^2 - y^2 = 16$
5. $y = \sin x$
6. $x = e^z$
7. $y^2 - z^2 = 49$
8. $z = \cosh x$
9. $x = \tan y$
10. $y = x$
11. $y^2 + z^2 = 4$
12. $z = \sinh x$
13. $4x^2 - 9z^2 = 36$
14. $z^2 - 16y = 0$
15. $2x - 3y = 6$
16. $x = \cos z$
17. $9x^2 + 4y^2 = 36$
18. $z^2 + x^2 = 4$
19. $y^2 - 9z^2 = 16$
20. $z^2 - 9y = 0$

FIG. 12-8   FIG. 12-9

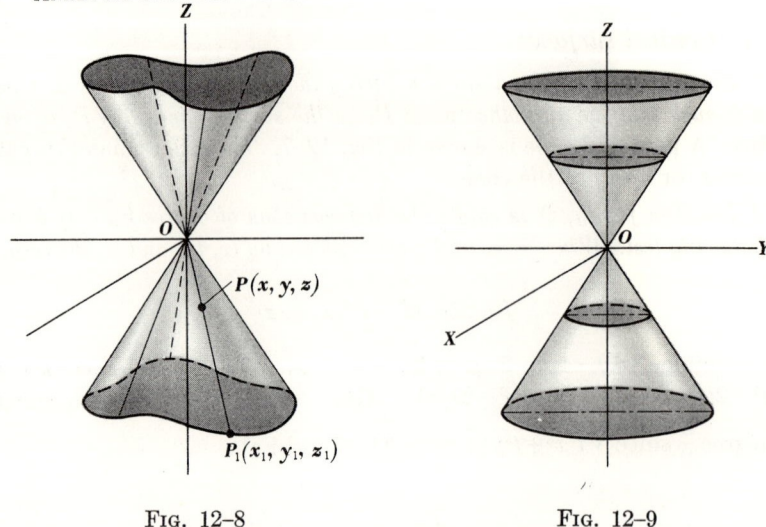

EXAMPLE 12-4. Show that $4x^2+y^2-z^2=0$ is an equation of a cone and sketch its graph.

Solution. Since the equation is homogeneous, it represents a cone having the vertex at the origin. If we rewrite the equation in the form $4x^2+y^2=z^2$, we see that the cross-section on any plane for which z is constant is an ellipse. The cone is shown in Fig. 12-9.

EXAMPLE 12-5. Find an equation of the right circular cone obtained by revolving the line whose equations in the yz-plane are $2z+3y-6=0$ and $x=0$ about the y-axis.

Solution. Let $P=(x, y, z)$ be any point on the surface, as indicated in Fig. 12-10, and let the point associated with it on the line be $L=(0, y, z_1)$,

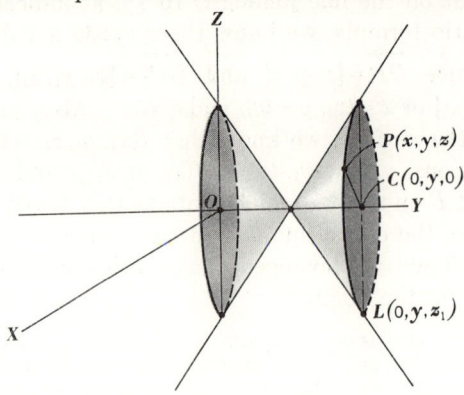

FIG. 12-10

where $z_1 \leq 0$. Then, if the center of the circle through P and L is $C = (0, y, 0)$, it follows that $|\overrightarrow{CP}| = |\overrightarrow{CL}| = -z_1$. Since $\overrightarrow{CP} = [x, 0, z]$ and $|\overrightarrow{CP}| = \sqrt{x^2 + z^2}$, we have

(1) $$-z_1 = \sqrt{x^2 + z^2}$$

Also, since L is on the line,

(2) $$2z_1 + 3y - 6 = 0$$

Eliminating z_1 from equations (1) and (2), we obtain

(3) $$2\sqrt{x^2 + z^2} + 3y - 6 = 0$$

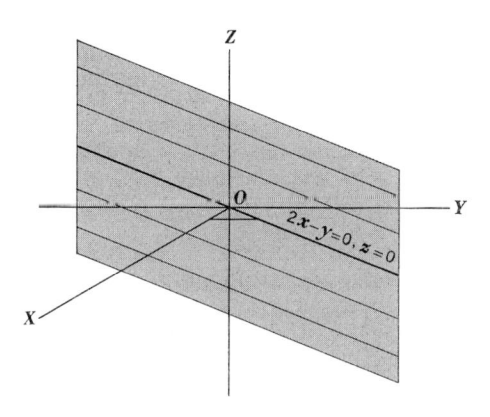

FIG. 12-11

Rationalizing this result, we find an equation of the surface to be

(4) $$4(x^2 + z^2) = (6 - 3y)^2$$

We know that equation (4) represents a cone; yet, it is not homogeneous.

Note: We might have chosen $L = (0, y, z_1)$ where $z_1 \geq 0$. If we had done so, what changes would have occurred in equations (1), (2), (3), and (4)?

EXAMPLE 12-6. Find an equation of a surface which is both a cylinder and a cone.

Solution. Consider the equation $2x - y = 0$. It is a single equation and therefore represents a surface. Since there are only two variables in the equation, the locus is a cylinder. Also, the equation is homogeneous; and, hence, it represents a cone with the vertex at the origin. Since it is a linear equation, it also represents a plane. See Fig. 12-11.

EXERCISES

For each equation state if it represents a cone, a cylinder, or a surface of revolution; and draw its graph:

1. $x^2 - y^2 = z^2$
2. $y^2 = 4z$
3. $x^2 - y^2 + x - y - 6 = 0$
4. $xy = yz$
5. $y^2 + z^2 = 16$
6. $4x^2 + z^2 = 16$
7. $z^2 + 4y = 0$
8. $x^2 - 4yz = 0$
9. $x - 3y + 3 = 0$
10. $9z^2 + 9y^2 = x^2$
11. $x^2 + y^2 = 25$
12. $z^2 + 4y^2 = 4$
13. $xy = 5$
14. $z^2 - y^2 = 0$
15. $xz + yz = 0$
16. $z + 2y - 6 = 0$
17. $x^2 + y^2 = z^2$
18. $4y^2 + 9z^2 = 36$
19. $x^2 = 4y$
20. $y^2 - z^2 = 16x^2$

12–5. Quadric surfaces

A surface whose equation is of the second degree is called a quadric surface. The general equation of such a surface is

$$Ax^2 + By^2 + Cz^2 + Dxy + Exz + Fyz + Gx + Hy + Iz + J = 0 \quad (12\text{–}15)$$

where not all the coefficients A, B, C, D, E, and F are zero. By means of a translation and a rotation of the axes this equation can be simplified. In the sections that follow we shall discuss in detail several of the simpler forms of the second-degree equation.

12–6. The spheres

Consider the set of points $P(x, y, z)$, Fig. 12–12, which have the property that $|\overrightarrow{CP}| = r$, where C is a fixed point with coordinates x_0, y_0, and z_0. The surface which is generated by P is called a *sphere*; the point C is called the *center*, and r is the *radius* of the sphere. Since $\overrightarrow{CP} = [x - x_0, y - y_0, z - z_0]$, it follows that

$$|\overrightarrow{CP}| = \sqrt{(x - x_0)^2 + (y - y_0)^2 + (z - z_0)^2} = r$$

or
$$(x - x_0)^2 + (y - y_0)^2 + (z - z_0)^2 = r^2 \quad (12\text{–}16)$$

This is known as the *standard equation* of a sphere.

If the center C of the sphere is at the origin, an equation of the sphere is

$$x^2 + y^2 + z^2 = r^2 \quad (12\text{–}17)$$

If we intersect the sphere with a plane whose equation is $z = k$, we

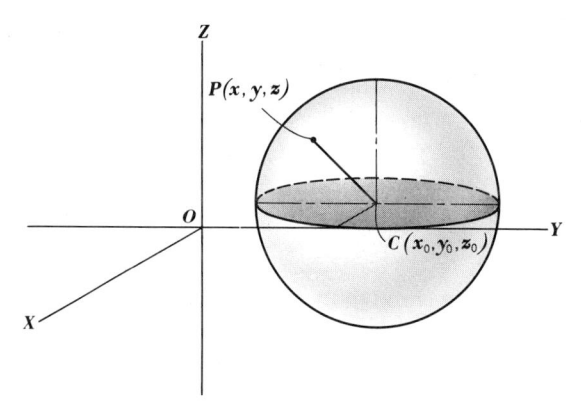

FIG. 12-12

obtain a cross-section whose equations are $x^2+y^2=r^2-k^2$ and $z=k$. This is a circle if $|k|<r$ and is a point if $|k|=r$; there is no locus, or there is an imaginary circle, if $|k|>r$. We speak of the plane section when $z=0$ as the *trace* of the surface in the xy-plane. In our example, the trace is the circle $x^2+y^2=r^2$ and $z=0$. By a similar argument we see that the cross-sections with $x=k$ or with $y=k$ are also circles when $|k|<r$, single points when $|k|=r$, and imaginary circles when $|k|>r$. The trace in the yz-plane is the circle $y^2+z^2=r^2$ and $x=0$; and the trace in the xz-plane is the circle $x^2+z^2=r^2$ and $y=0$.

Let us now consider the equation

$$x^2+y^2+z^2+Dx+Ey+Gz+F=0 \qquad (12\text{-}18)$$

If we complete the squares on the x, y, and z terms, we obtain

$$x^2+Dx+\frac{D^2}{4}+y^2+Ey+\frac{E^2}{4}+z^2+Gz+\frac{G^2}{4}=\frac{D^2+E^2+G^2}{4}-F$$

or

$$\left(x+\frac{D}{2}\right)^2+\left(y+\frac{E}{2}\right)^2+\left(z+\frac{G}{2}\right)^2=\frac{D^2+E^2+G^2-4F}{4} \qquad (12\text{-}19)$$

This is the standard equation of a sphere in which $x_0=-\frac{D}{2}$, $y_0=-\frac{E}{2}$, $z_0=-\frac{G}{2}$, and $r=\frac{\sqrt{D^2+E^2+G^2-4F}}{2}$. Hence, an equation of the second degree in which x^2, y^2, and z^2 have the same coefficient and in which there is no xy-term, no xz-term, or no yz-term represents a sphere. We agree to speak of the case when no locus exists, that is, when $D^2+E^2+G^2-4F$ is negative, as an imaginary sphere.

Equation (12-18) is called the *general equation* of a sphere. If $r>0$, the sphere has a real locus; if $r=0$, the locus is a point-sphere; and, if r is imaginary, there is no locus.

408 ANALYTIC GEOMETRY OF SPACE

EXAMPLE 12-7. Find an equation of the sphere that has its center at $(-1, 2, 3)$ and a radius equal to 5.

Solution. Using the standard equation of the sphere, we have
$$(x+1)^2 + (y-2)^2 + (z-3)^2 = 25$$

EXAMPLE 12-8. Find the center and radius of the sphere whose equation is $2x^2 + 2y^2 + 2z^2 - 3x + y - 4z - 3 = 0$.

Solution. Dividing through by 2 and completing the square, we obtain
$$x^2 - \frac{3}{2}x + \frac{9}{16} + y^2 + \frac{1}{2}y + \frac{1}{16} + z^2 - 2z + 1 = \frac{9}{16} + \frac{1}{16} + 1 + \frac{3}{2}$$
or
$$\left(x - \frac{3}{4}\right)^2 + \left(y + \frac{1}{4}\right)^2 + (z-1)^2 = \frac{50}{16}$$

The center of the sphere is the point $\left(\frac{3}{4}, -\frac{1}{4}, 1\right)$, and the radius is $\frac{5\sqrt{2}}{4}$.

EXERCISES

1. Find an equation of each of the following spheres having the given center and radius:
   (a) $C = (-1, 2, 3); r = 4$
   (b) $C = (0, 0, 0); r = 6$
   (c) $C = (3, 1, -2); r = 1$
   (d) $C = (-1, -1, 0); r = 3$
   (e) $C = (2, 0, -3); r = 5$
   (f) $C = (0, -4, 1); r = 10$

2. Find the center and radius of each of the following spheres:
   (a) $x^2 + y^2 + z^2 - 2x + 4y - 6z - 11 = 0$
   (b) $2x^2 + 2y^2 + 2z^2 - 4x + 6z - 3 = 0$
   (c) $x^2 + y^2 + z^2 + 2y - 4z - 4 = 0$
   (d) $3x^2 + 3y^2 + 3z^2 - x + 7y + 3z - 3 = 0$
   (e) $x^2 + y^2 + z^2 - 6x + 4z - 36 = 0$

3. Find an equation of the sphere having the line segment joining the two points $(5, 2, -1)$ and $(-3, 4, 7)$ as a diameter.

4. A sphere has its center in the plane $3x + y - z - 2 = 0$ and passes through the three points $(2, 1, 3)$, $(1, -1, 2)$ and $(-1, 3, -1)$. Find its equation.

5. Find an equation of the sphere which has its center on the line $3x + 6 = 2y - 3 = 3z$ and passes through the points $(2, -1, 1)$ and $(1, 3, 2)$.

6. A sphere has its center at the point $(-2, 3, 1)$ and touches the plane $2x - y + 2z - 7 = 0$. What is its equation? Hint: The absolute distance from the plane to the point is the radius.

7. A sphere passes through the point $(1, 1, -3)$ and is tangent to the

plane $x - 2y - 2z - 7 = 0$ at the point $(3, -1, -1)$. What is its equation?

8. A sphere passes through the points $(1, -1, 2)$ and $(2, 1, 1)$. If its center lies on the line $x = y + 3 = z + 1$, what is its equation?

9. A sphere is tangent to the plane $2x - y + 2z + 3 = 0$ and has its center at the point $(3, 1, 5)$. Find its equation.

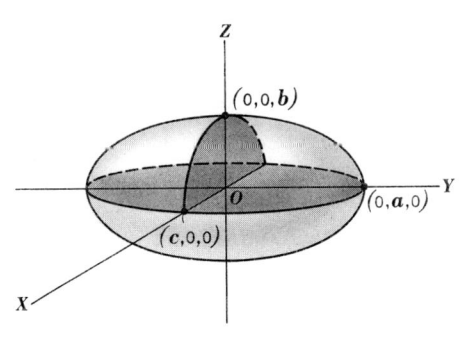

Fig. 12-13

12-7. The ellipsoids

Consider a surface whose equation is

$$\frac{y^2}{a^2} + \frac{z^2}{b^2} + \frac{x^2}{c^2} = 1 \tag{12-20}$$

where a, b, and c are positive numbers. The locus of this equation, or of one which can be reduced to this form by transformation of axes, is called an *ellipsoid*. Let us assume that $a > b > c$, as shown is Fig. 12-13. The trace in the xy-plane is the ellipse represented by the equations $\frac{y^2}{a^2} + \frac{x^2}{c^2} = 1$ and $z = 0$. The trace in the xz-plane is the ellipse whose equations are $\frac{z^2}{b^2} + \frac{x^2}{c^2} = 1$ and $y = 0$. The trace in the yz-plane is the ellipse with equations $\frac{y^2}{a^2} + \frac{z^2}{b^2} = 1$ and $x = 0$.

The cross-section of the ellipsoid on any plane for which $z = k$ is an ellipse whose equations are

$$\frac{y^2}{a^2} + \frac{x^2}{c^2} = 1 - \frac{k^2}{b^2} \text{ and } z = k$$

If $|k| < b$, the ellipse is real and not degenerate; if $|k| = b$, the locus is a point-ellipse; and, if $|k| > b$, the locus is an imaginary ellipse.

In the same way we find that the cross-section of the ellipsoid on a plane for which $y = k$ or on a plane for which $x = k$ is also an ellipse.

The surface represented by equation (11–20) is symmetric with respect to the x-axis, the y-axis, the z-axis, and the origin.

We will now consider several special types of ellipsoids.

(a) When $a = b = c$, the ellipsoid is a sphere.

(b) When $a = b \neq c$, or $b = c \neq a$, or $a = c \neq b$, the ellipsoid is a surface of revolution. If the ellipsoid of revolution is obtained by revolving the ellipse whose equations are $\frac{y^2}{a^2} + \frac{z^2}{b^2} = 1$ and $x = 0$ ($a > b$) about its *major* axis, it is called a *prolate spheroid*. The equation of this surface is

$$\frac{y^2}{a^2} + \frac{z^2}{b^2} + \frac{x^2}{b^2} = 1 \qquad (12\text{–}21)$$

If the ellipsoid is obtained by revolving an ellipse about its *minor* axis, the surface is called an *oblate spheroid*. In our example, the equation is

$$\frac{x^2}{a^2} + \frac{y^2}{a^2} + \frac{z^2}{b^2} = 1 \qquad (12\text{–}22)$$

In order to discuss a surface, we find the following characteristics: the intercepts, i.e., the x-, y-, and z-coordinates of the points where the surface intersects the axes; the traces; and the cross-sections with planes perpendicular to the axes. Then we draw the graph.

EXAMPLE 12–9. Discuss the surface whose equation is $z^2 + 4y^2 + 9x^2 = 36$.

Solution. *Intercepts:* If $y = z = 0$, then $x = \pm 2$; if $x = z = 0$, then $y = \pm 3$; if $x = y = 0$, then $z = \pm 6$.

Traces: The trace on the yz-plane is the ellipse $z^2 + 4y^2 = 36$ and $x = 0$; that on the xz-plane is the ellipse $z^2 + 9x^2 = 36$ and $y = 0$; and that on the xy-plane is the ellipse $4y^2 + 9x^2 = 36$ and $z = 0$.

Cross-sections: For $x = k$, where $|k| < 2$, the cross-sections are ellipses; for $y = k$, where $|k| < 3$, the cross-sections are ellipses; and, for $z = k$, where $|k| < 6$, they are ellipses. The surface is the ellipsoid shown in Fig. 12–14.

EXAMPLE 12–10. Find an equation of the prolate spheroid that has for its generating curve the ellipse whose equations are

$$\frac{y^2}{a^2} + \frac{z^2}{b^2} = 1 \text{ and } x = 0 \text{ (where } a > b)$$

Solution. This is a surface of revolution in which the ellipse is revolved about the y-axis (major axis). It is shown in Fig. 12–15.

Let $P = (x, y, z)$ be any point on the surface and let $L = (0, y, z_1)$, where $z_1 \geq 0$, be the point on the generating curve associated with P. Then,

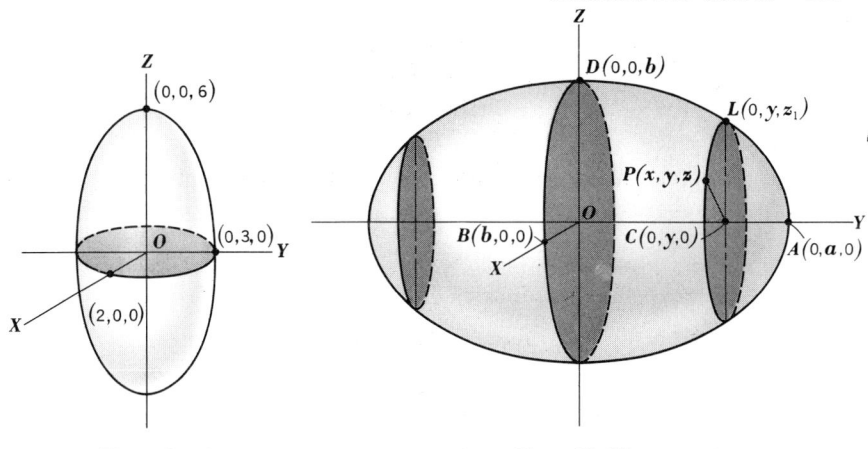

SURFACES AND CURVES 411

if the center of the circle through P and L is at $C = (0, y, 0)$, we know that $|\overrightarrow{CL}| = |\overrightarrow{CP}| = z_1$. Also, since $\overrightarrow{CP} = [x, 0, z]$ and $|CP| = \sqrt{x^2+z^2}$,

(1) $$z_1 = \sqrt{x^2+z^2}$$

Since L is on the generating curve

(2) $$\frac{y^2}{a^2} + \frac{z_1^2}{b^2} = 1$$

Eliminating z_1 from equations (1) and (2), we obtain as an equation of the prolate spheroid:

$$\frac{y^2}{a^2} + \frac{z^2+x^2}{b^2} = 1$$

EXERCISES

1. Discuss and sketch each of the following surfaces:
   (a) $x^2 + 4y^2 + 16z^2 = 144$
   (b) $9x^2 + y^2 + 4z^2 = 36$
   (c) $16x^2 + 9y^2 + 4z^2 = 144$
   (d) $4x^2 + 4y^2 + 9z^2 = 36$
   (e) $x^2 + 9y^2 + 9z^2 = 81$
   (f) $4x^2 + 9y^2 + 4z^2 = 1$
   (g) $5x^2 + 25y^2 + 25z^2 = 25$
   (h) $x^2 + y^2 + 4z^2 = 4$

2. In each of the following exercises, derive an equation of the surface of revolution obtained by revolving the given plane curve about the given axis:
   (a) $9x^2 + 4y^2 = 36$ and $z = 0$; the x-axis
   (b) $9x^2 + 4y^2 = 36$ and $z = 0$; the y-axis
   (c) $5x^2 + 3z^2 - 15 = 0$ and $y = 0$; the z-axis
   (d) $y^2 + 4z^2 - 4 = 0$ and $x = 0$; the y-axis

(e) $x^2+z^2-16=0$ and $y=0$; the z-axis
(f) $25y^2+z^2=625$ and $x=0$; the y-axis
(g) $4x^2+3y^2=12$ and $z=0$; the x-axis
(h) $25y^2+z^2=625$ and $x=0$; the z-axis

3. Which of the surfaces in Exercise 2 are oblate spheroids and which are prolate spheroids?

12-8. The elliptic paraboloids

Consider the surface whose equation is

$$\frac{x^2}{a^2}+\frac{y^2}{b^2}=pz \qquad (12\text{-}23)$$

where a and b are greater than 0 and $p\neq 0$. First we shall assume that $p>0$. The trace in the xy-plane is the point-ellipse whose equations are $\frac{x^2}{a^2}+\frac{y^2}{b^2}=0$ and $z=0$. The trace in the xz-plane is the parabola whose equations are $x^2=pa^2z$ and $y=0$. The trace in the yz-plane is the parabola whose equations are $y^2=b^2pz$ and $x=0$.

The cross-sections at the intersections of the surface with the planes for which $z=k$ are ellipses whose equations are

$$\frac{x^2}{a^2}+\frac{y^2}{b^2}=pk \text{ and } z=k$$

If $k>0$, the ellipses are real; if $k=0$, the locus is a point-ellipse; and, if $k<0$, the locus is an imaginary ellipse.

The cross-sections at the intersections with the planes for which $x=k$ are parabolas whose equations are

$$\frac{y^2}{b^2}=pz-\frac{k^2}{a^2} \text{ and } x=k$$

The cross-sections at the intersections with the planes for which $y=k$ are also parabolas whose equations are

$$\frac{x^2}{a^2}=pz-\frac{k^2}{b^2} \text{ and } y=k$$

If $p<0$, the discussion is the same as that just given, but the cross-sections for which $z=k$ are real when $k\leq 0$.

The surface represented by equation (12-23) is symmetric with respect to the z-axis. Since the cross-sections of the surface are either ellipses or parabolas, as shown in Fig. 12-16, the surface is called an *elliptic paraboloid*.

If $a=b$, the surface is a surface of revolution obtained by revolving the parabola whose equations are $y^2=a^2pz$ and $x=0$ about the z-axis. If $p=0$ in equation (12-23), the result is an equation of the z-axis.

Fig. 12-16

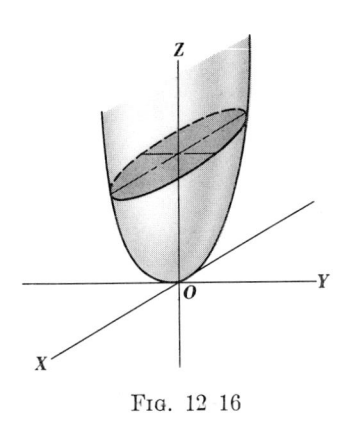

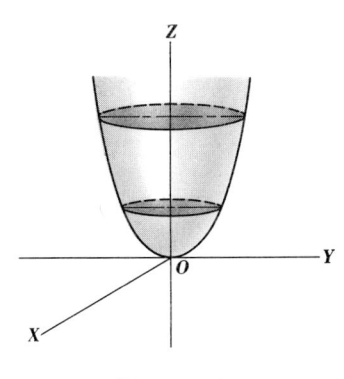

Fig. 12-17

EXAMPLE 12-11. Discuss the surface whose equation is $4y^2 + 9x^2 = 36z$.

Solution. Intercepts: The intercepts are 0 since the surface intersects each axis only at the origin.

Traces: The trace on the xy-plane is the point-ellipse whose equations are $4y^2 + 9x^2 = 0$ and $z = 0$. That on the xz-plane is the parabola whose equations are $x^2 = 4z$ and $y = 0$. That on the yz-plane is the parabola whose equations are $y^2 = 9z$ and $x = 0$. The surface is an *elliptic paraboloid*, as shown in Fig. 12-17.

EXERCISES

1. Discuss each of the following surfaces:
   (a) $4x^2 + 9y^2 = 4z$
   (b) $9x^2 + 4z^2 = 36y$
   (c) $z^2 + 4y^2 + 12x = 0$
   (d) $3x^2 + z^2 - 27y = 0$
   (e) $2z^2 + y^2 - 18x = 0$
   (f) $4x^2 + 9y^2 = -72z$

2. Derive the equation of each of the following surfaces of revolution obtained by revolving the given plane curve about the given axis:
   (a) $x^2 - 6y = 0$ and $z = 0$; the x-axis
   (b) $x^2 - 6y = 0$ and $z = 0$; the y-axis
   (c) $4y^2 + 9z = 0$ and $x = 0$; the y-axis
   (d) $4y^2 + 9z = 0$ and $x = 0$; the z-axis
   (e) $2x^2 + 25z = 0$ and $y = 0$; the x-axis
   (f) $6z^2 - 15y = 0$ and $x = 0$; the z-axis
   (g) $9y^2 - 24x = 0$ and $z = 0$; the y-axis
   (h) $5y^2 + 12x = 0$ and $z = 0$; the x-axis

414   ANALYTIC GEOMETRY OF SPACE

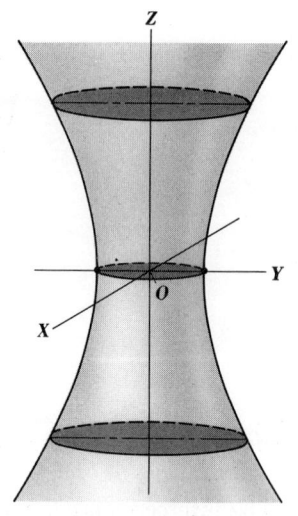

FIG. 12-18

12-9. The hyperboloids

(a) *The hyperboloid of one sheet.* The next surface to be considered is one whose equation is

$$\frac{y^2}{a^2}+\frac{x^2}{b^2}-\frac{z^2}{c^2}=1 \qquad (12\text{-}24)$$

where a, b, and c are greater than 0. Such a surface is called a *hyperboloid of one sheet*. It is shown in Fig. 12-18. This surface intersects the y-axis at $(0, a, 0)$, intersects the x-axis at $(b, 0, 0)$, and does not intersect the z-axis at all.

Its trace on the xy-plane is the ellipse whose equations are $\frac{y^2}{a^2}+\frac{x^2}{b^2}=1$ and $z=0$; that on the yz-plane is the hyperbola whose equations are $\frac{y^2}{a^2}-\frac{z^2}{c^2}=1$ and $x=0$; and that on the xz-plane is the hyperbola whose equations are $\frac{x^2}{b^2}-\frac{z^2}{c^2}=1$ and $y=0$.

The cross-sections at the intersections of the surface with the planes for which $z=k$ are ellipses whose equations are $\frac{y^2}{a^2}+\frac{x^2}{b^2}=1+\frac{k^2}{c^2}$ and $z=k$. As $|k|$ increases, the ellipses increase in size.

The cross-sections at the intersections with the planes for which $x=k$ are hyperbolas whose equations are $\frac{y^2}{a^2}-\frac{z^2}{c^2}=1-\frac{k^2}{b^2}$ and $x=k$. If $|k|>b$, a line parallel to the z-axis is the transverse axis; if $|k|=b$, the hyperbola degenerates into a pair of lines.

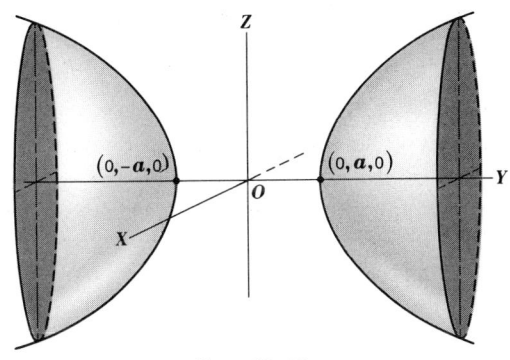

FIG. 12-19

Similarly, the cross-sections at the intersections with the planes for which $y = k$ are hyperbolas whose equations are $\frac{x^2}{b^2} - \frac{z^2}{c^2} = 1 - \frac{k^2}{a^2}$ and $y = k$. If $|k| > a$, a line parallel to the z-axis is the transverse axis; if $|k| < a$, a line parallel to the x-axis is the transverse axis; if $|k| = a$, the hyperbola again degenerates into a pair of lines.

When $a = b$, the surface is a *hyperboloid of revolution* of one sheet.

(b) *The hyperboloid of two sheets.* The equation of another type of hyperboloid is

$$\frac{y^2}{a^2} - \frac{x^2}{b^2} - \frac{z^2}{c^2} = 1 \tag{12-25}$$

where a, b, and c are greater than 0. This surface has two disconnected parts and is called a *hyperboloid of two sheets*. Such a surface is shown in Fig. 12-19. It intersects the y-axis at the points $(0, \pm a, 0)$ and does not intersect the x-axis or the z-axis at all. Hence, the y-intercepts are a and $-a$, and there is no x-intercept or z-intercept.

The trace on the xy-plane is the hyperbola whose equations are $\frac{y^2}{a^2} - \frac{x^2}{b^2} = 1$ and $z = 0$; that on the yz-plane is the hyperbola whose equations are $\frac{y^2}{a^2} - \frac{z^2}{c^2} = 1$ and $x = 0$; and there is no trace in the xz-plane.

The cross-sections at the intersections of the surface with the planes for which $y = k$ are ellipses whose equations are $\frac{x^2}{b^2} + \frac{z^2}{c^2} = \frac{k^2}{a^2} - 1$ and $y = k$. If $|k| > a$, the ellipses are real; if $|k| < a$, the ellipses are imaginary; and, if $|k| = a$, the locus is a point-ellipse.

On planes for which $x = k$, the cross-sections are hyperbolas whose equations are $\frac{y^2}{a^2} - \frac{z^2}{c^2} = 1 + \frac{k^2}{b^2}$ and $x = k$. For each value of k, a line parallel to the y-axis is the transverse axis.

On planes for which $z = k$, the cross-sections are hyperbolas whose equations are $\dfrac{y^2}{a^2} - \dfrac{x^2}{b^2} = 1 + \dfrac{k^2}{c^2}$ and $z = k$. For each value of k, a line parallel to the y-axis is the transverse axis.

When $b = c$, the hyperboloid is a surface of revolution.

When the equation of a hyperboloid is written so that the right-hand side is a positive constant, the surface has one sheet if only one minus sign appears on the left side, while it consists of two sheets if two minus signs appear on the left side.

If a hyperbola is revolved about its conjugate axis, the surface generated is a hyperboloid of revolution of one sheet. If a hyperbola is revolved about its transverse axis, the surface generated is a hyperboloid of two sheets.

EXAMPLE 12-12. Discuss the equation $4x^2 + 9y^2 - z^2 = 36$.

Solution. Since the right-hand side of the equation is a positive constant, and since only one minus sign appears on the left side, the surface is a hyperboloid of one sheet. Dividing through by 36, we obtain
$$\frac{x^2}{9} + \frac{y^2}{4} - \frac{z^2}{36} = 1.$$

The intercepts are $x = \pm 3$ and $y = \pm 2$.

The trace on the xy-plane is the ellipse whose equations are $4x^2 + 9y^2 = 36$ and $z = 0$; that on the xz-plane is the hyperbola whose equations are $4x^2 - z^2 = 36$ and $y = 0$; and that on the yz-plane is the hyperbola whose equations are $9y^2 - z^2 = 36$ and $x = 0$.

The cross-sections at the intersections of the surface with planes for which $z = k$ are ellipses. The cross-sections at the intersections of the surface with planes for which $x = k$ or for which $y = k$ are hyperbolas. The surface is shown in Fig. 12-20.

EXAMPLE 12-13. Discuss the equation $9z^2 - 4y^2 - 16x^2 = 144$.

Solution. Since the right-hand side of the equation is a positive constant and two minus signs appear on the left-hand side, the surface is a hyperboloid of two sheets. Dividing each term by 144, we obtain
$$\frac{z^2}{16} - \frac{y^2}{36} - \frac{x^2}{9} = 1$$

The intercepts are $z = \pm 4$. There are no intercepts on the x-axis and y-axis.

On the xy-plane, the trace is imaginary; the trace on the yz-plane is the hyperbola whose equations are $\dfrac{z^2}{16} - \dfrac{y^2}{36} = 1$ and $x = 0$; and that on the

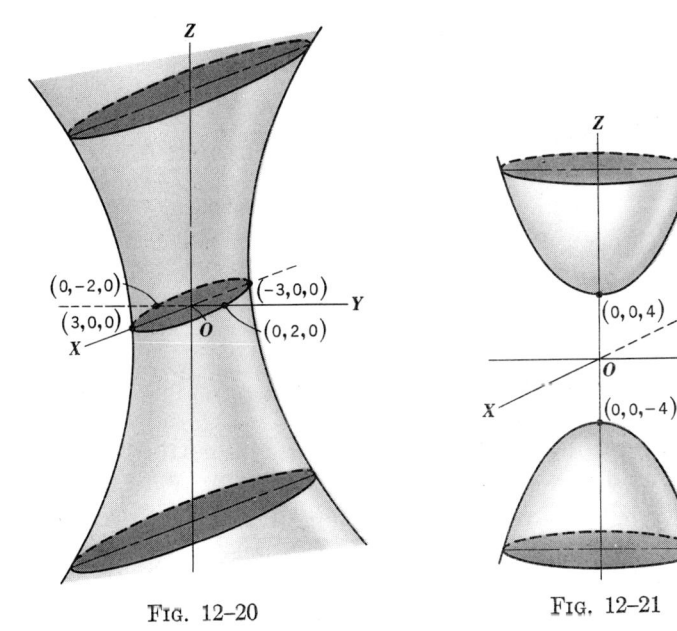

FIG. 12-20

FIG. 12-21

xz-plane is the hyperbola whose equations are $\dfrac{z^2}{16} - \dfrac{x^2}{9} = 1$ and $y = 0$.

At the intersections with planes for which $z = k$, the cross-sections are ellipses. When $|k| > 4$, the ellipses are real; when $|k| < 4$, the ellipses are imaginary; and when $|k| = 4$, the locus is a point-ellipse. On planes for which $x = k$, the cross-sections are hyperbolas. On planes for which $y = k$, the cross-sections are hyperbolas. The surface is shown in Fig. 12-21.

EXAMPLE 12-14. Find an equation of the surface of revolution obtained by revolving the hyperbola whose equations are $y^2 - 4z^2 = 4$ and $x = 0$ about the y-axis.

Solution. Let $P = (x, y, z)$ be any point on the surface; and, as shown in Fig. 12-22, let the point associated with it on the hyperbola be $L = (0, y, z_1)$, where $z_1 \geq 0$. Then, if the center of the circle through P and L is $C = (0, y, 0)$, we know that $|\overrightarrow{CP}| = |\overrightarrow{CL}| = z_1$. But $\overrightarrow{CP} = [x, 0, z]$ and $|\overrightarrow{CP}| = \sqrt{x^2 + z^2}$. Hence,

(1) $$z_1 = \sqrt{x^2 + z^2}$$

Also, since L is on the curve,

(2) $$y^2 - 4z_1^2 = 4$$

Eliminating z_1 from equations (1) and (2), we obtain the following equation of the surface:

$$y^2 - 4(x^2 + z^2) = 4$$

The surface is a hyperboloid of two sheets.

418   ANALYTIC GEOMETRY OF SPACE

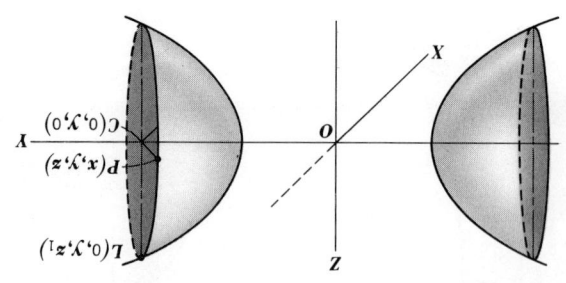

FIG. 12-22

EXERCISES

1. Discuss the surfaces whose equations are:
   (a) $4x^2 + 9y^2 - 9z^2 = 36$
   (b) $36y^2 - 16x^2 + 9z^2 = 144$
   (c) $4x^2 + 9y^2 - z^2 = 36$
   (d) $9x^2 + 4z^2 - 16y^2 = 144$
   (e) $16y^2 + 9z^2 - 4x^2 = 36$
   (f) $4z^2 + 9y^2 - x^2 - 64 = 0$
   (g) $4x^2 - y^2 - 16z^2 = 16$
   (h) $y^2 - 2z^2 + 4x^2 + 16 = 0$
   (i) $9z^2 - 16x^2 - y^2 - 144 = 0$
   (j) $5x^2 - 15z^2 + y^2 + 25 = 0$
   (k) $y^2 + 3x^2 - 9z^2 + 27 = 0$
   (l) $4z^2 - x^2 + 4y^2 - 16 = 0$

2. In each of the following exercises derive an equation of the surface of revolution obtained by revolving the given plane curve about the given axis:
   (a) $2x^2 - 3y^2 = 4$ and $z = 0$; the y-axis
   (b) $x^2 - z^2 = 9$ and $y = 0$; the x-axis
   (c) $4y^2 - 9z^2 = 36$ and $x = 0$; the y-axis
   (d) $16z^2 - 25x^2 = 25$ and $y = 0$; the z-axis
   (e) $3z^2 - 5y^2 = 15$ and $x = 0$; the y-axis
   (f) $3z^2 - 5y^2 = 15$ and $x = 0$; the z-axis
   (g) $y^2 - x^2 = 1$ and $z = 0$; the x-axis
   (h) $y^2 - x^2 = 1$ and $z = 0$; the y-axis

12-10. The hyperbolic paraboloid

Consider now a surface whose equation is of the form

$$\frac{y^2}{a^2} - \frac{x^2}{b^2} = pz \qquad (12\text{-}26)$$

where a and b are greater than 0. This surface is called a *hyperbolic paraboloid*. Assume first that $p > 0$. The surface then has the form shown in Fig. 12-23.

Each of the three intercepts is 0, since the surface intersects each axis only at the origin.

Its trace on the xy-plane is a pair of lines whose equations are $\frac{y}{a} - \frac{x}{b} = 0$

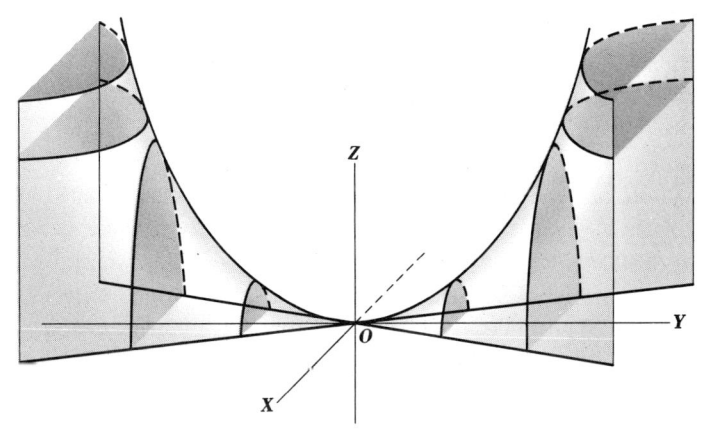

FIG. 12-23

and $z = 0$, and $\frac{y}{a} + \frac{x}{b} = 0$ and $z = 0$; that on the xz-plane is the parabola whose equations are $x^2 = -pb^2 z$ and $y = 0$; and that on the yz-plane is the parabola whose equations are $y^2 = a^2 pz$ and $x = 0$.

On the planes for which $z = k$, the cross-sections are hyperbolas whose equations are $\frac{y^2}{a^2} - \frac{x^2}{b^2} = pk$ and $z = k$. If $k > 0$, a line parallel to the y-axis is the transverse axis. If $k < 0$, a line parallel to the x-axis is the transverse axis.

On the planes for which $x = k$, the cross-sections are parabolas whose equations are $\frac{y^2}{a^2} = pz + \frac{k^2}{b^2}$ and $x = k$.

On the planes for which $y = k$, the cross-sections are parabolas whose equations are $\frac{x^2}{b^2} = -pz + \frac{k^2}{a^2}$ and $y = k$.

If $p < 0$, the equation of the hyperbolic paraboloid may be written in the form

$$\frac{x^2}{b^2} - \frac{y^2}{a^2} = -pz \qquad (12\text{-}27)$$

where $-p > 0$. The student should show that, for a given value of p, the surface represented by equation (12-27) is the image in the xy-plane of the surface represented by equation (12-26) when $p > 0$.

If $p = 0$, the surface degenerates into a pair of planes whose equations are $\frac{y}{a} - \frac{x}{b} = 0$ and $\frac{y}{a} + \frac{x}{b} = 0$.

A hyperbolic paraboloid is never a surface of revolution.

420   ANALYTIC GEOMETRY OF SPACE

EXAMPLE 12-15. Discuss the surface whose equation is $4y^2 - x^2 = 16z$.

Solution. This surface is a hyperbolic paraboloid.

Each of its intercepts is 0, since the curve intersects each axis only at the origin.

Its trace on the xy-plane is the lines whose equations are $2y - x = 0$ and $z = 0$, and $2y + x = 0$ and $z = 0$; that on the xz-plane is the parabola whose equations are $x^2 = -16z$ and $y = 0$; and that on the yz-plane is the parabola whose equations are $y^2 = 4z$ and $x = 0$.

The cross-sections of the surface at the intersections with planes for which $z = k$ are hyperbolas. The cross-sections at planes for which $x = k$ or for which $y = k$ are parabolas. See Fig. 12-23.

EXERCISES

Discuss each of the following surfaces:

1. $9x^2 - y^2 = 4z$
2. $4x^2 - 16z^2 + 25y = 0$
3. $y^2 - z^2 = x$
4. $16y^2 - 9z^2 = -144x$
5. $2z^2 - 5x^2 = 10y$
6. $25y^2 - x^2 = 100z$

12-11. Ruled surfaces

If a surface has the property that, through every point on it, there passes at least one line which is contained entirely within the surface, the surface is called a *ruled surface*. The lines that make up the surface are called *rulings*. Obvious examples of ruled surfaces are cylinders and cones. Other examples of ruled surfaces are the hyperboloid of one sheet, the hyperbolic paraboloid, and the surface called an elliptic wedge, as will be shown here.

The equation of the hyperboloid of one sheet has been given as

$$\frac{x^2}{a^2} + \frac{y^2}{b^2} - \frac{z^2}{c^2} = 1$$

Let us rewrite this equation in the form

$$\frac{x^2}{a^2} - \frac{z^2}{c^2} = 1 - \frac{y^2}{b^2}, \text{ or } \left(\frac{x}{a} - \frac{z}{c}\right)\left(\frac{x}{a} + \frac{z}{c}\right) = \left(1 - \frac{y}{b}\right)\left(1 + \frac{y}{b}\right)$$

Now we will consider the line whose equations are

$$k_1\left(\frac{x}{a} - \frac{z}{c}\right) = k_2\left(1 - \frac{y}{b}\right) \text{ and } k_2\left(\frac{x}{a} + \frac{z}{c}\right) = k_1\left(1 + \frac{y}{b}\right)$$

where k_1 and k_2 are arbitrary constants and both are not 0. Multiplying these two equations together, we obtain

$$k_1 k_2 \left(\frac{x^2}{a^2} - \frac{z^2}{c^2}\right) = k_1 k_2 \left(1 - \frac{y^2}{b^2}\right)$$

Unless $k_1 = 0$ or $k_2 = 0$, this equation reduces to

$$\frac{x^2}{a^2} - \frac{z^2}{c^2} = 1 - \frac{y^2}{b^2}$$

which is the equation of the original surface. Hence, it follows that any point whose coordinates satisfy the equations of the line must also satisfy the equation of the surface. In case $k_1 = 0$ or $k_2 = 0$, the line again lies on the surface. Conversely, for each point on the surface, we may determine the ratio of k_1 to k_2 or of k_2 to k_1 and so obtain an equation of a line through that point lying on the surface.

We might have chosen the line whose equations are

$$k_1'\left(\frac{x}{a} - \frac{z}{c}\right) = k_2'\left(1 + \frac{y}{b}\right) \text{ and } k_2'\left(\frac{x}{a} + \frac{z}{c}\right) = k_1'\left(1 - \frac{y}{b}\right)$$

By reasoning similar to that just given, we can show that this line also lies on the surface. Hence, it follows that, through every point on *this* surface, it is possible to draw *two* lines each of which is contained in the surface. For this reason it is said that the hyperboloid of one sheet is a *doubly* ruled surface. Further analysis shows that the two straight lines on the surface through a selected point are distinct and that no other straight line can be drawn on the surface through the point.

The equation previously given for the hyperbolic paraboloid is

$$\frac{x^2}{a^2} - \frac{y^2}{b^2} = 2pz$$

We can rewrite this equation in the form

$$\left(\frac{x}{a} - \frac{y}{b}\right)\left(\frac{x}{a} + \frac{y}{b}\right) = 2pz$$

Hence, the equations of lines that lie on the surface are:

$$k_1\left(\frac{x}{a} + \frac{y}{b}\right) = k_2 2p \text{ and } k_2\left(\frac{x}{a} - \frac{y}{b}\right) = k_1 z$$

$$k_1'\left(\frac{x}{a} - \frac{y}{b}\right) = k_2' 2p \text{ and } k_2'\left(\frac{x}{a} + \frac{y}{b}\right) = k_1' z$$

where both arbitrary constants in an equation are not 0. To see this, we form the products of the equations of the lines, obtaining

$$k_1 k_2 \left(\frac{x^2}{a^2} - \frac{y^2}{b^2}\right) = k_1 k_2 2pz$$

or
$$k_1'k_2'\left(\frac{x^2}{a^2} - \frac{y^2}{b^2}\right) = k_1'k_2'2pz$$

From either of these equations we obtain $\frac{x^2}{a^2} - \frac{y^2}{b^2} = 2pz$, which is the equation of the surface. As in the case of the hyperboloid of one sheet, for every point on the hyperbolic paraboloid we may determine the ratio of k_1 to k_2 or of k_1' to k_2' and hence find two rulings of the surface through a selected point. This surface is also a *doubly* ruled surface.

As a final example of a ruled surface, consider the surface whose equation is

$$y^2z^2 - a^2y^2 + b^2x^2 = 0 \qquad (12\text{-}28)$$

This surface is called an *elliptic wedge*. It is shown in Fig. 12–24.

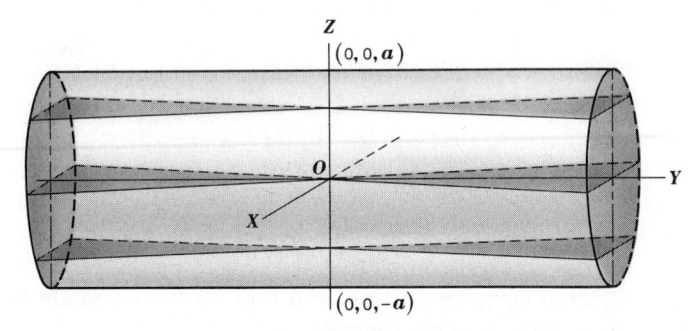

Fig. 12–24

Its intercepts on the x-axis and the y-axis are 0. If $x = y = 0$, then z may be any number.

Its trace on the yz-plane is the lines whose equations are $z = a$ and $x = 0$, $z = -a$ and $x = 0$, and $y = 0$ and $x = 0$. The trace on the xz-plane is the line $x = 0$ and $y = 0$. That on the xy-plane is the lines whose equations are $bx + ay = 0$ and $z = 0$, and $bx - ay = 0$ and $z = 0$.

The cross-sections of the surface at its intersections with planes for which $y = c$ are ellipses whose equations are $c^2z^2 + b^2x^2 = a^2c^2$ and $y = c$ when $|c| \neq |b|$, and are circles when $|c| = |b|$.

The plane sections for which $z = c$ are real lines if $|c| \leq |a|$ and are imaginary lines if $|c| > |a|$, except for the points $(0, 0, c)$. The plane sections for which $x = c$ are fourth-degree curves whose equations are $y^2z^2 - a^2y^2 = -b^2c^2$ and $x = c$.

Now let us show that this surface is a ruled surface. Consider the system of lines whose equations are

(1) $\qquad bx \pm \sqrt{a^2 - k^2}\, y = 0$ and $z = k$ (where $|k| \leq |a|$)

If we eliminate k from equations (1), we obtain

(2) $$bx \pm \sqrt{a^2 - z^2}\, y = 0$$

Then, on rationalizing equation (2) we obtain the equation of the surface. Hence, the lines represented by equations (1) lie on the surface.

Let $P_1 = (x_1, y_1, z_1)$ be any point on the surface. Then, if $k = z_1$, equations (1) become

(3) $$bx + \sqrt{a^2 - z_1^2}\, y = 0 \text{ and } z = z_1$$
and
(3') $$bx - \sqrt{a^2 - z_1^2}\, y = 0 \text{ and } z = z_1$$

When $x = x_1$ and $y = y_1$, *either* equation (3) *or* equation (3') is satisfied, since P_1 lies on the surface. Hence, there is *one* line on the surface through each point P_1 of the surface. When $P_1 = (0, 0, z_1)$ is the selected point on the surface, both lines represented by equation (1) contain it; when $P_1 = (0, y_1, a)$ is the selected point, the lines coincide.

The elliptic wedge is also known as a *right conoid*. It is symmetric with respect to the x-axis, y-axis, z-axis, and origin.

EXERCISES

Verify that each of the following is a ruled surface:

1. $z = xy$
2. $9x^2 - 4y^2 = z$
3. $9x^2 + 4y^2 - z^2 = 36$
4. $y^2z^2 - 4y^2 + 9x^2 = 0$

12–12. Curves in space

(a) *Cartesian equations*. We have already seen that a line can be represented algebraically by the equations of two intersecting planes. So a curve which happens to be the intersection of two surfaces can be represented algebraically by the equations of the surfaces. Thus, a line can be represented by the equations $a_1 x + b_1 y + c_1 z + d_1 = 0$ and $a_2 x + b_2 y + c_2 z + d_2 = 0$. Since a circle is the intersection of a sphere and a plane, the equations $x^2 + y^2 + z^2 = a^2$ and $z = 0$ are equations of a circle. In general, $f_1(x, y, z) = 0$ and $f_2(x, y, z) = 0$ are equations of the curve of intersection of the surfaces whose equations are $f_1(x, y, z) = 0$ and $f_2(x, y, z) = 0$, respectively, provided the surfaces do intersect.

If each of two cylinders contains a certain space curve and its generators are parallel to a coordinate axis, these cylinders are called the *projecting cylinders* of the space curve. Equations of the projecting cylinders are obtained by eliminating each variable in turn from the equations of the curve. This procedure follows immediately since, when one of the variables is eliminated, the resulting equation represents a cylinder whose generators are parallel to the axis of the missing variable. This surface contains the curve, and its equation is also the equation of the projection of the curve on the coordinate plane. Hence, the name *projecting cylinder* is used.

EXAMPLE 12-16. Find equations of the projecting cylinders of the circle whose equations are $x^2 + y^2 + z^2 = 4$ and $z = 0$.

Solution. One of the cylinders is the plane $z = 0$. The equation of the other cylinder is $x^2 + y^2 = 4$.

EXAMPLE 12-17. Find equations of the projecting cylinders of the curve whose equations are $x^2 + y^2 + z^2 = 16$ and $2x^2 + y^2 - z^2 = 0$.

Solution. Eliminating x, we obtain $y^2 + 3z^2 = 32$; eliminating y, we obtain $2z^2 - x^2 = 16$; and eliminating z, we obtain $3x^2 + 2y^2 = 16$. These are equations of the projecting cylinders of the space curve.

(b) *Parametric equations.* If x, y, and z are expressed as functions of a parameter t, the point $P = (x, y, z)$ will move on a curve as t varies. A special case has already been considered, namely, the straight line, whose parametric equations were studied in Section 11-2. In this case, x, y, and z are expressed linearly in terms of t. In general, if we have $x = f(t)$, $y = g(t)$, and $z = h(t)$, where $f(t)$, $g(t)$, and $h(t)$ are functions of t only, the equations are called *parametric equations* of a curve in space.

An an example, consider the curve whose equations are $x = a \cos t$, $y = a \sin t$, and $z = ct$. This curve is called a *circular helix* and has the form of a spiral wound about a right circular cylinder whose equation is $x^2 + y^2 = a^2$. The student should verify this by sketching a graph of the curve.

Cartesian equations of the curve are obtained from parametric equations by eliminating the parameter. So, in the case of the helix, we find a pair of cartesian equations to be $x^2 + y^2 = a^2$ and $x = a \cos \dfrac{z}{c}$.

12-13. Sketching the curve of intersection of two surfaces

Purpose. The purpose of this section is to illustrate the technique of sketching the curve of intersection of two surfaces by utilizing the projecting cylinders of the curve.

Discussion. In Section 12-12 it was pointed out that if two surfaces intersect, the curve of intersection can be represented algebraically by the equations of the two surfaces. Further, by eliminating each variable in turn (or by eliminating the parameter if the curve is represented by parametric equations), we can find the equations of the three projecting cylinders of the curve. Since these cylinders also pass through the curve, the problem is reduced to that of sketching the curve of intersection of any two of the three projecting cylinders.

SURFACES AND CURVES 425

EXAMPLE 12–18. Sketch the curve of intersection of the surfaces:
$x^2 + 5y^2 - 8y + z - 4 = 0$ and $x^2 + 3y^2 - 8y - z + 4 = 0$.

Solution.

(a) Eliminating first x and then z, we obtain the equations of two of the projecting cylinders:

and
$$y^2 = -(z-4)$$
$$\frac{x^2}{4} + \frac{(y-1)^2}{1} = 1$$

(b) Draw the traces of the projecting cylinders. (See Fig. 12–25.)

(c) Let P be a plane parallel to the generators of both cylinders; in this case, P is a plane of the form $y = k$. (Note that this plane will be

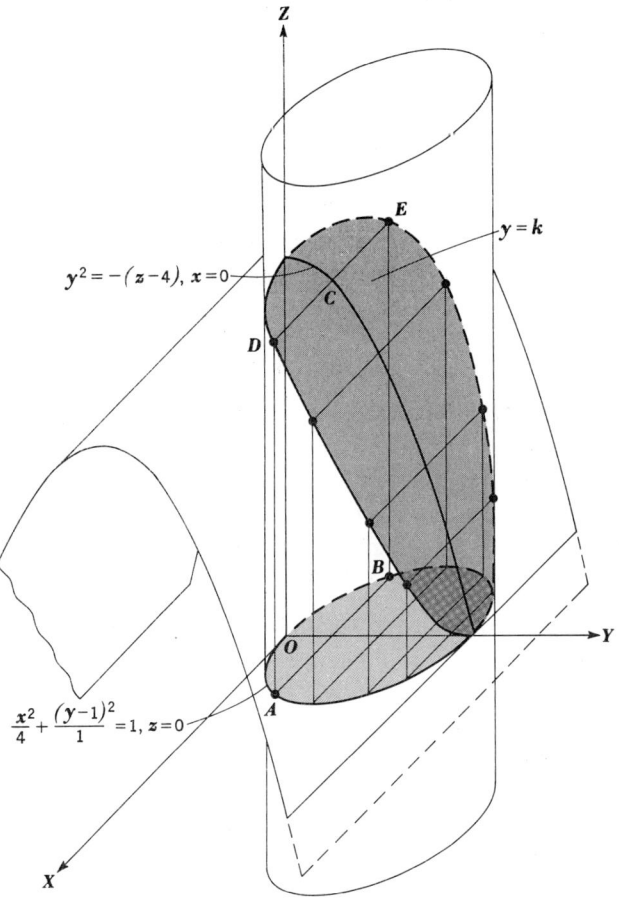

FIG. 12–25

perpendicular to the axis of the variable which is common to the equations of both cylinders.) The plane P will intersect the traces at the points A, B, and C.

Now, in order to locate points on the curve of intersection, we observe that a line through point A parallel to the z-axis (line AD) will be a generator of the elliptical cylinder. Likewise, a line through point C parallel to the x-axis (line CD) will be a generator of the parabolic cylinder. Since point D is on both generators, it must be on both cylinders and, thus, on the curve of intersection of the cylinders. Similarly, point E is on the curve.

(d) To find additional points on the curve, we consider additional planes parallel to P and repeat the process.

Any two of the three projecting cylinders may be used for this technique. However, in this case, we note that eliminating y would necessitate solving for y in terms of x and z in one equation and substituting the result in the other, yielding an unwieldy equation:

$$x^2 + 3\left[\frac{4 \pm \sqrt{16-5(x^2+z-4)}}{5}\right]^2 - 8\left[\frac{4 \pm \sqrt{16-5(x^2+z-4)}}{5}\right] - z + 4 = 0$$

This result points out the fact that the choice of projecting cylinders will sometimes depend upon the algebra involved in finding their equations.

EXERCISES

1. Sketch each of the following curves and find equations of the projecting cylinders:
   (a) $y^2 = 4x$ and $z^2 = 4x$
   (b) $x^2 + y^2 + z^2 = 16$ and $x = 0$
   (c) $x^2 + 2y^2 + 4z^2 = 16$ and $x^2 + y^2 - z^2 = 8$
   (d) $4x^2 + y^2 = z$ and $x^2 - y^2 - z^2 = 4$

2. Find cartesian equations of each of the following curves and sketch it:
   (a) $x = \cos t$, $y = \sin t$, and $z = \sin t$
   (b) $x = t^2$, $y = t$, and $z = t + 1$
   (c) $x = 2 \sin^2 t$, $y = 3 \cos^2 t$, and $z = \sin t$
   (d) $x = t$, $y = t^2$, and $z = t^3$

12–14. Transformation of the axes

(a) *Translation.* Sometimes, after a coordinate system has been established, it is found expedient to select a new point in space to be the origin. If the new axes are parallel, respectively, to the original axes, the

SURFACES AND CURVES   427

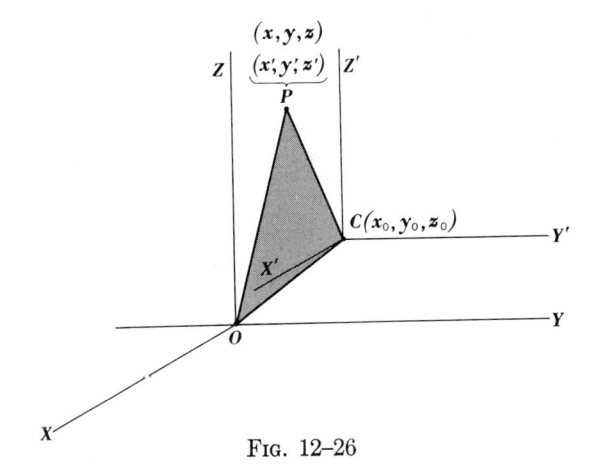

FIG. 12-26

new ones are said to be related to the old ones by a translation. In Fig. 12–26, let $C = (x_0, y_0, z_0)$ be a new origin and let P be a given point having coordinates x, y, and z with respect to the original axes and having coordinates x', y', and z' with respect to the new axes. Then, using vectors, we have $\overrightarrow{OP} = \overrightarrow{OC} + \overrightarrow{CP}$, where $\overrightarrow{OP} = [x, y, z]$, $\overrightarrow{OC} = [x_0, y_0, z_0]$, and $\overrightarrow{CP} = [x', y', z']$. Hence,

$$[x, y, z] = [x_0, y_0, z_0] + [x', y', z']$$

or

$$x = x_0 + x', \; y = y_0 + y', \text{ and } z = z_0 + z' \qquad (12\text{-}29)$$

Equations (12–29) are the formulas for a *translation* of axes.

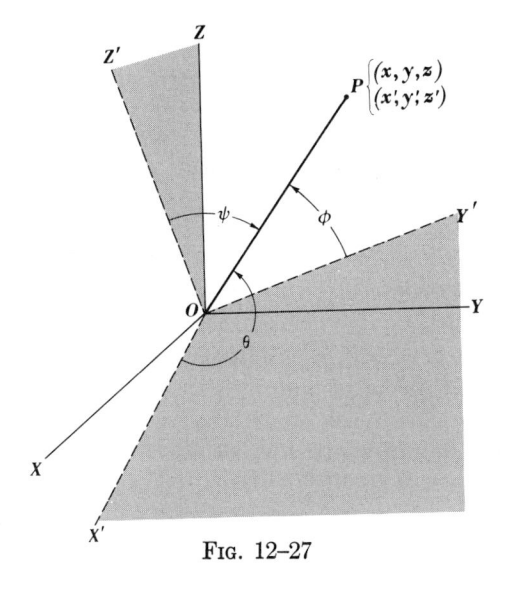

FIG. 12-27

(b) *Rotation.* Let a new system of axes have the same origin as the original system and assume that the two systems have the same character (that is, both are right-handed), as indicated in Fig. 12-27. The second system is said to be related to the first by a *rotation*. Let us designate the direction cosines of OX', OY', and OZ' with respect to the original axes by the notations (l_1, m_1, n_1), (l_2, m_2, n_2), and (l_3, m_3, n_3), respectively. Let P be a point that is different from the origin and whose coordinates are x, y, and z with respect to the original system of axes and are x', y', and z' with respect to the new system of axes. Let the direction cosines of OP be l, m, and n with respect to the original axes and l', m', and n' with respect to the new axes. Then if θ is the angle from OX' to OP, it follows that $\cos \theta = ll_1 + mm_1 + nn_1$. Multiplying through by $|OP|$, we get $|OP| \cos \theta = |OP|ll_1 + |OP|mm_1 + |OP|nn_1$. Since $\cos \theta = l'$, $|OP| \cos \theta = x'$, $|OP| l = x$, $|OP| m = y$, and $|OP| n = z$, we have

$$x' = l_1 x + m_1 y + n_1 z$$

In exactly the same way, if ϕ is the angle between OP and OY', it follows that $\cos \phi = ll_2 + mm_2 + nn_2$. Multiplying through by $|OP|$, we obtain $|OP| \cos \phi = |OP| ll_2 + |OP| mm_2 + |OP| nn_2$. Since $|OP| \cos \phi = y'$, we have

$$y' = l_2 x + m_2 y + n_2 z$$

If ψ is the angle between OP and OZ', it follows that $\cos \psi = ll_3 + mm_3 + nn_3$. Multiplying through by $|OP|$, and noting that $|OP| \cos \psi = z'$, we obtain

$$z' = l_3 x + m_3 y + n_3 z$$

So, our formulas for rotation are

$$\begin{aligned} x' &= l_1 x + m_1 y + n_1 z \\ y' &= l_2 x + m_2 y + n_2 z \\ z' &= l_3 x + m_3 y + n_3 z \end{aligned} \quad (12\text{-}30)$$

It is important to note that equations (12-30) also hold when P is the origin. Often it is necessary to know the values of the coordinates x, y, and z in terms of x', y', and z'. We know that l_1 is the cosine of the angle between OX' and OX, that l_2 is the cosine of the angle between OY' and OX, and that l_3 is the cosine of the angle between OZ' and OX. Hence, l_1, l_2, and l_3 are direction cosines of OX with respect to the new axes. In the same way, m_1, m_2, and m_3 are direction cosines of OY with respect to the new axes, and n_1, n_2, and n_3 are direction cosines of OZ. Hence, $l_1^2 + l_2^2 + l_3^2 = 1$, $m_1^2 + m_2^2 + m_3^2 = 1$, and $n_1^2 + n_2^2 + n_3^2 = 1$. Since the axes are mutually perpendicular to each other, it follows also that $l_1 m_1 + l_2 m_2 + l_3 m_3 = 0$, $l_1 n_1 + l_2 n_2 + l_3 n_3 = 0$, and $m_1 n_1 + m_2 n_2 + m_3 n_3 = 0$.

If we multiply equations (12–30) by l_1, l_2, and l_3, respectively, and add the results, we obtain
$$x = l_1 x' + l_2 y' + l_3 z'$$
By multiplying equations (12–30) by m_1, m_2, and m_3, respectively, and adding the products, we obtain
$$y = m_1 x' + m_2 y' + m_3 z'$$
Finally, by multiplying equations (12–30) by n_1, n_2, and n_3, respectively, and adding the results, we obtain
$$z = n_1 x' + n_2 y' + n_3 z'$$
Hence, we have the following second set of formulas for rotation of the axes:

$$\begin{aligned} x &= l_1 x' + l_2 y' + l_3 z' \\ y &= m_1 x' + m_2 y' + m_3 z' \\ z &= n_1 x' + n_2 y' + n_3 z' \end{aligned} \tag{12-31}$$

12–15. Non-rectangular coordinate systems

(a) *Cylindrical coordinates.* In this system we name a point P by means of the polar coordinates in the xy-plane of the projection of P on the xy-plane and by means of the z-coordinate of P. As indicated in Fig. 12–28, we use the notation $P = (\rho, \theta, z)$.

The relations between the coordinates x, y, and z and the coordinates ρ, θ, and z are:

$$x = \rho \cos \theta, \; y = \rho \sin \theta, \text{ and } z = z \tag{12-32}$$

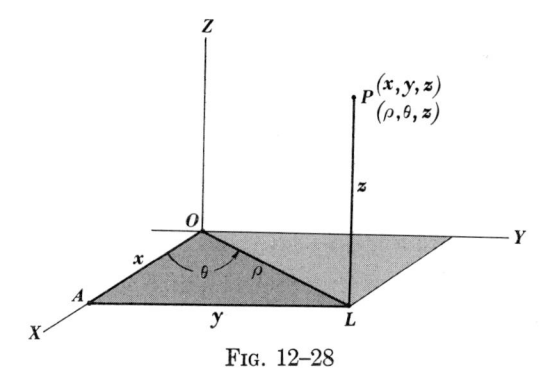

FIG. 12–28

EXAMPLE 12–19. What does the equation of the sphere having its center at the origin and having a radius equal to a become after transformation to cylindrical coordinates?

Solution. Since an equation of the sphere in cartesian coordinates is $x^2+y^2+z^2 = a^2$, and since $\rho^2 = x^2+y^2$, we obtain
$$\rho^2 + z^2 = a^2$$

EXAMPLE 12-20. What is an equation in cylindrical coordinates for the right circular cylinder whose cartesian equation is $x^2+y^2 = a^2$?

Solution. Since $\rho^2 = x^2+y^2$, we know at once that $\rho = a$ or $\rho = -a$. Either equation may be taken to represent the surface.

EXAMPLE 12-21. What is a cartesian equation of the surface whose equation in cylindrical coordinates is $2\rho \cos\theta - 3\rho \sin\theta + 3z = 4$?

Solution. Since $x = \rho \cos\theta$ and $y = \rho \sin\theta$, we have $2x - 3y + 3z - 4 = 0$, which represents a plane.

(b) *Spherical coordinates.* Let P be any point in space. Also, in Fig. 12-29, let $|OP| = r$, let θ be the angle which OP makes with the positive z-axis, and let ϕ be the angle which the projection of OP on the xy-plane makes with the positive x-axis. The spherical coordinates of P are r, ϕ, and θ, where $0° \leq \theta \leq 180°$.

To find the relations between the coordinates x, y, and z and the coordinates r, ϕ, and θ, we proceed as follows:

$$x = \overline{OA} = |OC| \cos\phi = r \sin\theta \cos\phi \text{ (since } |OC| = r \sin\theta\text{)}$$
$$y = \overline{AC} = |OC| \sin\phi = r \sin\theta \sin\phi$$
$$z = r \cos\theta$$

Hence,

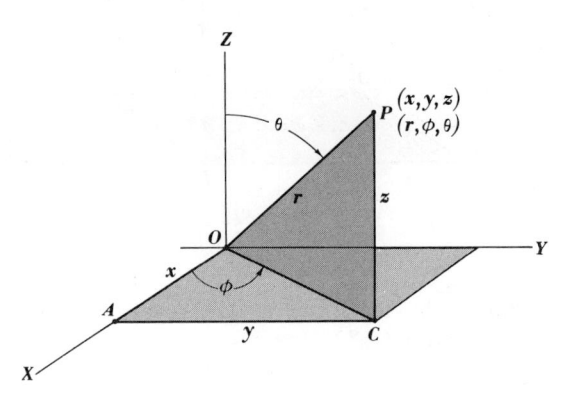

$$x = r \sin\theta \cos\phi, \ y = r \sin\theta \sin\phi, \text{ and } z = r \cos\theta \qquad (12\text{-}33)$$

FIG. 12-29

EXAMPLE 12–22. Express in spherical coordinates an equation of the sphere with its center at the origin and a radius equal to a.

Solution. Since all points on a sphere are equidistant from the center, it follows that an equation of the sphere is

$$r = a$$

EXERCISES

1. Using spherical coordinates, find an equation of a right circular cone having its vertex at the origin and having the z-axis as its axis.

2. Using spherical coordinates, find an equation of a plane that contains the z-axis and makes an angle ϕ_0 with the x-axis.

3. Identify each of the following surfaces and sketch it:
   (a) $\rho^2 = a^2 \cos 2\theta$
   (b) $\rho = a(1 - \cos \theta)$
   (c) $\rho = a \cos \theta$
   (d) $\rho = \dfrac{3}{1 + 2 \sin \theta}$
   (e) $\rho = a \cos 2\theta$
   (f) $\rho = a \sin \theta$
   (g) $\rho = \dfrac{2}{1 - \cos \theta}$
   (h) $\rho = \dfrac{4}{2 - \cos \theta}$

4. Find, in spherical coordinates, an equation of a sphere having its center at the origin and its radius equal to 10.

5. Using cylindrical coordinates, find an equation of the sphere in Exercise 4.

6. Transform each of the following equations to cylindrical coordinates:
   (a) $x^2 + y^2 - z^2 = 0$
   (b) $x^2 + y^2 + z^2 = 16$
   (c) $x^2 + y^2 = 4z$
   (d) $x^2 + y^2 = 16$
   (e) $a^2 y^2 = a^2 - z$
   (f) $az = a^2 - x^2 - y^2$

REVIEW EXERCISES

1. Derive an equation of the surface of revolution obtained by revolving each of the following curves about the given axis:
   (a) $x^2 + y^2 = 16$ and $z = 0$; the x-axis
   (b) $y^2 = 16z$ and $x = 0$; the y-axis
   (c) $y^2 = 16z$ and $x = 0$; the z-axis
   (d) $9x^2 - 16y^2 = 144$ and $z = 0$; the x-axis
   (e) $9x^2 - 16y^2 = 144$ and $z = 0$; the y-axis
   (f) $4z^2 + y^2 = 16$ and $x = 0$; the z-axis
   (g) $2x - 3y = 6$ and $z = 0$; the x-axis

432 ANALYTIC GEOMETRY OF SPACE

2. Discuss and sketch each of the following surfaces:

(a) $x^2 + z^2 = 16$
(b) $y^2 = 4x$
(c) $\rho = -4$
(d) $r = 4$
(e) $4x^2 + y^2 + 9z^2 = 144$
(f) $4x^2 + y^2 - 9z^2 = 144$
(g) $4x^2 - y^2 - 9z^2 = 144$
(h) $x^2 + y^2 = z^2$
(i) $4x^2 - y^2 = z^2$
(j) $x^2 - y^2 = z^2$
(k) $x^2 + y^2 = 16z$
(l) $4x^2 - z^2 + 16y = 0$
(m) $4x^2 + 4y^2 + 25z^2 = 100$
(n) $4x^2 + 25y^2 + 25z^2 = 100$
(o) $\rho \cos \theta = 4$
(p) $x^2 + y^2 + z^2 = 9$
(q) $x^2 - z^2 = 4y^2$
(r) $r = 2$
(s) $y^2 - 4z = 0$
(t) $3x^2 + 2y^2 + z^2 = 6$
(u) $y^2 + 2z^2 = x^2$
(v) $y^2 + 2z^2 - x^2 = 8$
(w) $\rho = 4 \sin \theta$
(x) $\rho^2 = \cos 2\theta$
(y) $\rho \cos \theta - 2\rho \sin \theta + z = 5$
(z) $\rho = a(1 - \cos \theta)$

3. Identify all the ruled surfaces in Exercise 2.

4. Find the rectangular coordinates of each of the following points given in cylindrical coordinates:

(a) $\left(4, \dfrac{\pi}{6}, 3\right)$
(b) $\left(3, -\dfrac{\pi}{2}, -4\right)$
(c) $(-2, \pi, -3)$
(d) $\left(1, \dfrac{3\pi}{4}, 0\right)$
(e) $\left(6, -\dfrac{7\pi}{6}, 2\right)$
(f) $(4, 120°, -1)$

5. Find the rectangular coordinates of each of the following points given in spherical coordinates:

(a) $(2, 60°, 30°)$
(b) $(4, 30°, 45°)$
(c) $\left(5, -\dfrac{\pi}{2}, \dfrac{\pi}{4}\right)$
(d) $\left(1, \dfrac{5\pi}{6}, \dfrac{3\pi}{4}\right)$
(e) $(10, 120°, 60°)$
(f) $(6, -210°, 30°)$

6. Find the center and radius of the circle whose equations are $x^2 + y^2 + z^2 = 16$ and $x - 2y + 2z + 9 = 0$.

7. Sketch each of the following curves and eliminate the parameter:

(a) $x = \sin^2 t$, $y = \cos^2 t$, and $z = \cos t$
(b) $x = t$, $y = 2t^2$, and $z = 3t^3$
(c) $x = \sin t$, $y = 1 - \cos t$, and $z = \cos t$
(d) $x = e^t$, $y = e^{-t}$, and $z = e^t + e^{-t}$
(e) $x = \cosh^2 t$, $y = \sinh^2 t$, $z = \sinh t$

Appendixes

Appendix A

Brief Introduction to Matrix Theory

A–1. Introduction

The theory of matrices is important in solving problems in physics and engineering. It is also becoming increasingly useful in its applications to business problems, economics, and linear programming. Because of this, we are introducing the concept here. There are places in *Analytic Geometry* where the matrix notation can sometimes be used to advantage. This is particularly true in rotation problems. The advantage is that the notation remains the same regardless of the dimensions of the space being studied. However, we shall make the discussion here as brief and concise as possible.

A rectangular array of numbers enclosed in brackets and satisfying properties 1 and 2 listed below is called a *matrix*. If the array has m rows and n columns, it is called an m by n matrix and the numbers m and n are said to be its dimensions. The numbers between the brackets are called *elements*. So

$$\begin{bmatrix} a_{11} & a_{12} & a_{13} & a_{14} \\ a_{21} & a_{22} & a_{23} & a_{24} \\ a_{31} & a_{32} & a_{33} & a_{34} \end{bmatrix}$$

is a 3 by 4 matrix. The first dimension *always* determines the number of rows while the second dimension determines the number of columns. The element in the second row and the third column is a_{23}.

Property 1: Two matrices are said to be equal provided they are of *like* dimensions and they consist of *equal* elements in corresponding positions. So

$$\begin{bmatrix} a_{11} & a_{12} & a_{13} \\ a_{21} & a_{22} & a_{23} \end{bmatrix} = \begin{bmatrix} b_{11} & b_{12} & b_{13} \\ b_{21} & b_{22} & b_{23} \end{bmatrix}$$

provided $a_{ij} = b_{ij}$ for all i and j where i is the row and j is the column that contains the a_{ij} element.

Property 2: If two matrices are of like dimensions their *sum* is defined as the matrix obtained by adding together corresponding elements. So

$$\begin{bmatrix} a_{11} & a_{12} & a_{13} \\ a_{21} & a_{22} & a_{23} \end{bmatrix} + \begin{bmatrix} b_{11} & b_{12} & b_{13} \\ b_{21} & b_{22} & b_{23} \end{bmatrix} = \begin{bmatrix} a_{11}+b_{11} & a_{12}+b_{12} & a_{13}+b_{13} \\ a_{21}+b_{21} & a_{22}+b_{22} & a_{23}+b_{23} \end{bmatrix}$$

A matrix whose dimensions are equal, that is $m = n$, is called a *square matrix*. So

$$\begin{bmatrix} a_{11} & a_{12} \\ a_{21} & a_{22} \end{bmatrix}$$

is a 2 by 2 (written 2×2) square matrix.

A 1×2 matrix $[a_{11} \; a_{12}]$ may be thought of as a plane vector. A 2×1 matrix $\begin{bmatrix} a_{11} \\ a_{21} \end{bmatrix}$ may also be thought of as a vector whose scalar components are a_{11} and a_{21}. We sometimes think of a 2×2 matrix as consisting of two row vectors and two column vectors.

Matrices which have the same number of rows and columns will be said to be of the *same dimension* or *type*. From the definition for addition, it follows that matrices of unlike type cannot be added.

We shall define the multiplication of a matrix by a *scalar* to be the matrix obtained by multiplying *each* of the elements of the original matrix by the scalar. So if

$$\mathbf{A} = \begin{bmatrix} a_{11} & a_{12} \\ a_{21} & a_{22} \end{bmatrix}†$$

then

$$t\mathbf{A} = \begin{bmatrix} ta_{11} & ta_{12} \\ ta_{21} & ta_{22} \end{bmatrix}$$

EXAMPLE A-1: If $\mathbf{A} = \begin{bmatrix} 3 \\ 1 \end{bmatrix}$ and $\mathbf{B} = \begin{bmatrix} -2 \\ 3 \end{bmatrix}$, find $2\mathbf{A} - 3\mathbf{B}$.

Solution. Since $2\mathbf{A} = \begin{bmatrix} 6 \\ 2 \end{bmatrix}$ and $-3\mathbf{B} = \begin{bmatrix} 6 \\ -9 \end{bmatrix}$, it follows that

$$2\mathbf{A} - 3\mathbf{B} = \begin{bmatrix} 12 \\ -7 \end{bmatrix}$$

EXAMPLE A-2: Given $\mathbf{A} = \begin{bmatrix} 2 & -3 \\ 1 & 4 \end{bmatrix}$ and $\mathbf{B} = \begin{bmatrix} x+y & -3 \\ 1 & x-y \end{bmatrix}$ and $\mathbf{A} = \mathbf{B}$, find x and y.

Solution. Since two matrices are equal when their corresponding elements are equal, we must have $x + y = 2$ and $x - y = 4$. Solving these simultaneously, we find that $x = 3$ and $y = -1$.

Definition. A matrix whose elements are all zero is defined to be a *zero matrix*. An $n \times m$ zero matrix will be symbolized by $0_{n \times m}$. So

$$0_{3 \times 2} = \begin{bmatrix} 0 & 0 \\ 0 & 0 \\ 0 & 0 \end{bmatrix}$$

---

† When using a capital letter to represent a matrix, it will be written in boldface type.

The zero matrix has the important property that

$$A + 0 = A = 0 + A$$

So
$$\begin{bmatrix} a_{11} & a_{12} \\ a_{21} & a_{22} \end{bmatrix} + \begin{bmatrix} 0 & 0 \\ 0 & 0 \end{bmatrix} = \begin{bmatrix} a_{11}+0 & a_{12}+0 \\ a_{21}+0 & a_{22}+0 \end{bmatrix} = \begin{bmatrix} a_{11} & a_{12} \\ a_{21} & a_{22} \end{bmatrix}$$

Definition. The *additive inverse* or the *negative* of a matrix A is the matrix whose elements are the negatives of the elements of A and is indicated by $-A$. So if

$$A = \begin{bmatrix} a_{11} & a_{12} \\ a_{21} & a_{22} \end{bmatrix},$$

we have
$$-A = \begin{bmatrix} -a_{11} & -a_{12} \\ -a_{21} & -a_{22} \end{bmatrix}$$

The additive inverse has the property that

$$A + (-A) = 0$$

Next we shall explain a method for multiplying matrices. If the number of columns of A is the same as the number of rows of B, we define the ij element of the product matrix AB to be equal to the *sum of the products of the elements of the ith row of A by the corresponding elements of the jth column of B*. Another way to say this is that the ij element is equal to the *dot product* of the ith row vector of A by the jth column vector of B. So if

$$A = \begin{bmatrix} a_{11} & a_{12} \\ a_{21} & a_{22} \end{bmatrix} \quad \text{and} \quad B = \begin{bmatrix} b_{11} & b_{12} & b_{13} \\ b_{21} & b_{22} & b_{23} \end{bmatrix}$$

then the element in the second row and third column of the product is $a_{21}b_{13} + a_{22}b_{23}$. The entire product is

$$AB = \begin{bmatrix} a_{11}b_{11}+a_{12}b_{21} & a_{11}b_{12}+a_{12}b_{22} & a_{11}b_{13}+a_{12}b_{23} \\ a_{21}b_{11}+a_{22}b_{21} & a_{21}b_{12}+a_{22}b_{22} & a_{21}b_{13}+a_{22}b_{23} \end{bmatrix}$$

We term AB the product of B by A.

EXAMPLE A-3: Given $B = \begin{bmatrix} 1 & 2 \\ 3 & 1 \\ -1 & 1 \end{bmatrix}$ and $A = \begin{bmatrix} 2 & 5 \\ 1 & 2 \end{bmatrix}$, find the product of A by B.

Solution. Since we want the product of A by B, we must find BA. So

$$BA = \begin{bmatrix} 1 & 2 \\ 3 & 1 \\ -1 & 1 \end{bmatrix} \begin{bmatrix} 2 & 5 \\ 1 & 2 \end{bmatrix} = \begin{bmatrix} 1\cdot2+2\cdot1 & 1\cdot5+2\cdot2 \\ 3\cdot2+1\cdot1 & 3\cdot5+1\cdot2 \\ -1\cdot2+1\cdot1 & -1\cdot5+1\cdot2 \end{bmatrix} = \begin{bmatrix} 4 & 9 \\ 7 & 17 \\ -1 & -3 \end{bmatrix}$$

If the number of columns of A is *not* equal to the number of rows of B, we do not attempt to define the product AB.

438 APPENDIXES

It is important to note that matrix multiplication is not, in general, *commutative*. That is, in general,

$$\mathbf{AB} \neq \mathbf{BA}$$

EXAMPLE A-4: Show that the matrices $\mathbf{A} = \begin{bmatrix} -1 & 2 \\ 3 & 4 \end{bmatrix}$ and $\mathbf{B} = \begin{bmatrix} 2 & -1 \\ 1 & 3 \end{bmatrix}$ are not commutative.

Solution.

$$\mathbf{AB} = \begin{bmatrix} -1 & 2 \\ 3 & 4 \end{bmatrix} \begin{bmatrix} 2 & -1 \\ 1 & 3 \end{bmatrix} = \begin{bmatrix} 0 & 7 \\ 10 & 9 \end{bmatrix}$$

while

$$\mathbf{BA} = \begin{bmatrix} 2 & -1 \\ 1 & 3 \end{bmatrix} \begin{bmatrix} -1 & 2 \\ 3 & 4 \end{bmatrix} = \begin{bmatrix} -5 & 0 \\ 8 & 14 \end{bmatrix}$$

EXAMPLE A-5: Show that $\mathbf{A} = \begin{bmatrix} \sqrt{3}/2 & -1/2 \\ 1/2 & \sqrt{3}/2 \end{bmatrix}$ and $\mathbf{B} = \begin{bmatrix} \sqrt{3}/2 & 1/2 \\ -1/2 & \sqrt{3}/2 \end{bmatrix}$ are commutative matrices.

Solution.

$$\mathbf{AB} = \begin{bmatrix} \sqrt{3}/2 & -1/2 \\ 1/2 & \sqrt{3}/2 \end{bmatrix} \begin{bmatrix} \sqrt{3}/2 & 1/2 \\ -1/2 & \sqrt{3}/2 \end{bmatrix} = \begin{bmatrix} 1 & 0 \\ 0 & 1 \end{bmatrix}$$

and

$$\mathbf{BA} = \begin{bmatrix} \sqrt{3}/2 & 1/2 \\ -1/2 & \sqrt{3}/2 \end{bmatrix} \begin{bmatrix} \sqrt{3}/2 & -1/2 \\ 1/2 & \sqrt{3}/2 \end{bmatrix} = \begin{bmatrix} 1 & 0 \\ 0 & 1 \end{bmatrix}$$

Since $\mathbf{AB} = \mathbf{BA}$, the matrices are commutative.

We have seen that if $\mathbf{A}$ is a $p \times q$ matrix, then $\mathbf{B}$ must be a $q \times r$ matrix in order that the product matrix $\mathbf{AB}$ may be defined. The resulting matrix will then have p rows and r columns and will be a $p \times r$ matrix. Thus the product of a 2×2 matrix and a 2×3 matrix is a 2×3 matrix; the product of a 1×3 matrix and a 3×2 matrix is a 1×2 matrix; the product of a 1×3 and a 2×2 matrix is *not* defined.

Definition. The transpose $\mathbf{A}^*$ of a matrix $\mathbf{A}$ is defined to be the matrix obtained by interchanging the rows and the columns of $\mathbf{A}$.

Hence the transpose of $\mathbf{A} = \begin{bmatrix} a_{11} & a_{12} \\ a_{21} & a_{22} \end{bmatrix}$ is

$$\mathbf{A}^* = \begin{bmatrix} a_{11}^* & a_{12}^* \\ a_{21}^* & a_{22}^* \end{bmatrix} = \begin{bmatrix} a_{11} & a_{21} \\ a_{12} & a_{22} \end{bmatrix}$$

where $a_{11}^* = a_{11}$, $a_{21}^* = a_{12}$, $a_{12}^* = a_{21}$, $a_{22}^* = a_{22}$. It should be apparent that the interchange of rows and columns corresponds to an interchange of indices on the elements.

EXAMPLE A-6: Given $\mathbf{A} = \begin{bmatrix} 2 & 1 \\ -1 & 4 \\ 3 & 5 \end{bmatrix}$, find $\mathbf{A}^*$.

Solution. From the definition, we have

$$\mathbf{A}^* = \begin{bmatrix} 2 & -1 & 3 \\ 1 & 4 & 5 \end{bmatrix}$$

EXAMPLE A-7: Find the transpose of the matrix $\mathbf{B} = \begin{bmatrix} 1 & -2 \\ 3 & 4 \end{bmatrix}$.

Solution. The transpose of $\mathbf{B}$ is

$$\mathbf{B}^* = \begin{bmatrix} 1 & 3 \\ -2 & 4 \end{bmatrix}$$

EXERCISES

1. Given $\begin{bmatrix} -1 & 2 \\ 3 & 1 \\ 4 & -5 \end{bmatrix}$, find a_{32}, a_{22}, and a_{12}.

2. Given $\begin{bmatrix} 2 & -1 & 1 \\ 0 & 3 & -2 \\ 3 & 1 & 4 \end{bmatrix}$, find a_{21}, a_{32}, a_{13}, and a_{33}.

3. Given $\begin{bmatrix} 2 & -3 & 1 \\ 1 & 5 & -4 \end{bmatrix}$, find a_{23}, a_{12}, and a_{22}.

4. Find the value of x and y if the following matrices are equal:

$$\begin{bmatrix} 2 & x-y \\ -1 & 2x+3y \end{bmatrix} = \begin{bmatrix} 2 & 3 \\ -1 & -2 \end{bmatrix}$$

5. If $\mathbf{A} = \begin{bmatrix} 2 & 3 & -1 \\ 1 & 0 & -2 \end{bmatrix}$ and $\mathbf{B} = \begin{bmatrix} -1 & -2 & 0 \\ 3 & 1 & 4 \end{bmatrix}$, find $\mathbf{A} - \mathbf{B}$; $-3\mathbf{B}$; $3\mathbf{A} + 2\mathbf{B}$.

6. What is the additive inverse of the matrix $\mathbf{A} = \begin{bmatrix} -2 & 3 \\ 1 & -1 \\ 4 & 2 \end{bmatrix}$?

7. Find the product matrix in each of the following cases:

(a) $\begin{bmatrix} 2 & 1 \\ 3 & -2 \\ 1 & -3 \end{bmatrix} \begin{bmatrix} 1 & -1 & 2 \\ 3 & 2 & -1 \end{bmatrix}$

(f) $\begin{bmatrix} 1 & 0 & 0 \\ 0 & 1 & 0 \\ 0 & 0 & 1 \end{bmatrix} \begin{bmatrix} 2 \\ -3 \\ 3 \end{bmatrix}$

(b) $[u_1 \; u_2 \; u_3] \begin{bmatrix} v_1 \\ v_2 \\ v_3 \end{bmatrix}$

(g) $\begin{bmatrix} 0 & 0 & 0 \\ 0 & 0 & 0 \end{bmatrix} \begin{bmatrix} 1 & 1 & 3 \\ -1 & 2 & 5 \\ 2 & -1 & -4 \end{bmatrix}$

(c) $\begin{bmatrix} 3 & -2 \\ 1 & 4 \end{bmatrix} \begin{bmatrix} 1 & -3 \\ -2 & 1 \end{bmatrix}$

(h) $[2 \; -1 \; 3] \begin{bmatrix} 1 \\ -1 \\ -2 \end{bmatrix}$

(d) $\begin{bmatrix} 2 & 1 \\ -3 & -1 \end{bmatrix} \begin{bmatrix} -2 & 4 \\ 0 & 3 \end{bmatrix}$ (i) $\begin{bmatrix} 1 & 0 & 0 \\ 0 & 1 & 0 \\ 0 & 0 & 1 \end{bmatrix} \begin{bmatrix} 2 & -1 & 3 \\ 1 & 4 & 2 \\ 3 & 0 & 5 \end{bmatrix}$

(e) $\begin{bmatrix} 2 & 0 & -1 \\ 3 & 1 & 0 \\ -1 & 3 & -2 \end{bmatrix} \begin{bmatrix} 1 & 1 & 2 \\ -1 & 3 & 1 \\ 0 & 2 & -4 \end{bmatrix}$ (j) $\begin{bmatrix} 2 & -1 & 3 \\ 1 & 4 & 2 \\ 3 & 0 & 5 \end{bmatrix} \begin{bmatrix} 1 & 0 & 0 \\ 0 & 1 & 0 \\ 0 & 0 & 1 \end{bmatrix}$

8. Given $\mathbf{A} = \begin{bmatrix} 2 & -1 \\ 3 & 1 \end{bmatrix}$ and $\mathbf{B} = \begin{bmatrix} -1 & 3 \\ 2 & 4 \end{bmatrix}$, find $\mathbf{AB}$. Is $\mathbf{AB} = \mathbf{BA}$?

9. Given $\mathbf{A} = \begin{bmatrix} 2 & -1 \\ 3 & 1 \\ 1 & 4 \end{bmatrix}$, $\mathbf{I}_3 = \begin{bmatrix} 1 & 0 & 0 \\ 0 & 1 & 0 \\ 0 & 0 & 1 \end{bmatrix}$, and $\mathbf{I}_2 = \begin{bmatrix} 1 & 0 \\ 0 & 1 \end{bmatrix}$, show that

$$\mathbf{AI}_2 = \mathbf{I}_3 \mathbf{A} = \mathbf{A}$$

10. Find numbers x, y, z such that $[x \ y \ z] \begin{bmatrix} 2 & -1 & 1 \\ 0 & 3 & 2 \\ 1 & -1 & 1 \end{bmatrix} = [1 \ -2 \ 1]$.

11. Find r, s, and t if

$$[r \ s \ t] \begin{bmatrix} 1 & 1 & 0 \\ -1 & 2 & 1 \\ 3 & -1 & 2 \end{bmatrix} = [-1 \ 3 \ 2]$$

12. Find the transpose of each of the following matrices:

(a) $[u_1 \ u_2 \ u_3]$ (d) $\begin{bmatrix} 1 & 0 & 0 \\ 0 & 1 & 0 \\ 0 & 0 & 1 \end{bmatrix}$

(b) $\begin{bmatrix} 2 & -1 & 3 \\ 0 & 1 & -1 \\ 4 & 2 & -3 \end{bmatrix}$ (e) $\begin{bmatrix} -1 \\ 3 \\ 2 \end{bmatrix}$

(c) $\begin{bmatrix} 3 & 2 \\ 5 & -4 \end{bmatrix}$ (f) $\begin{bmatrix} x_1 & y_1 & z_1 \\ x_2 & y_2 & z_2 \end{bmatrix}$

A-2. Square matrices

In this section we shall discuss some properties of *square* matrices. The matrix $0 = \begin{bmatrix} 0 & 0 \\ 0 & 0 \end{bmatrix}$ is the *additive identity* of order 2. This means that

$$\mathbf{A} + 0 = 0 + \mathbf{A} = \mathbf{A}$$

for $\mathbf{A}$ *any* 2×2 matrix.

So

$$\begin{bmatrix} a_{11} & a_{12} \\ a_{21} & a_{22} \end{bmatrix} + \begin{bmatrix} 0 & 0 \\ 0 & 0 \end{bmatrix} = \begin{bmatrix} 0 & 0 \\ 0 & 0 \end{bmatrix} + \begin{bmatrix} a_{11} & a_{12} \\ a_{21} & a_{22} \end{bmatrix} = \begin{bmatrix} a_{11} & a_{12} \\ a_{21} & a_{22} \end{bmatrix}.$$

If two matrices **A** and **B** have the property that $\mathbf{A}+\mathbf{B}=0$, then **B** is said to be the additive inverse of **A**. It can be easily determined that if $\mathbf{A} = \begin{bmatrix} a_{11} & a_{12} \\ a_{21} & a_{22} \end{bmatrix}$, its additive inverse is $\mathbf{B} = \begin{bmatrix} -a_{11} & -a_{12} \\ -a_{21} & -a_{22} \end{bmatrix}$.

The matrix $\mathbf{I}_2 = \begin{bmatrix} 1 & 0 \\ 0 & 1 \end{bmatrix}$ is the multiplicative identity matrix of order 2. This means that
$$\mathbf{AI}_2 = \mathbf{I}_2\mathbf{A} = \mathbf{A}$$
for **A** any 2×2 matrix. So
$$\begin{bmatrix} a_{11} & a_{12} \\ a_{21} & a_{22} \end{bmatrix} \begin{bmatrix} 1 & 0 \\ 0 & 1 \end{bmatrix} = \begin{bmatrix} 1 & 0 \\ 0 & 1 \end{bmatrix} \begin{bmatrix} a_{11} & a_{12} \\ a_{21} & a_{22} \end{bmatrix} = \begin{bmatrix} a_{11} & a_{12} \\ a_{21} & a_{22} \end{bmatrix}$$

Associated with every square matrix is a *determinant* (see Section 1-5) but the two must not be confused. A square *matrix* is merely an ordered square array of numbers, whereas a *determinant* is a square array of numbers representing a specific number.

Thus if $\mathbf{A} = \begin{bmatrix} a_{11} & a_{12} \\ a_{21} & a_{22} \end{bmatrix}$, the determinant associated with **A** is

$$\text{Determinant } A = \begin{vmatrix} a_{11} & a_{12} \\ a_{21} & a_{22} \end{vmatrix} = a_{11}\, a_{22} - a_{12}\, a_{21}$$

If
$$\mathbf{B} = \begin{bmatrix} b_{11} & b_{12} & b_{13} \\ b_{21} & b_{22} & b_{23} \\ b_{31} & b_{32} & b_{33} \end{bmatrix} \quad \text{then} \quad \text{Determinant } B = \begin{vmatrix} b_{11} & b_{12} & b_{13} \\ b_{21} & b_{22} & b_{23} \\ b_{31} & b_{32} & b_{33} \end{vmatrix}$$

where determinant $B = b_{11}\, b_{22}\, b_{33} - b_{11}\, b_{23}\, b_{32} - b_{12}\, b_{21}\, b_{33} + b_{12}\, b_{23}\, b_{31} + b_{13}\, b_{21}\, b_{32} - b_{13}\, b_{22}\, b_{31}$.

It is important to note carefully that the value of a determinant may be zero, even when the associated matrix is not the zero matrix. So $\begin{bmatrix} 3 & -6 \\ -2 & 4 \end{bmatrix} \neq 0_{2 \times 2}$, while $\begin{vmatrix} 3 & -6 \\ -2 & 4 \end{vmatrix} = 0$. Also, it is possible to have two nonzero matrices whose *product* is the zero matrix. For example, if $\mathbf{A} = \begin{bmatrix} 2 & -1 \\ -4 & 2 \end{bmatrix}$ and $\mathbf{B} = \begin{bmatrix} 1 & 3 \\ 2 & 6 \end{bmatrix}$, then $\mathbf{AB} = \begin{bmatrix} 0 & 0 \\ 0 & 0 \end{bmatrix}$. Is **BA** also equal to the zero matrix?

A system of equations can be represented in a condensed form by using matrices. For example, consider the system of equations
$$a_1 x + b_1 y = c_1$$
$$a_2 x + b_2 y = c_2$$

If we let **X** be the column matrix $\begin{bmatrix} x \\ y \end{bmatrix}$, **D** the matrix of coefficients $\begin{bmatrix} a_1 & b_1 \\ a_2 & b_2 \end{bmatrix}$, and **C** the column matrix $\begin{bmatrix} c_1 \\ c_2 \end{bmatrix}$, then we can write

$$\mathbf{DX} = \mathbf{C}$$

The same set of equations can also be written as

$$(\mathbf{DX})^* = \mathbf{C}^* \quad \text{or} \quad \mathbf{X}^*\mathbf{D}^* = \mathbf{C}^*$$

Note the reverse order of the matrices; that is, $(\mathbf{DX})^* = \mathbf{X}^*\mathbf{D}^*$.

To see this, we have $\mathbf{X}^* = [x \ y]$, $\mathbf{D}^* = \begin{bmatrix} a_1 & a_2 \\ b_1 & b_2 \end{bmatrix}$, and $\mathbf{C}^* = [c_1 \ c_2]$. Hence,

$$[x \ y] \begin{bmatrix} a_1 & a_2 \\ b_1 & b_2 \end{bmatrix} = [c_1 \ c_2]$$

Definition. Given a square matrix **A**, if a square matrix **B** exists such that

$$\mathbf{AB} = \mathbf{BA} = \mathbf{I}$$

then **B** is defined to be the *multiplicative inverse matrix* of **A** and is written $\mathbf{A}^{-1}$.

If **A** is any square matrix whose determinant is *not* zero, the inverse matrix $\mathbf{A}^{-1}$ always exist. To find $\mathbf{A}^{-1}$ let us first recall some properties of determinants that were discussed in Section 1-8.

Definition. A co-factor of an element is $(-1)^{p+q}$ times the determinant which remains when the row and column that contain the element have been deleted, where p is the number of the row and q is the number of the column containing the element.

Theorem 1: *The value of a determinant is equal to the sum of the products of the elements of any row (or column) and their corresponding co-factors.*

Theorem 2: *The sum of the products of the elements of any row (or column) and the co-factors of the elements of any other row (or column) is zero.*

If $\mathbf{A} = \begin{bmatrix} a_{11} & a_{12} \\ a_{21} & a_{22} \end{bmatrix}$, then co-$\mathbf{A} = \begin{bmatrix} A_{11} & A_{12} \\ A_{21} & A_{22} \end{bmatrix}$ is the matrix whose elements are the co-factors of the corresponding elements of **A**. That is $A_{11} = a_{22}$, $A_{12} = -a_{21}$, $A_{21} = -a_{12}$, $A_{22} = a_{11}$. So co-$\mathbf{A} = \begin{bmatrix} a_{22} & -a_{21} \\ -a_{12} & a_{11} \end{bmatrix}$. We also have

$$(\text{co-}\mathbf{A})^* = \begin{bmatrix} A_{11} & A_{21} \\ A_{12} & A_{22} \end{bmatrix} = \begin{bmatrix} a_{22} & -a_{12} \\ -a_{21} & a_{11} \end{bmatrix}$$

Now

$$\mathbf{A}^{-1} = \frac{(\text{co-}\mathbf{A})^*}{\text{determinant } A}$$

To see this, form

$$AA^{-1} = \begin{bmatrix} a_{11} & a_{12} \\ a_{21} & a_{22} \end{bmatrix} \frac{\begin{bmatrix} a_{22} & -a_{12} \\ -a_{21} & a_{11} \end{bmatrix}}{\text{determinant } A} = \frac{\begin{bmatrix} \text{determinant } A & 0 \\ 0 & \text{determinant } A \end{bmatrix}}{\text{determinant } A}$$

$$= \begin{bmatrix} 1 & 0 \\ 0 & 1 \end{bmatrix} = I_{2 \times 2}$$

The student should verify that $A^{-1}A = I_{2 \times 2}$.

We shall illustrate the above by using a 3×3 matrix.

Let $A = \begin{bmatrix} 2 & -1 & 1 \\ 1 & 3 & -1 \\ 0 & 2 & 1 \end{bmatrix}$, then co-$A = \begin{bmatrix} 5 & -1 & 2 \\ 3 & 2 & -4 \\ -2 & 3 & 7 \end{bmatrix}$ and (co-A)*

$= \begin{bmatrix} 5 & 3 & -2 \\ -1 & 2 & 3 \\ 2 & -4 & 7 \end{bmatrix}$ Also determinant $A = 13$. Therefore,

$$A^{-1} = \frac{\begin{bmatrix} 5 & 3 & -2 \\ -1 & 2 & 3 \\ 2 & -4 & 7 \end{bmatrix}}{13}$$

Let us check that $AA^{-1} = I_{3 \times 3}$. We have

$$AA^{-1} = \begin{bmatrix} 2 & -1 & 1 \\ 1 & 3 & -1 \\ 0 & 2 & 1 \end{bmatrix} \frac{\begin{bmatrix} 5 & 3 & -2 \\ -1 & 2 & 3 \\ 2 & -4 & 7 \end{bmatrix}}{13} = \frac{\begin{bmatrix} 13 & 0 & 0 \\ 0 & 13 & 0 \\ 0 & 0 & 13 \end{bmatrix}}{13} = \frac{13 \begin{bmatrix} 1 & 0 & 0 \\ 0 & 1 & 0 \\ 0 & 0 & 1 \end{bmatrix}}{13} = I_{3 \times 3}$$

The student should verify that $A^{-1}A = I_{3 \times 3}$.

EXERCISES

1. Find the additive inverse of each of the following matrices:

    (a) $\begin{bmatrix} 2 & 1 \\ -1 & 4 \\ 3 & 2 \end{bmatrix}$ (b) $\begin{bmatrix} 3 & 1 \\ -2 & 5 \end{bmatrix}$ (c) $\begin{bmatrix} a_{11} & a_{12} & a_{13} \\ a_{21} & a_{22} & a_{23} \\ a_{31} & a_{32} & a_{33} \end{bmatrix}$

2. Find the determinants associated with each of the following matrices:

    (a) $\begin{bmatrix} 2 & 3 \\ 4 & -5 \end{bmatrix}$ (b) $\begin{bmatrix} 1 & 0 & -1 \\ 3 & 2 & 1 \\ -1 & 1 & 4 \end{bmatrix}$ (c) $\begin{bmatrix} 3 & 1 & -2 \\ 0 & 2 & 5 \\ -1 & -2 & 1 \end{bmatrix}$

3. Find the multiplicative inverses of each of the matrices in Exercise 2 by each of two methods.

4. Given $A = \begin{bmatrix} a_{11} & a_{12} \\ a_{21} & a_{22} \end{bmatrix}$, $B = \begin{bmatrix} b_{11} & b_{12} \\ b_{21} & b_{22} \end{bmatrix}$, $C = \begin{bmatrix} c_{11} & c_{12} \\ c_{21} & c_{22} \end{bmatrix}$, prove that:

(a) $A(B+C) = AB + AC$. (This is known as the *distributive law* for matrices.)

(b) $A(BC) = (AB)C$. (This is known as the *associative law* for matrix multiplication.)

A-3. Matrices as operators

We conclude these remarks on matrices by exhibiting their use as operators in geometry.

(a) Reflection in the x-axis. Let $P = (x_0, y_0)$ be any point in the plane and $P_1 = (x_1, y_1)$ its image in the x-axis. See Fig. A-1.

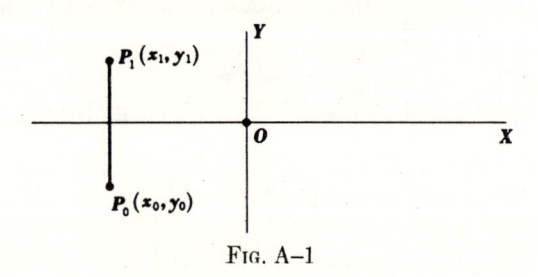

Fig. A-1

We have
$$x_1 = x_0$$
$$y_1 = \quad -y_0 \quad (A\text{-}1)$$

If we let $P_1 = \begin{bmatrix} x_1 \\ y_1 \end{bmatrix}$, $A = \begin{bmatrix} 1 & 0 \\ 0 & -1 \end{bmatrix}$, and $P_0 = \begin{bmatrix} x_0 \\ y_0 \end{bmatrix}$, we could have written equation (A-1) as

$$\begin{bmatrix} x_1 \\ y_1 \end{bmatrix} = \begin{bmatrix} 1 & 0 \\ 0 & -1 \end{bmatrix} \begin{bmatrix} x_0 \\ y_0 \end{bmatrix} \quad \text{or} \quad P_1 = AP_0$$

Here we can think of A as a reflection matrix in the x-axis since using it as an operator, any point P_0 is sent into its image in the x-axis.

(b) Reflection in the y-axis. From Fig. A-2, we see that
$$x_1 = -x_0$$
$$y_1 = \quad y_0 \quad (A\text{-}2)$$

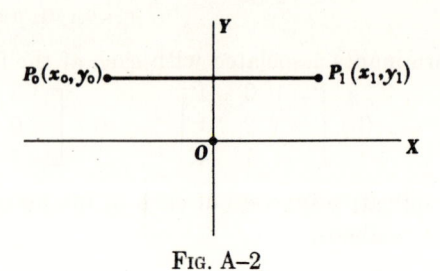

Fig. A-2

APPENDIX A   445

If we let $\mathbf{B} = \begin{bmatrix} -1 & 0 \\ 0 & 1 \end{bmatrix}$, in matrix notation equation (A-2) becomes

$$\mathbf{P}_1 = \mathbf{B}\,\mathbf{P}_0$$

The matrix **B** is a reflection operator in the y-axis.

(c) Reflection in the origin. In this case, we have

$$\begin{aligned} x_1 &= -x_0 \\ y_1 &= \;\; -y_0 \end{aligned} \tag{A-3}$$

If $\mathbf{C} = \begin{bmatrix} -1 & 0 \\ 0 & -1 \end{bmatrix}$, we have

$$\mathbf{P}_1 = \mathbf{C}\,\mathbf{P}_0$$

C is the matrix operator sending a point through a 180° turn about the origin. See Fig. A-3.

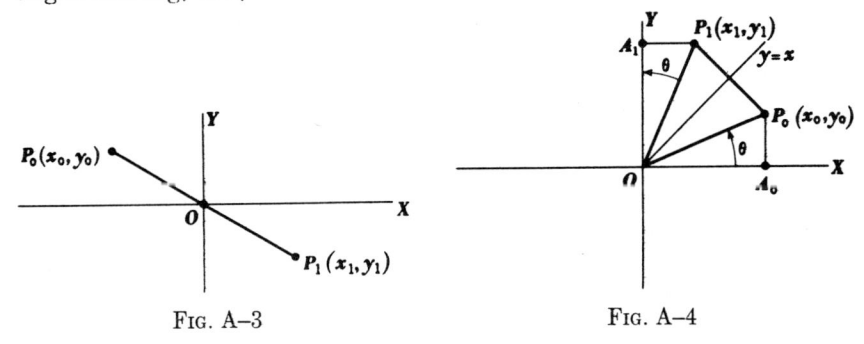

FIG. A-3   FIG. A-4

(d) Reflection in the line $y = x$. P_1 is the image of P_0 in the line $y = x$. See Fig. A-4.

The triangles OA_0P_0 and OA_1P_1 are congruent. Why? Hence,

$$\begin{aligned} x_1 &= y_0 \\ y_1 &= x_0 \end{aligned} \tag{A-4}$$

If we let $\mathbf{D} = \begin{bmatrix} 0 & 1 \\ 1 & 0 \end{bmatrix}$, we have

$$\mathbf{P}_1 = \mathbf{D}\,\mathbf{P}_0$$

The matrix operator **D** sends a point (x_0, y_0) into its image in the line $y = x$.

EXAMPLE A-7. Use the matrix operators **A**, **B**, **C**, and **D** to find the image of the point $(-3, 2)$ in the x-axis, the y-axis, the origin, and the line $y = x$, respectively.

Solution. Reflection in the x-axis. Using **A** we have

$$\begin{bmatrix} x_1 \\ y_1 \end{bmatrix} = \begin{bmatrix} 1 & 0 \\ 0 & -1 \end{bmatrix} \begin{bmatrix} -3 \\ 2 \end{bmatrix} = \begin{bmatrix} -3 \\ -2 \end{bmatrix}$$

Thus $x_1 = -3$ and $y_1 = -2$.

Reflection in the y-axis. Using **B** we have
$$\begin{bmatrix} x_1 \\ y_1 \end{bmatrix} = \begin{bmatrix} -1 & 0 \\ 0 & 1 \end{bmatrix} \begin{bmatrix} -3 \\ 2 \end{bmatrix} = \begin{bmatrix} 3 \\ 2 \end{bmatrix}$$
Thus $x_1 = 3$, $y_1 = 2$.

Reflection in the origin. Using **C** we have
$$\begin{bmatrix} x_1 \\ y_1 \end{bmatrix} = \begin{bmatrix} -1 & 0 \\ 0 & -1 \end{bmatrix} \begin{bmatrix} -3 \\ 2 \end{bmatrix} = \begin{bmatrix} 3 \\ -2 \end{bmatrix}$$
Thus $x_1 = 3$, $y_1 = -2$.

Reflection in line $y = x$. Using **D** we have
$$\begin{bmatrix} x_1 \\ y_1 \end{bmatrix} = \begin{bmatrix} 0 & 1 \\ 1 & 0 \end{bmatrix} \begin{bmatrix} -3 \\ 2 \end{bmatrix} = \begin{bmatrix} 2 \\ -3 \end{bmatrix}$$
Thus $x_1 = 2$, $y_1 - 3$.

As an exercise, operate with **A**, **B**, **C**, and **D** on the plane vector $\begin{bmatrix} x \\ y \end{bmatrix}$.

(e) *Translation of the axes.* In Section 6-2, we studied *translation of the axes* and saw that this was obtained by the linear transformation
$$x = x' + h \qquad y = y' + k \tag{A-5}$$
where (x, y) are the coordinates of a point with respect to an original set of axes, (x', y') are the coordinates of the same point with respect to a new set of axes, and (h,k) are the coordinates of the new origin with respect to the original axes. See Fig. A-5.

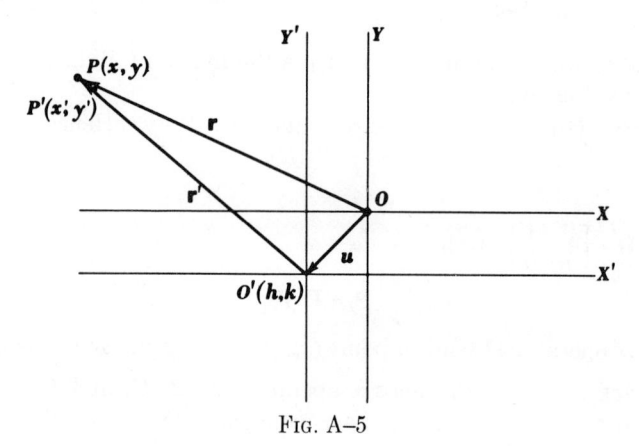

Fig. A-5

If we let $\mathbf{r} = [x, y]$ be the row matrix representing the position vector $\overrightarrow{OP}$, $\mathbf{r}' = [x', y']$ be the row matrix representing the position vector $\overrightarrow{O'P'}$, and $\mathbf{u} = [h, k]$ be the row matrix representing the vector $\overrightarrow{OO'}$, then
$$\mathbf{r} = \mathbf{u} + \mathbf{r}' \tag{A-6}$$
gives us the matrix formula for a translation of axes.

(f) Rotation of the axes. We conclude this brief discussion of matrix theory by discussing the matrix as an operator for a rotation of the axes.

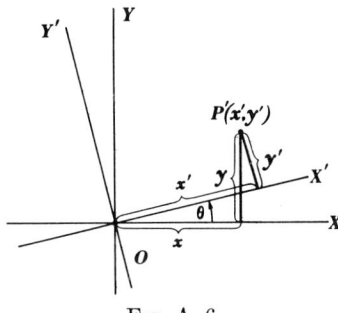

Fig. A-6

In Section 6-5, we saw that the linear transformation

$$x = x' \cos \theta - y' \sin \theta$$
$$y = x' \sin \theta + y' \cos \theta \qquad \text{(A-7)}$$

are the equations for the rotation of the axes in the plane. See Fig. A-6.

Letting $\mathbf{r} = [x, y]$, $\mathbf{r}' = [x', y']$, and $\mathbf{R} = \begin{bmatrix} \cos \theta & -\sin \theta \\ \sin \theta & \cos \theta \end{bmatrix}$, we can write equations (A-7) as

$$\mathbf{r}^* = \mathbf{R}\mathbf{r}'^* \qquad \text{(A-8)}$$

R is called a *rotation matrix*. If we multiply both sides of equation (A-8) on the left by $\mathbf{R}^{-1}$, we obtain:

$$\mathbf{R}^{-1}\mathbf{r}^* = \mathbf{R}^{-1}\mathbf{R}\mathbf{r}'^*$$

or

$$\mathbf{r}'^* = \mathbf{R}^{-1}\mathbf{r}^* \qquad \text{(A-9)}$$

This becomes:

$$\begin{bmatrix} x' \\ y' \end{bmatrix} = \begin{bmatrix} \cos \theta & \sin \theta \\ -\sin \theta & \cos \theta \end{bmatrix} \begin{bmatrix} x \\ y \end{bmatrix}$$

and

$$x' = x \cos \theta + y \sin \theta$$
$$y' = -x \sin \theta + y \cos \theta \qquad \text{(A-10)}$$

The student should note that

$$\mathbf{R}\mathbf{R}^{-1} = \begin{bmatrix} \cos \theta & -\sin \theta \\ \sin \theta & \cos \theta \end{bmatrix} \begin{bmatrix} \cos \theta & \sin \theta \\ -\sin \theta & \cos \theta \end{bmatrix} = \begin{bmatrix} 1 & 0 \\ 0 & 1 \end{bmatrix} = \mathbf{I}_{2 \times 2}$$

In like manner, we have

$$\mathbf{R}^{-1}\mathbf{R} = \mathbf{I}_{2 \times 2}$$

EXAMPLE A-8. What do the coordinates of the point $(-1, 2)$ become after the axes have been rotated through an angle of 30°?

Solution. Using equation (A-9) we have

$$\begin{bmatrix} x' \\ y' \end{bmatrix} = \begin{bmatrix} \cos 30° & \sin 30° \\ -\sin 30° & \cos 30° \end{bmatrix} \begin{bmatrix} -1 \\ 2 \end{bmatrix}$$

or

$$x' = -\frac{\sqrt{3}}{2} + 1 \text{ and } y' = \frac{1}{2} + \sqrt{3}$$

Equation (A-8) also represents a rotation in a space of any dimensions. We shall illustrate this with a space of three dimensions.

Consider a set of axes X, Y, Z that have, after a rotation, become the set X', Y', Z'. Let (l_1, m_1, n_1), (l_2, m_2, n_2), and (l_3, m_3, n_3) be the direction cosines of the X' axis, the Y' axis and the Z' axis, with respect to the original axes.

Then $\mathbf{u} = [l_1\ m_1\ n_1]$, $\mathbf{v} = [l_2\ m_2\ n_2]$, and $\mathbf{w} = [l_3\ m_3\ n_3]$ are unit vectors on the X' axis, the Y' axis, and the Z' axis, respectively. Now let

$$\mathbf{R} = [\mathbf{u}^*\ \mathbf{v}^*\ \mathbf{w}^*] \quad \text{or} \quad \mathbf{R} = \begin{bmatrix} l_1 & l_2 & l_3 \\ m_1 & m_2 & m_3 \\ n_1 & n_2 & n_3 \end{bmatrix}$$

Hence, if $\mathbf{r} = [xyz]$ and $\mathbf{r}' = [x'y'z']$, then $\mathbf{r}^* = \mathbf{R}\mathbf{r}'^*$ gives

$$x = l_1 x' + l_2 y' + l_3 z'$$
$$y = m_1 x' + m_2 y' + m_3 z'$$
$$z = n_1 x' + n_2 y' + n_3 z'$$

But this is the linear transformation for a rotation of axes that was given in equation (12-31) of Chapter 12.

In the same way $\mathbf{r}'^* = \mathbf{R}^{-1}\mathbf{r}^*$ gives

$$x' = l_1 x + m_1 y + n_1 z$$
$$y' = l_2 x + m_2 y + n_2 z$$
$$z' = l_3 x + m_3 y + n_3 z$$

This is the set of equations given in (12-30).

Thus it is apparent that the matrix notation for a rotation of axes as given by equations (A-8) or (A-9) is a simple way of representing rotations in any dimensions.

Appendix B

Answers to Selected Exercises

CHAPTER 1

Sec. 1–3

3. (a) $\{d,e,f,g\}$; (c) $\{(a,c),(a,d),(b,c),(b,d),(c,c),(c,d)\}$;
(e) $\{e,f,g\}$; (g) $\{d,e,f,g\}$.

4. $(A \cap B) \cup (A \cap C)$; $(A \cap B') \cup (C \cap A')$; $(A \cap B' \cap C) \cup (B \cap A' \cap C')$.

Sec. 1–5

2. (a) $\dfrac{9}{5}$; (c) $\dfrac{192}{25}$; (e) $\dfrac{t+2}{r-s}$. **3.** (b) $\dfrac{3 \pm \sqrt{17}}{4}$; (d) $0, -5$; (f) $\dfrac{1 \pm \sqrt{13}}{2}$.

4. (a) real; (c) complex; (e) real. **5.** -2.

7. (a) 8; (b) -7.

9. (a) $\dfrac{3 \pm \sqrt{9 + 8x - 24x^3}}{2x}$; (c) $1 \pm \sqrt{-(9x^2 + 36x + 32)}$.

10. (b) $\dfrac{1}{2}y, -2y$.

Sec. 1–6

1. 2^5; 3^6; $\dfrac{1}{5}$; 3^4; 1; 2^6; 2; $\dfrac{1}{6}$; $\dfrac{8}{27}$; 4; $\dfrac{2}{3}$; $\dfrac{1}{4}$; 4; 3^8.

3. $a^2 b^{3/2}$ **4.** $\dfrac{3\sqrt{2}}{2}$.

Sec. 1–7

1. $\log_2 8 = 3$; $\log_9 3 = \dfrac{1}{2}$; $\log_2 \dfrac{1}{4} = -2$; $\log_{16} 8 = \dfrac{3}{4}$.

3. (a) 3; (c) 8; (e) -3; (g) 1000; (i) $a^{2/3}$.

7. (a) 5880; (c) 732; (e) 460; (f) 0.1403.

Sec. 1–8

1. (a) 11; (c) 0.
2. (a) 21; (c) 0.
3. (a) 10; (c) 261.

Sec. 1–9

1. (a) $-\frac{3}{2}$; (c) $\frac{-1 \pm \sqrt{-11}}{2}$.
2. (b) $x = \frac{14}{17}, y = \frac{15}{17}$.
3. (a) $x = \frac{1}{4}, y = \frac{17}{4}, z = \frac{31}{4}$; (c) $x = \frac{5}{11}, y = \frac{20}{11}, z = \frac{13}{11}$.
4. (a) $x = 1, y = -2, z = 2, w = -1$.

Sec. 1–10

1. (a) $(2,2), \left(\frac{8}{9}, -\frac{4}{3}\right)$; (c) $\left(\frac{12\sqrt{97}}{97}, -\frac{18\sqrt{97}}{97}\right), \left(\frac{-12\sqrt{97}}{97}, \frac{18\sqrt{97}}{97}\right)$;
   (e) $(0,3), \left(\frac{33}{8}, -\frac{9}{8}\right)$.
2. (a) $(3,2), (3,-2), (-3,2), (-3,-2)$; (c) $(\sqrt{3},i), (\sqrt{3},-i), (-\sqrt{3},i), (-\sqrt{3},-i)$;
   (e) $(5,1), (-1,-5), (1,5), (-5,-1)$.

Sec. 1–11

1. F, T, T, T, F, T, T, T.
2. 6; 2; 7; 2; −3; 1; 8; −3.
6. (a) $x < \frac{7}{3}$; (c) $x > -\frac{11}{2}$; (e) $x \neq 0$.
7. (a) 2, −2; (c) 2, −6; (e) 5, −9.

Sec. 1–12

3. The graph is the set of points defined by:

   (a) $-2 < x < 2$; (c) $1 < x < 3$; (e) $x > 1$; (g) $-\frac{7}{3} < x < -1$; (h) $\frac{1}{3} \leq x \leq 3$.

Sec. 1–13

1. (a) Yes, Domain = $\{1,2,4,5\}$, Range = $\{2,5,-2,0\}$;
   (c) No, Domain = $\{1,4,5\}$, Range = $\{2,1,5,0\}$;
   (e) Yes, Domain = $\{x \mid x \epsilon R\}$, Range = $\{y \mid y \geq 0 \text{ and } y \epsilon R\}$;
   (g) Yes, Domain = $\{x \mid x \epsilon R\}$, Range = $\{1\}$;
   (i) No, Domain = $\{x \mid x \epsilon R\}$, Range = $\{y \mid y \epsilon R\}$.
2. (a) 6; (c) 0; (e) $3(a+h)$.

3. Yes. **4. (a)** 6; **(c)** 2; **(e)** $a^2 + 2ah + h^2 + 2$. **5.** No.

6. No. Domain = $\{x \mid -1 \leq x \leq 1\}$, Range = $\{y \mid -1 \leq y \leq 1\}$.

8. (a) Domain = $\{x \mid x \epsilon R\}$, Range = $\{0\}$.
   (c) Domain = $\{x \mid x \epsilon R\}$, Range = $\{y \mid y \epsilon R\}$.

9. (a) $y = -x^2$; **(c)** $y = \sqrt{1-x^2}$ and $y = -\sqrt{1-x^2}$; **(e)** $y = 2x$.

10. (a) $x = \sqrt{-y}$ and $x = -\sqrt{-y}$; **(b)** $x = -\frac{1}{4}y^2$;
   (d) $x = \frac{1}{y}$; **(f)** $x = -\frac{y^2+4}{2y}$.

Sec. 1–14

1. Conditional. **3.** Identity. **5.** Identity. **7.** Identity.

CHAPTER 2
Sec. 2–2

5. $(-a, b)$, $(a, -b)$, and $(-a, -b)$. **7.** A vertical line.

9. (a) Horizontal line 3 units below the x-axis; **(c)** a horizontal line. **11.** A line bisecting quadrants II and IV.

13. $(b, 0)$ and (b, b) or $(-b, 0)$ and $(-b, b)$ or $\left(\frac{-b}{2}, \frac{b}{2}\right)$ and $\left(\frac{b}{2}, \frac{b}{2}\right)$; three solutions for a given b.

15. $\left(\frac{a}{\sqrt{2}}, 0\right)$, $\left(0, \frac{a}{\sqrt{2}}\right)$, $\left(-\frac{a}{\sqrt{2}}, 0\right)$, and $\left(0, -\frac{a}{\sqrt{2}}\right)$.

16. The set of points lying on the circumference of a circle having center at $(0, 0)$ and radius equal to the given number.

17. The set of points whose coordinates are $(k, 2k)$, for k any real number. The locus is a line.

18. The set of points to the left of, and on a vertical line six units to the right of, the y-axis.

19. The set of points above and on the horizontal line two units below the x-axis.

20. The set of points to the left of the vertical line six units to the right of the y-axis and above the horizontal line two units below the x-axis.

21. Three solutions: $(a, 0)$, $(0, a)$; $(2a, 0)$, $(a, -a)$; $(0, 2a)$, $(-a, a)$.

22. The set of points to the right of, and on the vertical line three units to the left of, the y-axis and below and on the x-axis.

23. The set of points inside the circle having center at the origin and radius 4.

24. The set of points contained between the two vertical lines, each of which is three units from the y-axis.

25. (a) The set of points on and between the two vertical lines each of which is four units from the y-axis and on and between the horizontal lines each of which is two units from the x-axis.
   (b) The set of points lying between the vertical lines each of which is three units from the y-axis and also between the horizontal lines each of which is one unit from the x-axis.

(c) The set of points lying outside of the rectangle formed by two vertical lines two units from the y-axis and two horizontal lines three units from the x-axis.

(d) The set of points contained within the strip bounded by the vertical lines each of which is one unit from the y-axis, above the horizontal line three units above the x-axis, and below the horizontal line three units below the x-axis.

26. The set of points lying on the two lines bisecting the four quadrants.
27. The set of points lying outside the area bounded by the circle of radius five and center at (0, 0).

Sec. 2–3

1. (a) $(-5, -4)$; (c) $(-5, 4)$. 3. $(-1, -2)$ and $(1, 2)$.
5. $\{(-2,-3),(-4,0),(0,4)\}$; (c) $\{(2,1),(-2,3),(-2,-1),(-4,0)\}$.
8. $\left\{(-d,c),\ (d,-c),\ \left(-\dfrac{a+b}{2},\dfrac{a-b}{2}\right),\ \left(\dfrac{b-a}{2},\dfrac{a+b}{2}\right),\ (-c,-d),\ (d,c)\right\}$.

Sec. 2–4

3. (a) 7 and -1; (c) 3 and -8; (e) -4 and 10.

Sec. 2–5

1. (a) $[5, -9]$; (c) $[-7, -1]$; (e) $[7, 2]$. 3. (a) $(5, 2)$; (c) $(-3, -4)$; (e) $(-3, 4)$.
4. (a) $(-5, 6)$; (c) $(-7, 1)$; (e) $(7, -6)$.

Sec. 2–6

1. (a) $3\sqrt{5}$; (c) 5; (e) $2\sqrt{10}$; (g) $\sqrt{106}$; (i) 9.
2. (a) $\sqrt{13}+5+\sqrt{2}$; (c) $\sqrt{65}+3\sqrt{2}+\sqrt{17}$; (e) $\sqrt{17}+\sqrt{13}+2\sqrt{10}$.
5. $(-5, -5)$ or $(10+5\sqrt{3}, 10-5\sqrt{3})$. 7. 20 square units.

Sec. 2–7

1. Sum of their squares is 1. 3. (a) No.
4. (a) Yes, for the unit segment; (b) not generally.
5. No. The *directions* are different.
7. (a) $[1, 0]$ if x is positive; $[-1, 0]$ if x is negative.
8. (a) $\left[-\dfrac{4}{5}, -\dfrac{3}{5}\right]$; (c) $\left[\dfrac{7}{\sqrt{130}}, \dfrac{9}{\sqrt{130}}\right]$; (e) $\left[\dfrac{8}{\sqrt{113}}, -\dfrac{7}{\sqrt{113}}\right]$.

Sec. 2–8

1. $\sqrt{13}$. 2. (b) $[-4, 0]$ and 4; (d) $[7, -7]$ and $\sqrt{98}$; (f) $[-1, 3]$ and $\sqrt{10}$.
3. (b) No. 4. (b) $\sqrt{5}\left[\dfrac{2}{\sqrt{5}}, \dfrac{1}{\sqrt{5}}\right]$; (d) $5\left[-\dfrac{3}{5}, -\dfrac{4}{5}\right]$.
8. $\mathbf{u} = u_1\mathbf{i} + u_2\mathbf{j}$.
11. (a) $[-6, 14]$, $2\sqrt{58}$; (b) $[-6, -9]$, $3\sqrt{13}$; (c) $[-4, 7]$, $\sqrt{65}$;
    (d) $[-6, -20]$, $2\sqrt{109}$; (e) $\left[\dfrac{2}{3}, 1\right]$, $\dfrac{1}{3}\sqrt{13}$; (f) $[17, -16]$, $\sqrt{545}$;
    (g) $[-7, -3]$, $\sqrt{58}$; (h) $[0, 1]$, 1.
12. $-\mathbf{i}+3\mathbf{j}$; $2\mathbf{i}+2\mathbf{j}$; $4\mathbf{i}-3\mathbf{j}$; $\mathbf{i}+\mathbf{j}$; $-2\mathbf{i}-3\mathbf{j}$.

13. (a) $\left(-\frac{1}{\sqrt{10}}, \frac{3}{\sqrt{10}}\right)$; (b) $\left(\frac{3}{\sqrt{13}}, -\frac{2}{\sqrt{13}}\right)$; (c) $(1, 0)$; (d) $(0, -1)$;

(e) $\left(-\frac{3}{5}, -\frac{4}{5}\right)$; (f) $\left(\frac{5}{\sqrt{34}}, \frac{3}{\sqrt{34}}\right)$; (g) $(-1, 0)$; (h) $\left(-\frac{2}{\sqrt{29}}, \frac{5}{\sqrt{29}}\right)$.

Sec. 2–10

1. (a) $\frac{21}{\sqrt{13}\sqrt{82}}, \frac{1}{\sqrt{13}\sqrt{2}}$, and $\frac{4}{\sqrt{41}}$; (c) $\frac{-9}{\sqrt{10}\sqrt{13}}, \frac{19}{\sqrt{41}\sqrt{10}}$, and $\frac{22}{\sqrt{13}\sqrt{41}}$;
(e) $-\frac{1}{5\sqrt{2}}, \frac{5}{\sqrt{2}\sqrt{37}}$, and $\frac{27}{5\sqrt{37}}$.

2. $k = \frac{7}{5}$. 4. $(5, 5)$. 6. $\frac{1 \pm \sqrt{65}}{2}$ or $\frac{19}{3}$ or $\frac{-11}{3}$. 7. -5.

14. The sets (a), (b), (c), and (e). 15. (a) $x = -\frac{11}{7}$; (b) $y = 3$.

Sec. 2–12

1. $\left(\frac{9}{8}, \frac{41}{8}\right)$. 3. $\left(\frac{22}{5}, \frac{8}{5}\right)$ or $\left(\frac{26}{3}, -\frac{16}{3}\right)$. 5. $\frac{\sqrt{101} + \sqrt{61} + \sqrt{50}}{2}$.

7. $(-7, 6)$. 10. $(4, 0), (6, 6)$, and $(-6, 2)$. 15. 1 to 1.

17. $\left(\frac{11}{4}, 0\right)$ or $(0, -11)$. 19. -4 to 1.

Sec. 2–13

(a) $[2, 3]$; (c) $[1, -4]$; (e) $\left[-\frac{4}{5}, \frac{3}{5}\right]$; (g) $[-\Delta y, \Delta x]$.

Sec. 2–14

2. (b) $a_1 a_2 + b_1 b_2$. 3. (b) Bar product. 4. (b) 15.

5. (a) $\frac{37}{2}$; (c) $\frac{45}{2}$. 7. (a) $\frac{6}{5}$; (b) $-\frac{10}{3}$.

11. (a) $\frac{49}{2}$ square units; (b) $\frac{67}{2}$ square units; (c) $\frac{55}{2}$ square units.

Review Exercises

1. (a) 5; (b) $(4, 3)$. 6. (a) $(3, 4)$; (c) $(-3, 4)$. 7. $(5, 6)$.
9. $(12, -7)$. 11. $(6, -4 \pm \sqrt{2})$ or $(6, -5)$ or $-1 \pm \sqrt{17}$. 13. $(0, 10)$.
15. (c) $[-9, 12]$ and $|15|$. 16. (b) $[-2, 3]$; (d) $[-7, 6]$; (f) $[3, 0]$; (h) $[5, 3]$
17. (a) $(-8, 17)$; (b) -3 to 4. 19. $(-15, -3)$. 21. 1 to 1.
23. $(-13, -2)$ or $(17, 10)$. 25. Rectangle.
37. $\mathbf{i} = -\frac{1}{5}\mathbf{u} - \frac{2}{5}\mathbf{v}, \mathbf{j} = \frac{1}{5}\mathbf{u} - \frac{3}{5}\mathbf{v}$.

Chapter 3

Sec. 3–2

1. (a) [4, 2]; (c) [−1, 2]; (e) [2, 0].
2. (a) $\left[\frac{2}{\sqrt{5}}, \frac{1}{\sqrt{5}}\right]$; (c) $\left[-\frac{1}{\sqrt{5}}, \frac{2}{\sqrt{5}}\right]$; (e) [1, 0].
3. (a) $\frac{1}{2}$; (c) −2; (e) 0. 5. (a) [5, 3]; (c) [1, 4]; (e) [1, 0].
7. (a) $\left[-\frac{2}{\sqrt{29}}, \frac{5}{\sqrt{29}}\right]$; (c) $\left[\frac{2}{\sqrt{5}}, -\frac{1}{\sqrt{5}}\right]$.
8. (a) $-\frac{5}{2}$; (c) $-\frac{1}{2}$.

Sec. 3–3*

1. (a) $x = 3 - 7t$, $y = 5 - 4t$; (c) $x = 3 - 3t$, $y = 0 + 4t$; (e) $x = 4 - 7t$, $y = -2$; (g) $x = 3 + 4t$, $y = 4$; (i) $x = 1 + 4t$, $y = 1 + t$.
2. (a) [2, −3] and (−1, 2); (c) [0, 3] and (2, 0); (e) [−1, −3] and (−3, 2); (g) [1, 0] and (0, 0); (i) [2, 1] and (−5, 3).
5. (a) $x = t$, $y = 0$; (b) $x = 0$, $y = t$.
6. (a) ∥ (b), (e) ∥ (h); (a) ⊥ (c), (b) ⊥ (c), (f) ⊥ (g), (d) ⊥ (h).
8. If $\frac{a_1}{a_2} = \frac{c_1}{c_2}$, lines are parallel. If $a_1 a_2 + c_1 c_2 = 0$, lines are perpendicular.
9. (a) $x = -1 + 2t$, $y = 3 + 5t$; (b) $x = 3 - t$, $y = -2 - 2t$; (c) $x = -3 + t$, $y = -2 + 4t$; (d) $x = 4 + 4t$, $y = 5 + 3t$; (e) $x = 1 + 2t$, $y = -1 + t$; (f) $x = -2 + 5t$ $y = 5 + 3t$; (g) $x = 3$, $y = 2 + t$; (h) $x = -1 + t$, $y = 4$.

Sec. 3–4

2. (a) $x - 4y + 11 = 0$; (c) $6x - 5y - 23 = 0$; (e) $y = 4$; (g) $x = 3$; (i) $2x - y = 0$.
3. (a) $3x + y - 5 = 0$; (c) $x - 2y + 12 = 0$; (e) $2x + 3y + 8 = 0$; (g) $x - 2y - 8 = 0$; (i) $x + y = 6$.
6. (a) $x + 2y - 5 = 0$; (b) $x - y - 3 = 0$; (c) $2x + y - 11 = 0$.

Sec. 3–5

2. (a) [3, −2]; (c) [4, 3]; (e) [0, 1]; (g) [1, 0].
3. (a) $\left[\frac{3}{\sqrt{13}}, -\frac{2}{\sqrt{13}}\right]$; (c) $\left[\frac{4}{5}, \frac{3}{5}\right]$; (e) [0, 1]; (g) [1, 0].
4. Coincident. 6. $a = 2$, $b = -3$, and $c = 11$.
10. For AB, $5x - 3y = 0$; for AC, $2x + y = 0$; for BC, $3x - 4y + 11 = 0$.
11. (b) [0, 1]; (d) $\left[\frac{4}{5}, \frac{3}{5}\right]$; (f) [1, 0]; (h) $\left[-\frac{3}{\sqrt{10}}, \frac{1}{\sqrt{10}}\right]$.

---

*Answers not unique for exercises 1, 5, and 9.

12. (b) none; (d) $\frac{3}{4}$; (f) 0; (h) $-\frac{1}{3}$.

13. (a) [3, −4]; (b) [4, 3]; (c) $\frac{3}{4}$; (d) $\left[\frac{4}{5}, \frac{3}{5}\right]$; (e) (−4, 0), (0, 3).

14. (a) [5, −12]; [12, 5]; $\frac{5}{12}$; $\left[\frac{12}{13}, \frac{5}{13}\right]$; $\left(\frac{-12}{5}, 0\right)$, (0, 1).

   (b) [2, −3]; [3, 2]; $\frac{2}{3}$; $\left[\frac{3}{\sqrt{13}}, \frac{2}{\sqrt{13}}\right]$; (3, 0), (0, −2).

   (c) [3, 4]; [−4, 3]; $-\frac{3}{4}$; $\left[-\frac{4}{5}, \frac{3}{5}\right]$; (−4, 0), (0, −3).

15. (a) Set of points to the left of the line $x - 3y = 0$.
   (c) Set of points to the left of the line $3x + 2y = 0$.
   (d) Set of points to the right of the line $x + y - 4 = 0$.
   (e) Set of points to the left of $2x - 3y + 7 = 0$.
   (f) Set of points to the left of and on the line $5x - 3y + 1 = 0$.
   (h) Set of points below and on the line $y - 1 = 0$.
   (i) Set of points to the right of and on the line $x = -1$.
   (j) Set of points to the right of and on the line $4x - 3y - 6 = 0$.

16. (a) $2x - 3y + 14 = 0$; (h) $3x + y - 9 = 0$; (c) $x + 2y + 9 = 0$; (d) $5x - 2y - 19 = 0$;
   (e) $x - y + 2 = 0$; (f) $8x + 6y + 7 = 0$.

Sec. 3–6

1. (a) $\left[-\frac{2}{\sqrt{13}}, \frac{3}{\sqrt{13}}\right]$; (c) [0, 1]; (e) $\left[\frac{4}{\sqrt{41}}, \frac{5}{\sqrt{41}}\right]$; (g) $\left[\frac{4}{5}, -\frac{3}{5}\right]$.

2. (a) $-\frac{3}{2}$; (c) none; (e) $\frac{5}{4}$; (g) $-\frac{3}{4}$.

3. (a) $\left[\frac{3}{\sqrt{13}}, \frac{2}{\sqrt{13}}\right]$; (c) [0, 1]; (e) $\left[\frac{1}{\sqrt{2}}, \frac{1}{\sqrt{2}}\right]$; (g) $\left[\frac{12}{13}, \frac{5}{13}\right]$.

4. (a) $\frac{2}{3}$; (c) none; (e) 1; (g) $\frac{5}{12}$. **5.** (a) No; (c) yes.

6. (a) $-\frac{1}{\sqrt{26}}$; (c) 0; (e) 0.

10. (a) $\frac{2\sqrt{5}}{5}$; (b) $-\frac{1}{2}$; (c) 13 square units.
   (d) Both the sum and the product are undefined since tan $P_2P_1P_3$ is undefined.

11. (a) Cosines of angles are: $\frac{4}{\sqrt{65}}$; $\frac{1}{\sqrt{65}}$; $\frac{4}{5}$. (b) Tangents of angles are: $\frac{7}{4}$; 8; $\frac{3}{4}$.
   (c) $\frac{28}{3}$ square units.

12. $\frac{-17}{6}$.

Sec. 3–10

1. (a) $x-3y+7=0$; (c) $3x-y+3=0$; (e) $3x-7y-21=0$; (g) $2x-5y-10=0$;
  (i) $2x+y-1=0$.
2. (a) $\dfrac{x}{-3}+\dfrac{y}{2}=1$ and $y=\dfrac{2x}{3}+2$; (c) $\dfrac{x}{-2}+\dfrac{y}{10}=1$ and $y=\dfrac{5x}{7}+\dfrac{10}{7}$.
4. $s=-\dfrac{2}{9}$ or -2. (Here *area* is taken positive.)
10. (a) $\sqrt{3}\,x+y-20=0$; (c) $y=-1$; (e) $\sqrt{3}\,x-y-8=0$; (g) $x+\sqrt{3}\,y+4=0$.
11. (a) $-\dfrac{3}{5}x+\dfrac{4}{5}y=\dfrac{12}{5}$; (c) $\dfrac{3}{5}x+\dfrac{4}{5}y+\dfrac{12}{5}$; (e) $-\dfrac{5}{13}x+\dfrac{12}{13}y=5$; (g) $x=\dfrac{5}{2}$.
  (i) $-\dfrac{x}{\sqrt{10}}+\dfrac{3y}{\sqrt{10}}=0$ or $\dfrac{x}{\sqrt{10}}-\dfrac{3y}{\sqrt{10}}=0$.
12. $x+y=7$.   13. $x+y+5=0$.
14. $x+2y+6=0$; $x+8y-12=0$.   17. $x-y+8=0$.
18. $x-2y+12=0$.   19. $3x+2y-13=0$.
20. $4x+3y=25$.   21. $\sqrt{3}\,x-2y+6=0$.
22. $2x-3y=11$; $x+2y+5=0$.

Sec. 3–11

1. (a) $3x-2y+5=0$; (c) $2x-y-5=0$.
2. (a) $[2,-3]$; (c) $[3,4]$; (e) $[1,-1]$.
3. (a) $[3,2]$; (c) $[4,-3]$; (e) $[1,1]$.
4. (a) $2x-3y-1=0$ and $3x+2y-8=0$; (c) $x-y-1=0$ and $x+y+3=0$;
  (e) $y+1=0$ and $x=6$; (g) $3x+2y-5=0$ and $2x-3y+1=0$; (i) $x-3y-10=0$ and $3x+y-10=0$.
5. (a) For AB, $2x-y-2=0$; for AC, $x+2y-1=0$; for BC, $3x+y-13=0$;
  (c) $x+2y-6=0$; $2x-y-7=0$; $x-3y-1=0$; (e) $x=1$ and $y=0$; $x=5$ and $y=-2$; $x=3$ and $y=4$.
6. (b) $x+5y=0$, $2x-y-4=0$, and $3x+4y-4=0$;
  (d) $x+2y-22=0$, $4x-3y-22=0$, and $5x-y=0$.
7. (a) $x+4y-7=0$, $4x+3y-2=0$, and $7x+2y-23=0$;
  (c) $8x-2y-5=0$, $3x-4y-14=0$, and $4x-14y-51=0$;
  (e) $x=-1$ and $y=2$; $x=3$ and $y=1$; $x=5$ and $y=-6$.
9. $x+y-2=0$, $x+y+2=0$, $x-y+2=0$, $x-y-2=0$.
10. $x+2y-8=0$, $x+2y+8=0$.   11. $2x+y+6=0$, $2x+y-6=0$.
12. $ax+by=a^2+b^2$.   13. $3x+2y=28$.
14. $(2,-2)$.   15. $\left(\dfrac{10}{7},\dfrac{34}{7}\right)$.
16. $cx-(a+b)y=0$.
17. The set of points common to the lines $2x-3y-4=0$ and $2x-3y+4=0$.

Sec. 3–12

3. (a) $\dfrac{23}{5}$; (c) $\dfrac{19}{\sqrt{5}}$; (e) $\dfrac{24}{5}$.   4. (a) $\dfrac{2}{\sqrt{13}}$; (c) $\left|\dfrac{c_2-c_1}{\sqrt{a^2+b^2}}\right|$.
5. (b) $\dfrac{17}{\sqrt{26}}$, $\dfrac{17}{\sqrt{13}}$, and $\dfrac{17}{5}$; (d) $\dfrac{44}{\sqrt{73}}$, $\dfrac{44}{\sqrt{65}}$, and $\dfrac{11}{\sqrt{2}}$.

APPENDIX B 457

6. (b) $\frac{17}{2}$; (d) 22.

8. $\left(0, \frac{13}{2}\right); \left(0, -\frac{7}{2}\right)$.
9. $3x+4y+5=0,\ 3x+4y-5=0$.
10. $x-2y-2\sqrt{5}=0,\ x-2y+2\sqrt{5}=0$. 11. $2x+3y-6\pm3\sqrt{13}=0$.
12. $3x+4y-42=0,\ 3x+4y+18=0$. 13. $x-3y-3=0,\ x-3y+17=0$.
14. $3x+4y-3=0,\ 4x+3y-18=0$. 15. $4x+3y-3=0,\ 4x+3y-23=0$.
16. $x+y-11=0,\ x+y+7=0$. 17. $x-y+1=0$.
19. (a) The set of all points in the plane to the left of the line $x-y=0$.
    (b) The set of all points in the plane to the left of the line $2x+3y=0$.
    (c) The set of all points in the plane to the right of the line $x+2y=0$.
    (d) The set of all points in the plane to the left of the line $x-y-4=0$.
    (e) The set of all points in the plane to the left of the line $2x+3y-2=0$.
    (f) The set of all points in the plane to the right of the line $x+2y-3=0$.
    (g) The set of all points in the plane to the left of and on the line $3x-4y+2=0$.
    (h) The set of all points in the plane to the left of and on the line $2x-3y+11=0$.
    (i) The set of all points in the plane to the right of the line $3x+4y-12=0$.

20. (a) The set of all points in the plane contained in the area to the left of the line $x+y=0$ and to the right of the line $x-y=0$.
    (b) The set of all points in the plane contained in the area to the right of the line $2x-y-4=0$ and to the right of the line $3x+4y-12=0$.
    (c) The set of all points in the plane contained in the area to the left of the line $x-2y+6=0$ and to the right of the line $2x+3y-6=0$.

Sec. 3–14

1. $-\frac{10}{3}$. 3. $\frac{4}{5}$. 5. $3x-4y+13=0$ and $4x+3y-16=0$.
7. (a) $\left(\frac{a+b}{3}, \frac{c}{3}\right)$; (c) $\left(\frac{a}{2}, \frac{b^2+c^2-ab}{2c}\right)$.

Sec. 3–15

1. $\frac{x-2y+6}{1} = \pm\frac{3x+y+9}{\sqrt{2}}$. 3. $2x+4y-15=0$.
5. $x-y-5=0,\ 2x+2y+1=0$. 6. $9x-27y+61=0$.
7. $18x-18y-17=0$. 8. $4x-4y+7=0;\ -\frac{4}{3}$.
9. $9x-3y+14=0$. 10. $\left(-\frac{4}{11}, 0\right)$.

Sec. 3–16

1. (a) $y=3x+k$; (c) $y+1=s(x-2)$; (e) $2x-y+c=0$; (g) $y=sx$; (i) $y=-x+k$.
2. (a) all non-vertical lines through $(0, -2)$; (c) all horizontal lines; (e) all vertical lines; (g) all lines 4 units from the origin which are either vertical or above the origin; (i) all lines with slope $-\frac{3}{2}$; (k) all non-horizontal lines through $(0, 3)$.
3. (a) $y=sx-1$; (b) $3x-4y+k=0$; (c) $2x-y=k$; (d) $2x-3y=k$; (e) $x+ky+5=0$; (f) $x\cos\alpha + y\sin\alpha = 4$; (g) $y-5=s(x+2)$; (h) $y=-5x+k$; (i) $12x+5y=k$.

458 APPENDIXES

4. (a) all horizontal lines; (b) all non-horizontal lines through (2, 0); (c) all lines with slope -1; (d) all vertical lines; (e) all lines with slope $\frac{2}{3}$; (f) all non-vertical lines through (0, 2); (g) all non-vertical lines through (0, -5); (h) all non-vertical lines through (-1, -3); (i) all vertical lines; (j) all horizontal lines; (k) all lines with slope 1; (l) all non-vertical lines through (1, 0).

Sec. 3–17

1. $x = 0$. 2. (b) $7x - 13y + 25 = 0$; (d) $22x + 7y + 2 = 0$.
3. (b) $24x - 5y - 110 = 0$; (d) $2x - 61y - 114 = 0$.
4. (b) $15x + 5y - 32 = 0$; (d) $5x - 6 = 0$.
5. (b) $5x + 3y + 110 = 0$; (d) $x = -16$. 6. (b) $x - 6y + 24 = 0$.
9. The two lines $5x - 12y - 40 = 0$ and $5x - 12y + 64 = 0$.
10. (a) $4x + y - 3 = 0$; (b) $x = 0$; (c) $3x + 7y - 21 = 0$; (d) $x - y + 3 = 0$; (e) $12x + 5y - 15 = 0$; (f) $2x - y + 3 = 0$; (g) $y = 3$; (h) $x = 0$.

Review Exercises

3. $x = 2 - 3t$ and $y = -3 + 7t$.*
4. (b) $x - 3y + 5 = 0$; (d) $3x - y - 2 = 0$. 6. (a) $7x + 3y - 2 = 0$.
7. (a) $x + 2y + 7 = 0$; (c) $y = 5$. 8. (a) $\left[\frac{3}{\sqrt{13}}, \frac{2}{\sqrt{13}}\right]$.
9. (a) $-\frac{3}{4}$.
10. (a) $y = \frac{3}{4}x + \frac{7}{4}$; $\frac{x}{-\frac{7}{3}} + \frac{y}{\frac{7}{4}} = 1$; $\frac{3}{5}x + \frac{4}{5}y - \frac{7}{5} = 0$.
11. (a) [3, 4]; (c) [1, -1]. 12. (a) $-\frac{3}{4}$; (c) 1.
14. (a) $x - 7y + 22 = 0$; (c) $7x + y - 56 = 0$; (e) $\left(\frac{19}{2}, \frac{9}{2}\right)$; (g) $3\sqrt{2}$; (i) $x - y - 2 = 0$.
15. (b) $4x + 10y = 5$; (d) $5x - 2y - 34 = 0$; (f) $5x - (\sqrt{29} + 2)y + 1 - 2\sqrt{29} = 0$; (h) $\frac{35}{2}$; (j) $\frac{5}{2}$.
16. $x + 2y - 10 = 0$ and $2x - y + 5 = 0$. 18. $2x - y + 1 = 0$.
20. (a) $t_1(x - 2y + 6) + t_2(3x - y + 2) = 0$; (c) $15x - 20y + 58 = 0$; (e) $5x - 5y + 14 = 0$.
24. -17 to 8.
26. $\sqrt{3}\,x - y - 4\sqrt{3} + 5 = 0$ and $\sqrt{3}\,x + y - 4\sqrt{3} - 5 = 0$.
28. $3x + 4y - 27 = 0$ and $3x + 4y + 3 = 0$.
29. (b) 11; (d) $\begin{matrix}2x - y = 11 \\ x + 2y = 3\end{matrix}$, $\begin{matrix}5x + 3y = 0 \\ 3x - 5y = 0\end{matrix}$, $\begin{matrix}x + 5y - 22 = 0 \\ 5x - y - 6 = 0\end{matrix}$;
(f) $\frac{-2x + y}{\sqrt{5}} = \frac{5x + 3y - 22}{\sqrt{34}}$, $\frac{2x - y}{\sqrt{5}} = \frac{x + 5y}{\sqrt{26}}$, and $\frac{5x + 3y - 22}{\sqrt{17}} = -\frac{x + 5y}{\sqrt{13}}$.
31. $7x + 56y + 72 = 0$ and $8x - y - 57 = 0$.
33. $8x + y - 12 = 0$ and $2x + y - 6 = 0$.
35. (a) $x + by + 2 = 0$; (c) $y = sx$; (e) $y = sx + 5$; (g) $x - 2y + c = 0$.
36. $x + 2y - 4 = 0$.

* Answers not unique.

Chapter 4

Sec. 4–1

1. (a) $(x-2)^2+(y+1)^2=25$; (c) $(x-2)^2+(y-4)^2=1$; (e) $(x-3)^2+(y+2)^2=36$.
2. (a) $(x+1)^2+(y-3)^2=58$; (c) $(x-3)^2+y^2=25$.
3. $\left(x-\frac{1}{2}\right)^2+(y-4)^2=\frac{13}{4}$.
4. (b) $(x-4)^2+(y-4)^2=16$ or $(x-4)^2+(y+4)^2=16$ or $(x+4)^2+(y+4)^2=16$ or $(x+4)^2+(y-4)^2=16$.
5. (a) $(x-6)^2+(y-4)^2=36$; (c) $(x+1)^2+(y-3)^2=\frac{9}{25}$.
6. (b) $(-2,-1)$ and 5; (d) $(0,-3)$ and 3; (f) $(0,0)$ and $\sqrt{7}$.
7. (a) $(x\pm 5)^2+(y\pm 5)^2=25$ (all possible combinations of signs);
    (b) $(x-3)^2+(y+1)^2=1$; (c) $(x+4)^2+(y+2)^2=16$; (d) $(x+1)^2+(y+4)^2=\frac{484}{25}$.
8. $(x+1)^2+(y+2)^2=8$.   9. $(x-2)^2+(y+3)^2=4$.
10. $(x-5)^2+(y-5)^2=25$, $(x-1)^2+(y-1)^2=1$.

Sec. 4–2

1. $(1,-2)$ and 2.   3. $\left(\frac{3}{4},-1\right)$ and $\frac{\sqrt{33}}{4}$.   5. $(0,0)$ and $\frac{\sqrt{30}}{3}$.
7. $(2,-3)$ and 4.   9. $\left(1,-\frac{3}{2}\right)$ and $\frac{\sqrt{13}}{2}$.
11. $(1,-2)$ and 0 (point-circle).   13. Imaginary circle.
15. $2x+3y-31=0$.   16. $(x+4)^2+(y+4)^2=16$, $(x-1)^2+(y+1)^2=1$.
17. $(x-10)^2+(y-10)^2=100$, $(x-2)^2+(y-2)^2=4$.
18. (1) $y = -2 + \sqrt{4-(x-1)^2}$; half circle above the line $y = -2$.
    $y = -2 - \sqrt{4-(x-1)^2}$; half circle below the line $y = -2$.

    (3) $y = \dfrac{-4+\sqrt{33-(4x-3)^2}}{4}$; half circle above the line $y = -1$.

    $y = \dfrac{-4-\sqrt{33-(4x-3)^2}}{4}$; half circle below the line $y = -1$.

    (5) $y = \dfrac{\sqrt{30-3x^2}}{3}$; half circle above the line $y = 0$.

    $y = \dfrac{-\sqrt{30-3x^2}}{3}$; half circle below the line $y = 0$.

    (7) $y = -3 + \sqrt{4-(x-2)^2}$; half circle above the line $y = -3$.
    $y = -3 - \sqrt{4-(x-2)^2}$; half circle below the line $y = -3$.

19. (1) $x = 1 + \sqrt{4-(y+2)^2}$; half circle right of the line $x = 1$.
    $x = 1 - \sqrt{4-(y+2)^2}$; half circle left of the line $x = 1$.

    (3) $x = \dfrac{3+\sqrt{33-16(y+1)^2}}{4}$; half circle right of the line $x = \dfrac{3}{4}$.

    $x = \dfrac{3-\sqrt{33-16(y+1)^2}}{4}$; half circle left of the line $x = \dfrac{3}{4}$.

    (5) $x = \dfrac{\sqrt{30-3y^2}}{3}$; half circle right of the line $x = 0$.

    $x = \dfrac{-\sqrt{30-3y^2}}{3}$; half circle left of the line $x = 0$.

    (7) $x = 2 + \sqrt{4-(y+3)^2}$; half circle right of the line $x = 2$.
    $x = 2 - \sqrt{4-(y+3)^2}$; half circle left of the line $x = 2$.

460 APPENDIXES

Sec. 4–3

1. (a) $x^2+y^2-12x+6y+20=0$; (c) $5x^2+5y^2-11x-11y-10=0$;
   (e) $x^2+y^2-2x+4y=0$; (g) $27x^2+27y^2+35x-7y-442=0$.
2. $7x^2+7y^2-29x-37y=0$.
4. (a) $x^2+y^2+5y-15=0$; (c) $(x+2)^2+(y-2)^2=4$;
   (e) $(x-4)^2+(y+11)^2=25$ or $(x+2)^2+(y+3)^2=25$; (g) $(x-2)^2+(y+3)^2=\dfrac{961}{25}$.
5. $(x-6)^2+y^2=5$. 6. $x^2+(y+3)^2=17$.
7. $(x-1)^2+(y-3)^2=2$, $(x-3)^2+(y-5)^2=2$.
8. $(x+5)^2+(y-10)^2=90$, $(x-1)^2+(y+8)^2=90$.
9. $(x-3)^2+(y-5)^2=25$.
10. $(x-2)^2+(y-2)^2=8$, $(x-2)^2+(y+14)^2=200$.
11. $(x+1)^2+(y+4)^2=52$.
12. $(x-5)^2+(y-1)^2=45$, $(x+7)^2+(y-7)^2=45$.
13. $x^2+y^2+2x-2y+1=0$.

Sec. 4–6

1. $\sqrt{13}$. 3. $\sqrt{5}$. 5. Point lies inside.

Sec. 4–7

1. (a) $3x-5y+1=0$; (c) $3x-29y-7=0$.
2. (a) $(3,-4)$ and $\left(\dfrac{63}{13},-\dfrac{16}{13}\right)$; (b) $(-3,5)$ and $\left(\dfrac{24}{17},\dfrac{40}{17}\right)$;
   (c) $(-4,-2)$ and $(2,6)$; (d) $(-9,8)$ and $(5,-6)$;
   (e) $\left(\dfrac{-5-2\sqrt{39}}{2},\dfrac{2+\sqrt{39}}{2}\right)$; (f) $(-3,2)$ and $\left(-\dfrac{29}{5},-\dfrac{32}{5}\right)$.
3. (a) $(-2,-1)$;* (c) $\left(\dfrac{7}{3},0\right)^*$. 4. (a) $\left(-\dfrac{22}{3},\dfrac{88}{9}\right)$.

Sec. 4–8

1. (a) $3x+\sqrt{7}\,y=16$ and $\sqrt{7}\,x-3y=0$; (c) $y=-5$ and $x=0$;
   (e) $4x+5y=-41$ and $5x-4y=0$.
3. (a) $y=1$ and $x=3$; (c) $x+3y+9=0$ and $3x-y+7=0$;
   (e) $x-y-4=0$ and $x+y-2=0$.
5. (a) $x-2y-3=0$ and $2x+y-1=0$; (c) $2x-5y-19=0$ and $5x+2y-4=0$;
   (e) $3x-4y+18=0$ and $4x+3y-1=0$.

Sec. 4–9

1. (a) $3x-4y\pm25=0$; (c) $y=\sqrt{15}\,x\pm8$; (e) $y=-2x\pm5$.
2. (a) $3x-4y\pm25=0$; (c) $8x-15y\pm68=0$; (e) $y=+8$ and $y=-8$.
3. (a) $3x+4y\pm40=0$; (c) $12x-5y\pm195=0$; (e) $x=9$ and $x=-9$.
5. (a) $x-2y\pm15=0$; (b) $3x+7y\pm58=0$; (c) $5x+2y\pm58=0$.

* Answers not unique.

Sec. 4–10

1. (a) $(x-2)^2+(y+3)^2=r^2$; (c) $(x-2)^2+(y-k)^2=16$; (e) $(x-h)^2+(y+4)^2=9$.
2. (a) $15x^2+15y^2+4x-6y-265=0$; (c) $25x^2+25y^2-64x+96y=0$; (e) $x^2+y^2=16$; (g) $x^2+y^2+4x-6y-41=0$.
3. (b) $49x^2+49y^2+114x-224y-204=0$; (d) $2x^2+2y^2-y-6=0$; (f) $x^2+y^2-10x+17y+2=0$; (h) $7x^2+7y^2-26x+42y-8=0$.

Review Exercises

1. (a) $(x+1)^2+(y-2)^2=25$; (c) $(x+3)^2+(y+3)^2=9$; (e) $x^2+\left(y+\dfrac{5}{2}\right)^2=\dfrac{145}{4}$; (g) $(x-2)^2+(y-4)^2=25$ or $(x-2)^2+(y+4)^2=25$; (i) $2x^2+2y^2-7x+3=0$; (k) $(x-7)^2+(y+5)^2=5$.
2. (b) $x^2+y^2+5x-3y+4=0$.
4. $\left(\dfrac{5}{4}, -\dfrac{1}{3}\right)$; point lies inside circles.
6. (x_1, y_1) lies inside circle; outside circle; on the circle.
7. (b) $23x^2+23y^2+x-83y-130=0$.
9. (a) $x-2y+5=0$ and $2x+y=0$; (c) $x+3y-4=0$ and $3x-y-2=0$ or $2x-6y-23=0$ and $6x+2y+1=0$.
11. $(1, 1)$ and $\left(\dfrac{1}{13}, \dfrac{21}{13}\right)$.
13. (a) $x^2+y^2-80x-74y+72=0$; (c) $x^2+y^2-6x-2=0$.
14. $(x+7)^2+y^2=9$.
16. (a) $x=4\cos\theta$, $y=4\sin\theta$; (c) $x=7\cos\theta$, $y=7\sin\theta$;* (e) $x+3=2\cos\theta$, $y-2=2\sin\theta$.*
18. $x^2+y^2+4x+6y-108=0$. 19. $x^2+y^2-2x+4y-35=0$.
20. (a) $x-2y+12=0$, $2x+y-21=0$; (b) $3x+7y-30=0$, $7x-3y+46=0$; (c) $5x+2y-37=0$, $2x-5y-38=0$.

Supplementary Exercises

6. (a) $F=\dfrac{17}{2}$; (c) $E=\dfrac{11}{4}$. 7. (b) $D=\dfrac{2}{5}$ and $F=\dfrac{19}{10}$.
8. $x^2+y^2-32x+43y+25=0$.
9. (b) $3x-5y-9=0$; (d) $x-4y+4=0$; (f) $y=2$.

CHAPTER 5

Sec. 5–2

1. $y^2+8x-4y-4=0$. 3. $3x^2-y^2-32x+64=0$.
5. $8x^2+9y^2-44x+56=0$. 7. $4x^2-5y^2-76y-260=0$.
9. $48x^2-12xy+43y^2+544x+140y+1316=0$. 11. $x^2+6y-9=0$.

---

* Answers not unique.

Sec. 5-5

1. $y^2 = -4ax$. 3. $x^2 = -4ay$. 4. (b) $x^2 = -12y$; (d) $y^2 = -16x$; (f) $y^2 = 24x$.
5. (b) $y^2 = -12x$; (d) $x^2 = 20y$. 6. (b) $y^2 = -12x$; (d) $x^2 = -16y$.
7. (b) $y^2 = -25x$. 8. (b) $x^2 = -16y$ and $4y^2 = -x$.

Sec. 5-6

4. (a) $x^2 - 6x - 8y + 25 = 0$; (c) $9x^2 + 12xy + 4y^2 - 24x + 36y - 36 = 0$;
   (e) $x^2 + 4x + 12y - 44 = 0$; (g) $y^2 + 24x + 8y - 176 = 0$.
5. $x^2 + y^2 - 5ay = 0$. 7. Area $= 16a^2$; perimeter $= 8a + 8a\sqrt{2}$.

Sec. 5-9

2. (a) $y^2 - 2y - 8x + 17 = 0$.
3. (a) $x^2 - 2x - 4y + 5 = 0$; (c) $x^2 - 4xy + 4y^2 + 14x + 32y + 49 = 0$;
   (d) $4x^2 + 4xy + y^2 - 30x + 20y + 65 = 0$.
4. (a) $x^2 = 16y$, $x^2 = -16y$.
5. (a) $(a^2 + b^2)(x^2 + y^2) = (ax + by + c)^2$; (b) $(y-k)^2 = 4a(x-h)$, $(y-k)^2 = -4a(x-h)$.
6. (b) $x^2 = 48y$. 11. $(0, 0)$, $(4a, 4a)$.
13. $y^2 + 16x - 8y = 0$. 14. $x^2 + 24y = 0$. 15. $x^2 = 4ay$.
16. The part of $x^2 + 10y - 25 = 0$ for which $y \leq 2$, together with the part of $x^2 - 2y - 1 = 0$ for which $y \geq 2$.
17. (d) $y = \dfrac{\sqrt{74x}}{2}$; vertex at $(0,0)$; upper half of the parabola; Domain $= [0, \infty)$; Range $= [0, \infty)$. $y = \dfrac{-\sqrt{74x}}{2}$; lower half of the parabola; Domain $= [0, \infty)$; Range $= (-\infty, 0]$.
19. $40{,}000\sqrt{3}$ ft.; $10{,}000$ ft.
21. $\dfrac{9}{2}$ ft.
23. $\dfrac{27}{8}$ in.

Sec. 5-10

1. $\dfrac{x^2}{a^2} + \dfrac{y^2}{b^2} = 1$. 3. $\dfrac{y^2}{a^2} + \dfrac{x^2}{b^2} = 1$.
5. (a) $\dfrac{y^2}{8} + \dfrac{x^2}{4} = 1$; (c) $\dfrac{x^2}{35} + \dfrac{y^2}{10} = 1$; (e) $\dfrac{y^2}{25} + \dfrac{x^2}{16} = 1$.

Sec. 5-11

1. $\dfrac{x^2}{100} + \dfrac{y^2}{36} = 1$. 3. $\dfrac{x^2}{81} + \dfrac{y^2}{45} = 1$. 5. $\dfrac{y^2}{25} + \dfrac{x^2}{16} = 1$.
7. $\dfrac{3x^2}{100} + \dfrac{y^2}{25} = 1$. 9. $\dfrac{y^2}{64} + \dfrac{x^2}{48} = 1$.

APPENDIX B   463

Sec. 5–12
4. $9x^2 + 25y^2 = 225$.   5. $7x^2 + 16y^2 = 112$.
6. $11y^2 + 36x^2 = 396$.   7. $3y^2 + 4x^2 = 12$.

Sec. 5–13
1. $a = 3$, $b = 2$, $c = \sqrt{5}$, $e = \dfrac{\sqrt{5}}{3}$, $\dfrac{a}{e} = \dfrac{9}{\sqrt{5}}$, $(\pm\sqrt{5}, 0)$, $(\pm 3, 0)$, $(0, \pm 2)$, $x = \pm \dfrac{9}{\sqrt{5}}$.

3. $a = \sqrt{15}$, $b = \sqrt{5}$, $c = \sqrt{10}$, $e = \sqrt{\dfrac{2}{3}}$, $\dfrac{a}{e} = \dfrac{3\sqrt{5}}{\sqrt{2}}$, $(0, \pm\sqrt{10})$, $(0, \pm\sqrt{15})$, $(\pm\sqrt{5}, 0)$, $y = \pm \dfrac{3\sqrt{5}}{\sqrt{2}}$.

5. $a = 6$, $b = 4$, $c = \sqrt{20}$, $e = \dfrac{\sqrt{5}}{3}$, $\dfrac{a}{e} = \dfrac{18}{\sqrt{5}}$, $(0, \pm\sqrt{20})$, $(0, \pm 6)$, $(\pm 4, 0)$, $y = \pm \dfrac{18}{\sqrt{5}}$

7. $a = 3$, $b = \sqrt{2}$, $c = \sqrt{7}$, $e = \dfrac{\sqrt{7}}{3}$, $\dfrac{a}{e} = \dfrac{9}{\sqrt{7}}$, $(0, \pm\sqrt{7})$, $(0, \pm 3)$, $(\pm\sqrt{2}, 0)$, $y = \pm \dfrac{9}{\sqrt{7}}$.

9. (1) $y = \dfrac{2}{3}\sqrt{9 - x^2}$; upper half of the ellipse; Domain = $[-3, 3]$; Range = $[0, 2]$.

$y = -\dfrac{2}{3}\sqrt{9 - x^2}$; lower half of the ellipse; Domain = $[-3, 3]$; Range = $[-2, 0]$.

(3) $y = \sqrt{15 - 3x^2}$; upper half of the ellipse; Domain = $[-\sqrt{5}, \sqrt{5}]$; Range = $[0, \sqrt{15}]$.
$y = -\sqrt{15 - 3x^2}$; lower half of the ellipse; Domain = $[-\sqrt{5}, \sqrt{5}]$; Range = $[-\sqrt{15}, 0]$.

(5) $y = \dfrac{3}{2}\sqrt{16 - x^2}$; upper half of the ellipse; Domain = $[-4, 4]$; Range = $[0, 6]$.

$y = -\dfrac{3}{2}\sqrt{16 - x^2}$; lower half of the ellipse; Domain = $[-4, 4]$; Range = $[-6, 0]$.

(7) $y = \dfrac{3}{2}\sqrt{4 - 2x^2}$; upper half of the ellipse; Domain = $[-\sqrt{2}, \sqrt{2}]$; Range = $[0, 3]$.

$y = -\dfrac{3}{2}\sqrt{4 - 2x^2}$; lower half of the ellipse; Domain = $[-\sqrt{2}, \sqrt{2}]$; Range = $[-3, 0]$.

Sec. 5–14
2. (a) $\dfrac{y^2}{36} + \dfrac{x^2}{16} = 1$; (c) $\dfrac{y^2}{36} + \dfrac{x^2}{9} = 1$; (e) $\dfrac{x^2}{81} + \dfrac{y^2}{25} = 1$; (g) $\dfrac{x^2}{25} + \dfrac{y^2}{9} = 1$.

3. (a) $\dfrac{x^2}{256} + \dfrac{y^2}{112} = 1$; (c) $\dfrac{y^2}{144} + \dfrac{x^2}{80} = 1$; (e) $\dfrac{4y^2}{81} + \dfrac{x^2}{9} = 1$; (g) $\dfrac{x^2}{64} + \dfrac{y^2}{9} = 1$; (i) $\dfrac{y^2}{64} + \dfrac{x^2}{28} = 1$;
(k) $\dfrac{9x^2}{625} + \dfrac{y^2}{25} = 1$; (m) $\dfrac{y^2}{100} + \dfrac{x^2}{36} = 1$.

4. (a) $\dfrac{5x^2}{108} + \dfrac{7y^2}{432} = 1$; (c) $\dfrac{x^2}{33} + \dfrac{2y^2}{33} = 1$.   5. $e = \dfrac{1}{2}$.

10. $3x^2 + 4y^2 = k$, $k > 0$.   11. $a = \sqrt{2}\, b$, $b = c$.

13. If center of base is at origin and end points on x-axis, then $k^2 x^2 + y^2 = a^2 k^2$, where k^2 is the positive constant.
14. Approximately 94,553 miles and 91,447 miles.
15. 0.762.

Sec. 5–16

1. (a) $\dfrac{x^2}{9}+\dfrac{y^2}{1}=1$; (c) $\dfrac{x^2}{64}+\dfrac{y^2}{4}=1$. 2. $x=8\cos\theta$ and $y=5\sin\theta$.
3. $x=12\cos\theta$, $y=6\sin\theta$.

Sec. 5–18

2. $\dfrac{x^2}{a^2}-\dfrac{y^2}{b^2}=1$. 4. $\dfrac{y^2}{a^2}-\dfrac{x^2}{b^2}=1$. 6. (a) $\dfrac{y^2}{2}-\dfrac{x^2}{2}=1$; (c) $\dfrac{y^2}{24}-\dfrac{x^2}{12}=1$.

Sec. 5–20

1. (a) $a=2$, $b=2$, $c=2\sqrt{2}$, $e=\sqrt{2}$, $\dfrac{a}{e}=\sqrt{2}$, $(\pm 2,0)$, $(\pm 2\sqrt{2},0)$, $x=\pm\sqrt{2}$, $x\pm y=0$;
(c) $a=2$, $b=4$, $c=2\sqrt{5}$, $e=\sqrt{5}$, $\dfrac{a}{e}=\dfrac{2}{\sqrt{5}}$, $(0,\pm 2)$, $(0,\pm 2\sqrt{5})$, $y=\pm\dfrac{2}{\sqrt{5}}$, $x\pm 2y=0$;
(e) $a=4$, $b=3$, $c=5$, $e=\dfrac{5}{4}$, $\dfrac{a}{e}=\dfrac{16}{5}$, $(0,\pm 4)$, $(0,\pm 5)$, $y=\pm\dfrac{16}{5}$, $4x\pm 3y=0$.
(g) $a=3\sqrt{3}$, $b=3$, $c=6$, $e=\dfrac{2}{\sqrt{3}}$, $\dfrac{a}{e}=\dfrac{9}{2}$, $(\pm 3\sqrt{3},0)$, $(\pm 6,0)$, $x=\pm\dfrac{9}{2}$, $\sqrt{3}\,y\pm x=0$.

2. (a) $\dfrac{x^2}{100}-\dfrac{y^2}{36}=1$; (c) $\dfrac{9y^2}{64}-\dfrac{9x^2}{80}=1$; (e) $\dfrac{x^2}{16}-\dfrac{y^2}{9}=1$; (g) $\dfrac{3y^2}{16}-\dfrac{x^2}{16}=1$; (i) $\dfrac{y^2}{16}-\dfrac{x^2}{20}=1$;
(k) $\dfrac{y^2}{25}-\dfrac{16x^2}{225}=1$; (m) $\dfrac{9y^2}{4}-\dfrac{x^2}{4}=1$.

7. $3x^2-y^2=3$. 8. $\dfrac{\sqrt{13}}{3}$. 9. Hyperbola. 10. $5y^2-4x^2=80$.
11. $y^2-11x^2=5$. 12. The same foci.
13. The same asymptotes. If $k<0$, y-axis is transverse; if $k>0$, x-axis is transverse; if $k=0$, the equation represents a pair of lines.
15. Same as the graph of the two hyperbolas $x^2-y^2=1$ and $x^2-y^2=-1$.

Sec. 5–21

4. $7x^2-9y^2=63$. 5. $8y^2-x^2=32$. 6. $5y^2-4x^2=80$. 7. $\dfrac{x^2}{25}-\dfrac{y^2}{24}=1$.

Sec. 5–25

2. (a) $\dfrac{y^2}{64}-\dfrac{x^2}{9}=1$; (c) $\dfrac{x^2}{16}-\dfrac{y^2}{9}=1$; (e) $\dfrac{x^2}{16}-\dfrac{y^2}{9}=1$; (g) $\dfrac{y^2}{9}-\dfrac{x^2}{27}=1$; (i) $\dfrac{y^2}{81}-\dfrac{x^2}{63}=1$;
(k) $\dfrac{y^2}{36}-\dfrac{x^2}{48}=1$; (m) $\dfrac{y^2}{36}-\dfrac{x^2}{9}=1$; (o) $\dfrac{y^2}{16}-\dfrac{x^2}{20}=1$; (q) $\dfrac{x^2}{36}-\dfrac{y^2}{540}=1$; (s) $x^2-y^2=4$;
(u) $\dfrac{x^2}{36}-\dfrac{y^2}{64}=1$; (w) $\dfrac{3x^2}{20}-\dfrac{7y^2}{20}=1$; (y) $\dfrac{x^2}{5}-\dfrac{4y^2}{5}=1$.

3. $5y^2-4x^2=20$. 4. $16x^2-9y^2=144$.
5. $3y^2-x^2=12$. 6. $5y^2-9x^2=36$.
7. $9x^2-25y^2=k$. 8. $y^2-x^2=k^2$.
9. $x\pm 2\sqrt{2}\,y=0$; $2\sqrt{2}\,x\pm y=0$.

Sec. 5–28

2. (a) $\dfrac{x^2}{9}-\dfrac{y^2}{1}=1$; (c) $\dfrac{x^2}{1}-\dfrac{y^2}{4}=1$; (e) $\dfrac{y^2}{1}-\dfrac{x^2}{25}=1$.

Review Exercises

2. (a) $\dfrac{y^2}{36} - \dfrac{x^2}{64} = 1$; (c) $x^2 = -16y$; (e) $x^2 = 12y$; (g) $\dfrac{y^2}{5/2} - \dfrac{x^2}{5/3} = 1$; (i) $\dfrac{x^2}{5} - \dfrac{y^2}{5} = 1$.

6. $20\sqrt{3}$ feet. 9. $\dfrac{1}{2}$. 11. $x^2 = 4ky$. 14. $\dfrac{17x^2}{50} - \dfrac{17y^2}{18} = 1$. 17. $\dfrac{2\sqrt{e^2-1}}{2-e^2}$. 29. $b = c$.

Chapter 6

Sec. 6–2

1. (a) $(4, -5)$; (c) $(3, 4)$; (e) $(-2, 4)$; (g) $(5, -5)$.
2. (a) $(5, -1)$; (c) $(4, 1)$; (e) $(x'+h, y'+k)$; (g) $(-2, -4)$.
3. (a) $x'^2 + y'^2 - 4 = 0$; (c) $x'^2 - 3y' = 0$; (e) $x'^2 - 4y'^2 - 16 = 0$; (g) $x'^2 - 9y'^2 + 9 = 0$.

Sec. 6–3

1. $Ax'^2 + Ey' = 0$. 3. $9x'^2 + 4y'^2 - 144 = 0$. 5. $y'^2 + 4x' = 0$.
7. $x'^2 + 4y'^2 - 16 = 0$. 9. $9x'^2 - 16y'^2 - 11 = 0$.
11. $2x'^2 - 3y'^2 - 48 = 0$. 13. $25x'^2 + 9y'^2 - 225 = 0$.
15. $4x'^2 - y'^2 + 1 = 0$.
17. (4) $y = \dfrac{15 + 3\sqrt{25 - (x+2)^2}}{5}$; half of the ellipse above the line $y = 3$;

    Domain $= [-7, 3]$; Range $= [3, 6]$.

    $y = \dfrac{15 - 3\sqrt{25 - (x+2)^2}}{5}$; half of the ellipse below the line $y = 3$;
    Domain $= [-7, 3]$; Range $= [0, 3]$.

    (10) $\dfrac{(x-1)^2}{8} - \dfrac{(y-3)^2}{18} = -1$; $y = \dfrac{6 + 3\sqrt{8 + (x-1)^2}}{2}$; half of the hyperbola above the line $y = 3$; Domain $= (-\infty, \infty)$; Range $= [3 + 3\sqrt{2}, \infty)$.

    $y = \dfrac{6 - 3\sqrt{8 + (x-1)^2}}{2}$; half of the hyperbola below the line $y = 3$;
    Domain $= (-\infty, \infty)$; Range $= (-\infty, 3 - 3\sqrt{2}]$.

Sec. 6–4

2. (a) $\dfrac{(x-1)^2}{16} + \dfrac{(y-4)^2}{12} = 1$; (c) $(y-2)^2 = 6\left(x - \dfrac{3}{2}\right)$.
   (e) $\dfrac{(y+2)^2}{4} + \dfrac{(x-5)^2}{3} = 1$; (g) $(y-3)^2 = 16(x-4)$ or $(y-3)^2 = -16(x-4)$;
   (i) $(x-8)^2 = -36(y+3)$; (k) $\dfrac{(y-3)^2}{36} - \dfrac{(x-5)^2}{16} = 1$; (m) $\dfrac{(y-4)^2}{64} + \dfrac{(x-2)^2}{28} = 1$;
   (o) $(x-3)^2 = 20(y+1)$; (q) $(x-5)^2 = 8(y+1)$; (s) $\dfrac{(y-2)^2}{16} - \dfrac{(x+2)^2}{20} = 1$;
   (u) $\dfrac{(y-4)^2}{25} - \dfrac{(x-2)^2}{75} = 1$.

Sec. 6–5

1. (a) $\left(\dfrac{3\sqrt{3}-1}{2}, \dfrac{-3-\sqrt{3}}{2}\right)$; (c) $\left(\dfrac{\sqrt{2}}{2}, \dfrac{3\sqrt{2}}{2}\right)$; (e) $\left(\dfrac{-2-5\sqrt{3}}{2}, \dfrac{2\sqrt{3}-5}{2}\right)$.
2. (a) $2\sqrt{2}\, y'^2 - 3x' + 3y' = 0$; (c) $9x'^2 + 4y'^2 - 36 = 0$; (e) $y'^2 - 1 = 0$; (g) $14x'^2 + y'^2 - 5 = 0$.

Sec. 6–6

1. $(1-\sqrt{3})x'^2+(1+\sqrt{3})y'^2+8=0$. **3.** $50x''^2+25y''^2-7=0$.
5. $9x''^2-16y''^2-144=0$. **7.** $y''^2-4x''^2+64=0$. **9.** $x''^2-4\sqrt{2}\,y''=0$.
11. (a) $x'^2-y'^2=4$, hyperbola; (b) $4(x'^2-y'^2)+1=0$, hyperbola;
(c) $x'^2+4y'^2=4$, ellipse; (d) $7x'^2+y'^2=14$, ellipse;
(e) $2x'^2-y'^2=2$, hyperbola; (f) $4y'^2-x'^2+16=0$, hyperbola;
(g) $x'^2-y'^2=a^2$, hyperbola; (h) $2x'^2+y'^2=4$, ellipse;
(i) $x'^2-4y'^2=8$, hyperbola; (j) $2x'^2+y'^2=2$, ellipse;
(k) $11x'^2+y'^2=44$, ellipse; (l) $26x'^2+y'^2=144$, ellipse;
(m) $4y'^2-3\sqrt{2}x'+3\sqrt{2}y'=0$, parabola; (n) $y'^2=1$ (two parallel lines);
(o) $2x'^2-3y'^2=4$, hyperbola; (p) $14x'^2+y'^2=5$, ellipse;
(q) $4x'^2+9y'^2=36$, ellipse; (r) $x'^2+9y'^2=9$, ellipse;
(s) $9x'^2-4y'^2=4$, hyperbola; (t) $5\sqrt{5}y'^2-6x'-12y'=0$, parabola.

Sec. 6–9

1. Hyperbola, degenerate. **3.** Circle, degenerate. **5.** Hyperbola, non-degenerate.
7. Ellipse, degenerate. **9.** Hyperbola.

Sec. 6–11

1. (a) $x^2-20xy+16y^2-3x+30y-4=0$; (c) $3x^2-10xy-7y^2-9x+7y=0$;
(d) $61x^2-166xy-114y^2-257x+508y+196=0$
2. (a) $x^2+2xy+y^2-3x-3y=0$ (degenerate);
(c) $\begin{cases} x^2+4xy+4y^2-3x-6y+2=0 \\ x^2+2xy+y^2+y-x-2=0; \end{cases}$ (d) $\begin{cases} x^2-2xy+y^2-x-y=0 \\ x^2-10xy+25y^2+23x-73y=0. \end{cases}$
3. (a) $x^2+10xy-7y^2+28=0$; (c) $13x^2+17xy+10y^2-40=0$;
(d) $8x^2+16xy+5y^2-5=0$.
4. (a) $3x^2+23y^2-6x+22y-45=0$; (c) $4x^2+y^2-4x-4y-5=0$; (d) $y^2-x-2y=0$.
5. (a) $y^2+2x-3y-2=0$; (c) $7y^2+6x-19y+6=0$; (d) $3y^2-2x-5y-6=0$.
6. (a) $x^2-11x-6y=0$; (c) $3x^2+13x+10y-6=0$; (d) $7x^2-33x-6y+38=0$.

Sec. 6–12

1. (a) $(3,0)$; (b) $e=1$; (c) $y=-1$; (d) none; (e) $y=\dfrac{(x-3)^2}{4}$;

(f) The locus of a point (x,y) moving in the plane so that its distance from the fixed point (focus), $(3,1)$, is always equal to its distance from the line $y=-1$ (directrix).

3. (a) $(-3,1)$; (b) $a=3$, $b=2$, $c=\sqrt{5}$, $e=\dfrac{\sqrt{5}}{3}$; $\dfrac{a}{e}=\dfrac{9\sqrt{5}}{5}$; (c) $x=-3\pm\dfrac{9\sqrt{5}}{5}$;

(d) none; (e) $y=\dfrac{3+2\sqrt{9-(x+3)^2}}{3}$ and $y=\dfrac{3-2\sqrt{9-(x+3)^2}}{3}$; (f) The locus of a point moving in the plane so that its distance from $(-3+\sqrt{5},1)$ is always equal to $\dfrac{\sqrt{5}}{3}$ times its distance from the line $x=-3+\dfrac{9\sqrt{5}}{5}$. A similar statement could be made using the focus $(-3-\sqrt{5},1)$ and the directrix $x=-3-\dfrac{9\sqrt{5}}{5}$;

(g) The locus of a point (x,y) such that the sum of its distances from the points $(-3+\sqrt{5},1)$ and $(-3-\sqrt{5},1)$ is always equal to 6.

Chapter 7

Sec. 7-2
2. (a) $(-5, 240°)$, $(-5, -120°)$, or $(5, -300°)$; (c) $(-6, 0°)$ or $(6, -180°)$;
(e) $\left(-2, \frac{5\pi}{4}\right), \left(-2, -\frac{3\pi}{4}\right)$, or $\left(2, -\frac{7\pi}{4}\right)$; (g) $\left(-3, \frac{\pi}{2}\right), \left(3, -\frac{\pi}{2}\right)$, or $\left(-3, -\frac{3\pi}{2}\right)$;
(i) $(6, 60°)$, $(-6, 240°)$, or $(6, -300°)$; (k) $(-1, -90°)$, $(-1, 270°)$, or $(1, -270°)$;
(m) $\left(4, \frac{4\pi}{3}\right), \left(-4, \frac{\pi}{3}\right)$, or $\left(-4, -\frac{5\pi}{3}\right)$; (o) $(0, \theta)$, for any θ.

Sec. 7-3
1. (a) $\rho = 4$ or $\rho = -4$; (c) $\rho^2 \cos 2\theta = -4$; (e) $\rho^2 \cos^2 \theta + \rho^2 = 16$;
(g) $\rho^2 \cos^2 \theta - 32\rho \sin \theta - 256 = 0$.
2. (a) $y = 4$; (c) $x^2 + y^2 = 16$; (e) $x^2+y^2=x$; (g) $3x^2-y^2+12x+9=0$.

Sec. 7-4
1. $y = -4$. 3. $\sqrt{3}\,x + y - 8 = 0$. 5. $y = -1$.
7. $y = -3$. 8. $x + y = 3\sqrt{2}$.
9. $x + \sqrt{3}y + 4 = 0$. 10. $x = -4$.
11. $3y - 5$. 12. $x - \sqrt{3}y + 10 = 0$.
13. $\sqrt{3}\,x - y + 6 = 0$. 14. $x - y - \sqrt{2} = 0$.
15. $y = -\sqrt{3}\,x$. 16. $y = 2x$.
17. $y = 0$. 18. $x = 0$.
19. $5y + 3x - 15 = 0$. 20. $4x - y - 4 = 0$.
21. $y + x = 0$. 22. $x - \sqrt{3}\,y = 0$.
23. $x - y - 2 = 0$. 24. $2x + y + 2 = 0$.
25. $2x - 1 = 0$. 26. $\sqrt{2}y + 1 = 0$.
27. $2x - 3y - 6 = 0$. 28. $6x - y + 4 = 0$.

Sec. 7-5
1. (a) $\rho = -2a \cos \theta$; (c) $\rho = r$. 2. (a) $\rho = -12 \cos \theta$; (c) $\rho = 8 \sin \theta$.
4. (a) $x^2 + y^2 + 3y = 0$; (c) $x^2 + y^2 - 2x + 3y = 0$; (e) $x^2 + y^2 - 2y = 0$.
5. (a) $\rho = 6 \cos \theta$; (c) $\rho - 3 \cos \theta + \sin \theta = 0$; (e) $\rho - 2 \cos \theta + 4 \sin \theta = 0$;
(g) $\rho^2 - 8\rho \cos \theta - 2\rho \sin \theta + 6 = 0$.
6. (a) $\left(\frac{3}{2}, -1\right), r = \frac{\sqrt{13}}{2}$; (b) $\left(-1, \frac{3}{2}\right), r = \frac{\sqrt{13}}{2}$;
(c) $(-1, 2), r = \sqrt{11}$; (d) $(2, 3), r = \sqrt{3}$;
(e) $(3, -1), r = \sqrt{14}$; (f) $(-1, 2), r = \sqrt{5}$;
(g) $(-2, 3), r = \sqrt{15}$.
10. $\rho^2 - 2k\rho \sin \theta + k^2 - r^2 = 0$.

Sec. 7-9
1. (a) $(x^2 + y^2 - 2x)^2 = x^2 + y^2$; (c) $(x^2 + y^2 + x)^2 = 9x^2 + 9y^2$;
(e) $x^4 + y^4 + 2x^2y^2 - x(x^2 - 3y^2) = 0$; (g) $(x^2 + y^2)^2 = 4x^2 - 4y^2$;

(i) $(x^2+y^2)^3 = 4x^4$; **(k)** $x^2+y^2+4y = 0$; **(m)** $x^2+y^2 = 9\left(\tan^{-1}\frac{y}{x}\right)^2$;
(o) $y^2+4x-4 = 0$; **(q)** $3x^2-y^2-4x+1 = 0$; **(s)** $4x^2+3y^2-8y-16 = 0$;
(u) $(x^2+y^2-2x)^2 = 4(x^2+y^2)$; **(w)** $(x^2+y^2)^3 = 4(x^2-y^2)^2$;
(y) $(x^2+y^2)^5 = 144x^2y^2(x^2-y^2)^2$; **(aa)** $(x^2+y^2)^2 = -16(x^2-y^2)$; **(ac)** $y = 2$;
(ae) $2y+x = 0$; **(ag)** $(x^2+y^2)\left(\tan^{-1}\frac{y}{x}\right)^2 = 16$; **(ai)** $9x^2+24y-16 = 0$;
(ak) $9x^2+8y^2-4y-4 = 0$.

Sec. 7–10

1. (b) $\rho = \dfrac{ep}{1+e\cos\theta}$.

2. (a) Parabola;
 (c) parabola;
 (e) hyperbola;
 (g) ellipse.

4. (a) Foci: $(0, 0°)$, $\left(\dfrac{3}{2}, 0°\right)$; Vertices: $\left(-\dfrac{3}{2}, 0°\right)$, $(3, 0°)$;

Directrices: $\rho\cos\theta = -6$, $\rho\cos\theta = \dfrac{15}{2}$; $8x^2+9y^2-12x-36 = 0$.

(b) Foci: $(0, 0°)$, $\left(\dfrac{10}{3}, 0°\right)$; Vertices: $(5, 0°)$, $\left(-\dfrac{5}{3}, 0\right)$;

Directrices: $\rho\cos\theta = \dfrac{25}{3}$, $\rho\cos\theta = -5$; $3x^2+4y^2-10x-25 = 0$.

(c) Foci: $(0, 0°)$, $\left(\dfrac{27}{4}, 90°\right)$; Vertices: $\left(\dfrac{9}{4}, 90°\right)$, $\left(\dfrac{9}{2}, 90°\right)$;

Directrices: $\rho\sin\theta = 3$, $\rho\sin\theta = \dfrac{15}{4}$; $8y^2-x^2-54y+81 = 0$.

(d) Foci: $(0, 0°)$, $\left(-\dfrac{64}{15}, 90°\right)$; Vertices: $\left(-\dfrac{8}{5}, 90°\right)$, $\left(-\dfrac{8}{3}, 90°\right)$;

Directrices: $\rho\sin\theta = -2$, $\rho\sin\theta = -\dfrac{34}{15}$; $15y^2-x^2+64y+64 = 0$.

(e) Foci: $(0, 0°)$, $(2, 0°)$; Vertices: $\left(\dfrac{1}{2}, 0°\right)$, $\left(\dfrac{3}{2}, 0°\right)$;

Directrices: $4\rho\cos\theta = 3$, $4\rho\cos\theta = 5$; $12x^2-4y^2-24x+9 = 0$.

(f) Foci: $(0, 0°)$, $(-3, 90°)$; Vertices: $\left(\dfrac{3}{2}, 90°\right)$, $\left(\dfrac{9}{2}, 270°\right)$;

Directrices: $\rho\sin\theta = -\dfrac{15}{2}$, $\rho\sin\theta = \dfrac{9}{2}$; $16x^2+12y^2+36y-81 = 0$.

Sec. 7–11

1. $\rho = a\sec\theta + b$, for P; $\rho = a\sec\theta - b$, for P'.
3. $\rho = a(1+\cos\theta)$.
5. $2\rho = 2a\sin\theta + a$.

Sec. 7–12

1. $\left(\frac{3\sqrt{2}}{2}, 45°\right)$ and the pole.

3. $(1, 0°)$, $\left(-\frac{1}{2}, 120°\right)$, $\left(-\frac{1}{2}, 240°\right)$, and the pole.

5. $(2, 0°)$ and $(0, 90°)$.

7. $(2, 15°)$, $(2, 75°)$, $(2, 105°)$, $(2, 165°)$, $(2, 195°)$, $(2, 255°)$, $(2, 285°)$, and $(2, 345°)$.

9. $(2, 2 \text{ radians})$, $\left(\frac{2}{1-k\pi}, 2-2k\pi\right)$, and $\left(\frac{4}{2-\{2k+1\}\pi}, 2-\{2k+1\}\pi\right)$, where k is any integer.

11.

k	θ	$\rho = \theta$	$\theta + \{2k+1\}\pi$	$\rho = \pi + \theta$
0	$-\pi$	$-\pi$	0	π
1	-2π	-2π	π	2π
2	-3π	-3π	2π	3π
.	.	.	.	.
.	.	.	.	.
.	.	.	.	.

The curves intersect on the polar axis and 180°-axis and at the pole.

Review Exercises

1. (a) $(2, 210°)$, $(2, -150°)$, or $(-2, -330°)$; (c) $(-3, 60°)$, $(3, 240°)$, or $(-3, -300°)$;
   (e) $\left(\frac{1}{2}, \pi\right)$ or $\left(-\frac{1}{2}, 0\right)$; (g) $\left(-1, \frac{\pi}{2}\right)$, $\left(1, -\frac{\pi}{2}\right)$, or $\left(-1, -\frac{3\pi}{2}\right)$;
   (i) $(1, -90°)$, $(1, 270°)$, or $(-1, -270°)$.

2. $\rho = -12 \sin \theta$.   4. $\rho = \frac{4e}{1 + e \cos \theta}$.

6. (a) $(x^2 + y^2 - 3x)^2 = 4x^2 + 4y^2$; (c) $(x^2 + y^2)^3 = (x^2 - y^2)^2$; (e) $(x^2 + y^2)^2 = -16(x^2 - y^2)$;
   (g) $y^2 + 4x - 4 = 0$; (i) $5x^2 - 4y^2 + 12x + 4 = 0$; (k) $x^2 - 3y^2 - 16y - 16 = 0$;
   (m) $(x^2 + y^2)^3 = x^2$; (o) $(x^2 + y^2)(2x^2 + 2y^2 - 1)^2 = x^2$.

7. (a) $\rho = \pm 4$; (c) $\rho^2 \sin 2\theta = -16$; (e) $\rho = 4 \tan \theta \sec \theta$; (g) $\rho \sin \theta = -4$; (i) $\rho^2 = 2 \sin 2\theta$.

8. (a) $(3, 60°)$ and $(3, 300°)$; (c) $(0, 0°)$, $\left(\frac{\sqrt{3}}{2}, 60°\right)$, and $\left(-\frac{\sqrt{3}}{2}, 300°\right)$;
   (e) $\left(\frac{1}{\sqrt{5}}, \text{Tan}^{-1}\frac{1}{2}\right)$ and the pole; (g) $\left(\frac{4}{5}, \text{Sin}^{-1}\frac{2}{5}\right)$, $\left(\frac{4}{5}, \pi - \text{Sin}^{-1}\frac{2}{5}\right)$, and the pole.

Chapter 8

Sec. 8–2

	Intercepts	Symmetry	Asymptotes
16.	$(0, 0)$, $(3, 0)$, $(-3, 0)$	origin	none
17.	$(0, 0)$, $(3, 0)$, $(-3, 0)$	origin	none
18.	$(0, 0)$	y-axis	none
19.	$(0, 0)$	y-axis	none

470 APPENDIXES

20.	(0, 0), (9, 0)	none	none
21.	(0, 0), (1, 0), (−1, 0)	origin	none
22.	(0, 0), (1, 0), (−1, 0)	origin	none
23.	(−2, 0)	none	$x=0; y=6$
24.	(0, 1)	none	$x=3; y=0$
25.	(3, 0)	none	$x=0; y=2$
26.	(1, 0)	x-axis	$x=0; y=\pm\sqrt{6}$
27.	(0, 3)	y-axis	$y=0$
28.	(0, −1)	y-axis	$x=+3; x-3; y=1$
29.	None	y-axis	$x=0; y=0$
30.	None	x-axis, y-axis	$x=0; y=0$
31.	(0, 2)	none	$x=-4; y=0$
32.	(0, 0)	none	$x=2; y=0$
33.	(0, 0), (2, 0), (−2, 0)	x-axis, y-axis	none
34.	(0, 0), (−2, 0), (2, 0)	x-axis, y-axis	none
35.	(0, 0)	x-axis	$x=-1, y=\pm 2$
36.	(0, 1)	none	$x=1, x=-2, x=2; y=0$

Sec. 8–5

7. (a) $x = \dfrac{(2k-1)\pi}{2}$, k an integer; (b) $x = k\pi$, k an integer;
(c) $x = \dfrac{(2k-1)\pi}{2}$, k an integer; (d) $x = k\pi$, k an integer.

Sec. 8–7

1. $\dfrac{2\pi}{3}$, 1, and none. 3. 2π, none, and none. 5. 6π, none, and none. 7. π, 3, and π.
9. π, none, and π. 11. 2π, 1, and $-\dfrac{\pi}{2}$. 13. $\dfrac{2\pi}{3}$, 2, and none. 15. $\dfrac{4\pi}{3}$, 1, and none.

Sec. 8–8

1. (a) $-\dfrac{\pi}{2}$; (c) $\dfrac{2\pi}{3}$; (e) $\dfrac{5\pi}{6}$; (g) $\dfrac{2\pi}{3}$; (i) $\dfrac{\pi}{6}$.

Sec. 8–9, par. (a)

1. $x = rt - a\sin t$, $y = r - a\cos t$.

Sec. 8–9

1. $x - y - 2 = 0$, line. 2. $x - y + 1 = 0$, line.
3. $b^2x^2 - a^2y^2 = a^2b^2$, hyperbola. 4. $y^2 = (x-1)(5-x)^2$.
5. $y^2 = (1-x)(2-x)^2$. 6. $y^2 = (1-x^2)(1-4x^2)^2$.
7. $y = 1 - 4x + 2x^2$, parabola. 8. $b^2x^2 + a^2y^2 = a^2b^2$, ellipse.
9. $y = x^2 - 4x$, parabola. 10. $x - 3y + 6 = 0$, line.
11. $x - y - 1 = 0$, line. 12. $x^{2/3} + y^{2/3} = a^{2/3}$, asteroid.
13. $x^2 + 2y^2 - 2xy = 1$, ellipse. 14. $x^2 - 2y = 1$, parabola.

Chapter 9

Sec. 9–2

3. In the xz-plane.

5. (a) Plane containing the z-axis and all points for which $y = x$; (c) plane parallel to the y-axis and containing points for which $2x + z - 4 = 0$; (e) plane parallel to the xy-plane and 4 units above it.

6. $(-x, -y, -z)$; one.

8. $(0,0,0)$, $(a,0,0)$, $(0,a,0)$, $(a,a,0)$, $(0,0,a)$, $(a,0,a)$, $(0,a,a)$, and (a,a,a). There are three other possible positions.

10. (a) $\Delta x = -4$, $\Delta y = 3$, and $\Delta z = -5$; (c) $\Delta x = -3$, $\Delta y = 6$, and $\Delta z = 0$; (e) $\Delta x = 5$, $\Delta y = 0$, and $\Delta z = -2$.

Sec. 9–5

1. (a) $\sqrt{18} + \sqrt{17} + \sqrt{29}$; (c) $\sqrt{14} + \sqrt{30} + \sqrt{66}$.

5. (a) $(3, -2, 1)$, $\sqrt{6}$, and $\left[\dfrac{2}{\sqrt{6}}, \dfrac{1}{\sqrt{6}}, -\dfrac{1}{\sqrt{6}}\right]$;

(c) $(-3, 2, 7)$, $\sqrt{29}$, and $\left[-\dfrac{3}{\sqrt{29}}, \dfrac{2}{\sqrt{29}}, \dfrac{4}{\sqrt{29}}\right]$.

6. No. **8.** $(2, -3, 0)$ or $(6, 1, 4)$. **10.** $l = \pm\dfrac{\sqrt{3}}{4}$. **12.** $90°$; yes. **13.** $\cos^{-1}\left(\pm\dfrac{1}{\sqrt{3}}\right)$.

14. $\sqrt{y^2 + z^2}$, to the x-axis; $\sqrt{x^2 + z^2}$, to the y-axis; $\sqrt{x^2 + y^2}$, to the z-axis.

15. $(-x, y, z)$, image in x-plane; $(x, -y, z)$, image in the y-plane; $(x, y, -z)$, image in the z-plane.

16. $(x, -y, -z)$ image in x-axis; $(-x, y, -z)$, image in y-axis; $(-x, -y, z)$, image in z-axis.

Sec. 9–6

2. (a) $\sqrt{14}$; (c) $\sqrt{26}$.

4. (a) $|u + v| = \sqrt{17}$ and $|u - v| = \sqrt{17}$; (c) $|u + v| = \sqrt{6}$ and $|u - v| = \sqrt{26}$.

5. (a) $|3u| = \sqrt{126}$; (c) $|\tfrac{1}{2}u| = \dfrac{\sqrt{14}}{2}$.

6. (b) $\sqrt{11}\left[\dfrac{1}{\sqrt{11}}, -\dfrac{1}{\sqrt{11}}, \dfrac{3}{\sqrt{11}}\right]$. **9.** No.

10. (a) $\left[\dfrac{2}{\sqrt{14}}, \dfrac{3}{\sqrt{14}}, \dfrac{1}{\sqrt{14}}\right]$; (c) $\left[\dfrac{3}{\sqrt{29}}, \dfrac{4}{\sqrt{29}}, \dfrac{2}{\sqrt{29}}\right]$.

11. $u = u_1 i + u_2 j + u_3 k$.

12. (a) $-i + 3j + 2k$; (b) $2i + 5j - 3k$; (c) $3j + 2k$; (d) $4i - 2j + 5k$.

13. (a) $6i - 4j + 8k$, $2\sqrt{29}$; (b) $5i - 8j + 8k$, $\sqrt{153}$;

(c) $-i + 10j - 10k$, $\sqrt{201}$; (d) $6i + 9j - 3k$, $\sqrt{126}$;

(e) $-4i + 3j - 5k$, $5\sqrt{2}$; (f) $11i + 10j + 2k$, 15.

Sec. 9–7

2. (a) -8.

3. (a) $\cos \angle P_1 = \dfrac{14}{\sqrt{590}}$, $\cos \angle P_2 = \dfrac{45}{\sqrt{59}\sqrt{41}}$, and $\cos \angle P_3 = -\dfrac{4}{\sqrt{410}}$;

(c) $\cos \angle P_1 = -\dfrac{5}{\sqrt{33}}$, $\cos \angle P_2 = \dfrac{8}{\sqrt{66}}$, and $\cos \angle P_3 = \dfrac{2\sqrt{2}}{3}$.

6. $(5, 0, -2)$. **8.** $x = -\dfrac{1}{4}$ and $z = -\dfrac{1}{4}$. **10.** 5 or $-\dfrac{8}{3}$.

Sec. 9–9

1. $\left(\dfrac{27}{5}, \dfrac{7}{5}, \dfrac{7}{5}\right)$. **3.** $\left(-\dfrac{2}{3}, \dfrac{17}{3}, -3\right)$. **5.** $(1, 2, -5)$.

8. $(1, -4, -1)$, $(13, 6, -11)$, and $(-3, 0, 9)$.

Review Exercises

1. In the yz-plane. **3.** On the x-axis. **5.** $(0, 0, 0)$.

11. On or between two planes perpendicular to the x-axis and each at a distance of 4 units from the origin.

13. On a plane perpendicular to the x-axis at a distance x from the origin.

15. $\sqrt{x^2 + y^2}$. **17.** $[3, -4, 2]$. **19.** $(-1, 5, -11)$.

21. $\left(-\dfrac{3}{5}, 0, 0\right)$. **25.** $\left(\dfrac{1}{2}, \dfrac{1}{2}, 0\right)$, $\left(\dfrac{1}{2}, 0, \dfrac{1}{2}\right)$, $\left(\dfrac{1}{2}, 0, 0\right)$, $\left(0, \dfrac{1}{2}, \dfrac{1}{2}\right)$, $\left(0, \dfrac{1}{2}, 0\right)$, and $\left(0, 0, \dfrac{1}{2}\right)$.

27. $P = (0, 7, -5)$, ratio is -2.

28. $A = \left(\dfrac{7}{2}, 0, \dfrac{11}{2}\right)$, ratio is $-\dfrac{3}{5}$.

29. $\left(\dfrac{7}{3}, -\dfrac{1}{3}, 0\right)$; $\left(\dfrac{11}{5}, 0, -\dfrac{1}{5}\right)$; $\left(0, \dfrac{11}{2}, -\dfrac{7}{2}\right)$.

CHAPTER 10

Sec. 10–1

6. (a) $3x + y + 2z + 7 = 0$; (c) $5x - 2y + 3z + 1 = 0$.

7. (b) $(1, 0, 0)$, $(0, -3, 0)$, and $\left(0, 0, \dfrac{1}{2}\right)$.

8. (a) $[1, -1, 2]$; (c) $[1, -2, 3]$; (e) $[1, -3, 2]$.

9. $x + 4y - 4z - 3 = 0$. **11.** $2y - z + 5 = 0$. **13.** $2x + 3z - 5 = 0$.

15. $x - 2y - z = 0$. **16.** (b) $x = 1$. **17.** $x = 0$, $y = 0$, and $z = 0$.

Sec. 10–2

1. (a) $2x + y - 3z - 17 = 0$; (c) $x - 2y + 2z - 5 = 0$. **2.** (a) -4; (c) $-\dfrac{3}{2}$.

Sec. 10–3

1. (a) $\dfrac{x}{-\dfrac{d}{a}} + \dfrac{y}{-\dfrac{d}{b}} + \dfrac{z}{-\dfrac{d}{c}} = 1$; (c) $\dfrac{x}{-\dfrac{8}{3}} + \dfrac{y}{\dfrac{8}{5}} + \dfrac{z}{4} = 1$; (e) $\dfrac{x}{-3} + \dfrac{y}{-3} + \dfrac{z}{\dfrac{3}{2}} = 1$.

2. (a) $\dfrac{x}{3} + \dfrac{y}{4} + \dfrac{z}{-1} = 1$; (c) $\dfrac{x}{2} + \dfrac{3y}{4} - \dfrac{2z}{1} = 1$.

Sec. 10–5

1. (a) i; (c) 0; (e) 6i.
3. (a) $-i + 7j + 5k$. (c) $2i - 14j - 10k$.
5. $\pm \frac{1}{7}(3i - 2j + 6k)$.
7. $\pm \frac{\sqrt{3}}{3}(i + j + k)$.
9. $15\sqrt{3}$ square units.
11. $\frac{3}{5}i + \frac{4}{5}j$.

Sec. 10–6

5. (a) 11; (c) $\frac{3}{2}$. 6. $[4, -2, 5]$.
10. (a) $5\sqrt{3}$ square units; (b) $3\sqrt{59}$ square units;
 (c) $\sqrt{(u_2q_3 - u_3q_2)^2 + (u_3q_1 - u_1q_3)^2 + (u_1q_2 - u_2q_1)^2}$.
11. $(-3, -1, -2)$, $2\sqrt{146}$ square units.
12. $(10, -2, -1)$, $\sqrt{1714}$ square units.
13. (a) 15 cubic units; (b) 21 cubic units; (c) 1 cubic unit. 14. i; j; $-k$. 15. 1; -1.
16. (a) $\overrightarrow{P_1P_2} = [-1, -2, 1]$, $\overrightarrow{P_1P_3} = [-3, -2, 7]$; (b) $u = [-12, 4, -4]$.
17. $u + v = i + j + 2k$, $u - v = 3i + 5j - 4k$, $-2u = -4i - 6j + 2k$,
 $|u + v| = \sqrt{6}$, $|u - v| = 5\sqrt{2}$, $|-2u| = 2\sqrt{14}$.

Sec. 10–7

7. (a) $3\sqrt{10}$; (b) $5\sqrt{26}$; (c) 20; (d) 20; (e) $-40i - 20j + 20k$;
 (f) $35i - 35j + 35k$.
10. (a) $11x + 5y + 13z - 30 = 0$.

Sec. 10–8

1. (a) $x + y + z - 1 = 0$; (c) $2x - 3y + 2z + 7 = 0$; (e) $x = 0$; (g) $x - 4y - 3z - 5 = 0$.
2. $x + 5y + 3z - 10 = 0$. 4. $11x + 7y - 5z + 14 = 0$. 7. $3x + y + 2z - 28 = 0$.
8. $3x - y - 5z + 2 = 0$. 9. $3x - 2y + 4z - 12 = 0$.
10. $3x - 2y + z - 14 = 0$. 11. $x - 6y - 5z + 19 = 0$.
12. (a) $6x - 8y - 4z - 95 = 0$; (b) $12x - 16y - 8z + 13 = 0$; (c) $9x - 12y - 6z + 46 = 0$.

Sec. 10–9

1. (a) $\frac{12}{\sqrt{26}}$; (c) $\frac{12}{\sqrt{14}}$. 2. $\frac{6}{\sqrt{14}}$. 4. $x - 5y + 6z - 12 = 0$.
6. $2x - 5y + 3z - 18 = 0$ and $4x + y - z + 6 = 0$.

7. $x+2y-3z-6=0$ and $3x+z=0$.
9. (a) Set of points in space lying on the positive side of the plane $2x-3y+z-6=0$;
   (b) Set of points in space lying on the negative side of as well as on the plane $x-y-2z-4=0$;
   (c) Set of points in space lying on the negative side of the plane $x-y=0$;
   (d) Set of points in space lying on the positive side as well as on the plane $z-5=0$;
   (e) Set of points in space on negative side of plane $2y-3z-6=0$;
   (f) Set of points on the positive side or on the plane $2x-3y-4=0$.
11. $(0, 3, 0)$ and $(0, -9, 0)$. 12. $\left(0, 0, -\frac{7}{2}\right)$ and $\left(0, 0, \frac{17}{2}\right)$.
13. $2x-2y+z-15=0$, $2x-2y+z+3=0$.
14. $6x-2y+3z-28=0$, $6x-2y+3z+28=0$.
15. $3x+2y-6z+21=0$, $3x+2y-6z-21=0$.

Sec. 10–11

1. (a) $\dfrac{5}{\sqrt{84}}$; (c) $\dfrac{11}{\sqrt{132}}$.
2. (b) $2x+7y-5z+18=0$; (d) $2x-3y+z-6=0$; (f) $x+y-z+3=0$; (h) $3x-2y-3=0$.
3. (a) $t_1(3x+2y-z-4)+t_2(x-y+z+5)=0$; (c) $19x+6y-z=0$; (e) $x-y+z+5=0$;
   (g) $11x+4y-z-2=0$; (i) $5x+z+6=0$. 5. $44x-16y+9z-27=0$.

Review Exercises

1. $8x-3z-12=0$. 3. $5x-4y-5z+13=0$. 5. $z=0$.
7. (a) Plane goes through the origin; (b) plane is parallel to the y-axis;
   (c) plane is perpendicular to the x-axis; (d) plane is parallel to the z-axis and goes through the origin.
9. $2x-4y-5z+7=0$. 12. $\dfrac{7}{\sqrt{88}}$.
13. (c) $5y+5z-24=0$; (e) $13x+11y-2z-6=0$.
14. (b) $x+y-1=0$; (d) there isn't any, but the plane $2y-z=0$ contains the x-axis and therefore the point $(4, 0, 0)$; (f) $x+y-1=0$.
15. $3x+y-30=0$, $x-3y+10=0$, $z-1=0$. 16 $\dfrac{\sqrt{195}}{14}$.
17. Cosine of angle made with xy-plane is $\dfrac{1}{\sqrt{6}}$. Cosine of angle made with yz-plane is $\dfrac{1}{\sqrt{6}}$. 18. $k=2$.

Chapter 11

Sec. 11–1

1. (a) $[4, -3, -1]$ and $\left[\dfrac{4}{\sqrt{26}}, -\dfrac{3}{\sqrt{26}}, -\dfrac{1}{\sqrt{26}}\right]$; (c) $[-4, 0, 3]$ and $\left[-\dfrac{4}{5}, 0, \dfrac{3}{5}\right]$.
2. (a) $n=\dfrac{\sqrt{164}}{15}$; (c) $n=\dfrac{\sqrt{39}}{10}$. 3. Yes. 5. Yes.

Sec. 11-4

1. (a) $x = -3+5t$, $y = -2+t$, and $z = 1+2t$; (c) $x = -2-3t$, $y = -t$, and $z = 3+4t$.
2. (a) $x = 5-t$, $y = -1+3t$, and $z = 2+2t$.
3. (a) $\dfrac{x}{3} = \dfrac{y+1}{-5} = \dfrac{z-2}{3}$; (c) $\dfrac{x-2}{-1} = \dfrac{z-1}{3}$ and $y = -1$.
4. (a) $\dfrac{x-2}{-1} = \dfrac{y+1}{1} = \dfrac{z-3}{2}$; $x = 2-t$, $y = -1+t$, and $z = 3+2t$;
   (c) $\dfrac{x+5}{2} = \dfrac{y-3}{-1} = \dfrac{z-2}{4}$; $x = -5+2t$, $y = 3-t$, and $z = 2+4t$.
5. $\dfrac{x-3}{3} = \dfrac{y-2}{1} = \dfrac{z-1}{-1}$.
7. (a) $[-5, 3, 1]$, $x = -5t - 23$, $y = -3t - 17$, $z = 1+t$;
   (b) $[-7, -8, -5]$, $x = -1 - 7t$, $y = -2 - 8t$, $z = -5t$.
   (c) $[1, 7, 1]$, $x = t$, $y = 7t - 10$, $z = -2+t$;
   (d) $[2, -4, -3]$, $x = -2+2t$, $y = 10 - 4t$, $z = -3 - 3t$.
8. $3x + 4y - 2z - 20 = 0$ and $x - 2y - z + 3 = 0$. 10. $2x - 3z - 1 = 0$.
12. $x = 3$ and $3y + z = 0$. 14. $\dfrac{x+1}{15} = \dfrac{y-2}{-50} = \dfrac{z-3}{37}$. 15. (b) $\dfrac{22}{\sqrt{574}}$.

18. $\begin{vmatrix} x - x_1 & y - y_1 & z - z_1 \\ u_1 & u_2 & u_3 \\ v_1 & v_2 & v_3 \end{vmatrix} = 0$. 20. $3x - 2y + 7z + 8 = 0$.

Sec. 11-5

1. $(2, 1, -1)$. 3. Collinear. 5. Intersect in parallel lines.

Sec. 11-6

1. $(0, 0, 0)$. 2. A line of intersection containing all points $(0, -3k, -3k)$ for any k.
3. Two planes coincide. The solution set consists of all points on the line $2x - 3y + z = 0$, $x + 2y - z = 0$. This set can be written $(k, 3k, 7k)$ for any k.
4. $(4k, -k, 5k)$ for any k; the planes intersect in a line.
5. The three planes are coincident. The solution set consists of all points common to the plane.

Sec. 11-7

1. (a) $x = -1 + 4t_1 + 3t_2$, $y = 2 - t_1 + 2t_2$, and $z = 3 - 2t_1 - 4t_2$;
   (c) $x = 5 - 4t_1 - 2t_2$, $y = 2 - t_1$, and $z = 3 - t_1 - 4t_2$.
2. (a) $x = 2 - 2t_1$, $y = -\dfrac{4}{3}t_1 + \dfrac{4}{3}t_2$, and $z = \dfrac{2}{3}t_2$; (c) $x = 2 - t_1$, $y = t_1 - t_2$, and $z = -t_2$.

Review Exercises

1. $x + 2y - 1 = 0$. 3. $5x - 2y - 4z - 1 = 0$.
5. (a) $3x - y + 2z - 8 = 0$ and $x + y - z + 2 = 0$; (b) $x = \dfrac{3}{2} - t$, $y = -\dfrac{7}{2} + 5t$, and $z = 4t$.
7. (a) $\sqrt{117}$; (b) $P_4 = (-3, -2, 2)$.
9. (a) $x + 4y + 3z - 1 = 0$; (b) $\dfrac{24}{\sqrt{26}}$.
11. $5x + y + 2z + 18 = 0$ and $x + 3y - 4z + 6 = 0$.

Chapter 12

Sec. 12–2

3. (a) $256(y^2+z^2) = x^4$; (c) $x^2+y^2+z^2 = 16$; (e) $9y^2 - 4x^2 - 4z^2 = 36$; (g) $x^2+z^2 = (\sin^{-1} y)^2$; (i) $x^2+z^2 = (\log_e y)^2$; (k) $x^2+z^2+4y^2 = 16$.

Sec. 12–4

1. Cone. **3.** Cylinder. **5.** Cylinder and surface of revolution. **7.** Cylinder.
9. Plane, cylinder, and cone.
11. Cylinder and surface of revolution. **12.** Cylinder. **13.** Cylinder.
14. Cylinder, cone, two planes. **15.** Two planes. **16.** Plane, cylinder, and cone.
17. Cone and surface of revolution. **18.** Cylinder. **19.** Cylinder. **20.** Cone.

Sec. 12–6

1. (a) $(x+1)^2+(y-2)^2+(z-3)^2 = 16$; (c) $(x-3)^2+(y-1)^2+(z+2)^2 = 1$; (e) $(x-2)^2+y^2+(z+3)^2 = 25$.
2. (a) $(1, -2, 3)$ and 5; (c) $(0, -1, 2)$ and 3; (e) $(3, 0, -2)$ and 7.
3. $(x-1)^2+(y-3)^2+(z-3)^2 = 33$.
4. $\left(x-\dfrac{33}{74}\right)^2+\left(y-\dfrac{104}{74}\right)^2+\left(z-\dfrac{55}{74}\right)^2 = \dfrac{42014}{5476}$.
5. $\left(x+\dfrac{8}{3}\right)^2+\left(y-\dfrac{1}{2}\right)^2+\left(z+\dfrac{2}{3}\right)^2 = \dfrac{965}{36}$.
6. $(x+2)^2+(y-3)^2+(z-1)^2 = 16$.
7. $x^2+y^2+z^2-10y-10z-31 = 0$.
8. $x^2+y^2+z^2-5x+y-3z+6 = 0$.
9. $x^2+y^2+z^2-6x-2y-10z-1 = 0$.

Sec. 12–7

2. (a) $9x^2+4y^2+4z^2 = 36$; (c) $5x^2+5y^2+3z^2-15 = 0$; (e) $x^2+y^2+z^2-16 = 0$; (g) $4x^2+3y^2+3z^2-12 = 0$.

Sec. 12–8

2. (a) $x^4-36y^2-36z^2 = 0$; (c) $16y^4 = 81(x^2+z^2)$; (e) $4x^4 = 625(y^2+z^2)$; (g) $81y^4 = 576(x^2+z^2)$.

Sec. 12–9

2. (a) $2x^2+2z^2-3y^2 = 4$; (c) $4y^2-9x^2-9z^2 = 36$; (e) $3x^2+3z^2-5y^2 = 15$; (g) $y^2+z^2-x^2 = 1$.

Sec. 12–12

1. (a) $y^2 = 4x$, $z^2 = 4x$, and $z^2-y^2 = 0$; (c) $y^2-5z^2 = 8$, $5x^2+6y^2 = 48$, and $x^2-16z^2 = 0$.
2. (a) $x^2+y^2 = 1$ and $y = z$; (c) $3x+2y-6 = 0$ and $x = 2z^2$.

Sec. 12–15

1. $\theta = k$. **4.** $r = 10$. **5.** $\rho^2+z^2 = 100$. **6.** (a) $\rho = \pm z$; (c) $\rho^2 = 4z$; (e) $a^2 \rho^2 \sin^2 \theta = a^2 - z$.

Review Exercises

1. (a) $x^2+y^2+z^2=16$; (c) $x^2+y^2=16z$; (e) $9x^2+9z^2-16y^2=144$; (g) $(2x-6)^2=9y^2+9z^2$.
2. (a) Cylinder; (b) cylinder; (c) cylinder; (d) sphere; (e) ellipsoid;
   (f) hyperboloid of one sheet; (g) hyperboloid of two sheets; (h) cone;
   (i) cone; (j) cone; (k) paraboloid of revolution; (l) paraboloid;
   (m) ellipsoid of revolution; (n) ellipsoid of revolution; (o).plane; (p) sphere;
   (q) cone; (r) sphere; (s) cylinder; (t) ellipsoid; (u) cone; (v) hyperboloid;
   (w) cylinder; (x) cylinder; (y) plane; (z) cylinder.
3. (a), (b), (c), (f), (h), (i), (j), (l), (o), (q), (s), (u), (v), (w), (x), (y), (z).
4. (a) $(2\sqrt{3}, 2, 3)$; (c) $(2, 0, -3)$; (e) $(-3\sqrt{3}, 3, 2)$.
5. (a) $\left(\dfrac{1}{2}, \dfrac{\sqrt{3}}{2}, \sqrt{3}\right)$; (c) $\left(0, -\dfrac{5\sqrt{2}}{2}, \dfrac{5\sqrt{2}}{2}\right)$; (e) $\left(-\dfrac{5\sqrt{3}}{2}, \dfrac{15}{2}, 5\right)$.
6. $(-1, 2, -2)$ and $\sqrt{7}$. 7. (b) $y=2x^2$ and $z=3x^3$; (d) $xy=1$ and $z=x+y$.

Index

Abscissa, 39
Absolute value, 24
Addition of vectors, 56, 344
Alternating product, 72, 367
Amplitude, 313
Angle
  between two lines, 94
  between two vectors, 58
  lag, 313
  lead, 313
  phase, 313
  vectorial, 262
Apex, 162
Arch, parabolic, 172
Area, 71
Astroid, 325
Asymptotes, 198, 241, 302
Auxiliary circles, 192
Axis
  conjugate, 197
  major, 182
  minor, 182
  normal, 99
  of parabola, 166
  transverse, 196

Bar product, 71
Base of logarithms, 11
Bisectors of an angle, 115

Cardioid, 281
Cartesian geometry, 39
Cartesian product, 5
Catenary, 332
Characteristic, 244
Circle, 127, 269
Coefficient vector, 90
Co-factor, 13
Complement, 4
Complementary vector, 70
Components, 46
Concurrent lines, 112
Conditional equation, 32
Cone, 162, 403

Conic sections, 162
Conjugate axis, 197
Conjugate hyperbolas, 206
Conoid, 423
Coordinates
  cylindrical, 429
  polar, 261
  rectangular, 39, 335
  spherical, 430
Cosine formula, 59, 348
Cosine law, 37
Cramer's rule, 19
Cross product, 361
Curtate cycloid, 322
Cycloid, 322
Cylinder, 401
Cylindrical coordinates, 429

Degree, 32
Determinant, 12
Difference of two vectors, 56, 345
Directed segment, 44
Direction cosines
  of a line, 79, 381
  of a segment, 52, 341
  of a vector, 56, 345
Direction number equation of a line, 86
Direction numbers of a line, 79, 381
Directrix, 163, 178
Discriminant, 7, 246
Distance formula
  between two lines in space, 388
  between two points, 48, 340
  from line to point, 107
  from plane to point, 373
Domain, 27, 224
Dot product, 60, 348

Eccentricity, 163
Ellipse, 178
Ellipsoid, 409
Elliptic paraboloid, 412
Elliptic wedge, 422
Empty set, 4

Epicycloid, 323
Equilateral hyperbola, 207, 241
Equivalent equations, 137
Explicitly, 29
Exponential curves, 306
Exponents, 9
Extent, 301

Focal chord, 166
Focal radii of conics, 184, 202
Focus, 163
Function, 27

General definition of a conic, 163
General equation
  of circle, 130
  of conic, 165
  of line, 89, 384
  of quadric surface, 406
Generator, 162, 425
Graphs, 295, 317

Homogeneous equation, 390
Hyperbola
  asymptotes of, 198
  conjugate, 206
  equilateral, 207
  rectangular, 207
Hyperbolic functions, 330
Hyperbolic paraboloid, 418
Hyperboloid
  of one sheet, 414
  of revolution, 415
  of two sheets, 415
Hypocycloid, 324

Identities, 35
Identity equation, 31
Image, 42
Implicitly, 29
Inclination, 81
Increment, 46
Indeterminate, 301
Inequality, 23
Intercept equation
  of a line, 97
  of a plane, 360
Intercepts, 91, 272, 301
Intersecting lines, 110
Intersecting planes, 388
Intersection, 4
Invariants, 243
Inverse trigonometric functions, 38, 317
Involute of a circle, 325

Latus rectum, 166, 188, 205
Lemniscate, 279

Length of segment, 44
Length of tangent, 139
Limacon, 278
Line, 79, 267, 381
Line of centers, 143
Linear equation
  general, 89
  one unknown, 6
  three unknowns, 19
  two unknowns, 18
Logarithmic curves, 308
Logarithms, 10

Magnitude of a vector, 55, 344
Major axis, 182
Matrices, 435
Mid-point formula, 68, 352
Minor axis, 182

Normal axis, 100
Normal equation of a line, 100
Normal intercepts, 100
Normal vector, 100, 357
Null set, 4
Number interval, 25

Octant, 336
Ordered pair, 4
Ordinate, 39
Origin, 39, 335
Orthogonal circles, 160
Ovals of Cassini, 329

Parabola, 162, 166
Paraboloid
  elliptic, 412
  hyperbolic, 418
  of revolution, 412
Parameter, 83
Parametric equations
  of circle, 156
  of curve in space, 424
  of cycloid, 321
  of ellipse, 193
  of epicycloid, 323
  of hyperbola, 210
  of hypocycloid, 324
  of involute of circle, 325
  of line, 83, 383
  of parabola, 173
  of plane, 392
  of strophoid, 326
  of witch, 328
Pencil
  of circles, 151
  of conics, 251
  of lines, 120
  of planes, 377

Period, 310
Plane, 356
Plane sections of a cone, 162
Point circle, 131
Point-slope equation of line, 97
Polar axis, 261
Polar coordinates, 261
Polar equation, 265, 277
Polar line, 160
Pole, 261
Polynomial, 295
Product
  alternating, 72, 367
  bar, 72
  cross, 361
  dot, 60, 348
  scalar triple, 367
  symmetric, 60, 348
  vector, 361
  vector triple, 369
Projecting cylinders, 424
Projection, 44, 336
Prolate cycloid, 323
Proper subset, 5

Quadrant, 39
Quadratic equation, 7
Quadric surface, 406

Radian, 32
Radical axis, 141
Radius vector, 262
Range, 27, 224
Ratio of a segment, 65, 350
Rectangular hyperbola, 207
Relation, 26
Rotation of axes, 231, 428
Ruled surface, 420
Ruling, 420

Scalar components, 46, 339
Scalar product, 60, 348
Scalar triple product, 367
Secant curve, 312
Set, 3
Set notation, 3, 256
Sets of lines, 118
Simple equations
  of circle, 128
  of ellipse, 178
  of hyperbola, 194
  of parabola, 167
Sine curve, 310
Sine law, 37
Slope, 81

Slope-intercept equation of line, 99
Sphere, 406
Spherical coordinates, 430
Spheroid, 410
Spiral
  Archimedes, 281
  hyperbolic, 281
  lituus, 281
Standard equation
  of circle, 128
  of ellipse, 226
  of hyperbola, 227
  of parabola, 225
Strophoid, 326
Subset, 4
Sum of two vectors, 56, 344
Surface, 396
Surface of revolution, 396
Symmetric equation of a line, 87, 384
Symmetric product of two vectors, 60
Symmetry
  axis of, 42
  center of, 42
  with respect to a line, 42, 136, 272
  with respect to a point, 42, 136, 275

Tangent curve, 311
Tangent formula, 95
Tangent line, 139, 146, 148
Trace, 409, 425
Transcendental curves, 295
Transformation formulas, 264
Translation of the axes, 217, 426
Transverse axis, 197
Triangle, area of, 71
Trigonometric curves, 310
Trigonometric functions, 34

Union, 4
Unit circle, 58
Unit vector, 56, 345
Universal set, 4

Vector
  parallel, 61
  perpendicular, 62
  in plane, 55
  in space, 343
Vector product, 361
Vectorial angle, 262
Vertex, 162, 166
Vertices, 178, 182

Witch of Agnesi, 328

Zero vector, 56